LABORATORY MANUAL FOR

HUMAN ANATOMY & PHYSIOLOGY

THIRD EDITION

581
605   due
4/20/17

TERRY R. MARTIN

Kishwaukee College

CONTRIBUTORS

Brian H. Kipp

Cynthia Prentice-Craver

Phillip D. Snider, Jr.

Mc
Graw
Hill
Education

LABORATORY MANUAL FOR HUMAN ANATOMY & PHYSIOLOGY: CAT VERSION, THIRD EDITION

Published by McGraw-Hill Education, 2 Penn Plaza, New York, NY 10121. Copyright © 2016 by McGraw-Hill Education. All rights reserved. Printed in the United States of America. Previous editions © 2013 and 2010. No part of this publication may be reproduced or distributed in any form or by any means, or stored in a database or retrieval system, without the prior written consent of McGraw-Hill Education, including, but not limited to, in any network or other electronic storage or transmission, or broadcast for distance learning.

Some ancillaries, including electronic and print components, may not be available to customers outside the United States.

This book is printed on acid-free paper.

3 4 5 6 7 8 9 0 RMN/RMN 1 0 9 8 7 6

ISBN 978-0-07-802430-6
MHID 0-07-802430-7

Senior Vice President, Products & Markets: *Kurt L. Strand*
Vice President, General Manager, Products & Markets: *Marty Lange*
Vice President, Content Design & Delivery: *Kimberly Meriwether David*
Managing Director: *Michael S. Hackett*
Brand Manager: *Amy Reed*
Director, Product Development: *Rose Koos*
Product Developer: *Fran Simon*
Marketing Manager: *Jessica Cannavo*
Digital Product Developer: *John J. Theobald*
Director, Content Design & Delivery: *Linda Avenarius*
Program Manager: *Angela R. FitzPatrick*
Content Project Managers: *Vicki Krug/Sherry L. Kane*
Buyer: *Sandy Ludovissy*
Design: *Matthew Backhaus*
Content Licensing Specialists: *John Leland/Beth Thole*
Cover Image: *Ross Martin*
Art Studio and Compositor: *MPS North America LLC*
Printer: *R. R. Donnelley*

All credits appearing on page or at the end of the book are considered to be an extension of the copyright page.

The Internet addresses listed in the text were accurate at the time of publication. The inclusion of a website does not indicate an endorsement by the authors or McGraw-Hill Education, and McGraw-Hill Education does not guarantee the accuracy of the information presented at these sites.

Some of the laboratory experiments included in this text may be hazardous if materials are handled improperly or if procedures are conducted incorrectly. Safety precautions are necessary when you are working with chemicals, glass test tubes, hot water baths, sharp instruments, and the like, or for any procedures that generally require caution. Your school may have set regulations regarding safety procedures that your instructor will explain to you. Should you have any problems with materials or procedures, please ask your instructor for help.

www.mhhe.com

CONTENTS

*These exercises are available in the eBook via Connect Anatomy & Physiology and also online for instructor distribution; see Instructor Resources via Connect Library tab.

PREFACE

In**TOUCH** | WITH **Anatomy & Physiology Lab Courses**

Author Terry Martin's forty years of teaching anatomy and physiology courses, authorship of three laboratory manuals, and active involvement in the Human Anatomy and Physiology Society (HAPS) drove his determination to create a laboratory manual with an innovative approach that would benefit students. The *Laboratory Manual for Human Anatomy & Physiology* includes a cat version, and a fetal pig version. Each of these versions includes sixty-two laboratory exercises, three supplemental labs found online, and six cat, or fetal pig, dissection labs. A main version with no dissection exercises and a rat version are also available. All versions are written to work well with any anatomy and physiology text.

Martin Lab Manual Series . . .
In**TOUCH** WITH **Anatomy & Physiology Lab Courses**

▶ Anatomy and Physiology REVEALED® icons are found in figure legends. These icons indicate that there is a direct link to APR available in the eBook provided with Connect® for this title.

▶ Incorporates **learning outcomes and assessments** to help students master important material!

▶ **Pre-Lab** assignments are printed in the lab manual. They will help students be more prepared for lab and save instructors time during lab.

▶ **Clear, concise** writing style facilitates more thorough understanding of lab exercises.

FIGURE 24.5 APR Label the deep back muscles of a cadaver, using the terms provided. The trapezius, latissimus dorsi, rhomboids, and serratus posterior have been removed.

▶ **BIOPAC©** exercises use hardware and software for data acquisition, analysis, and recording.

▶ **NEW!** Exercise 31A Reaction Time: BIOPAC Exercise. This new lab will enhance the physiology concepts of the nervous system.

▶ **Ph.I.L.S. 4.0** physiology lab simulations interspersed throughout make otherwise difficult and expensive experiments a breeze through digital simulations.

▶ Cadaver images from **Anatomy & Physiology REVEALED®** **(APR)** are incorporated throughout the lab. Cadaver images help students make the connection from specimen to cadaver.

▶ **Micrographs** incorporated throughout the lab aid students' visual understanding of difficult topics.

▶ **Instructor's Guide** is annotated for quick and easy use by adjuncts and is available online.

FEATURES OF THIS LABORATORY MANUAL

InTOUCH WITH Student Needs

▶ The procedures are clear, concise, and easy to follow. Relevant lists and summary tables present the contents efficiently. Histology micrographs and cadaver photos are incorporated in the appropriate locations within the associated labs.

▶ The pre-lab section includes quiz questions. It also directs the student to carefully read the introductory material and the entire lab to become familiar with its contents. If necessary, a textbook or lecture notes might be needed to supplement the concepts.

▶ **Terminologia Anatomica** is used as the source for universal terminology in this laboratory manual. Alternative names are included when a term is introduced for the first time.

▶ Laboratory assessments immediately follow each laboratory exercise.

▶ Histology photos are placed within the appropriate laboratory exercise.

▶ A section called "Study Skills for Anatomy and Physiology" is located in the front of this laboratory manual. This section was written by students enrolled in a Human Anatomy and Physiology course.

▶ Critical Thinking Activities are incorporated within most of the laboratory exercises to enhance valuable critical thinking skills that students need throughout their lives.

▶ Cadaver images are incorporated with dissection labs.

InTOUCH WITH Instructor Needs

▶ The instructor will find digital assets for use in creating customized lectures, visually enhanced tests and quizzes, and other printed support material.

▶ A correlation guide for **Anatomy & Physiology REVEALED® (APR)** and the entire lab manual is available. Contact your McGraw-Hill Learning Technology Representative. Cadaver images from APR are included within many of the laboratory exercises.

▶ Some unique labs included are "Scientific Method and Measurements," "Chemistry of Life," "Fetal Skeleton," "Surface Anatomy," "Diabetic Physiology," and "Genetics."

▶ The annotated instructor's guide for *Laboratory Manual for Human Anatomy and Physiology* describes the purpose of the laboratory manual and its special features, provides suggestions for presenting the laboratory exercises to students, instructional approaches, a suggested time schedule, and annotated figures and assessments. It contains a "Student Safety Contract" and a "Student Informed Consent Form."

▶ Each laboratory exercise can be completed during a single laboratory session.

InTOUCH WITH Educational Needs

▶ Learning outcomes with icons (O) have matching assessments with icons (A) so students can be sure they have accomplished the laboratory exercise content. Outcomes and assessments include all levels of learning skills: remember, understand, apply, analyze, evaluate, and create.

▶ Assessment rubrics for entire laboratory assessments are included in Appendix 2.

InTOUCH WITH Technology

▶ Anatomy & Physiology REVEALED® 3.0

Detailed cadaver photographs blended together with a state-of-the-art layering technique provide a uniquely interactive dissection experience.

▶ Physiology Interactive Lab Simulations (Ph.I.L.S. 4.0) is included with the lab manual. Eleven lab simulations are interspersed throughout the lab manual. The correlation guide for all of the simulations is included in Appendix 3.

▶ BIOPAC Systems, Inc. BIOPAC© exercises are included on four different body systems. BIOPAC© systems use hardware and software for data acquisition, analysis, and recording of information for an individual.

ENGAGING PRESENTATION MATERIALS FOR LECTURE AND LAB

connect®
|ANATOMY & PHYSIOLOGY

McGraw-Hill Connect® Anatomy & Physiology is a digital teaching and learning environment that saves students and instructors time while improving performance over a variety of critical outcomes.

- From in-site tutorials, to tips and best practices, to live help from colleagues and specialists–you're never left alone to maximize Connect's potential.
- Instructors have access to a variety of resources including assignable and gradable interactive questions based on textbook images, case study activities, tutorial videos, and more.
- Digital images, PowerPoint slides, and instructor resources are also available through Connect.
- Digital Lecture Capture: Get Connected. Get Tegrity. Capture your lectures for students. Easy access outside of class anytime, anywhere, on just about any device.

Visit www.mcgrawhillconnect.com.

> "…(We) have reduced the time we spend on grading by 90 percent and student test scores have risen, on average, 10 points since we began using Connect!"
>
> —*William Hoover,*
> Bunker Hill Community College

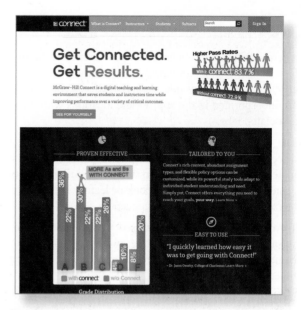

Save time with auto-graded assessments and tutorials

Fully editable, customizable, auto-graded interactive assignments using high-quality art from the lab manual, and animations and videos from a variety of sources take you way beyond multiple choice. Assignable content is available for every Learning Outcome in the lab manual. Extremely high quality content, includes case study modules, animated learning modules, and more!

Generate powerful data related to student performance based on question tagging for Learning Outcomes, HAPS topics and outcomes, specific topics, Bloom's level, and more.

LearnSmart Labs is a superadaptive simulated lab experience that brings meaningful scientific exploration to students. Through a series of adaptive questions, LearnSmart Labs identifies a student's knowledge gaps and provides resources to quickly and efficiently close those gaps. Once students have mastered the necessary basic skills and concepts, they engage in a highly realistic simulated lab experience that allows for mistakes and the execution of the scientific method.

The primary goal of LearnSmart Prep is to help students who are unprepared to take college-level courses. Using superadaptive technology, the program identifies what a student doesn't know and then provides "teachable moments" designed to mimic the office hour experience. When combined with a personalized learning plan, an unprepared or struggling student has all the tools needed to quickly and effectively learn the foundational knowledge and skills necessary to be successful in a college-level course.

What You've Only Imagined

The Future of Custom Publishing is Here.

Introducing McGraw-Hill Create™—a new, self-service website that allows you to quickly and easily create custom course materials by drawing upon McGraw-Hill Education's comprehensive, cross disciplinary content and other third party resources.

- Select, then arrange the content in a way that makes the most sense for your course
- Combine material from different sources and even upload your own content
- Choose the best format for your students-print or eBook
- Edit and update your course materials as often as you'd like

GUIDED TOUR THROUGH AN EXERCISE

The laboratory exercises include a variety of special features that are designed to stimulate interest in the subject matter, to involve students in the learning process, and to guide them through the planned activities. These features include the following:

Purpose of the Exercise   The purpose provides a statement about the intent of the exercise—that is, what will be accomplished.

Learning Outcomes   The learning outcomes list what a student should be able to do after completing the exercise. Each learning outcome will have matching assessments indicated by the corresponding icon (A) in the laboratory exercise or the laboratory assessment.

Materials Needed   This section lists the laboratory materials that are required to complete the exercise and to perform the demonstrations and learning extensions.

Safety   A list of safety guidelines is included inside the front cover. Each lab session that requires special safety guidelines has a safety section. Your instructor might require some modifications of these guidelines.

Introduction   The introduction describes the subject of the exercise or the ideas that will be investigated. It includes all of the information needed to perform the laboratory exercise.

Pre-Lab   The pre-lab includes quiz questions and directs the student to carefully read introductory material and examine the entire laboratory contents after becoming familiar with the topics from a textbook or lecture. After successfully answering the pre-lab questions, the student is prepared to become involved in the laboratory exercise.

Procedure   The procedure provides a set of detailed instructions for accomplishing the planned laboratory activities. Usually these instructions are presented in outline form so that a student can proceed efficiently through the exercise in stepwise fashion.

The procedures include a wide variety of laboratory activities and, from time to time, direct the student to complete various tasks in the laboratory assessments.

There are also separate procedures in 11 labs that utilize Ph.I.L.S. 4.0.

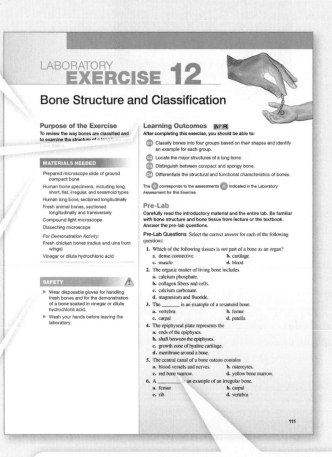

Demonstration Activities Demonstration activities appear in separate boxes. They describe specimens, specialized laboratory equipment, or other materials of interest that an instructor may want to display to enrich the student's laboratory experience.

DEMONSTRATION ACTIVITY

Examine a fresh chicken bone and a chicken bone that has been soaked for several days in vinegar or overnight in dilute hydrochloric acid. Wear disposable gloves for handling these bones. This acid treatment removes the inorganic salts from the bone extracellular matrix. Rinse the bones in water and note the texture and flexibility of each (fig. 12.8a). The bone becomes soft and flexible without the support of the inorganic salts with calcium.

Learning Extension Activities Learning extension activities also appear in separate boxes. They encourage students to extend their laboratory experiences. Some of these activities are open-ended in that they suggest the student plan an investigation or experiment and carry it out after receiving approval from the laboratory instructor. Some of the figures are illustrated as line art or in grayscale. This will allow colored pencils to be used as a visual learning activity to distinguish various structures.

LEARNING EXTENSION ACTIVITY

Repeat the demonstration of diffusion using a petri dish filled with ice-cold water and a second dish filled with very hot water. At the same moment, add a crystal of potassium permanganate to each dish and observe the circle as before. Record your findings in the data table in Part A of the laboratory assessment. What difference do you note in the rate of diffusion in the two dishes? How do you explain this difference? _____

_____
_____
_____
_____

one that has
121°C/250°F
protein and
extracellular
ttle and frag-
ers. A living

Illustrations Diagrams similar to those in a textbook often are used as aids for reviewing subject matter. Other illustrations provide visual instructions for performing steps in procedures or are used to identify parts of instruments or specimens. Micrographs are included to help students identify microscopic structures or to evaluate student understanding of tissues.

Laboratory Assessments A laboratory assessment form to be completed by the student immediately follows each exercise. These assessments include various types of review activities, spaces for sketches of microscopic objects, tables for recording observations and experimental results, and questions dealing with the analysis of such data.

As a result of these activities, students will develop a better understanding of the structural and functional characteristics of their bodies and will increase their skills in gathering information by observation and experimentation. By completing all of the assessments, students will be able to determine if they were able to accomplish all of the learning outcomes.

LABORATORY
ASSESSMENT
12

Name _____
Date _____
Section _____

The Ⓐ corresponds to the indicated Learning Outcome(s) Ⓞ found at the beginning of the Laboratory Exercise.

Bone Structure and Classification

PART A: Assessments
Complete the following statements: (Note: Questions 1–6 pertain to bone classification by shape.)

1. A bone that has a wide surface is classified as a(an) _____ bone.
2. The bones of the wrist are examples of _____ bones.
3. The bone of the thigh is an example of a(an) _____ bone.
4. Vertebrae are examples of _____ bones.
5. The patella (kneecap) is an example of a special type of short bone called a _____ bone.
6. The bones of the skull that form a protective covering for the brain are examples of _____ bones.
7. Distinguish between the epiphysis and the diaphysis of a long bone. _____
8. Describe where cartilage is found on the surface of a long bone. _____
9. Describe where the periosteum is found on the surface of a long bone. _____

Histology Histology photos placed within the appropriate exercise.

Osteon
Lamella
Bone extracellular matrix
Central canal
Lacuna (occupied by osteocyte in living bone)
Canaliculi

d tissues of the periosteum and those of the the endosteum. _____

_____

ote between compact bone and spongy bone? _____

_____

117

CHANGES TO THIS EDITION

Global Changes

- New design incorporated in the laboratory exercises and assessments.
- Revised icons for Learning Outcomes and matching Assessments.
- BIOPAC exercises rewritten and an additional exercise added.
- Ph.I.L.S. 4.0 laboratory exercises revised and placed within a new design section at the end of the Laboratory Assessments.
- Added APR icons.
- Increased italicized terminology in laboratory introductory materials.

LABORATORY EXERCISE	TOPIC	CHANGE
1	Introductory material Table 1.1	Added independent and dependent variables Improved depth
2	Fig. 2.1b (anterior body cavity) Procedure B (organ systems) Fig. 2.5 (anatomical position) Figs. 2.6 and 2.10 (anatomical planes) Figs. 2.8a and 2.8b (surface regions) Assessments: Part E	Added figure Expanded components Improved labels Revised and improved labels Revised and full color Expanded components
4	Fig. 4.7 (wet mount steps)	Redrawn figure
5	Introductory material Figs. 5.1 and 5.4 (composite cell) Table 5.1 (cellular structures and functions) Fig. 5.5	Improved depth Expanded content and labels Added table New micrograph
6	Procedure B (osmosis) Figs. 6.2 (osmosis), 6.3 (filtration), and 6.4 (water bath) Assessments: Parts A, B, C, and D	Improved directions Redrawn figures Expanded and improved directions
7	Fig. 7.3 (prophase cell) Figs. 7.4 and 7.5 (mitosis phases) Fig. 7.6a (chromosomes) Fig. 7.7a and d (mitosis phases)	Revised labels Redrawn figures New figure Improved leader lines
8	Fig. 8.1c,d, and j (epithelial tissues) Fig. 8.1g and k (epithelial tissues) Table 8.1 (epithelial tissues)	New micrographs Added micrographs Expanded components
9	Fig. 9.1a,b, and h (connective tissues)	New micrographs
10	Fig. 10.2 (nervous tissue) Tables 10.1 and 10.2 (muscle and nervous tissues)	Added figure and micrograph Updated and expanded components
11	Table 11.2 (skin glands) Fig. 11.4 (fingernail) Fig. 11.5d (light skin) Assessments: Parts A, B, and C	Added table Added figure Added micrograph Expanded components
12	Fig. 12.6 (bone tissue) Fig. 12.7 (bone marrow locations)	New micrograph Added figure
13	Fig. 13.1a and b (bones) Assessments: Part A (bones)	New figures Improved figures in color
14	Fig. 14.7 (bones of orbit) Fig. 14.8 (paranasal sinuses) Assessments: Part D	Added figure Redrawn figure Expanded components
15	Figs. 15.1 (vertebrae) and 15.5 (rib) Assessments: Part B (Fig. 15.8)	Redrawn figures New radiograph of neck
16	Fig. 16.1 (shoulder and upper limb) Fig. 16.2 (clavicle) Assessments: Part C (Figs. 16.8 and 16.10)	Redrawn figure Added figure New radiographs of elbow and hand

LABORATORY EXERCISE	TOPIC	CHANGE
17	Assessments: Part C (Figs. 17.6, 17.7, and 17.8)	New radiographs of pelvis, knee, and foot
19	Procedure A and Table 19.1 (joints) Figs. 19.2 (synovial joint) Fig. 19.5a (supination/pronation) Assessments: Part B (Fig. 19.6) Assessments: Part E (Fig. 19.7)	Expanded and improved depth New figure New photograph Redrawn joints in full color Redrawn in full color and clarified movements
20	Procedure (skeletal muscle structure and function) Fig. 20.2b (neuromuscular junction) Table 20.1 (various roles of muscles) Assessments: Parts B and C	Expanded and improved depth Added figure Updated terminology and descriptions Revised and expanded components
22	Fig. 22.5 (scalene muscles)	Added figure
23	New organization Fig. 23.1 and Table 23.1 (respiratory muscles) Assessments: Parts A, B, C, D, and E (Figs. 23.7 and 23.8)	Content divided into Procedures A and B Added figure and table New organization of components and added respiratory muscle figures
24	Table 24.1 (vertebral muscles) Fig. 24.1 (vertebral column muscles) Assessments: Part A (Fig. 24.6) Assessments: Part B	Revised and expanded components Expanded and relabeled figure New figure of abdominal wall Expanded components
25	Fig. 25.5 (deep quadriceps muscles) Fig. 25.8 (lateral leg) Assessments: Part D (Fig. 25.11)	Added figure Redrawn figure Added figure of leg muscles of cadaver
26	Assessments: Part C (Fig. 26.10)	Redrawn figure of surface features
27	Fig. 27.4 (nervous tissue)	New micrograph
28	Meninges, Spinal Cord, and Spinal Nerves Figs. 28.1(meninges), 28.3 (spinal cord), and 28.5 (spinal nerves) Figs. 28.2 (spinal nerves), 28.4 (spinal cord), 28.6 (spinal nerves), and 28.13 (dermatomes) Figs. 28.7, 28.8, 28.9, 28.10, 28.11, and 28.12 (nerve plexuses) Table 28.1 (nerve plexuses) Assessments: Parts A, B, C, and D	New order of topics and expanded depth Relabeled figures Added figures Added figures Added table New order of assessments; Part D added
29	Reflex Arc and Somatic Reflexes Introductory material Figs. 29.1 (stretch reflex) and 29.2 (withdrawal reflex) Assessments: Part A table	Revised title Improved organization and depth New figures with improved depth Reorganized categories
30	Procedure A (cranial meninges) and Fig. 30.1 (cranial meninges) Table 30.3 (clinical tests of cranial nerves) Figs. 30.10 (olfactory nerve), 30.11 (trigeminal nerve), and 30.12 (facial nerve) Assessments: Part B Assessments: Part F	Added material and added figure Added table Added figures of selected cranial nerves Expanded components Added assessments of cranial nerves
31A	Reaction Time: BIOPAC Exercise	Added lab exercise
32	Figs. 32.3 and 32.4 (sheep brain) Fig. 32.7 (frontal section human brain)	Redrawn figures Added figure
34	Introductory material (smell and taste) Fig. 34.1 (olfactory epithelium) Fig. 34.3 (taste buds)	Updated and expanded components New micrograph and revised labels Added micrograph

CHANGES TO THIS EDITION

LABORATORY EXERCISE	TOPIC	CHANGE
35	Fig. 35.2a (eye muscles) Figs. 35.6 and 35.8 Fig. 35.7 (retina)	Added figure Redrawn figures Revised legend and labels
36	Fig. 36.4 (accommodation) Fig. 36.6 (color blindness plate)	Added figure New figure
37	Fig. 37.1 (auditory pathways)	Added figure
38	Fig. 38.1 (vestibular pathways)	Added figure
39	Figs. 39.4 (pituitary) and 39.12 (pancreas) Fig. 39.11 (pancreas)	New micrographs Redrawn figure and revised labels
40	Introductory material (diabetic physiology) Fig. 40.1 (pancreas) Assessments: Part C	Updated and expanded components Added micrograph Expanded components
41	Introductory material (blood cells) Figs. 41.1 and 41.6 (blood smears) Figs. 41.2 (clinic lab) and 41.3 (RBC)	Updated and expanded components Redrawn figures Added figures
42	Introductory material (blood components) Procedure B and Fig. 42.2 (Tallquist color scale) Fig. 42.4 (blood clot)	Updated and expanded depth Added Tallquist directions and figure Added micrograph

43	Procedure A and Figs. 43.1, 43.2, and 43.3
	    (agglutination figures)
	Figs. 43.4 and 43.6 (blood typing)
	Assessments: Part A
		Updated and improved depth of agglutination
		    reactions and added figures
		Redrawn figures
		Expanded components

44	Figs. 44.2 (heart) and 44.7 (cardiac muscle) Procedure B (sheep heart dissection) Figs. 44.9, 44.10, 44.11, and 44.12 (sheep hearts) Assessments: Part C (sheep hearts)	Added figure and micrograph Rewritten and expanded depth Added photographs Updated and expanded components

45	Figs. 45.1 (heart sounds), 45.3 (ECG), and 45.5
	    (ECG electrodes)
	Procedure B (electrocardiogram)
	Assessments: Part D and Fig. 45.7 (cardiac
	    conduction)
		Redrawn figures
		
		Updated content and materials used
		Expanded content and added figure

47	Fig. 47.3 (frog) Fig. 47.5 (pulmonary circuit) and 47.9b (celiac trunk) Figs. 47.12 and 47.14 (limb veins)	Redrawn figure Added figures Revised and added labels
48	Fig. 48.1 (pulse locations) Figs. 48.4 (sphygmomanometer) Fig. 48.5 (taking blood pressure)	New figure Redrawn figure New photograph
49	Fig. 49.2 (lymphatic capillaries) Fig. 49.3 (lymphatic valve) Fig. 49.9 (thymus)	Added figure Added micrograph New micrograph
50	Fig. 50.2b (upper respiratory) Fig. 50.3c (larynx sagittal view) Fig. 50.10 (lung) Assessments: Part C	Added figure of cadaver Added figure New micrograph Expanded components
51	Fig. 51.2 (Boyle's Law) Fig. 51.3 (air movements)	Added figure Redrawn figure
53	Procedure A (control of breathing) Fig. 53.1 (respiratory control) Fig. 53.2 (chemoreceptors) Table 53.1 ((muscles of respiration)	Updated and expanded depth Updated new figure Redrawn figure Added table

LABORATORY EXERCISE	TOPIC	CHANGE
54	Figs. 54.8 (pancreas), 54.9 (liver), and 54.11a (microscopic liver)	Added figures
	Figs. 54.10 (pancreas) and 54.11b (liver)	Added micrographs
	Figs. 54.15 (small intestine) and 54.16 (large intestine)	Revised micrograph labels
	Assessments: Part D	Revised organization and expanded components
56	Fig. 56.2a (kidney)	Added figure
	Procedure B (renal blood vessels)	Updated and expanded depth
	Figs. 56.5b (renal medulla), 56.7 (ureter), and 56.8 (urinary bladder)	New micrographs
57	Introductory material and Procedure A (urinalysis)	Updated materials, new organization, and expanded components
	Figs. 57.1 (hydrometer) and 57.3 (urine sediments)	Redrawn figures
	Fig. 57.2 (reagent test strip)	Added figure
	Assessments: Part A	Revised table and expanded components
58	Figs. 58.8 and 58.9 (male structures)	Redrawn figures
59	Introductory material and Procedure A (female structures and functions)	Expanded depth
60	Meiosis, Fertilization, and Early Development	New title and order of components
	Introductory material and Procedure A (meiosis and fertilization)	Added meiosis, updated, and expanded components
	Fig. 60.1 (meiosis)	Added figure
	Procedure B (sea urchin)	Expanded content and materials used
	Assessments: Parts A, B, and C	New organization and expanded components
62	Fig. 62.3 (cat skeleton)	Added photograph
	Fig. 62.5	New APR cat muscle figure
	Fig. 62.10	Improved image and added muscle labels
	Figs. 62.21 and 62.24	Revised muscle labels
63	Assessments: Part C (Figs. 63.16 and 63.17)	Added APR cat cardiovascular figures
64	Figs. 64.3 and 64.5	Added APR cat respiratory figures
65	Fig. 65.2b	Added APR cat digestive figure
	Assessments: Part C (Fig. 65.5b)	Added APR cat digestive figure
66	Figs. 66.2 and 66.5	Added APR cat urinary figures
67	Figs. 67.2b and 67.4b	Added APR cat reproductive figures

ACKNOWLEDGMENTS

I value all the support and encouragement by the staff at McGraw-Hill Education, including Michael Hackett, Amy Reed, Jessica Cannavo, Fran Simon, Vicki Krug, and Sherry Kane. Special recognition is granted to Colin Wheatley for his insight, confidence, wisdom, warmth, and friendship.

I am grateful for the professional talent of contributors Brian Kipp, Cindy Prentice-Craver, and Phillip Snider Jr. for valuable updated contents to all of the laboratory exercises, including the Ph.I.L.S. Physiology Interactive Lab Simulations and the digital content. The BIOPAC® labs by Janet Brodsky have been updated, and an additional lab has been added.

I am particularly thankful to Dr. Norman Jenkins and Dr. David Louis, retired presidents of Kishwaukee College, and Dr. Thomas Choice, president of Kishwaukee College, for their support, suggestions, and confidence in my endeavors. I am appreciative for the expertise of Womack Photography for numerous contributions. The professional reviews of the nursing procedures were provided by Kathy Schnier. I am also grateful to Laura Anderson, Rebecca Doty, Michele Dukes, Troy Hanke, Jenifer Holtzclaw, Stephen House, Shannon Johnson, Brian Jones, Marissa Kannheiser, Morgan Keen, Marcie Martin, Angele Myska, Sparkle Neal, Bonnie Overton, Susan Rieger, Eric Serna, Robert Stockley, Shatina Thompson, Nancy Valdivia, Marla Van Vickle, Jana Voorhis, Joyce Woo, and DeKalb Clinic for their contributions. There have been valuable contributions from my students, who have supplied thoughtful suggestions and assisted in clarification of details.

To my son Ross, an art instructor, I owe gratitude for his keen eye, creative suggestions, and creative cover illustration. Foremost, I am appreciative to Sherrie Martin, my spouse and best friend, for advice, understanding, and devotion throughout the writing and revising.

Terry R. Martin
Kishwaukee College
21193 Malta Road
Malta, IL 60150

Reviewers

I would like to express my sincere gratitude to all reviewers of the laboratory manual who provided suggestions for its improvement. Their thoughtful comments and valuable suggestions are greatly appreciated. They include the following:

Andrew E. Accardi
  Central Carolina Technical College
Sharon Allen
  Reading Area Community College
Bert Atsma
  Union County College
Patricia Bernard
  Erie Community College
Carolyn J.W. Bunde
  Idaho State University
Claire M Carpenter
  Yakima Valley Community College
Alan Crandall
  Idaho State University
Iveta D. Dinbergs
  Middlesex Community College
Anna East Gilletly
  Central New Mexico Community College (CNM)
Danny J. Ebert
  Boise State University
Martha Eshleman
  Pulaski Technical College
David Evans
  Pennsylvania College of Technology, an affiliate of the Pennsylvania State University

Mitchell A. Freymiller
  University of Wisconsin–Eau Claire
Louis Giacinti
  Milwaukee Area Technical College
Gary Glaser
  Genesee Community College
Mark Haefele
  Community College of Denver
Clare Hays
  Metropolitan State College of Denver
D. J. Hennager
  Kirkwood Community College
Michael A. Hill
  Pennsylvania College of Technology
Anthony Holt
  University of Arkansas Community College at Morrilton
Michele Iannuzzi Sucich
  SUNY Orange
Alexander Ibe
  Weatherford College
Mary Katherine Lockwood
  University of New Hampshire
Allan L. Markezich
  Black Hawk College
Herbert L. Matthews
  Pulaski Technical College

Tiffany B. McFalls
  Elizabethtown Community & Technical College
Igor V. Oksov
  Union County College
Karen Payne
  Chattanooga State Community College
Michael W. Ruhl
  Vernon College
Jennifer Shaw
  Polk State College
Marina G. Smitherman
  Dalton State College
Paula S. Steiert
  Southwest Baptist University
Avani Thaci
  Malcolm X Jr. College
Eileen Walsh
  Westchester Community College
Sheila M. Wicks
  City Colleges of Chicago/RUSH University Department of Clinical Anatomy and Cell Biology/ University of Illinois Department of Anatomy & Kinesiology
Ronika R. Williams
  Coastal Bend College

DEDICATION

McGraw-Hill Education along with Brian Kipp, Cynthia Prentice-Craver, and Phillip Snider wish to dedicate this lab manual to Terry and Sherrie Martin.

On Tuesday, May 6, 2014, the ribbon cutting took place for the Terry & Sherrie Martin Health Careers Wing at Kishwaukee College. A highlight of the Terry & Sherrie Martin Health Careers wing is the Nursing Simulation Lab. According to Kishwaukee College:

High fidelity human patient simulators allow our students to have controlled, customized exposure to clinical events.

I consider it a true honor to have been able to work with Terry Martin for so many years. He is the epitome of an instructor, author, and friend. He and Sherrie have dedicated their lives to education. It's a true privilege to call them friends.

Fran Simon
Sr. Product Developer

Terry and Sherrie, Your dedication in teaching and support of your community is inspirational. Your hard work and generosity will be admired and genuinely appreciated by many for years to come. Congratulations!

Cindy Prentice-Craver

I am so proud to know Terry and Sherrie, who have done so much to help students become the health care professionals of tomorrow.

Ricki Lewis

Two of the most genuine people I have ever met. "I am truly blessed to have met and gotten to know Terry and Sherrie. The integrity that they both show is truly inspirational."

Phillip Snider Jr.

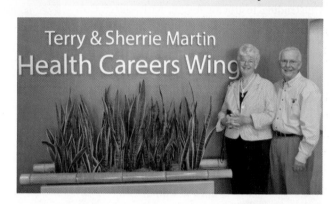

Terry and Sherrie Martin are two of those rare individuals who can take a sophisticated approach to solving a problem, yet keep things down to earth at the same time. I feel privileged to have them as colleagues and friends.

David Shier

Terry and Sherry Martin have always helped people achieve their dreams, as educators and now with this legacy of the Health Careers Wing at Kishwaukee. I am proud and inspired to have them as friends.

Jackie Butler

Terry's passion for teaching is infectious and evident in his writing. His clear and approachable writing style is a perfect reflection of his personality. It is an honor to work with him.

Brian Kipp

ABOUT THE AUTHOR

This laboratory manual series is by Terry R. Martin of Kishwaukee College. Terry's teaching experience of over forty years, his interest in students and love for college instruction, and his innovative attitude and use of technology-based learning enhance the solid tradition of his other well-established laboratory manuals. Among Terry's awards are the Kishwaukee College Outstanding Educator, Phi Theta Kappa Outstanding Instructor Award, Kishwaukee College ICCTA Outstanding Educator Award, Who's Who Among America's Teachers, Kishwaukee College Faculty Board of Trustees Award of Excellence, Continued Excellence Award for Phi Theta Kappa Advisors, and John C. Roberts Community Service Award. Terry's professional memberships include the National Association of Biology Teachers, Illinois Association of Community College Biologists, Human Anatomy and Physiology Society, former

TERRY R. MARTIN

Chicago Area Anatomy and Physiology Society (founding member), Phi Theta Kappa (honorary member), and Nature Conservancy. Terry revised the *Laboratory Manual to Accompany Hole's Human Anatomy and Physiology,* Fourteenth Edition, and revised the *Laboratory Manual to Accompany Hole's Essentials of Human Anatomy and Physiology*, Twelfth Edition. Terry teaches portions of EMT and paramedic classes and serves as a Faculty Consultant for Advanced Placement Biology examination readings. Terry has also been a faculty exchange member in Ireland. The author locally supports historical preservation, natural areas, scouting, and scholarship. We are pleased to have Terry continue the tradition of authoring laboratory manuals for McGraw-Hill Education.

The Editor

CONTRIBUTORS

BRIAN H. KIPP is an associate professor in the Biomedical Sciences Department at Grand Valley State University, in west Michigan. He earned a B.S. in Biological Science from Saginaw Valley State University in 1995 and a Ph.D. in Biology from Wayne State University in 2001. At WSU Brian's research focused on organic oxidation-reduction reactions and the generation of reactive oxygen species. Since coming to GVSU in 2002 he has been very active in the development of one- and two-semester anatomy and physiology courses and is responsible for creating the undergraduate pre-clinical A&P course sequence.

Brian has also been at the forefront of new and developing pedagogical methods of instruction. Since he began working with McGraw-Hill in 2007 he has worked with APR, authored digital content for other physiology books and LearnSmart, and served as a digital faculty consultant.

Brian has been married to his best friend Nicole since 2000, and they have three beautiful children who keep them very busy. Currently, Brian and two of his children are practitioners of the martial art of Tang Soo Do and will be testing for their black belts in 2016. Brian's other passion is the sport of powerlifting, where he has won state, national, and world championships.

CYNTHIA PRENTICE-CRAVER is a contributing author to this third edition of Terry Martin's laboratory manual series. At Chemeketa Community College in Oregon since 1994, Cynthia's professional passion has been teaching Human Anatomy and Physiology. Thirteen years ago, Cynthia pioneered the development of hybrid and all-online Human Anatomy and Physiology courses. Her M.S. in Curriculum and Instruction, along with undergraduate and graduate coursework in biological sciences, were instrumental in the effective results of these nontraditional courses. She is thrilled to be using the human cadaver lab at Chemeketa Community College in her teaching of Human Anatomy and Physiology. Cynthia's experience includes Life Sciences program chair for eight years, serving on faculty hiring and Nursing application committees, and serving as reviewer and advisor of textbooks and digital products. She is a member of the Human Anatomy and Physiology Society (HAPS). Beyond her professional pursuits, Cynthia's passions include being active outdoors, volunteering in local schools and in youth athletics, competing in numerous half-marathons, and raising her five boys who are now young men.

PHILLIP D. SNIDER JR. has been teaching Human Anatomy and Physiology at Gadsden State Community College for the last ten years. He also began teaching Human Gross Anatomy and Pathophysiology in 2008. He was instrumental in implementing the gross anatomy course and lab at his college, which distinguished it as the only community college in the state of Alabama to have two fully functional gross anatomy labs. Among Phillip's awards are Who's Who Among America's Teachers, Alabama Community College Systems Chancellor's Award inductee, Leadership Development Institute graduate at Gadsden State, and National Institute for Staff and Organizational Development Excellence Award. Phillip's professional memberships include Human Anatomy and Physiology Society (HAPS) and the National Institute for Staff and Organizational Development. Phillip also serves as a Deputy Coroner for Calhoun County, Alabama.

TO THE STUDENT

The exercises in this laboratory manual will provide you with opportunities to observe various anatomical structures and to investigate certain physiological phenomena. Such experiences should help you relate specimens, models, microscope slides, and your body to what you have learned in the lecture and read about in the textbook.

Frequent variations exist in anatomical structures among humans. The illustrations in the laboratory manual represent normal (normal means the most common variation) anatomy. Variations from normal anatomy do not represent abnormal anatomy unless some function is impaired.

The following list of suggestions and study skills may make your laboratory activities more effective and profitable.

1. Prepare yourself before attending the laboratory session by reading the assigned exercise and reviewing the related sections of the textbook and lecture notes as indicated in the pre-lab section of the laboratory exercise. Answer the pre-lab questions. It is important to have some understanding of what will be done in the lab before you come to class.

2. Be on time. During the first few minutes of the laboratory meeting, the instructor often will provide verbal instructions. Make special note of any changes in materials to be used or procedures to be followed. Also listen carefully for information about special techniques to be used and precautions to be taken.

3. Keep your work area clean and your materials neatly arranged so that you can locate needed items. This will enable you to proceed efficiently and will reduce the chances of making mistakes.

4. Pay particular attention to the purpose of the exercise, which states what you are to accomplish in general terms, and to the learning outcomes, which list what you should be able to do as a result of the laboratory experience. Then, before you leave the class, review the outcomes and make sure that you can perform all of the assessments.

5. Precisely follow the directions in the procedure and proceed only when you understand them clearly. Do not improvise procedures unless you have the approval of the laboratory instructor. Ask questions if you do not understand exactly what you are supposed to do and why you are doing it.

6. Handle all laboratory materials with care. Some of the materials are fragile and expensive to replace. Whenever you have questions about the proper treatment of equipment, ask the instructor.

7. Treat all living specimens humanely and try to minimize any discomfort they might experience.

8. Although at times you might work with a laboratory partner or a small group, try to remain independent when you are making observations, drawing conclusions, and completing the activities in the laboratory reports.

9. Record your observations immediately after making them. In most cases, such data can be entered in spaces provided in the laboratory assessments.

10. Read the instructions for each section of the laboratory assessment before you begin to complete it. Think about the questions before you answer them. Your responses should be based on logical reasoning and phrased in clear and concise language.

11. At the end of each laboratory period, clean your work area and the instruments you have used. Return all materials to their proper places and dispose of wastes, including glassware or microscope slides that have become contaminated with human blood or body fluids, as directed by the laboratory instructor. Wash your hands thoroughly before leaving the laboratory.

Study Skills for Anatomy and Physiology

Students have found that certain study skills worked well for them while enrolled in Human Anatomy and Physiology. Although everyone has his or her learning style, there are techniques that work well for most students. Using some of the skills listed here can make your course more enjoyable and rewarding.

1. **Time management:** Prepare monthly, weekly, and daily schedules. Include dates of quizzes, exams, and projects on the calendar. On your daily schedule, budget several short study periods. Daily repetition alleviates cramming for exams. Prioritize your tasks so that you still have time for work and leisure activities. Find an appropriate study atmosphere with minimum distractions.

2. **Note taking:** Look for the main ideas and briefly express them in your own words. Organize, edit, and review your notes soon after the lecture. Add textbook information to your notes as you reorganize them. Underline or highlight with different colors the important points, major headings, and key terms. Study your notes daily, as they provide sequential building blocks of the course content.

3. **Chunking:** Organize information into logical groups or categories. Study and master one chunk of information at

a time. For example, study the bones of the upper limb, lower limb, trunk, and head as separate study tasks.

4. **Mnemonic devices:** An *acrostic* is a combination of association and imagery to aid your memory. It is often in the form of a poem, rhyme, or jingle in which the first letter of each word corresponds to the first letters of the words you need to remember. **S**o **L**ong **T**op **P**art, **H**ere **C**omes **T**he **T**humb is an example of such a mnemonic device for remembering the eight carpals in a correct sequence. *Acronyms* are words formed by the first letters of the items to remember. *IPMAT* is an example of this type of mnemonic device to help you remember the phases of the cell cycle in the correct sequence. Try to create some of your own.

5. **Note cards/flash cards:** Make your own. Add labels and colors to enhance the material. Keep them with you; study them often and for short periods. Concentrate on a small number of cards at one time. Shuffle your cards and have someone quiz you on their content. As you become familiar with the material, you can set aside cards that don't require additional mastery.

6. **Recording and recitation:** An auditory learner can benefit by recording lectures and review sessions with a cassette recorder. Many students listen to the taped sessions as they drive or just before going to bed. Reading your notes aloud can help also. Explain the material to anyone (even if there are no listeners). Talk about anatomy and physiology in everyday conversations.

7. **Study groups:** Small study groups that meet periodically to review course material and compare notes have helped and encouraged many students. However, keep the group on the task at hand. Work as a team and alternate leaders. This group often becomes a support group.

Practice sound study skills during your anatomy and physiology endeavor.

The Use of Animals in Biology Education*

The National Association of Biology Teachers (NABT) believes that the study of organisms, including nonhuman animals, is essential to the understanding of life on Earth. NABT recommends the prudent and responsible use of animals in the life science classroom. NABT believes that biology teachers should foster a respect for life. Biology teachers also should teach about the interrelationship and interdependency of all things.

Classroom experiences that involve nonhuman animals range from observation to dissection. NABT supports these experiences so long as they are conducted within the long-established guidelines of proper care and use of animals, as developed by the scientific and educational community.

As with any instructional activity, the use of nonhuman animals in the biology classroom must have sound educational objectives. Any use of animals, whether for observation or dissection, must convey substantive knowledge of biology. NABT believes that biology teachers are in the best position to make this determination for their students.

NABT acknowledges that no alternative can substitute for the actual experience of dissection or other use of animals and urges teachers to be aware of the limitations of alternatives. When the teacher determines that the most effective means to meet the objectives of the class do not require dissection, NABT accepts the use of alternatives to dissection, including models and the various forms of multimedia. The Association encourages teachers to be sensitive to substantive student objections to dissection and to consider providing appropriate lessons for those students where necessary.

To implement this policy, NABT endorses and adopts the "Principles and Guidelines for the Use of Animals in Precollege Education" of the Institute of Laboratory Animals Resources (National Research Council). Copies of the "Principles and Guidelines" may be obtained from the ILAR (2101 Constitution Avenue, NW, Washington, DC 20418; 202-334-2590).

---

*Adopted by the Board of Directors in October 1995. This policy supersedes and replaces all previous NABT statements regarding animals in biology education.

EXERCISE 1

Scientific Method and Measurements

Purpose of the Exercise

To become familiar with the scientific method of investigation, learn how to formulate sound conclusions, and provide opportunities to use the metric system of measurements.

MATERIALS NEEDED

Meterstick

Calculator

Human skeleton

Learning Outcomes

After completing this exercise, you should be able to:

O1 Convert English measurements to the metric system, and vice versa.

O2 Calculate expected upper limb length and actual percentage of height from recorded upper limb lengths and heights.

O3 Apply the scientific method to test the validity of a hypothesis concerning the direct, linear relationship between human upper limb length and height.

O4 Design an experiment, formulate a hypothesis, and test it using the scientific method.

The **O** corresponds to the assessments **A** indicated in the Laboratory Assessment for this Exercise.

Pre-Lab

Carefully read the introductory material and examine the entire lab. Be familiar with the scientific method from lecture or the textbook. Answer the pre-lab questions.

Pre-Lab Questions Select the correct answer for each of the following questions:

1. To explain biological phenomena, scientists use a technique called
   a. the scientific method.   **b.** the scientific law.
   c. conclusions.   **d.** measurements.

2. Which of the following represents the correct sequence of the scientific method?
   a. analysis of data, conclusions, observations, experiment, hypothesis
   b. conclusions, experiment, hypothesis, analysis of data, observations
   c. observations, hypothesis, experiment, analysis of data, conclusions
   d. hypothesis, observations, experiment, analysis of data, conclusions

3. A hypothesis, verified continuously from experiments by many investigators, can become known as a
   a. control.   **b.** variable.
   c. valid result.   **d.** theory.

4. The most likely scientific unit for measuring the height of a person would be
   a. feet.   **b.** centimeters.
   c. inches.   **d.** kilometers.

5. Which of the following is *not* a unit of the metric system of measurements?
   a. centimeters   **b.** liters
   c. inches   **d.** millimeters

6. The hypothesis is formulated from the results of the experiment.
   a. True _____   **b.** False _____

7. A centimeter represents an example of a metric unit of length.
   a. True _____   **b.** False _____

Scientific investigation involves a series of logical steps to arrive at explanations for various biological phenomena. It reflects a long history of asking questions and searching for knowledge. This technique, called the *scientific method,* is used in all disciplines of science. It allows scientists to draw logical and reliable conclusions.

The scientific method begins with making *observations* related to the topic under investigation. This step commonly involves the accumulation of previously acquired information and/or your observations of the phenomenon. These observations are used to formulate a tentative explanation known as the *hypothesis.* An important attribute of a hypothesis is that it must be testable. The testing of the proposed hypothesis involves designing and performing a carefully controlled *experiment* to obtain data that can be used to support, reject, or modify the hypothesis. During the experiment to test the proposed hypothesis, it is important to be able to examine only a single changeable factor, known as a *variable.* An *independent variable* is one that can be changed, but is determined before the experiment occurs; a *dependent variable* is determined from the results of the experiment.

An *analysis of data* is conducted using sufficient information collected during the experiment. Data analysis may include organization and presentation of data as tables, graphs, and drawings. From the interpretation of the data analysis, *conclusions* are drawn. (If the data do not support the hypothesis, you must reexamine the experimental design and the data, and if needed develop a new hypothesis.) The final presentation of the information is made from the conclusions. Results and conclusions are presented to the scientific community for evaluation through peer reviews, presentations at professional meetings, and published articles. If many investigators working independently can validate the hypothesis by arriving at the same conclusions, the explanation can become

a *theory.* A theory serves as the explanation from a summary of known experiments and supporting evidence unless it is disproved by new information. The five components of the scientific method are summarized as

Observations
↓
Hypothesis
↓
Experiment
↓
Analysis of data
↓
Conclusions

Metric measurements are characteristic tools of scientific investigations. The English system of measurements is often used in the United States, so the investigator must make conversions from the English system to the metric system. Table 1.1 provides the conversion factors necessary to change from English to metric units.

PROCEDURE A: Using the Steps of the Scientific Method

This procedure represents a specific example of the order of the steps utilized in the scientific method. Each of the steps for this procedure will guide you through the proper sequence in an efficient pathway.

1. A correlation exists between the length of the upper and lower limbs and the height (stature) of an individual. For example, a person who has long upper limbs (the arm, forearm, and hand combined) tends to be tall. Make some visual observations of other people in your class to observe a possible correlation.

2. From such observations, the following hypothesis can be formulated: The length of a person's upper limb is equal to 0.4 (40%) of the height of the person. To test this hypothesis, perform the following experiment.

3. Use a meterstick (fig. 1.1) to measure an upper limb length of ten subjects. Place the meterstick in the axilla (armpit) and record the length in centimeters to the end of the longest finger (fig. 1.2). Obtain the height of

FIGURE 1.1   Metric ruler with metric lengths indicated. A meterstick length would be 100 centimeters (10 decimeters). (The image size is approximately to scale.)

Metric ruler

Table 1.1 Metric Measurement System and Conversions

Measurement	Unit & Abbreviation	Metric Equivalent	Conversion Factor Metric to English (approximate)	Conversion Factor English to Metric (approximate)
Length	1 kilometer (km)	1,000 (10^3) m	1 km = 0.62 mile	1 mile = 1.61 km
	1 meter (m)	100 (10^2) cm 1,000 (10^3) mm	1 m = 1.1 yards = 3.3 feet = 39.4 inches	1 yard = 0.9 m 1 foot = 0.3 m
	1 decimeter (dm)	0.1 (10^{-1}) m	1 dm = 3.94 inches	1 inch = 0.25 dm
	1 centimeter (cm)	0.01 (10^{-2}) m	1 cm = 0.4 inches	1 foot = 30.5 cm 1 inch = 2.54 cm
	1 millimeter (mm)	0.001 (10^{-3}) m 0.1 (10^{-1}) cm	1 mm = 0.04 inches	
	1 micrometer (μm)	0.000001 (10^{-6}) m 0.001 (10^{-3}) mm		
Mass	1 metric ton (t)	1,000 (10^3) kg	1 t = 1.1 ton	1 ton = 0.91 t
	1 kilogram (kg)	1,000 (10^3) g	1 kg = 2.2 pounds	1 pound = 0.45 kg
	1 gram (g)	1,000 (10^3) mg	1 g = 0.04 ounce	1 pound = 454 g 1 ounce = 28.35 g
	1 milligram (mg)	0.001 (10^{-3}) g		
Volume (liquids and gases)	1 liter (L)	1,000 (10^3) mL	1 L = 1.06 quarts	1 gallon = 3.78 L 1 quart = 0.95 L
	1 milliliter (mL)	0.001 (10^{-3}) L 1 cubic centimeter (cc or cm^3)	1 mL = 0.03 fluid ounce 1 mL = 1/5 teaspoon 1 mL = 15–16 drops	1 quart = 946 mL 1 fluid ounce = 29.6 mL 1 teaspoon = 5 mL
Time	1 second (s)	1/60 minute	same	same
	1 millisecond (ms)	0.001 (10^{-3}) s	same	same
Temperature	Degrees Celsius (°C)		°F = 9/5 °C + 32	°C = 5/9 (°F − 32)

FIGURE 1.2 Measurement of upper limb length.

each person in centimeters by measuring them without shoes against a wall (fig. 1.3). The height of each person can also be calculated by multiplying each individual's height in inches by 2.54 to obtain his/her height in centimeters. Record all your measurements in Part A of Laboratory Assessment 1.

4. The data collected from all of the measurements can now be analyzed. The expected (predicted) correlation between upper limb length and height is determined using the following equation:

Height $\times$ 0.4 = expected upper limb length

The observed (actual) correlation to be used to test the hypothesis is determined by

Length of upper limb/height = actual % of height

5. A graph is an excellent way to display a visual representation of the data. Plot the subjects' data in Part A of the laboratory assessment. Plot the upper limb length of each subject on the x-axis (independent variable) and the height of each person on the y-axis (dependent variable). A line is already located on the graph that represents a hypothetical relationship of 0.4 (40%) upper limb length compared to height. This is a graphic representation of the original hypothesis.

6. Compare the distribution of all of the points (actual height and upper limb length) that you placed on the graph with the distribution of the expected correlation represented by the hypothesis.

7. Complete Part A of the laboratory assessment.

PROCEDURE B: Design an Experiment

You have completed the steps of the scientific method with guidance directions in Procedure A. This procedure will allow for less guidance and more flexibility using the scientific method.

CRITICAL THINKING ACTIVITY

You have probably concluded that there is some correlation of the length of body parts to height. Often, when a skeleton is found, it is not complete. It is occasionally feasible to use the length of a single bone to estimate the height of an individual. Observe human skeletons and locate the humerus bone in an upper limb or the femur bone in a lower limb. Use your observations to identify a mathematical relationship between the length of the humerus or femur and height. Formulate a hypothesis that can be tested. Make measurements, analyze data, and develop a conclusion from your experiment. Complete Part B of the laboratory assessment.

FIGURE 1.3 Measurement of height.

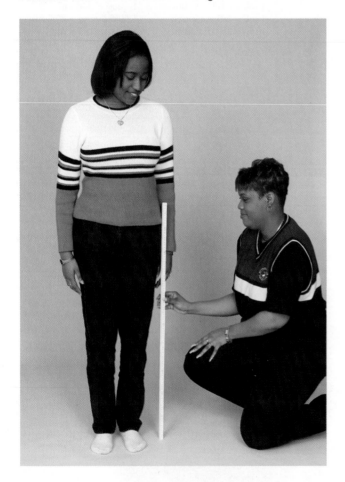

Name _____

Date _____

Section _____

The (A) corresponds to the indicated Learning Outcome(s) (O) found at the beginning of the Laboratory Exercise.

Scientific Method and Measurements

PART A: Assessments

1. Record measurements for the upper limb length and height of ten subjects. Use a calculator to determine the expected upper limb length and the actual percentage (as a decimal or a percentage) of the height for the ten subjects. Record your results in the following table. (A2)

Subject	Measured Upper Limb Length (cm)	Height* (cm)	Height × 0.4 = Expected Upper Limb Length (cm)	Actual % of Height = Measured Upper Limb Length (cm)/ Height (cm)
1.				
2.				
3.				
4.				
5.				
6.				
7.				
8.				
9.				
10.				

*The height of each person can be calculated by multiplying each individual's height in inches by 2.54 to obtain his/her height in centimeters.

2. Plot the distribution of data (upper limb length and height) collected for the ten subjects on the following graph. The line located on the graph represents the *expected* 0.4 (40%) ratio of upper limb length to measured height (the original hypothesis). (The x-axis represents upper limb length, and the y-axis represents height.) Draw a line of *best fit* through the distribution of points of the plotted data of the ten subjects. Compare the two distributions (expected line and the distribution line drawn for the ten subjects). **A3**

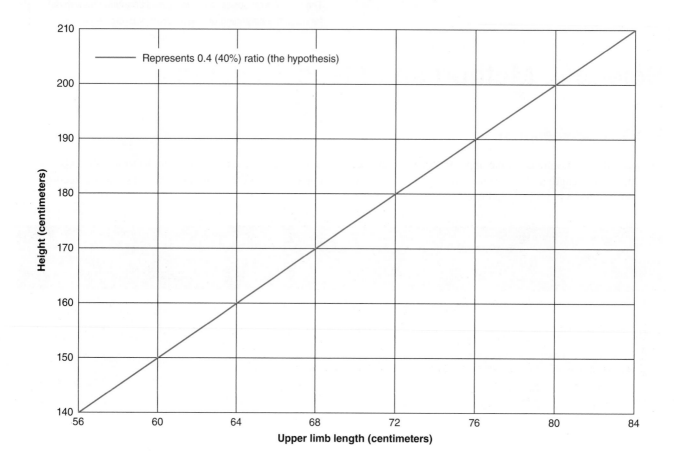

3. Does the distribution of the ten subjects' measured upper limb lengths support or reject the original hypothesis? _____ Explain your answer. **A3**

PART B: Assessments

1. Describe your observations of a possible correlation between the humerus or femur length and height.

2. Write a hypothesis based on your observations.

3. Describe the design of the experiment that you devised to test your hypothesis.

4. Place your analysis of the data in this space in the form of a table and a graph.

   a. Table:

b. Graph:

Humerus or femur length (centimeters)

(y-axis label: Height (centimeters))

5. Based on an analysis of your data, what can you conclude? Did these conclusions confirm or refute your original hypothesis? **A4**

6. Discuss your results and conclusions with classmates. What common conclusion can the class formulate about the correlation between the humerus or femur length and height? **A4**

Body Organization, Membranes, and Terminology

Purpose of the Exercise

To review the organizational pattern of the human body, to review its organ systems and the organs included in each system, and to become acquainted with the terms used to describe the relative position of body parts, body sections, and body regions.

MATERIALS NEEDED

Dissectible human torso model (manikin)

Variety of specimens or models sectioned along various planes

Learning Outcomes AP|R

After completing this exercise, you should be able to:

O1 Locate and name the major body cavities and identify the membranes associated with each cavity.

O2 Associate the organs and functions included within each organ system and locate the organs in a dissectible human torso model.

O3 Select the terms used to describe the relative positions of body parts.

O4 Differentiate the terms used to identify body sections and identify the plane along which a particular specimen is cut.

O5 Label body regions and associate the terms used to identify body regions.

The **O** corresponds to the assessments **A** indicated in the Laboratory Assessment for this Exercise.

Pre-Lab

Carefully read the introductory material and examine the entire lab. Be familiar with body cavities, membranes, organ systems, and body regions from lecture or the textbook. Answer the pre-lab questions.

Pre-Lab Questions Select the correct answer for each of the following questions:

1. The basis for communication in anatomy and physiology assumes
   a. the person is lying down.
   b. relative positions.
   c. anatomical position.
   d. the person is sleeping.

2. Which of the following is *not* a body cavity?
   a. diaphragm
   b. thoracic
   c. cranial
   d. abdominopelvic

3. The pericardium is associated with the
   a. lung.
   b. intestine.
   c. liver.
   d. heart.

4. The _____ plane divides the body into left and right sides.
   a. frontal
   b. cranial
   c. sagittal
   d. transverse

5. The abdominopelvic cavity can be subdivided into
   a. pleural cavities.
   b. pericardial cavities.
   c. quadrants.
   d. vertebral canals.

6. The larynx is part of the _____ system.
   a. urinary
   b. respiratory
   c. lymphatic
   d. nervous

7. The epigastric region is a portion of the _____ cavity.
   a. pelvic   b. pleural
   c. vertebral   d. abdominal
8. The brachial surface region pertains to the wrist.
   a. True _____   b. False ✓
9. A frontal plane divides the body into anterior and posterior parts.
   a. True ✓   b. False _____

The major features of the organization of the human body include certain body cavities. The *dorsal body cavity* includes a cranial cavity containing the brain and a vertebral canal (spinal cavity) containing the spinal cord. The *ventral body cavity* includes the *thoracic cavity,* which is subdivided into a mediastinum containing primarily the heart, esophagus, and trachea. The thoracic cavity also includes two pleural cavities, each surrounding a lung. Also included in the ventral body cavity is the *abdominopelvic cavity,* composed of an abdominal cavity and pelvic cavity. The entire abdominopelvic cavity is further subdivided into either nine regions or four quadrants. The large size of the abdominopelvic cavity, with its many visceral organs, warrants these further subdivisions into regions or quadrants for convenience and for accuracy in describing organ locations, injury sites, and pain locations.

Located within the ventral body cavities are thin serous membranes containing small amounts of a lubricating serous fluid. The double-layered membranes that secrete this fluid include the *pericardium, pleura,* and *peritoneum.* The pericardium is associated with the heart; the pleura is associated with a lung; the peritoneum is associated with the viscera located in the abdominopelvic cavity. The inner portion of each membrane attached to the organ is the visceral component, whereas the outer parietal portion forms an outer cavity wall. Several abdominopelvic organs, such as the kidneys, are located just behind (are retroperitoneal to) the parietal peritoneum, thus lacking a mesentery.

Although the human body functions as one entire unit, it is customary to divide the body into eleven body organ systems. In order to communicate effectively with each other about the body, scientists have devised anatomical terminology. Foremost in this task we use *anatomical position* as our basis for communication, including directional terms, body regions, and planes of the body. A person in anatomical position is standing erect, facing forward, with upper limbs at the sides and palms forward. This standard position allows us to describe relative positions of various body parts using such directional terms as left-right, anterior (ventral)-posterior (dorsal), proximal-distal, and superior-inferior. Body regions include certain surface areas, portions of limbs, and portions of body cavities. In order to study internal structures, often the body is depicted as sectioned into a sagittal plane, frontal plane, or transverse plane.

PROCEDURE A: Body Cavities and Membranes

This procedure outlines the body cavities by location. Certain cavities contain thin layers of cells, called membranes, that line the cavity and cover the organs within the cavity. The figures included in Procedure A will allow you to locate and identify the associated membrane location and name appropriate for the cavity and the organs involved.

1. Study figures 2.1, 2.2, and 2.3 to become familiar with body cavities and associated membranes.

FIGURE 2.1 AP|R   Major body cavities: (a) left lateral view; (b) anterior view.

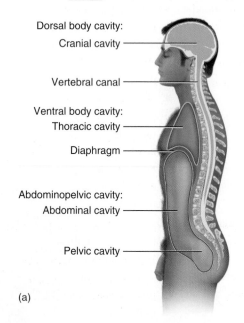

Dorsal body cavity:
Cranial cavity
Vertebral canal
Ventral body cavity:
Thoracic cavity
Diaphragm
Abdominopelvic cavity:
Abdominal cavity
Pelvic cavity

(a)

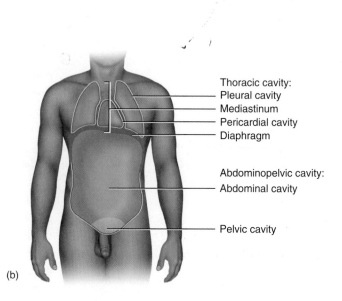

Thoracic cavity:
Pleural cavity
Mediastinum
Pericardial cavity
Diaphragm

Abdominopelvic cavity:
Abdominal cavity

Pelvic cavity

(b)

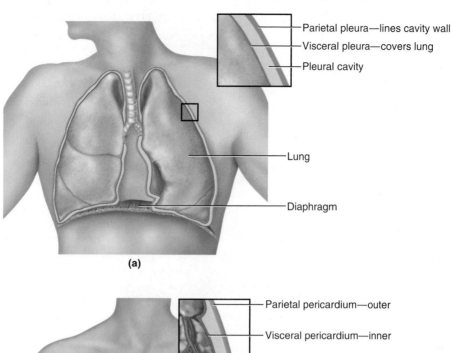

- Parietal pleura—lines cavity wall
- Visceral pleura—covers lung
- Pleural cavity

- Lung

- Diaphragm

(a)

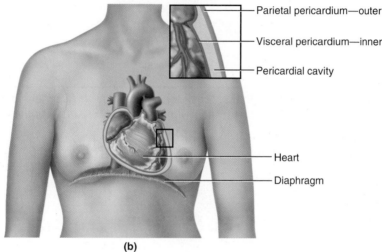

- Parietal pericardium—outer
- Visceral pericardium—inner
- Pericardial cavity

- Heart
- Diaphragm

(b)

2. Locate the following features on the dissectible human torso model (fig. 2.4):

body cavities

- cranial cavity—houses brain
- vertebral canal (spinal cavity)—houses spinal cord
- thoracic cavity
  - mediastinum—region between the lungs; includes pericardial cavity
  - pleural cavities (2)
- abdominopelvic cavity
  - abdominal cavity
  - pelvic cavity

diaphragm—separates thoracic and abdomino-pelvic cavities; functions in respiration

smaller cavities within the head

- oral cavity
- nasal cavity with connected sinuses
- orbital cavity
- middle ear cavity

membranes and cavities

- pleural cavity—associated with lungs
  - parietal pleura—lines cavity wall
  - visceral pleura—covers lungs
- pericardial cavity—associated with heart
  - parietal pericardium—covered by fibrous pericardium
  - visceral pericardium (epicardium)—covers heart
- peritoneal cavity—associated with abdominal organs
  - parietal peritoneum—lines cavity wall
  - visceral peritoneum—covers organs

3. Complete Part A of Laboratory Assessment 2.

FIGURE 2.3 AP|R   Serous membranes of the abdominal cavity are shown in this left lateral view.

Diaphragm
Liver
Pancreas
Stomach
Duodenum of small intestine
Large intestine
Small intestine
Mesentery
Urinary bladder
Rectum

Visceral peritoneum—covers organ

Peritoneal cavity
Greater omentum
Parietal peritoneum—lines cavity wall

Sagittal plane

FIGURE 2.4   Dissectible human torso model with body cavities, abdominopelvic quadrants, body planes, and major organs indicated. The abdominopelvic quadrants include right upper quadrant (RUQ), left upper quadrant (LUQ), right lower quadrant (RLQ), and left lower quadrant (LLQ).

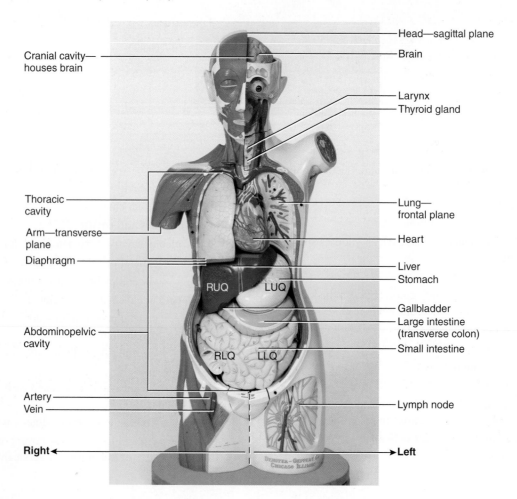

Head—sagittal plane
Brain
Larynx
Thyroid gland
Lung—frontal plane
Heart
Liver
Stomach
Gallbladder
Large intestine (transverse colon)
Small intestine
Lymph node

Cranial cavity—houses brain
Thoracic cavity
Arm—transverse plane
Diaphragm
Abdominopelvic cavity
Artery
Vein

RUQ   LUQ
RLQ   LLQ

Right ◄   ► **Left**

PROCEDURE B: Organ Systems

The eleven body systems are listed in this procedure, and a major function of each system is included. Representative major organs are listed within each of the systems. By using models and charts to locate each major organ included within the system, you will have an early overview of each system of the human body. More detailed organs and functions are covered within each system during later laboratory exercises.

1. Use the dissectible human torso model (fig. 2.4) to locate the following systems and their major organs:

    integumentary system—protection
    - skin—composed of epidermis and dermis
    - accessory organs—such as hair and nails

    skeletal system—support and protection
    - bones—in head, torso, and limbs
    - ligaments—connect bones to bones

    muscular system—movement
    - skeletal muscles—allow voluntary movements
    - tendons—connect muscles to bones

    nervous system—detects changes; interprets sensory information; stimulates muscles and glands
    - brain—within cranial cavity
    - spinal cord—extends through vertebral canal
    - nerves—conduct impulses into and from brain and spinal cord

    endocrine system—secretes hormones
    - pituitary gland—attached to base of brain
    - thyroid gland—anterior neck; inferior to larynx
    - parathyroid glands—four glands; embedded on posterior thyroid gland
    - adrenal glands—located superior to kidneys
    - pancreas—most in LUQ; influence blood sugar
    - ovaries—in females; produce reproductive hormones
    - testes—in males; produce testosterone
    - pineal gland—small gland within brain
    - thymus—within mediastinum

    cardiovascular system—transports gases, nutrients, and wastes
    - heart—muscular pump for blood
    - arteries—transport blood away from heart
    - veins—transport blood back to heart

    lymphatic system—immune cell production and return tissue fluid to blood
    - lymphatic vessels—carry lymph fluid
    - lymph nodes—along lymphatic vessels; help fight infections
    - thymus—within mediastinum
    - spleen—large organ in LUQ

    respiratory system—gas exchange between air and blood
    - nasal cavity—superior to mouth cavity
    - pharynx—passage for air superior to larynx
    - larynx—anterior neck; houses vocal cords
    - trachea—tube between larynx and bronchi
    - bronchi—airway tubes within lungs
    - lungs—large organs within thoracic cavity

    digestive system—food breakdown and absorption
    - mouth—contains tongue and teeth
    - tongue—for food manipulation
    - teeth—for biting and chewing food
    - salivary glands—secrete saliva into mouth
    - pharynx—passageway for food superior to esophagus
    - esophagus—tube from pharynx to stomach
    - stomach—between esophagus and small intestine; in LUQ
    - liver—produces bile; in RUQ
    - gallbladder—stores bile; in RUQ
    - pancreas—produces digestive enzymes; most in LUQ
    - small intestine—tube from stomach to large intestine
    - large intestine—tube from small intestine to anus

    urinary system—removes liquids and wastes from blood
    - kidneys—two large organs in upper adominopelvic cavity
    - ureters—tubes from kidneys to urinary bladder
    - urinary bladder—pelvic organ; stores urine
    - urethra—tube from urinary bladder to external opening

    male reproductive system—sperm production
    - scrotum—encloses testes
    - testes—produces sperm and hormones
    - penis—external reproductive organ
    - urethra—transports semen and urine

    female reproductive system—egg production and fetal development
    - ovaries—produce eggs and hormones
    - uterine tubes (oviducts; fallopian tubes)—transport eggs
    - uterus—muscular organ in pelvis; structure for fetal development
    - vagina—tube from uterus to external opening

2. Complete Part B of the laboratory assessment.

PROCEDURE C: Relative Positions, Planes, and Regions

This procedure illustrates and incorporates planes (sections), abdominopelvic subdivisions, and surface regions using the anatomical position of the human body. Communications and directions for anatomical study assume the person is in an anatomical position.

Anatomical planes (sections; cuts) of the entire body and individual organs allow a better understanding of the internal structure and function. Sectional anatomy is accomplished by cutting the body in standard ways to maximize the view of the particular organ. The anatomical planes include the *sagittal, frontal,* and *transverse* sections or cuts. Radiologic images by such techniques as a CT (computed tomography) scan or an MRI (magnetic resonance imaging) are examples of the advances and the importance of sectional anatomy for the diagnosis and treatment of medical situations.

The abdominopelvic area of the body is rather large and contains numerous viscera representing structural and functional parts of several body systems. People with abdominopelvic discomfort will often complain about it using terms like a "stomach ache." This is too general a term to be used for medical tests and procedures. Therefore, this abdominopelvic portion of the body is further subdivided into four *quadrants* and/or *nine regions*. Some medical disciplines prefer the use of quadrants, while others prefer the more specific use of the regions.

The general surface of the entire body is subdivided into regions using anatomical terms appropriate for that particular portion of the body. This becomes helpful for descriptions of injuries or pain locations that are more specific than using broader general terms such as head, neck, and upper limb, for example.

1. Observe the person standing in anatomical position (fig. 2.5). Anatomical terminology assumes the body is in anatomical position even though a person is often observed differently.
2. Study figures 2.6, 2.7, and 2.8 to become familiar with anatomical planes, abdominopelvic quadrants and regions, and body surface regions.
3. Examine the sectioned specimens on the demonstration table and identify the plane along which each is cut.
4. Complete Parts C, D, and E of the laboratory assessment.

FIGURE 2.5 AP|R Anatomical position with directional terms indicated. The body is standing erect, face forward, with upper limbs at the sides and palms forward. When the palms are forward (supinated) the radius and ulna in the forearm are nearly parallel. This results in an anterior view of the body. The relative positional terms are used to describe a body part's location in relation to other body parts.

FIGURE 2.6 AP|R Anatomical planes (sections) of the body.

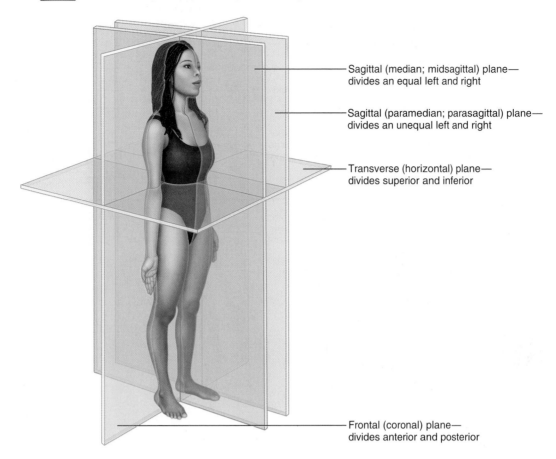

Sagittal (median; midsagittal) plane—
divides an equal left and right

Sagittal (paramedian; parasagittal) plane—
divides an unequal left and right

Transverse (horizontal) plane—
divides superior and inferior

Frontal (coronal) plane—
divides anterior and posterior

FIGURE 2.7 AP|R Abdominopelvic (a) quadrants and (b) regions. An area as large as the abdominopelvic cavity is subdivided into either four quadrants or nine regions for purposes of locating organs, injuries, or pain, or for performing medical procedures.

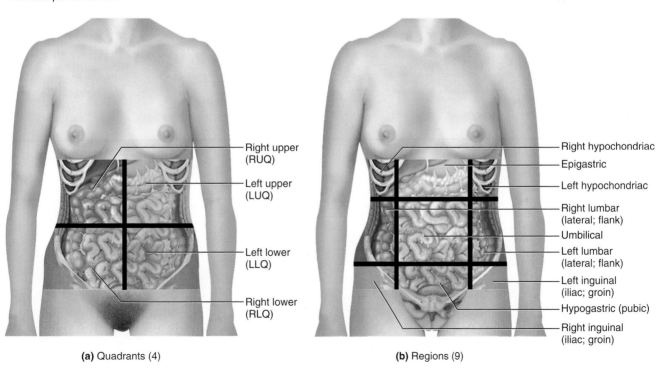

Right upper
(RUQ)

Left upper
(LUQ)

Left lower
(LLQ)

Right lower
(RLQ)

Right hypochondriac

Epigastric

Left hypochondriac

Right lumbar
(lateral; flank)

Umbilical

Left lumbar
(lateral; flank)

Left inguinal
(iliac; groin)

Hypogastric (pubic)

Right inguinal
(iliac; groin)

(a) Quadrants (4)

(b) Regions (9)

FIGURE 2.8 AP|R Diagrams of the body surface regions: (a) anterior regions; (b) posterior regions.

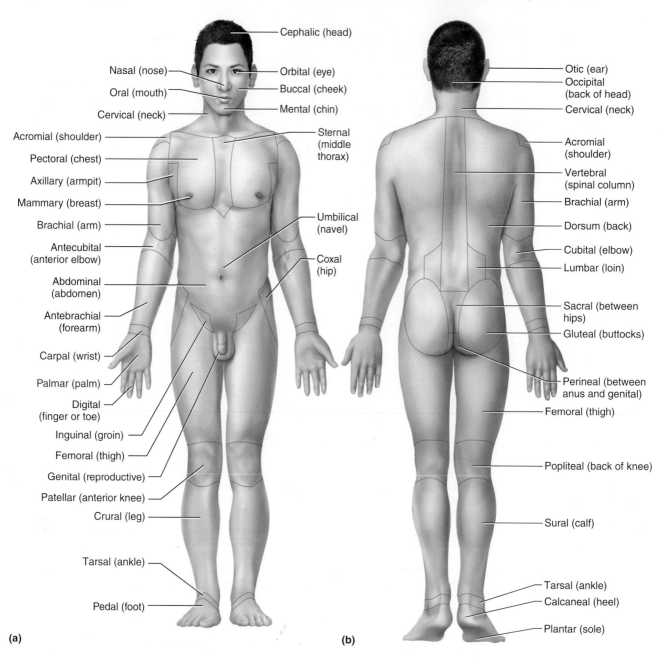

Cephalic (head)

Nasal (nose)
Oral (mouth)
Cervical (neck)

Orbital (eye)
Buccal (cheek)
Mental (chin)

Acromial (shoulder)
Pectoral (chest)
Axillary (armpit)
Mammary (breast)
Brachial (arm)
Antecubital (anterior elbow)
Abdominal (abdomen)
Antebrachial (forearm)
Carpal (wrist)
Palmar (palm)
Digital (finger or toe)
Inguinal (groin)
Femoral (thigh)
Genital (reproductive)
Patellar (anterior knee)
Crural (leg)
Tarsal (ankle)
Pedal (foot)

Sternal (middle thorax)

Umbilical (navel)

Coxal (hip)

Otic (ear)
Occipital (back of head)
Cervical (neck)
Acromial (shoulder)
Vertebral (spinal column)
Brachial (arm)
Dorsum (back)
Cubital (elbow)
Lumbar (loin)
Sacral (between hips)
Gluteal (buttocks)
Perineal (between anus and genital)
Femoral (thigh)
Popliteal (back of knee)
Sural (calf)
Tarsal (ankle)
Calcaneal (heel)
Plantar (sole)

(a)

(b)

ASSESSMENT

2

Name _____

Date _____

Section _____

The Ⓐ corresponds to the indicated Learning Outcome(s) Ⓞ found at the beginning of the Laboratory Exercise.

Body Organization, Membranes, and Terminology

PART A: Assessments

1. Match the body cavities in column A with the organs contained in the cavities in column B. Place the letter of your choice in the space provided. Ⓐ1 Ⓐ2

Column A	Column B
a. Abdominal cavity	_a._ **1.** Liver
b. Cranial cavity	_d._ **2.** Lungs
c. Pelvic cavity	_a._ **3.** Spleen
d. Thoracic cavity	_a._ **4.** Stomach
e. Vertebral canal (spinal cavity)	_b._ **5.** Brain
	_c._ **6.** Internal reproductive organs
	_c._ **7.** Urinary bladder
	_e._ **8.** Spinal cord
	_d._ **9.** Heart
	_c._ **10.** Small intestine

2. Label the body cavities in figure 2.9. Include an organ that would be located in each of the cavities, using the terms provided. Ⓐ1 Ⓐ2

FIGURE 2.9 Label body cavities 1–5. Add a representative organ located in the cavity, using the terms provided.

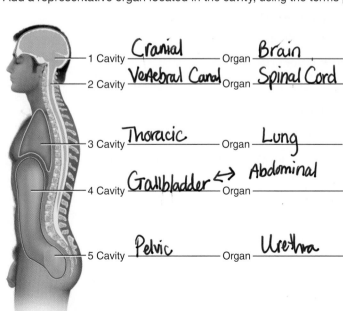

Terms:
Brain ✓
Gallbladder ✓
Lung ✓
Spinal cord ✓
Urethra ✓

1 Cavity _Cranial_ — Organ _Brain_
2 Cavity _Vertebral Canal_ — Organ _Spinal Cord_
3 Cavity _Thoracic_ — Organ _Lung_
4 Cavity _Gallbladder_ ↔ Organ _Abdominal_
5 Cavity _Pelvic_ — Organ _Urethra_

PART B: Assessments

Match the organ systems in column A with the principal functions in column B. Place the letter of your choice in the space provided. (A2)

Column A	Column B
a. Cardiovascular system	_c._ 1. The main system that secretes hormones
b. Digestive system	_m d._ 2. Provides an outer covering of the body for protection
c. Endocrine system	_h._ 3. Produces gametes (eggs and sperm)
d. Integumentary system	_g._ 4. Stimulates muscles to contract and interprets information from sensory organs
e. Lymphatic system	_j._ 5. Provides a framework and support for soft tissues and produces blood cells in red marrow
f. Muscular system	_i._ 6. Exchanges gases between air and blood
g. Nervous system	_e._ 7. Transports excess fluid from tissues to blood and produces immune cells
h. Reproductive system	_f._ 8. Involves contractions and creates most body heat
i. Respiratory system	_k._ 9. Removes liquid and wastes from blood and transports them to the outside of the body
j. Skeletal system	_b._ 10. Converts food molecules into forms that are absorbable
k. Urinary system	_a._ 11. Transports nutrients, wastes, and gases throughout the body

PART C: Assessments

Indicate whether each of the following sentences makes correct or incorrect usage of the word in boldface type (assume that the body is in the anatomical position). If the sentence is incorrect, in the space provided supply a term that will make it correct. (A3)

1. The mouth is **superior** to the nose. _false_

2. The stomach is **inferior** to the diaphragm. _true_

3. The trachea is **anterior** to the spinal cord. _true_

4. The larynx is **posterior** to the esophagus. _____

5. The heart is **medial** to the lungs. _____

6. The kidneys are **inferior** to the lungs. _true_

7. The hand is **proximal** to the elbow. _true_

8. The knee is **proximal** to the ankle. _false_

9. The thumb is the **lateral** digit of a hand. _true_

10. The popliteal surface region is **anterior** to the patellar region. _False_

CRITICAL THINKING ASSESSMENT

State the quadrant of the abdominopelvic cavity where the pain or sound would be located for each of the six conditions listed. In some cases, there may be more than one correct answer, and pain is sometimes referred to another region. This phenomenon, called *referred pain,* occurs when pain is interpreted as originating from some area other than the parts being stimulated. When referred pain is involved in the patient's interpretation of the pain location, the proper diagnosis of the ailment is more challenging. For the purpose of this exercise, assume the pain is interpreted as originating from the organ involved. **A2**

1. Stomach ulcer _____

2. Appendicitis _____

3. Bowel sounds _____

4. Gallbladder attack _____

5. Kidney stone in left ureter _____

6. Ruptured spleen _____

PART E: Assessments

Label figures 2.10 and 2.11.

FIGURE 2.10   Label the planes of the sectioned (cut) body part.   **A4**

1. Midsaggital
2. Coronal
3. Transverse

FIGURE 2.11   Label the indicated body surface regions 1–14.   A5

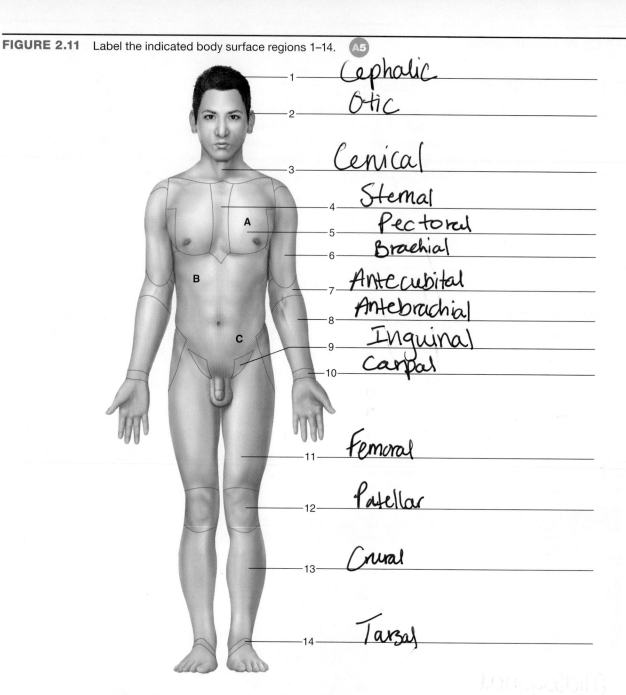

1 — Cephalic
2 — Otic
3 — Cervical
4 — Sternal
5 — Pectoral
6 — Brachial
7 — Antecubital
8 — Antebrachial
9 — Inguinal
10 — Carpal
11 — Femoral
12 — Patellar
13 — Crural
14 — Tarsal

CRITICAL THINKING ASSESSMENT

Examine the letters A, B, and C indicated on figure 2.11 that indicate a possible site of a deep puncture wound. Name the body cavity and the major internal organs that could be damaged from the injury.   A1   A2

Injury Site	Body Cavity	Internal Organ
A	Thoracic	Heart
B	Abdominopelvic	Liver
C	Pelvic	Small Intestine

LABORATORY EXERCISE 3

Chemistry of Life

Purpose of the Exercise

To review pH and basic categories of organic compounds, and to differentiate between types of organic compounds.

For pH Tests:

Chopped fresh red cabbage

Beaker (250 mL)

Distilled water

Tap water

Vinegar

Baking soda

Pipets for measuring

Laboratory scoop for measuring

Full-range pH test papers

7 assorted common liquids clearly labeled in closed bottles on a tray or in a tub

Droppers labeled for each liquid

For Organic Tests:

Test tubes

Test-tube rack

Test-tube clamps

China marker

Hot plate

Beaker for hot water bath (500 mL)

Pipets for measuring

Benedict's solution

Sudan IV dye

Biuret reagent (or 10% NaOH and 1% $CuSO_4$)

Iodine-potassium-iodide (IKI) solution

Egg albumin

10% glucose solution

Clear carbonated soft drink

10% starch solution

Potatoes for potato water

Distilled water

Vegetable oil

Brown paper

Numbered unknown organic samples

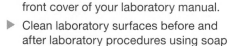

SAFETY

▶ Review all safety guidelines inside the front cover of your laboratory manual.

▶ Clean laboratory surfaces before and after laboratory procedures using soap and water.

▶ Use extreme caution when working with chemicals.

▶ Wear safety goggles.

▶ Wear disposable gloves while working with the chemicals.

▶ Take precautions to prevent chemicals from contacting your skin.

▶ Do not mix any of the chemicals together unless instructed to do so.

▶ Clean up any spills immediately and notify the instructor at once.

▶ Wash your hands before leaving the laboratory.

Learning Outcomes AP|R

After completing this exercise, you should be able to:

O1 Demonstrate pH values of various substances through testing methods.

O2 Determine categories of organic compounds with basic colorimetric tests.

O3 Discover the organic composition of an unknown solution.

The **O** corresponds to the assessments **A** indicated in the Laboratory Assessment for this Exercise.

Pre-Lab

Carefully read the introductory material and examine the entire lab. Be familiar with pH and organic molecules from lecture or the textbook. Answer the pre-lab questions.

Pre-Lab Questions Select the correct answer for each of the following questions:

1. The most basic unit of matter is
   - **a.** energy.
   - **b.** an element.
   - **c.** a molecule.
   - **d.** a cell.

2. Which number would represent a neutral pH?
   - **a.** 100
   - **b.** 0
   - **c.** 7
   - **d.** 14

3. The building block of a protein is
   - **a.** an amino acid.
   - **b.** a monosaccharide.
   - **c.** a lipid.
   - **d.** starch.

4. The building block of a carbohydrate is
   - **a.** glycerol.
   - **b.** an amino acid.
   - **c.** fatty acids.
   - **d.** a monosaccharide.

5. The test for starch uses which substance?
   - **a.** IKI (iodine solution)
   - **b.** Benedict's solution
   - **c.** Sudan IV dye
   - **d.** Biuret reagent

6. A positive test for lipids results in a
   a. pinkish color change.
   b. yellow color change.
   c. translucent stain.
   d. blue to red color change.
7. Which of the following pH numbers would represent the strongest acid?
   a. 1   b. 5
   c. 7   d. 14

The complexities of the human body arise from the organization and interactions of chemicals. Organisms are made of matter, and the most basic unit of matter is the chemical element. The smallest unit of an element is an atom, and two or more of those can unite to form a molecule. All processes that occur within the body involve chemical reactions—interactions between atoms and molecules. We breathe to supply oxygen to our cells for energy. We eat and drink to bring chemicals into our bodies that our cells need. Water fills and bathes all of our cells and allows an amazing array of reactions to occur, all of which are designed to keep us alive. Chemistry forms the very basis of life and thus forms the foundation of anatomy and physiology.

The *pH scale* is from 0 to 14 with the midpoint of 7.0 (pure water) representing a *neutral* solution. If the solution has a pH lower than 7.0, it represents an *acidic* solution with more hydrogen ions (H^+) than hydroxide ions (OH^-). A *basic* solution has a pH higher than 7.0 with more OH^- ions than H^+ ions. The weaker acids and bases are closer to 7, with the strongest acids near 0 and the strongest bases near 14. A seemingly minor change in the pH actually represents a more significant change as there is a tenfold difference in hydrogen ion concentrations with each whole number represented on the pH scale. Our cells can only function within minimal pH fluctuations.

The *organic molecules* (biomolecules) that are tested in this laboratory exercise include *carbohydrates, lipids,* and *proteins.* Carbohydrates are used to supply energy to our cells. The polysaccharide starch is composed of simple sugar (monosaccharide) building blocks. Lipids include fats, phospholipids, and steroids that provide some cellular structure and energy for cells. The building blocks of lipids include fatty acids and glycerol. Proteins compose important cellular structures, antibodies, and enzymes. Amino acids are the building blocks of proteins.

PROCEDURE A: The pH Scale

In this section, you will be able to determine the pH of various substances. The abbreviation pH is traditionally used to measure the power of H (hydrogen). The pH is a mathematical way to refer to the negative logarithm of the hydrogen ion (H^+) concentration in a solution. Each change of a number on the pH scale actually represents a tenfold increase or decrease of the hydrogen ion concentration.

The pH of various body fluids is very important in the homeostasis of a particular body system and the entire living organism. Blood, urine, saliva, and gastric secretions are examples of some body fluids with characteristic pH ranges for a healthy individual. Some body fluids like blood must fall within a very narrow pH range, while others like urine have a much wider normal range. In order to maintain normal ranges of the pH of body fluids, various buffering systems are involved to resist pH fluctuations. Some of these systems entail chemical buffers and others consist of physiological buffers to help regulate the pH levels. The acid-base balance is important since our bodies are composed predominantly of water. Various acid-base relationships are covered within the discussion of each body system.

1. Study figure 3.1, which shows the pH scale. Note the range of the scale and the pH value that is considered neutral.
2. **Cabbage water tests.** Many tests can determine a pH value, but among the more interesting are colorimetric tests in which an indicator chemical changes color when

FIGURE 3.1 AP|R   As the concentration of hydrogen ions (H^+) increases, a solution becomes more acidic and the pH value decreases. As the concentration of ions that combine with hydrogen ions (such as hydroxide ions) increases, a solution becomes more basic (alkaline) and the pH value increases. The pH values of some common substances are shown.

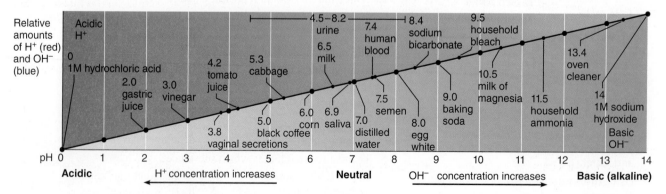

it reacts. Many plant pigments, especially anthocyanins (which give plants color, from blue to red), can be used as colorimetric pH indicators. One that works well is the red pigment in red cabbage.

    a. Prepare cabbage water to be used as a general pH indicator. Fill a 250 mL beaker to the 100 mL level with chopped red cabbage. Add water to make 150 mL. Place the beaker on a hot plate and simmer the mixture until the pigments come out of the cabbage and the water turns deep purple. Allow the water to cool. (You may proceed to step 3 while you wait for this to finish.)

    b. Label three clean test tubes: one for water, one for vinegar, and one for baking soda.

    c. Place 2 mL of cabbage water into each test tube.

    d. To the first test tube, add 2 mL of distilled water and swirl the mixture. Note and record the color in Part A of Laboratory Assessment 3.

    e. Repeat this procedure for test tube 2, adding 2 mL of vinegar and swirling the mixture. Note and record the results in Part A of the laboratory assessment.

    f. Repeat this procedure for test tube 3, adding one laboratory scoop of baking soda. Swirl the mixture. Note and record the results in Part A of the laboratory assessment.

  3. Testing with pH paper. Many commercial pH indicators are available. A very simple one to use is pH paper, which comes in small strips. Don gloves for this procedure so your skin secretions do not contaminate the paper and to protect you from the chemicals you are testing.

    a. Test the pH of distilled water by dropping one or two drops of distilled water onto a strip of pH paper, and note the color. Compare the color to the color guide on the container. Note the results of the test as soon as color appears, as drying of the paper will change the results. Record the pH value in Part A of the laboratory assessment.

    b. Repeat this procedure for tap water. Record the results.

    c. Individually test the various common substances found on your lab table. Record your results.

    d. Complete Part A of the laboratory assessment.

PROCEDURE B: Organic Molecules

In this section, you will perform some simple tests to check for the presence of the following categories of organic molecules (biomolecules): protein, sugar, starch, and lipid. Specific color changes will occur if the target compound is present. Please note the original color of the indicator being added so you can tell if the color really changed. For example, Biuret reagent and Benedict's solution are both initially blue, so an end color of blue would indicate no change. Use caution when working with these chemicals and follow all directions. Carefully label the test tubes and avoid contaminating any of your samples. Only the Benedict's test requires heating and a time delay for accurate results. Do not heat any

other tubes. To prepare for the Benedict's test, fill a 500 mL beaker half full with water and place it on the hot plate. Turn the hot plate on and bring the water to a boil. To save time, start the water bath before doing the Biuret test.

  1. Biuret test for protein. In the presence of protein, Biuret reagent reacts with peptide bonds and changes to violet or purple. A pinkish color indicates that shorter polypeptides are present. The color intensity is proportional to the number of peptide bonds; thus, the intensity reflects the length of polypeptides (amount of protein).

    a. Label six test tubes as follows: 1W, 1E, 1G, 1D, 1S, and 1P. The codes placed on the six test tubes used for the protein test include a number *1* referring to test 1 (biuret test for protein) followed by a letter. Letter *W* represents water (distilled); letter *E* represents egg (albumin); letter *G* represents glucose (10% solution); letter *D* represents drink (carbonated soft); letter *S* represents starch (solution); and letter *P* represents potato (juice). Similar codes are used for the tests for sugar (test 2) and starch (test 3) of Procedure B of this exercise.

    b. To each test tube, add 2 mL of one of the samples as follows:
      1W—2 mL distilled water
      1E—2 mL egg albumin
      1G—2 mL 10% glucose solution
      1D—2 mL carbonated soft drink
      1S—2 mL 10% starch solution
      1P—2 mL potato juice

    c. To each, add 2 mL of Biuret reagent, swirl the tube to mix it, and note the final color. [*Note:* If Biuret reagent is not available, add 2 mL of 10% NaOH (sodium hydroxide) and about 10 drops of 1% $CuSO_4$ (copper sulfate).] **Be careful—NaOH is very caustic.**

    d. Record your results in Part B of the laboratory assessment.

    e. Mark a "+" on any tubes with positive results and retain them for comparison while testing your unknown sample. Put remaining test tubes aside so they will not get confused with future trials.

  2. Benedict's test for sugar (monosaccharides). In the presence of sugar, Benedict's solution changes from its initial blue to a green, yellow, orange, or reddish color, depending on the amount of sugar present. Orange and red indicate greater amounts of sugar.

    a. Label six test tubes as follows: 2W, 2E, 2G, 2D, 2S, and 2P.

    b. To each test tube, add 2 mL of one of the samples as follows:
      2W—2 mL distilled water
      2E—2 mL egg albumin
      2G—2 mL 10% glucose solution
      2D—2 mL carbonated soft drink
      2S—2 mL 10% starch solution
      2P—2 mL potato juice

c. To each, add 2 mL of Benedict's solution, swirl the tube to mix it, and place all tubes into the boiling water bath for 3 to 5 minutes.

d. Note the final color of each tube after heating and record the results in Part B of the laboratory assessment.

e. Mark a "+" on any tubes with positive results and retain them for comparison while testing your unknown sample. Put remaining test tubes aside so they will not get confused with future trials.

3. Iodine test for starch. In the presence of starch, iodine turns a dark purple or blue-black color. Starch is a long chain formed by many glucose units linked together side by side. This regular organization traps the iodine molecules and produces the dark color.

a. Label six test tubes as follows: 3W, 3E, 3G, 3D, 3S, and 3P.

b. To each test tube, add 2 mL of one of the samples as follows:

3W—2 mL distilled water

3E—2 mL egg albumin

3G—2 mL 10% glucose solution

3D—2 mL carbonated soft drink

3S—2 mL 10% starch solution

3P—2 mL potato juice

c. To each, add 0.5 mL of IKI (iodine solution) and swirl the tube to mix it.

d. Record the final color of each tube in Part B of the laboratory assessment.

e. Mark a "+" on any tubes with positive results and retain them for comparison while testing your unknown sample. Put remaining test tubes aside so they will not get confused with future trials.

4. Tests for lipids.

a. Label two separate areas of a piece of brown paper as "water" or "oil."

b. Place a drop of water on the area marked "water" and a drop of vegetable oil on the area marked "oil."

c. Let the spots dry several minutes, then record your observations in Part C of the laboratory assessment. Upon drying, oil leaves a translucent stain (grease spot) on brown paper; water does not leave such a spot.

d. In a test tube, add 2 mL of water and 2 mL of vegetable oil and observe. Shake the tube vigorously; then let it sit for 5 minutes and observe again. Record your observations in Part B of the laboratory assessment.

e. Sudan IV is a dye that is lipid-soluble but not water-soluble. If lipids are present, the Sudan IV will stain them pink or red. Add a small amount of Sudan IV to the test tube that contains the oil and water. Swirl it; then let it sit for a few minutes. Record your observations in Part B of the laboratory assessment.

PROCEDURE C: Identifying Unknown Compounds

Now you will apply the information you gained with the tests in the previous section. You will retrieve an unknown sample that contains none, one, or any combination of the following types of organic compounds: protein, sugar, starch, or lipid. You will test your sample using each test from the previous section and record your results in Part C of the laboratory assessment.

1. Label four test tubes as follows: (1, 2, 3, and 4).

2. Add 2 mL of your unknown sample to each test tube.

3. Test for protein. Add 2 mL of Biuret reagent to tube 1. Swirl the tube and observe the color. Record your observations in Part C of the laboratory assessment.

4. Test for sugar. Add 2 mL of Benedict's solution to tube 2. Swirl the tube and place it in a boiling water bath for 3 to 5 minutes. Record your observations in Part C of the laboratory assessment.

5. Test for starch. Add several drops of iodine solution to tube 3. Swirl the tube and observe the color. Record your observations in Part C of the laboratory assessment.

6. Test for lipid. Add 2 mL of water to tube 4. Swirl the tube and note if there is any separation.

7. Add a small amount of Sudan IV to test tube 4, swirl the tube, and record your observations in Part C of the laboratory assessment.

8. Based on the results of these tests, determine if your unknown sample contains any organic compounds and, if so, what they are. Record and explain your identification in Part C of the laboratory assessment.

Name _____

Date _____

Section _____

The (A) corresponds to the indicated Learning Outcome(s) (O) found at the beginning of the Laboratory Exercise.

Chemistry of Life

PART A: Assessments

1. Results from cabbage water tests: (A1)

Substance	Cabbage Water	Distilled Water	Vinegar	Baking Soda
Color				
Acid, base, or neutral?				

2. Results from pH paper tests: (A1)

Substance Tested	Distilled Water	Tap Water	Sample 1: _____	Sample 2: _____	Sample 3: _____	Sample 4: _____	Sample 5: _____	Sample 6: _____	Sample 7: _____
pH Value									

3. Are the pH values the same for distilled water and tap water? _____

4. If not, what might explain this difference? _____

5. Draw the pH scale below, and indicate the following values: 0, 7 (neutral), and 14. Now label the scale with the names and indicate the location of the pH values for each substance you tested.

PART B: Assessments

1. **Biuret test results for protein.** Enter your results from the Biuret test for protein. A color change to purple indicates that protein is present; pink indicates that short polypeptides are present. (A2)

Tube	Contents	Color	Protein Present (+) or Absent (−)
1W	Distilled water		
1E	Egg albumin		
1G	Glucose solution		
1D	Soft drink		
1S	Starch solution		
1P	Potato juice		

2. **Benedict's test results for most sugars (monosaccharides).** Enter your results from the Benedict's test for sugars. A color change to green, yellow, orange, or red indicates that sugar is present. Note the color after the mixture has been heated for 3 to 5 minutes. **A2**

Tube	Contents	Color	Sugar Present (+) or Absent (–)
2W	Distilled water		
2E	Egg albumin		
2G	Glucose solution		
2D	Soft drink		
2S	Starch solution		
2P	Potato juice		

3. **Iodine test results for starch.** Enter your results from the iodine test for starch. A color change to dark blue or black indicates that starch is present. **A2**

Tube	Contents	Color	Sugar Present (+) or Absent (–)
3W	Distilled water		
3E	Egg albumin		
3G	Glucose solution		
3D	Soft drink		
3S	Starch solution		
3P	Potato juice		

4. **Lipid test results.** What did you observe when you allowed the drops of oil and water to dry on the brown paper? **A2**

_____

What did you observe when you mixed the oil and water together? _____

_____

What did you observe when you added the Sudan IV dye? **A2** _____

PART C: Assessments

1. What is the number of your sample? _____

2. Record the results of your tests on the unknown here: **A3**

Test Performed	Results
Biuret test for protein	
Benedict's test for sugars	
Iodine test for starch	
Water test for lipid	
Sudan IV test for lipid	

3. Based on these results, what organic compound(s) does your unknown sample contain? **A3** _____

_____

4. Explain your answer. **A3** _____

_____

CRITICAL THINKING ASSESSMENT

If a person were on a low-carbohydrate, high-protein diet, which of the six substances tested would the person want to increase? _____

Explain your answer. _____

_____

_____

_____

_____

NOTES

LABORATORY EXERCISE 4

Care and Use of the Microscope

Purpose of the Exercise

To become familiar with the major parts of a compound light microscope. Be able to use the microscope to observe small objects.

MATERIALS NEEDED

Compound light microscope
Lens paper
Microscope slides
Coverslips
Transparent plastic millimeter ruler
Prepared slide of letter *e*
Slide of three colored threads
Medicine dropper
Dissecting needle (needle probe)
Specimen examples for wet mounts
Methylene blue (dilute) or iodine-potassium-
    iodide stain

For Demonstration Activities:
Micrometer scale
Stereomicroscope (dissecting microscope)

Learning Outcomes

After completing this exercise, you should be able to:

O1 Differentiate the major functional parts of a compound light microscope.

O2 Calculate the total magnification produced by various combinations of eyepiece and objective lenses.

O3 Demonstrate proper use of the microscope to observe and measure small objects.

O4 Prepare a simple microscope slide and sketch the objects you observed.

The **O** corresponds to the assessments **A** indicated in the Laboratory Assessment for this Exercise.

Pre-Lab

Carefully read the introductory material and examine the entire lab. Visit www.mhhe.com/martinseries2 for LabCam videos. Answer the pre-lab questions.

Pre-Lab Questions Select the correct answer for each of the following questions:

1. The human eye cannot perceive objects less than
    a. one inch.   **b.** one millimeter.
    c. 0.1 millimeter.   **d.** one centimeter.

2. The objective lenses of the compound light microscope are attached to the
    a. stage.   **b.** base.
    c. body tube.   **d.** rotating nosepiece.

3. Which objective lens provides the least total magnification?
    a. scan   **b.** low power
    c. high power   **d.** oil immersion

4. The _____ increases or decreases the light intensity of the compound light microscope.
    a. eyepiece   **b.** stage
    c. adjustment knob   **d.** iris diaphragm

5. Basic lens cleaning is accomplished using
    a. water.   **b.** lens paper.
    c. a paper towel.   **d.** xylene.

6. When preparing a wet mount specimen for viewing, it should be covered with
    a. clear paper.   **b.** another glass slide.
    c. a coverslip.   **d.** transparent tape.

The human eye cannot perceive objects less than 0.1 mm in diameter, so a microscope is an essential tool for the study of small structures such as cells. The microscope used for this purpose is the *compound light microscope.* It is called compound because it uses two sets of lenses: an eyepiece, or ocular lens, and an objective lens system. The eyepiece lens system magnifies, or compounds, the image reaching it after the image is magnified by the objective lens system. Such an instrument can magnify images of small objects up to about 1000×.

PROCEDURE A: Microscope Basics

This procedure includes the terminology for the parts of the compound light microscope, the purpose of the parts, and how to manipulate the components for suitable image viewing. The proper storage of the microscope, lens-cleaning techniques, and magnification determination are also included.

1. Familiarize yourself with the following list of rules for care of the microscope:
   a. Keep the microscope under its *dustcover* and in a cabinet when it is not being used.
   b. Handle the microscope with great care. It is an expensive and delicate instrument. To move it or carry it, hold it by its *arm* with one hand and support its *base* with the other hand (fig. 4.1).
   c. Always store the microscope with the lowest power objective in place. Always start with this objective when using the microscope.
   d. To clean the lenses, rub them gently with *lens paper* or a high-quality cotton swab. If the lenses need additional cleaning, follow the directions in the "Lens-Cleaning Technique" section that follows.
   e. If the microscope has a substage lamp, be sure the electric cord does not hang off the laboratory table where someone might trip over it. The bulb life can be extended if the lamp is cool before the microscope is moved.

FIGURE 4.1 Major parts of a compound light microscope with a monocular (one eyepiece) body and a mechanical stage. Some microscopes are equipped with a binocular (two eyepieces) body.

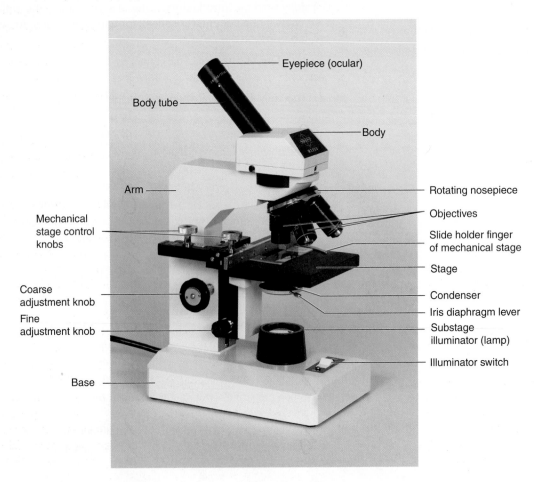

FIGURE 4.2 Scientist using a compound light microscope equipped with a binocular body and two eyepieces.

FIGURE 4.3 Microscope, showing the path of light through it.

Light path

LENS-CLEANING TECHNIQUE

1. Moisten one end of a high-quality cotton swab with one drop of lens cleaner. Keep the other end dry.
2. Clean the optical surface with the wet end. Dry it with the other end, using a circular motion.
3. Use a hand aspirator to remove lingering dust particles.
4. Start with the scanning objective and work upward in magnification, using a new cotton swab for each objective.
5. When cleaning the eyepiece, do not open the lens unless it is absolutely necessary.
6. Use alcohol for difficult cleaning.

  f. Never drag the microscope across the laboratory table.

  g. Never remove parts of the microscope or try to disassemble the eyepiece or objective lenses.

  h. If your microscope is not functioning properly, report the problem to your laboratory instructor immediately.

2. Observe a compound light microscope and study figure 4.1 to learn the names of its major parts. Your microscope might be equipped with a binocular body and two eyepieces (fig. 4.2). The lens system of a compound microscope includes three parts—the condenser, objective lens, and eyepiece (ocular).

Light enters this system from a *substage illuminator (lamp)* and is concentrated and focused by a *condenser* onto a microscope slide (fig. 4.3). The condenser, which contains a set of lenses, usually is kept in its highest position possible.

The *iris diaphragm,* located between the light source and the condenser, can be used to increase or decrease the intensity of the light entering the condenser and to control the contrast of the image. Locate the lever that operates the iris diaphragm beneath the *stage* and move it back and forth. Note how this movement causes the size of the opening in the diaphragm to change. (Some microscopes have a revolving plate called a disc diaphragm beneath the stage instead of an iris diaphragm. Disc diaphragms have different-sized holes to admit varying amounts of light.) Which way do you move the diaphragm to increase the light intensity (left or right)? _____ Which way to decrease it (left or right)? _____

After light passes through a specimen mounted on a microscope slide, it enters an *objective lens system.* This lens projects the light upward into the *body tube,* where it produces a magnified image of the object being viewed.

The *eyepiece (ocular) lens system* then magnifies this image to produce another image seen by the eye. Typically, the eyepiece lens magnifies the image ten times ($10\times$). Look for the number in the metal of the eyepiece that indicates its power (fig. 4.4). What is the eyepiece power of your microscope? _____

The objective lenses are mounted in a *rotating nosepiece* so that different magnifications can be achieved by rotating any one of several objective lenses into position above the specimen. Commonly, this set of lenses includes a scanning objective ($4\times$), a low-power objective ($10\times$), and a high-power objective, also called a high-dry-power objective (about $40\times$). Sometimes an

FIGURE 4.4 The powers of this 10× eyepiece (a) and this 40× objective (b) are marked in the metal. DIN is an international optical standard on quality optics. The 0.65 on the 40× objective is the numerical aperture, a measure of the light-gathering capabilities.

(a)

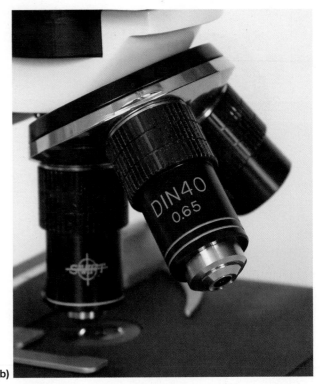

(b)

oil immersion objective (about 100×) is present. Look for the number printed on each objective that indicates its power. What are the objective lens powers of your microscope? _____

To calculate the *total magnification* achieved when using a particular objective, multiply the power of the eyepiece by the power of the objective used. Thus, the 10× eyepiece and the 40× objective produce a total magnification of 10 × 40, or 400×. See a summary of microscope lenses in table 4.1.

3. Complete Part A of Laboratory Assessment 4.

Table 4.1 Microscope Lenses

Objective Lens Name	Common Objective Lens Magnification	Common Eyepiece Lens Magnification	Total Magnification
Scan	4×	10×	40×
Low-power (LP)	10×	10×	100×
High-power (HP)	40×	10×	400×
Oil immersion	100×	10×	1,000×

Note: If you wish to observe an object under LP, HP, or oil immersion, locate and then center and focus the object first under scan magnification.

4. Turn on the substage illuminator and look through the eyepiece. You will see a lighted, circular area called the *field of view.*

You can measure the diameter of this field of view by focusing the lenses on the millimeter scale of a transparent plastic ruler. To do this, follow these steps:

a. Place the ruler on the microscope stage in the spring clamp of a slide holder finger on a *mechanical stage* or under the *stage (slide) clips.* (*Note:* If your microscope is equipped with a mechanical stage, it may be necessary to use a short section cut from a transparent plastic ruler. The section should be several millimeters long and can be mounted on a microscope slide for viewing.)

b. Center the millimeter scale in the beam of light coming up through the condenser, and rotate the scanning objective into position.

c. While you watch from the side to prevent the lens from touching anything, raise the stage until the objective is as close to the ruler as possible, using the *coarse adjustment knob* (fig. 4.5). (*Note:* The adjustment knobs on some microscopes move the body and objectives downward and upward for focusing.)

d. Look into the eyepiece and use the *fine adjustment knob* to raise the stage until the lines of the millimeter scale come into sharp focus.

e. Adjust the light intensity by moving the *iris diaphragm lever* so that the field of view is brightly

FIGURE 4.5 When you focus using a particular objective, you can prevent it from touching the specimen by watching from the side.

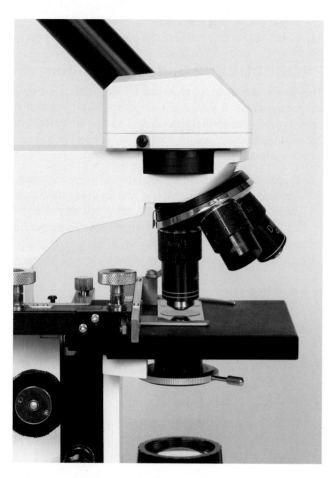

illuminated but comfortable to your eye. At the same time, take care not to overilluminate the field because transparent objects tend to disappear in bright light.

    f. Position the millimeter ruler so that its scale crosses the greatest diameter of the field of view. Also, move the ruler so that one of the millimeter marks is against the edge of the field of view.

    g. In millimeters, measure the distance across the field of view.

5. Complete Part B of the laboratory assessment.

6. Most microscopes are designed to be *parfocal*. This means that when a specimen is in focus with a lower-power objective, it will be in focus (or nearly so) when a higher-power objective is rotated into position. Always center the specimen in the field of view before changing to higher objectives.

    Rotate the low-power objective into position, and then look at the millimeter scale of the transparent plastic ruler. If you need to move the low-power objective to sharpen the focus, use the fine adjustment knob.

    Adjust the iris diaphragm so that the field of view is properly illuminated. Once again, adjust the millimeter ruler so that the scale crosses the field of view through its greatest diameter, and position the ruler so that a millimeter mark is against one edge of the field. Measure the distance across the field of view in millimeters.

7. Rotate the high-power objective into position, while you watch from the side, and then observe the millimeter scale on the plastic ruler. <u>All focusing using high-power magnification should be done only with the fine adjustment knob.</u> If you use the coarse adjustment knob with the high-power objective, you can accidentally force the objective into the coverslip and break the slide. This is because the *working distance* (the distance from the objective lens to the slide on the stage) is much shorter when using higher magnifications.

    Adjust the iris diaphragm for proper illumination. When using higher magnifications, more illumination usually will help you to view the objects more clearly. Measure the distance across the field of view in millimeters.

8. Locate the numeral 4 (or 9) on the plastic ruler and focus on it using the scanning objective. Note how the number appears in the field of view. Move the plastic ruler to the right and note which way the image moves. Slide the ruler away from you and again note how the image moves.

9. Observe a letter *e* slide. Note the orientation of the letter *e* using the scan, LP, and HP objectives. As you increase magnifications, note that the amount of the letter *e* shown is decreased. Move the slide to the left, and then away from you, and note the direction in which the observed image moves. If a *pointer* is visible in your field of view, you can manipulate the pointer to a location (structure) within the field of view by moving the slide or rotating the ocular lens.

10. Examine the slide of the three colored threads using the low-power objective and then the high-power objective. Focus on the location where the three threads cross. By using the fine adjustment knob, determine the order from top to bottom by noting which color is in focus at different depths. The other colored threads will still be visible, but they will be blurred. Be sure to notice whether the stage or the body tube moves up and down with the adjustment knobs of the microscope being used for this depth determination. The vertical depth of the specimen clearly in focus is called the *depth of field (focus)*.

CRITICAL THINKING ACTIVITY

What was the sequence of the three colored threads from top to bottom? Explain how you came to that conclusion.

_____

_____

_____

A compound light microscope is sometimes equipped with a micrometer scale mounted in the eyepiece. Such a scale is subdivided into 50 to 100 equal divisions (fig. 4.6). These arbitrary divisions can be calibrated against the known divisions of a micrometer slide placed on the microscope stage. Once the values of the divisions are known, the length and width of a microscopic object can be measured by superimposing the scale over the magnified image of the object.

Observe the micrometer scale in the eyepiece of the demonstration microscope. Focus the low-power objective on the millimeter scale of a micrometer slide (or a plastic ruler) and measure the distance between the divisions on the micrometer scale in the eyepiece. What is the distance between the finest divisions of the scale in micrometers? _____

Whenever specimens are examined, continue to use the fine adjustment focusing knob to determine relative depths of structures clearly in focus within cells, giving a three-dimensional perspective. The depth of field is less at higher magnifications.

11. Complete Parts C and D of the laboratory assessment.

PROCEDURE B: Slide Preparation

This procedure includes techniques for making your own temporary slides. If your microscope has an oil immersion objective lens, the special procedures and care are described. For viewing thicker, opaque specimens, a stereomicroscope demonstration activity is included.

1. Prepare several temporary *wet mounts,* using any small, transparent objects of interest, and examine the specimens using the low-power objective and then a high-power objective to observe their details. To prepare a wet mount, follow these steps (fig. 4.7):

   a. Obtain a precleaned microscope slide.

FIGURE 4.6 The divisions of a micrometer scale in an eyepiece can be calibrated against the known divisions of a micrometer slide.

Source: Courtesy of Swift Instruments, Inc., San Jose, California

Arbitrary divisions

Micrometer in eyepiece

Lines 0.01 mm apart

Micrometer slide

ULTRA LITE

FIGURE 4.7 Steps in the preparation of a wet mount.

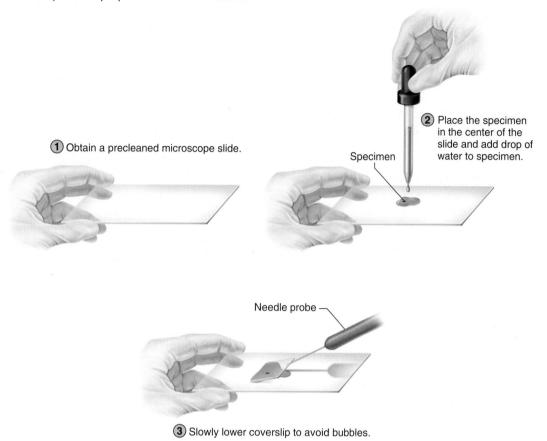

1. Obtain a precleaned microscope slide.

2. Place the specimen in the center of the slide and add drop of water to specimen.

Specimen

Needle probe

3. Slowly lower coverslip to avoid bubbles.

b. Place a tiny, thin piece of the specimen you want to observe in the center of the slide, and use a medicine dropper to put a drop of water over it. Consult with your instructor if a drop of stain might enhance the image of any cellular structures of your specimen. If the specimen is solid, you might want to tease some of it apart with dissecting needles. In any case, the specimen must be thin enough that light can pass through it. Why is it necessary for the specimen to be so thin?

_____

_____

c. Cover the specimen with a coverslip. Try to avoid trapping bubbles of air beneath the coverslip by slowly dragging the edge of the coverslip to the water or stain. When the edge of the coverslip makes contact with the water or stain, drop it from a 45-degree angle.

d. If your microscope has an inclination joint, do not tilt the microscope while observing wet mounts because the fluid will flow.

e. Place the slide under the stage (slide) clips or in the slide holder on a mechanical stage, and position the slide so that the specimen is centered in the light beam passing up through the condenser.

f. Focus on the specimen using the scanning objective first. Next focus using the low-power objective, and then examine it with the high-power objective.

2. If an oil immersion objective is available, use it to examine the specimen. To use the oil immersion objective, follow these steps:

a. Center the object you want to study under the high-power field of view.

b. Rotate the high-power objective away from the microscope slide, place a small drop of immersion oil on the coverslip, and swing the oil immersion objective into position. To achieve sharp focus, use the fine adjustment knob only.

c. You will need to open the iris diaphragm more fully for proper illumination. More light is needed because the oil immersion objective covers a very small lighted area of the microscope slide.

d. The oil immersion objective must be very close to the coverslip to achieve sharp focus, so care must be taken to avoid breaking the coverslip or damaging the objective lens. For this reason, never lower the objective when you are looking into the eyepiece.

A stereomicroscope (dissecting microscope) (fig. 4.8) is useful for observing the details of relatively large, opaque specimens. Although this type of microscope achieves less magnification than a compound light microscope, it has the advantage of producing a three-dimensional image rather than the flat, two-dimensional image of the compound light microscope. In addition, the image produced by the stereomicroscope is positioned in the same manner as the specimen, rather than being reversed and inverted, as it is by the compound light microscope.

Observe the stereomicroscope. The eyepieces can be pushed apart or together to fit the distance between your eyes. Focus the microscope on the end of your finger. Which way does the image move when you move your finger to the right? _____

When you move it away? _____

If the instrument has more than one objective, change the magnification to a higher power. Use the instrument to examine various small, opaque objects available in the laboratory.

Instead, always raise the objective to achieve focus, or prevent the objective from touching the coverslip by watching the microscope slide and coverslip from the side if the objective needs to be lowered. Usually when using the oil immersion objective, only the fine adjustment knob needs to be used for focusing.

3. When you have finished working with the microscope, remove the microscope slide from the stage and wipe any oil from the objective lens with lens paper or a high-quality cotton swab. Swing the scanning objective or the low-power objective into position. Wrap the electric cord around the base of the microscope and replace the dustcover.

4. Complete Part E of the laboratory assessment.

FIGURE 4.8 A stereomicroscope, also called a dissecting microscope.

Name _____

Date _____

Section _____

The corresponds to the indicated Learning Outcome(s) **o** found at the beginning of the Laboratory Exercise.

Care and Use of the Microscope

PART A: Assessments

Revisit Procedure A, number 2; then complete the following: **A2**

1. What total magnification will be achieved if the 10× eyepiece and the 10× objective are used? __*100x*__
2. What total magnification will be achieved if the 10× eyepiece and the 100× objective are used? __*1,000x*__

PART B: Assessments

Revisit Procedure A, number 2; then complete the following:

1. Sketch the millimeter scale as it appears under the scanning objective magnification. (The circle represents the field of view through the microscope.)

2. In millimeters, what is the diameter of the scanning field of view? **A3** _____

3. Microscopic objects often are measured in *micrometers*. A micrometer equals 1/1,000 of a millimeter and is symbolized by μm. In micrometers, what is the diameter of the scanning power field of view? **A3** _____

4. If a circular object or specimen extends halfway across the scanning field, what is its diameter in millimeters? **A3** _____

5. In micrometers, what is its diameter? **A3** _____

PART C: Assessments

Complete the following:

1. Sketch the millimeter scale as it appears using the low-power objective.

2. What do you estimate the diameter of this field of view to be in millimeters? **A3**

_____

3. How does the diameter of the scanning power field of view compare with that of the low-power field?

_____

4. Why is it more difficult to measure the diameter of the high-power field of view than that of the low-power field?

_____

_____

5. What change occurred in the intensity of the light in the field of view when you exchanged the low-power objective for the high-power objective? _____

6. Sketch the numeral 4 (or 9) as it appears through the scanning objective of the compound microscope.

7. What has the lens system done to the image of the numeral? (Is it right side up, upside down, or what?) _____

8. When you moved the ruler to the right, which way did the image move? _____

9. When you moved the ruler away from you, which way did the image move? _____

PART D: Assessments

1. Label the microscope parts in figure 4.9.

FIGURE 4.9 Identify the parts indicated on this compound light microscope.

1 —
2 —
3 —
4 —
5 —
6 —
7 —
8 —
9 —
10 —

2. Match the names of the microscope parts in column A with the descriptions in column B. Place the letter of your choice in the space provided. (A1)

Column A	Column B
a. Adjustment knob (coarse)	_____ 1. Increases or decreases the light intensity
b. Arm	_____ 2. Platform that supports a microscope slide
c. Condenser	_____ 3. Concentrates light onto the specimen
d. Eyepiece (ocular)	_____ 4. Causes stage (or objective lens) to move upward or downward
e. Field of view	_____ 5. After light passes through the specimen, it next enters this lens system
f. Iris diaphragm	
g. Nosepiece	_____ 6. Holds a microscope slide in position
h. Objective lens system	_____ 7. Contains a lens at the top of the body tube
i. Stage	_____ 8. Serves as a handle for carrying the microscope
j. Stage (slide) clip	_____ 9. Part to which the objective lenses are attached
	_____ 10. Circular area seen through the eyepiece

PART E: Assessments

Prepare sketches of the objects you observed using the microscope. For each sketch, include the name of the object, the magnification you used to observe it, and its estimated dimensions in millimeters and micrometers. (A4)

NOTES

Cell Structure and Function

Purpose of the Exercise

To review the structure and functions of major cellular components and to observe examples of human cells. To measure and compare the average cell's metabolic rate in individuals of different sizes (weight).

MATERIALS NEEDED

Animal cell model

Clean microscope slides

Coverslips

Flat toothpicks

Medicine dropper

Methylene blue (dilute) or iodine-potassium-iodide stain

Prepared microscope slides of human tissues

Compound light microscope

For Learning Extension Activities:

Single-edged razor blade

Plant materials such as leaves, soft stems, fruits, onion peel, and vegetables

Cultures of *Amoeba* and *Paramecium*

Ph.I.L.S. 4.0, Lesson 2 Metabolism: Size and Basal Metabolic Rate

Colored pencils

SAFETY

▶ Review all the safety guidelines inside the front cover.

▶ Clean laboratory surfaces before and after laboratory procedures.

▶ Wear disposable gloves for the wet-mount procedures of the cells lining the inside of the cheek.

▶ Work only with your own materials when preparing the slide of the cheek cells. Observe the same precautions as with all body fluids.

▶ Dispose of laboratory gloves, slides, coverslips, and toothpicks as instructed.

▶ Use the biohazard container to dispose of items used during the cheek cells procedure.

▶ Take precautions to prevent cellular stains from contacting your clothes and skin.

▶ Wash your hands before leaving the laboratory.

Learning Outcomes AP|R

After completing this exercise, you should be able to:

01 Name and locate the components of a cell.

02 Differentiate the functions of cellular components.

03 Prepare a wet mount of cells lining the inside of the cheek; stain the cells; and identify the plasma (cell) membrane, nucleus, and cytoplasm.

04 Examine cells on prepared slides of human tissues and identify their major components.

05 Explain the relationship of oxygen consumption with an animal's total weight.

06 Explain the relationship of oxygen consumption per gram of weight.

07 Explain the relationship of oxygen consumption and metabolic rate.

08 Calculate surface area–to–volume ratio and explain its significance to temperature regulation.

09 Integrate the concepts of weight, oxygen consumption, and metabolism and explain their application to the human body.

The ⓞ corresponds to the assessments Ⓐ indicated in the Laboratory Assessment for this Exercise.

Pre-Lab

Carefully read the introductory material and examine the entire lab. Be familiar with the basic structures and functions of a cell from lecture or the textbook. Answer the pre-lab questions.

Pre-Lab Questions Select the correct answer for each of the following questions:

1. Which of the following cellular structures is *not* easily visible with the compound light microscope?
   a. nucleus
   b. DNA
   c. cytoplasm
   d. plasma membrane

2. Which of the following cellular structures is located in the nucleus?
   a. nucleolus
   b. ribosomes
   c. mitochondria
   d. endoplasmic reticulum

3. The outer boundary of a cell is the
   a. mitochondrial membrane.   b. nuclear envelope.
   c. Golgi apparatus.   d. plasma membrane.

4. Microtubules, intermediate filaments, and microfilaments are components of
   a. vesicles.   b. the Golgi apparatus.
   c. the cytoskeleton.   d. ribosomes.

5. Easily attainable living cells observed in the lab are from
   a. inside the cheek.   b. blood.
   c. hair.   d. finger surface.

6. A slide of human cheek cells can be stained to observe _____ under the microscope.
   a. individual membrane proteins
   b. Golgi apparatus
   c. the plasma membrane and nucleus
   d. mitochondria

7. Cellular energy is stored in
   a. ER.   b. ATP.   c. DNA.   d. RNA.

8. The smooth ER possesses ribosomes.
   a. True _____   b. False _____

9. The nuclear envelope contains nuclear pores.
   a. True _____   b. False _____

Cells are the "building blocks" from which all parts of the human body are formed. Their arrangement and interactions result in the shape, organization, and construction of the body, and cells are responsible for carrying on its life processes. Using a compound light microscope and the proper stain, one can easily see the *plasma (cell) membrane,* the *cytoplasm,* and the *nucleus.* The cytoplasm is composed of a clear fluid, the *cytosol,* and numerous *cytoplasmic organelles* that are suspended in the cytosol.

The plasma membrane, composed of phospholipids, glycolipids, and glycoproteins, represents the cell boundary and functions in various methods of membrane transport. The plasma membrane might possess extensions as microvilli, cilia, or flagella. The nucleus is surrounded by a nuclear envelope, and many of the cytoplasmic organelles have membrane boundaries similar to the plasma membrane. Movements of substances across these membranes can be by passive processes, involving kinetic energy or hydrostatic pressure, or active processes, using the cellular energy of ATP (adenosine triphosphate).

Metabolism refers to all the chemical reactions within cells that use or release energy. How does the metabolic rate of cells in a man who is six feet tall compare to that of a man who is five feet nine? Surprisingly, an understanding of temperature regulation is required to answer this question. *Temperature regulation* is accomplished through a balance of heat production and heat loss. Heat is produced by cells through metabolism (second law of thermodynamics: For every chemical reaction, some energy is always lost as heat). The greater the number of cells or the more metabolically active the cells, the more heat that is produced. Heat is lost through the surface of the skin. The larger the surface area of the skin in proportion to body volume, the more heat that is lost. For a given volume of cells, the metabolic rate of cells is established and maintained to offset the heat that is lost by the surface of the skin (within limits). The more heat that is lost at the skin, the more metabolically active the cells must be to produce the heat required to keep the body warm.

PROCEDURE: Cell Structure and Function

The boundary of the nucleus is a double-layered *nuclear envelope,* which has nuclear pores allowing the passage of genetic information. The nucleus contains fine strands of *chromatin* that consist of DNA (deoxyribonucleic acid) and structural and regulatory proteins. Chromatin condenses to form chromosomes during cell division. The nucleus is often referred to as the "control center" of the cell because it contains the genetic material and is required for cell division. The *nucleolus* is a nonmembranous structure composed of RNA (ribonucleic acid) and protein and is a site of ribosomal formation.

Cytoplasmic organelles carry out specialized metabolic functions. The *endoplasmic reticulum (ER)* has numerous canals that serve in transporting molecules as proteins throughout the cytoplasm. *Ribosomes* synthesize proteins and are located free in the cytoplasm or on the surface of the endoplasmic reticulum *(rough ER).* If the ER lacks the ribosomes on the surface, it is called *smooth ER,* and it is involved in lipid synthesis but does not serve as a region of protein synthesis. The *Golgi apparatus (complex),* composed of flattened membranous sacs, is the site for packaging glycoproteins for transport and secretion. *Mitochondria* possess a double membrane and provide the main location for the cellular energy production of ATP. Mitochondria are often referred to as the "powerhouse" of a cell because of the energy production. *Lysosomes* are membranous sacs that contain intracellular digestive enzymes for destroying debris and worn-out organelles. Lysosomes are nicknamed "suicide sacs" because they have the ability to destroy the entire cell. *Vesicles* are membranous sacs produced by the cell that contain substances for storage or transport, or form from pinching off pieces of the plasma membrane. A *cytoskeleton* contains microtubules, intermediate filaments, and microfilaments that support cellular structures within the cytoplasm and are involved in cellular movements.

1. Study figures 5.1 and 5.2 and table 5.1.
2. Observe the animal cell model and identify its major structures.
3. Complete Part A of Laboratory Assessment 5.
4. Prepare a wet mount of cells lining the inside of the cheek. To do this, follow these steps:
   a. Gently scrape (force is not necessary and should be avoided) the inner lining of your cheek with the broad end of a flat toothpick.

FIGURE 5.1 AP|R The structures of a composite cell. The structures are not drawn to scale.

Apical cell surface
Microvilli
Microfilaments of cytoskeleton
Secretory vesicles
Fat droplet
Plasma (cell) membrane
Cytosol of cytoplasm
Golgi apparatus
Centrosome
Centrioles
Nuclear pores
Nucleus
Nucleolus
Chromatin
Lysosome
Peroxisome
Nuclear envelope
Mitochondrion
Ribosomes (free)
Smooth endoplasmic reticulum
Microtubules of cytoskeleton
Rough endoplasmic reticulum
Basal cell surface
Basement membrane

FIGURE 5.2 AP|R Structure of the plasma (cell) membrane.

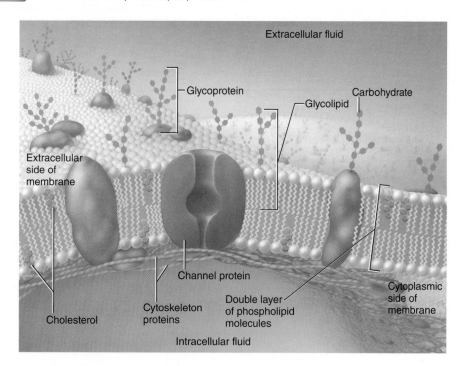

Extracellular fluid
Glycoprotein
Glycolipid
Carbohydrate
Extracellular side of membrane
Cholesterol
Cytoskeleton proteins
Channel protein
Double layer of phospholipid molecules
Cytoplasmic side of membrane
Intracellular fluid

Table 5.1 Cellular Structures and Functions

Cellular Component	Structure	Function
Plasma membrane	Bilayer of phospholipids, glycolipids, and glycoproteins	Cell boundary and involved in various membrane transport mechanisms
Microvilli	Numerous tiny folds containing microfilaments	Allow greater surface area for absorption
Cilia	Numerous short extensions containing microtubules	Move substances such as mucus
Flagellum	Long extension of sperm cell containing microtubules	Moves sperm
Nucleus	**Large structure**	**Center of cellular control**
Nuclear envelope	Double membrane boundary of nucleus with pores; continuous with rough ER	Passageway between nucleus and cytoplasm
Nucleolus	Nonmembranous body containing ribosomal RNA and protein	Location of ribosome formation
Chromatin	Strands of DNA and protein	DNA represents genetic control for protein synthesis
Cytoplasm	**Contents between plasma membrane and nucleus containing cytosol fluid and organelles**	**Location of numerous cellular processes**
Membrane-bound organelles	**Organelles possessing a membranous covering**	**Various functions depending upon organelle**
Rough endoplasmic reticulum	Extensive membrane network with attached ribosomes	Site of protein synthesis with ribosomes; cellular transport
Smooth endoplasmic reticulum	Extensive membrane network lacking ribosomes	Site of lipid synthesis; cellular transport
Mitochondria	Possess a double membrane with folded inner membrane	Synthesis of ATP
Golgi apparatus (complex)	Series of flattened sacs often near nucleus	Packaging site of glycoproteins for transport and secretion
Vesicles	Membranous sacs containing various substances	Store or transport cellular substances
Lysosomes	Sacs containing digestive enzymes	Sites that destroy cellular debris and worn-out organelles
Peroxisomes	Small sacs containing oxidative enzymes; abundant in liver and kidney cells	Detoxify harmful cellular substances
Non-membranous organelles	**Organelles that lack a membranous covering**	**Various functions depending upon organelle**
Ribosomes	Contain ribosomal RNA and protein; attached to rough ER or free in cytoplasm	Sites of protein synthesis
Centrosome	Area near nucleus containing a pair of centrioles	Formation center to produce microtubules and spindle fibers for cell division
Centrioles	Cylindrical bodies of microtubules within the centrosome of the cytoplasm	Form mitotic spindle during cell division
Cytoskeleton: microtubules	Hollow cylinders of proteins in cilia, flagella, and centrioles	Involved with support and motility within cell
Cytoskeleton: intermediate filaments	Various protein fibers	Provide internal cellular support for shape of cell
Cytoskeleton: microfilaments	Fine threads of a protein actin	Help maintain cell shape and change in shape; involved in some cell motility as muscle contraction

b. Stir the toothpick in a drop of water on a clean microscope slide and dispose of the toothpick as directed by your instructor.

c. Cover the drop with a coverslip.

d. Observe the cheek cells by using the microscope. Compare your image with figure 5.3. To report what you observe, sketch a single cell in the space provided in Part B of the laboratory assessment.

5. Prepare a second wet mount of cheek cells, but this time, add a drop of dilute methylene blue or iodine-potassium-iodide stain to the cells. Cover the liquid with a coverslip and observe the cells with the microscope. Add to your sketch any additional structures you observe in the stained cells.

6. Answer the questions in Part B of the laboratory assessment.

FIGURE 5.3 Iodine-stained cell from the inner cheek, as viewed through the compound light microscope using the high-power objective (400×).

Plasma membrane

Folds of the plasma membrane

Nucleus

Cytoplasm

7. Observe all safety guidelines for disposal of materials used during this procedure.
8. Using the microscope, observe each of the prepared slides of human tissues. To report what you observe, sketch a single cell of each type in the space provided in Part C of the laboratory assessment.
9. Complete Parts C and D of the laboratory assessment.

CRITICAL THINKING ACTIVITY

The cells lining the inside of the cheek are frequently removed for making observations of basic cell structure. The cells are from stratified squamous epithelium. Explain why these cells are used instead of outer body surface tissue.

_____

_____

_____

LEARNING EXTENSION ACTIVITY

Investigate the microscopic structure of various plant materials. To do this, prepare tiny, thin slices of plant specimens, using a single-edged razor blade. *(Take care not to injure yourself with the blade.)* Keep the slices in a container of water until you are ready to observe them. To observe a specimen, place it into a drop of water on a clean microscope slide and cover it with a cover-slip. Use the microscope and view the specimen using low- and high-power magnifications. Observe near the edges where your section of tissue is most likely to be one cell thick. Add a drop of dilute methylene blue or iodine-potassium-iodide stain, and note if any additional structures become visible. How are the microscopic structures of the plant specimens similar to the human tissues you observed? _____

_____

How are they different? _____

_____

LEARNING EXTENSION ACTIVITY

Prepare a wet mount of the *Amoeba* and *Paramecium* by putting a drop of culture on a clean glass slide. Gently cover with a clean coverslip. Observe the movements of the *Amoeba* with pseudopodia and the *Paramecium* with cilia. Try to locate cellular components such as the plasma (cell) membrane, nuclear envelope, nucleus, mitochondria, and contractile vacuoles. Describe the movement of the *Amoeba.*

_____

_____

Describe the movement of the *Paramecium.*

_____

_____

NOTES

Name _____

Date _____

Section _____

The Ⓐ corresponds to the indicated Learning Outcome(s) Ⓞ found at the beginning of the Laboratory Exercise.

Cell Structure and Function

PART A: Assessments

1. Label the cellular structures in figure 5.4. **A1**

FIGURE 5.4 Label the indicated cellular structures of this composite cell.

2. Match the cellular components in column A with the descriptions in column B. Place the letter of your choice in the space provided.　A2

Column A	Column B
a. Chromatin	_____ **1.** Loosely coiled fibers containing protein and DNA within nucleus
b. Cytoplasm	_____ **2.** Location of ATP production for cellular energy
c. Endoplasmic reticulum	_____ **3.** Small RNA-containing particles for the synthesis of proteins
d. Golgi apparatus (complex)	_____ **4.** Membranous sac formed by the pinching off of pieces of plasma membrane
e. Lysosome	
f. Microtubule	_____ **5.** Dense body of RNA and protein within the nucleus
g. Mitochondrion	_____ **6.** Part of the cytoskeleton involved in cellular movement
h. Nuclear envelope	_____ **7.** Composed of membrane-bound canals for tubular transport throughout the cytoplasm
i. Nucleolus	
j. Nucleus	_____ **8.** Occupies space between plasma membrane and nucleus
k. Ribosome	_____ **9.** Flattened, membranous sacs that package a secretion
l. Vesicle	_____ **10.** Membranous sac that contains digestive enzymes
	_____ **11.** Separates nuclear contents from cytoplasm
	_____ **12.** Spherical organelle that contains chromatin and nucleolus

PART B: Assessments

Complete the following:

1. Sketch a single cheek cell. Label the cellular components you recognize. Add any additional structures observed to your sketch after staining was completed. (The circle represents the field of view through the microscope.)　A3

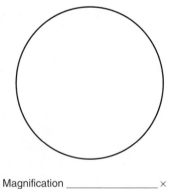

2. After comparing the wet mount and the stained cheek cells, describe the advantage gained by staining cells.

Magnification _____ ×

_____

_____

PART C: Assessments

Complete the following:

1. Sketch a single cell of each type you observed in the prepared slides of human tissues. Name the tissue, indicate the magnification used, and label the cellular components you recognize. **A4**

_____ ×

Tissue _____

_____ ×

Tissue _____

2. What do the various types of cells in these tissues have in common? _____

_____

3. What are the main differences you observed among these cells? _____

_____

_____

PART D: Assessments

Electron micrographs represent extremely thin slices of cells. Each micrograph in figure 5.5 contains a section of a nucleus and some cytoplasm. Compare the organelles shown in these micrographs with organelles of the animal cell model and figure 5.1.

Identify the structures indicated by the arrows in figure 5.5. **A1**

1. _____   6. _____

2. _____   7. _____

3. _____   8. _____

4. _____   9. _____

5. _____

FIGURE 5.5 Transmission electron micrographs of cellular components. The views are only portions of a cell. Magnifications: (a) 26,000×; (b) 10,000×. Identify the numbered cellular structures, using the terms provided. **A1**

(a)   Cytoplasm   Nucleus

(b)   Nucleus   Cytoplasm

Terms:

Chromatin (use 2 times)	Nucleolus
Endoplasmic reticulum	Plasma membrane
Mitochondrion (cross section)	Ribosomes (free)
Nuclear envelope (use 2 times)	

Answer the following questions after observing the transmission electron micrographs in figure 5.5.

11. What cellular structures were visible in the transmission electron micrographs that were not apparent in the cells you observed using the microscope? _____

_____

_____

LESSON 2 | # Metabolism
Size and Basal Metabolic Rate

The purpose of this exercise is to demonstrate how the size of an animal influences the metabolic rate of its body cells. **Metabolism** is the sum total of all the chemical reactions that occur in the body. Cells utilize fuel from the foods you eat and oxygen from the air you breathe to synthesize ATP. Carbon dioxide, water, and energy in the form of ATP *and* heat are produced in this reaction. The **basal metabolic rate (BMR)** of an animal is the amount of energy expended in a given amount of time while in a resting state. Warm-blooded animals rely on the production of heat to keep their body temperature within normal limits (homeostasis). Heat normally escapes the body surface toward the outside environment; therefore, the animal's cells must maintain a metabolic rate that produces enough heat to make up for the loss. Because heat moves through the surface of the body, the size of an animal, particularly its *surface area–to–volume ratio,* is a factor that influences BMR. Smaller animals have a greater surface area–to–volume ratio than larger animals, so they would lose heat more quickly and cells would have to work harder to make up for this heat loss. To determine the BMR of an animal, you *could* measure the amount of fuel (i.e., glucose) consumed, or you *could* measure the amount of carbon dioxide produced, or you *could* measure the amount of heat lost, but in this lab, you *will* indirectly measure the rate of oxygen consumption. In this laboratory experiment, you will weigh a mouse, place it in a chamber that contains CO_2 absorbent, and observe how quickly the mouse consumes oxygen by measuring the movement of a soap bubble along a tube every 15 seconds for up to 2 minutes. There are four mice of different sizes that you will be using in this experiment, and you will conduct a minimum of two trials for each mouse. You will not only look at the average oxygen consumed per minute, but you will also look at the average oxygen consumed per hour per gram of body weight for each mouse.

Before you begin, familiarize yourself with the following concepts (use your textbook as a reference):

■ Function of adenosine triphosphate (ATP)

■ Surface area–to–volume ratio of cells

■ Factors that affect metabolic rate (thyroid hormone, exercise, gender, size)

State your hypothesis regarding the size of an animal and the amount of oxygen consumed per hour per gram of body weight.

_____

_____

_____

_____

_____

1. Open Lesson 2: Metabolism: Size and Basal Metabolic Rate.

2. Read the objectives and introduction and take the pre-laboratory quiz.

3. After completing the pre-laboratory quiz, read through the wet lab.

4. The laboratory exercise will open when you click Continue after completing the wet lab (fig. 5.6).

5. Follow the instructions provided in the laboratory exercise and answer the following questions as you proceed.

6. Record the data from the Ph.I.L.S. *journal* of the individual trials for each mouse in table 5.2.

7. Which trial for all mice in the journal data shows the highest O_2 consumed (mL) per minute? _____ To what do you attribute this higher oxygen consumption in this particular trial?

_____

_____

_____

_____

_____

8. Record the data from the Ph.I.L.S. journal *Final Graph* in table 5.3.

FIGURE 5.6 Opening Screen for the Laboratory Exercise on Metabolism: Size and Basal Metabolic Rate.

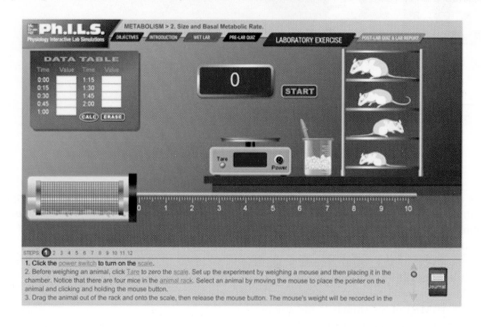

Table 5.2 Journal Data of Ph.I.L.S. 4.0, Lesson 2: Metabolism *Size and Basal Metabolic Rate*

Mouse	Weight (g)	Trial 1 (O_2 mL/min.)	Trial 2 (O_2 mL/min.)	Trial 3 (O_2 mL/min.)	Trial 4 (O_2 mL/min.)	Trial 5 (O_2 mL/min.)
1						
2						
3						
4						

Table 5.3 Results of Ph.I.L.S. 4.0, Lesson 2: Metabolism *Size and Basal Metabolic Rate*

Mouse	Weight (g)	Avg. O_2/min	Avg. O_2/h/g
1			
2			
3			
4			

9. Using a different color pencil for each final graph, graph the data you recorded in table 5.3 on the axis below.

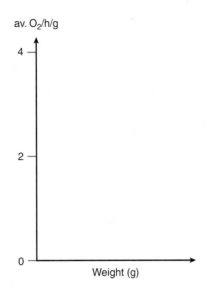

10. Make note of any pertinent observations here:

_____

_____

_____

_____

_____

11. Examine the graph on the left of the final graph. On the x-axis is weight (grams) and on the y-axis is average oxygen consumed per minute (av. O_2/m). As the weight of the mouse decreases, does the average oxygen consumed per minute decrease or increase? _____ Is the relationship between weight and oxygen consumed direct or inverse? **A5** _____

12. Examine the graph on the right of the final graph. Weight (grams) is on the x-axis and average oxygen consumed per hour per gram (of body weight) (av. O_2/h/g) is on the y-axis. As the weight of the mouse decreases, does the average oxygen consumed per hour per gram decrease or increase? _____ Is the relationship between weight and oxygen consumed per gram of body weight direct or inverse? **A6** _____

13. Based on the results of the graph on the right *and* the fact that oxygen consumed is an indirect measure of metabolic rate, as the weight of the mouse decreases, does the metabolic rate decrease or increase? _____ Is the relationship between weight and metabolic rate direct or inverse? **A7** _____

14. **Surface Area–to–Volume Ratio:** As a (spherical) structure increases in radius, the surface area increases at a rate of r^2 and the volume increases at a rate of r^3. If the radius of the sphere is 2, the calculated surface area $2^2 = 2 \times 2 = 4$ and the calculated volume $2^3 = 2 \times 2 \times 2 = 8$, and the ratio of surface area to volume is $4/8 = 0.5$. If r is 4, calculate the surface area, volume, and surface area/volume ratio. Place your answers in the following box. **A8**

$r = 4$

Surface area = _____   Volume = _____

Surface area/volume ratio = _____

15. As the (spherical) structure becomes larger, is the *ratio* becoming larger or smaller? (Hint: Compare the surface area/volume ratio between two spheres—one with a radius of 2 and the other with a radius of 4.) _____ Is the volume increasing at a faster rate than the surface area? **A8** _____

16. If the surface area–to–volume ratio is applied to the human body, where volume is the size (weight) of the body and surface area is the area of skin surface, as a person increases in size: **A8** **A9**

   a. What happens to the ratio of surface area/volume?

   _____

   b. Is volume (size) increasing at a faster rate than surface area (skin)? _____

   c. Is this a more favorable ratio for keeping the body warm? _____

   d. Will each individual cell have to be as metabolically active as in a larger individual to keep the body warm? _____ Why? _____

   _____

   e. Is the metabolic rate of individual cells faster or slower? _____

CRITICAL THINKING ASSESSMENT

17. Walt, a six-foot man, walks with a friend Albert, who is five-foot-nine.

   a. Would you predict that Walt, with the larger body, is consuming more oxygen than Albert?

   _____

   b. Would you predict that Walt is consuming more oxygen *per gram of weight* than Albert?

   _____

   c. Which man would have a higher *metabolic rate per gram of weight?* _____

   Why? _____

   _____

   _____

   _____

18. Which of the following factors would increase *or* decrease a person's metabolic rate?

   a. Hyperthyroidism would _____ metabolic rate.

   b. Hypothyroidism would _____ metabolic rate.

   c. Exercise would _____ metabolic rate.

   d. Being sedentary (a "couch potato") would _____ metabolic rate.

   e. In general, being a female as opposed to being a male would _____ metabolic rate.

   f. Experiencing stress would _____ metabolic rate.

19. Click the Post-Lab Quiz & Lab Report in the upper menu. Complete the ten questions in the Post-Lab Quiz.

20. Read the conclusion on the computer screen.

21. Print the Lab Report for Metabolism: Size and Basal Metabolic Rate.

LABORATORY
EXERCISE 6

Movements Through Membranes

Purpose of the Exercise

To become familiar with the physical processes by which substances move through membranes and to demonstrate those processes that are passive.

MATERIALS NEEDED

For Procedure A— Diffusion:

Petri dish

White paper

Forceps

Colored pencils

Potassium perman-ganate crystals

Millimeter ruler (thin and transparent)

For Procedure B— Osmosis:

Thistle tube

Molasses (or Karo dark corn syrup or 60% colored sucrose solution)

Ring stand and clamp

Beaker

Rubber band

Millimeter ruler

Selectively permeable (semipermeable) membrane (presoaked dialysis tubing of 1 5/16" or greater diameter)

For Procedure C— Hypertonic, Hypotonic, and Isotonic Solutions:

Test tubes

Marking pen

Test-tube rack

10 mL graduated cylinder

Medicine dropper

Uncoagulated animal blood

Distilled water

0.9% NaCl (aqueous solution)

3.0% NaCl (aqueous solution)

Clean microscope slides

Coverslips

Compound light microscope

For Procedure D— Filtration:

Glass funnel

Filter paper

Support stand and ring

Beaker

Powdered charcoal or ground black pepper

1% glucose (aqueous solution)

1% starch (aqueous solution)

Test tubes

10 mL graduated cylinder

Water bath (boiling water)

Benedict's solution

Iodine-potassium-iodide solution

Medicine dropper

For Alternative Activities:

Fresh chicken egg

Beaker

Laboratory balance

Spoon

Vinegar

Corn syrup (Karo)

Ph.I.L.S. 4.0, Lesson 1 Osmosis and Diffusion: Varying Extracellular Concentration

SAFETY

▶ Clean laboratory surfaces before and after laboratory procedures.

▶ Wear disposable gloves when handling chemicals and animal blood.

▶ Wear safety glasses when using chemicals.

▶ Dispose of laboratory gloves and blood-contaminated items as instructed.

▶ Wash your hands before leaving the laboratory.

Learning Outcomes AP|R

After completing this exercise, you should be able to:

01 Demonstrate the process of diffusion and identify examples of diffusion.

02 Explain diffusion by preparing and interpreting a graph.

03 Demonstrate the process of osmosis and identify examples of osmosis.

04 Distinguish among hypertonic, hypotonic, and isotonic solutions and examine the effects of these solutions on animal cells.

05 Demonstrate the process of filtration and identify examples of filtration.

06 Identify the percent transmittance of light through a blood sample placed in varying concentrations of NaCl.

07 Interpret the results to explain the relationship of percent transmittance and condition of the red blood cells (normal, shriveled or ruptured) in the blood sample.

08 Interpret the results to explain the condition of the red blood cells (normal, shriveled, or ruptured) to the relative concentration of NaCl solutions (isotonic, hypertonic, or hypotonic).

09 Relate plasma concentration and red blood cell integrity with events that occur in the human body.

The **O** corresponds to the assessments **A** indicated in the Laboratory Assessment for this Exercise.

Pre-Lab

Carefully read the introductory material and examine the entire lab. Be familiar with diffusion, osmosis, tonicity, and filtration from lecture or the textbook. Visit www.mhhe.com/martinseries2 for LabCam videos. Answer the pre-lab questions.

Pre-Lab Questions Select the correct answer for each of the following questions:

1. Which of the following membrane movements is *not* by a passive process?
   - **a.** osmosis
   - **b.** diffusion
   - **c.** active transport
   - **d.** filtration

2. Osmosis is a membrane passage process pertaining to
   - **a.** glucose solutions.
   - **b.** hydrostatic pressure.
   - **c.** saline solutions.
   - **d.** water.

3. Which of the following represents an isotonic solution to human cells?
   - **a.** 0.9% NaCl
   - **b.** 3% NaCl
   - **c.** 3% glucose
   - **d.** 0.9% glucose

4. Filtration requires _____ for movements through membranes.
   - **a.** hydrostatic pressure
   - **b.** active transport
   - **c.** pinocytosis
   - **d.** phagocytosis

5. Which of the following does *not* influence the rate of diffusion?
   - **a.** concentration gradient
   - **b.** size of petri dish
   - **c.** molecular size
   - **d.** temperature

6. Dialysis tubing serves as a _____ for the osmosis experiment.
   - **a.** cell wall
   - **b.** permeable membrane
   - **c.** selectively permeable
   - **d.** barrier membrane

7. Passive membrane passage requires ATP.
   - **a.** True _____
   - **b.** False _____

8. A cell will not swell or shrink when placed in an isotonic solution.
   - **a.** True _____
   - **b.** False _____

A plasma membrane functions as a gateway through which chemical substances and small particles may enter or leave a cell. Because of its bilayer of phospholipid molecules, the plasma membrane is *selectively permeable* in that it allows only certain substances to easily pass through it. These substances move through the membrane by *passive (physical) processes* such as diffusion, osmosis, and filtration, or by *active (physiological) processes* such as active transport, phagocytosis, or pinocytosis. Passive processes do not require the cellular energy of ATP, but occur due to molecular motion or hydrostatic pressure; active processes utilize ATP.

This laboratory exercise contains examples of passive membrane transport through either living plasma membranes or artificial membranes. *Diffusion* is the random motion of molecules from an area of higher concentration toward an area of lower concentration, thus moving down their concentration gradient. The rate of diffusion depends upon factors such as the concentration gradient, the molecular size, and the temperature. For example, an increase in temperature would cause an increase in molecular motion, or kinetic energy, resulting in an increase in the rate of diffusion. Facilitated diffusion is a carrier-mediated transport, which is not represented in this laboratory exercise.

A special case of passive membrane passage, called *osmosis,* occurs when water molecules diffuse through a selectively permeable membrane from an area of higher water concentration toward an area of lower water concentration. *Tonicity* refers to the osmotic pressure of a solution in relation to a cell. A solution is *isotonic* if the osmotic pressure is the same as the cell. A 5% glucose solution and a 0.9% normal saline solution serve as isotonic solutions to human cells. A *hypertonic* solution has a higher concentration of solutes than the cell; a *hypotonic* solution has a lower concentration of solutes than a cell. The osmosis of water will occur out of a cell when it is immersed in a hypertonic solution, and the cell will shrink (crenate). Conversely, the osmosis of water will occur into a cell when it is immersed in a hypotonic solution, and the cell will swell, and may burst (lyse).

Filtration is the movement of water and solutes through a selectively permeable membrane by hydrostatic pressure upon the membrane. The pressure gradient forces the water and solutes (filtrate) from the higher hydrostatic pressure area to a lower hydrostatic pressure area. For example, blood pressure provides the hydrostatic pressure for water and dissolved substances to move through capillary walls. Hypertension (high blood pressure) is a condition that would increase capillary filtration, as hydrostatic pressure is increased. This may lead to swelling (edema) in the tissues.

PROCEDURE A: Diffusion AP|R

Simple diffusion can occur in living organisms as well as in nonliving systems involving gases, liquids, and solids. In this procedure, diffusion will be examined in water without cells.

1. To demonstrate diffusion, refer to figure 6.1 as you follow these steps:
   - **a.** Place a petri dish, half filled with water, on a piece of white paper that has a millimeter ruler positioned on the paper. Wait until the water surface is still. Allow approximately 3 minutes.
     Note: The petri dish should remain level. A second millimeter ruler may be needed under the petri dish as a shim to obtain a level amount of the water inside the petri dish.
   - **b.** Using forceps, place one crystal of potassium permanganate near the center of the petri dish and near the millimeter ruler (fig. 6.1).
   - **c.** Measure the radius of the purple circle at 1-minute intervals for 10 minutes and record the results in Part A of Laboratory Assessment 6.
2. Complete Part A of the assessment.

FIGURE 6.1 To demonstrate diffusion, place one crystal of potassium permanganate in the center of a petri dish containing water. Place the crystal near the millimeter ruler (positioned under the petri dish).

Ruler used as a shim to maintain a level petri dish

LEARNING EXTENSION ACTIVITY

Repeat the demonstration of diffusion using a petri dish filled with ice-cold water and a second dish filled with very hot water. At the same moment, add a crystal of potassium permanganate to each dish and observe the circle as before. Record your findings in the data table in Part A of the laboratory assessment. What difference do you note in the rate of diffusion in the two dishes? How do you explain this difference? _____

_____

_____

_____

_____

PROCEDURE B: Osmosis AP|R

For the osmosis experiment, an artificial membrane will serve as a model for an actual plasma membrane. A layer of dialysis tubing that has been soaked for 30 minutes can easily be cut open because it becomes pliable. This membrane model works as a selectively (semipermeable; differentially) permeable membrane mimicking the plasma membrane of a cell.

When water moves across a selectively permeable membrane down its concentration gradient, it is termed osmosis. Penetrating solute particles will also move down their concentration gradients across a membrane if the solute is able to pass through the membrane. However, if the solute

particles are nonpenetrating because they are too large to pass across a selectively permeable membrane, only water will move across the selectively permeable membrane. Water will move toward the higher concentration of nonpenetrating particles. This will result in changes in the volume of water on a certain side of the membrane. This will create a volume change in relation to a cell or models used as simulated cells.

1. To demonstrate osmosis, refer to figure 6.2 as you follow these steps:
    a. One person plugs the tube end of a thistle tube with a finger.
    b. Another person then fills the bulb with molasses until it is about to overflow at the top of the bulb. Allow the molasses to enter the first centimeter of the stem, leaving the rest filled with trapped air.
    c. Cover the bulb opening with a single-thickness piece of moist selectively permeable membrane.
    d. Tightly secure the membrane in place with several wrappings of a rubber band.
    e. Immerse the bulb end of the tube in a beaker of water. If leaks are noted, repeat the procedures.
    f. Support the upright portion of the tube with a clamp on a support stand. Folded paper between the stem and the clamp will protect the thistle tube stem from breakage.
    g. Mark the meniscus level of the molasses in the tube. *Note:* The best results will occur if the mark of the molasses is a short distance up the stem of the thistle tube when the experiment starts.
    h. Measure the level changes after 10 minutes and 30 minutes and record the results in Part B of the laboratory assessment.

2. Complete Part B of the laboratory assessment.

ALTERNATE ACTIVITY

Eggshell membranes possess selectively permeable properties. To demonstrate osmosis using a natural membrane, soak a fresh chicken egg in vinegar for about 24 hours to remove the shell. Use a spoon to carefully handle the delicate egg. Place the egg in a hypertonic solution (corn syrup) for about 24 hours. Remove the egg, rinse it, and using a laboratory balance, weigh the egg to establish a baseline weight. Place the egg in a hypotonic solution (distilled water). Remove the egg and weigh it every 15 minutes for an elapsed time of 75 minutes. Explain any weight changes that were noted during this experiment.

_____

_____

FIGURE 6.2 (a) Fill the bulb of the thistle tube with molasses; (b) tightly secure a piece of selectively permeable (semipermeable) membrane over the bulb opening; and (c) immerse the bulb in a beaker of water. *Note:* These procedures require the participation of two people.

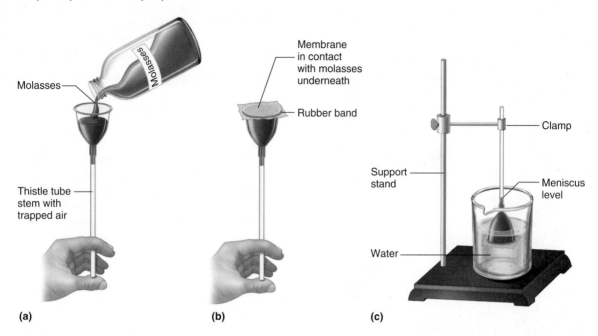

PROCEDURE C: Hypertonic, Hypotonic, and Isotonic Solutions AP|R

Applications of volume changes to cells from osmosis are addressed in this procedure. If a cell is placed in an isotonic solution with the same water and solute concentration, there will be no net osmosis and the cell will retain its normal size. If a cell is placed in a hypertonic solution, the cell will lose water and shrink; if a cell is in a hypotonic solution it will swell and burst.

1. To demonstrate the effect of hypertonic, hypotonic, and isotonic solutions on animal cells, follow these steps:
    a. Place three test tubes in a rack and mark them *tube 1, tube 2,* and *tube 3.* (*Note:* One set of tubes can be used to supply test samples for the entire class.)
    b. Using 10 mL graduated cylinders, add 3 mL of distilled water to tube 1; add 3 mL of 0.9% NaCl to tube 2; and add 3 mL of 3.0% NaCl to tube 3.
    c. Gently shake first, and then place 3 drops of fresh, uncoagulated animal blood into each of the tubes, and gently mix the blood with the solutions. Wait 5 minutes.
    d. Using three separate medicine droppers, remove a drop from each tube and place the drops on three separate microscope slides marked *1, 2,* and *3.*
    e. Cover the drops with coverslips and observe the blood cells, using the high power of the microscope.
2. Complete Part C of the laboratory assessment.

ALTERNATE ACTIVITY

Various substitutes for blood can be used for Procedure C. Onion, cucumber, and cells lining the inside of the cheek represent three possible options.

PROCEDURE D: Filtration AP|R

Filtration is another example of passive membrane passage. Unlike the previous experiments, hydrostatic pressure provides the mechanism for movements across a membrane instead of molecular motion. Water and solutes that pass through the membrane move simultaneously in the same direction. Solutes that are too large to cross the membrane remain on their original side of the membrane. Filtration occurs faster when the hydrostatic pressures increase.

1. To demonstrate filtration, follow these steps:
    a. Place a glass funnel in the ring of a support stand over an empty beaker. Fold a piece of filter paper in half and then in half again. Open one thickness of the filter paper to form a cone. Wet the cone, and place it in the funnel. The filter paper is used to demonstrate how movement across membranes is limited by the size of the molecules, but it does not represent a working model of biological membranes.
    b. Prepare a mixture of 5 cc (approximately 1 teaspoon) powdered charcoal (or ground black pepper) and equal amounts of 1% glucose solution and 1% starch

FIGURE 6.3 Apparatus used to illustrate filtration.

Support stand

Mixture

Filter paper

Funnel

Ring

Filtrate

FIGURE 6.4 Heat the filtrate and Benedict's solution in a boiling water bath for 2 minutes.

Test tube

Water bath

Filtrate and Benedict's solution

Hot plate

solution in a beaker. Pour some of the mixture into the funnel until it nearly reaches the top of the filter-paper cone. Care should be taken to prevent the mixture from spilling over the top of the filter paper. Collect the filtrate in the beaker below the funnel (fig. 6.3).

c. Test some of the filtrate in the beaker for the presence of glucose. To do this, place 1 mL of filtrate in a clean test tube and add 1 mL of Benedict's solution. Place the test tube in a water bath of boiling water for 2 minutes and then allow the liquid to cool

slowly. If the color of the solution changes to green, yellow, or red, glucose is present (fig. 6.4).

d. Test some of the filtrate in the beaker for the presence of starch. To do this, place a few drops of filtrate in a test tube and add 1 drop of iodine-potassium-iodide solution. If the color of the solution changes to blue-black, starch is present.

e. Observe any charcoal in the filtrate.

2. Complete Part D of the laboratory assessment.

NOTES

LABORATORY
ASSESSMENT
6

Name _____

Date _____

Section _____

The (A) corresponds to the indicated Learning Outcome(s) (O) found at the beginning of the Laboratory Exercise.

Movements Through Membranes

PART A: Assessments

Complete the following:

1. Enter data for changes in the movement of the potassium permanganate in the following table. If the Learning Extension Activity was performed, include those results in the table. (A1)

Elapsed Time	Radius of Purple Circle in Millimeters (mm) at Room Temperature	Learning Extension Activity: Radius of Purple Circle in Ice-Cold Water in mm	Learning Extension Activity: Radius of Purple Circle in Hot Water in mm
Initial			
1 minute			
2 minutes			
3 minutes			
4 minutes			
5 minutes			
6 minutes			
7 minutes			
8 minutes			
9 minutes			
10 minutes			

2. Prepare a graph that illustrates the diffusion distance of potassium permanganate in 10 minutes. Then, using a different colored pencil for ice-cold water and a third colored pencil for hot water, graph the data you obtained if you performed the Learning Extension Activity. (A2)

3. Interpret your graph. (A2) _____
_____
_____
_____
_____
_____
_____
_____
_____

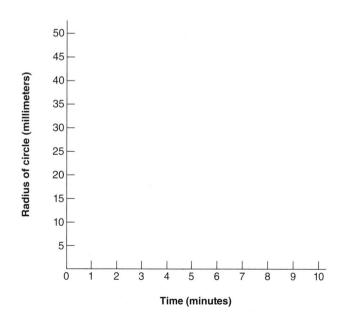

4. Using a tea bag placed in hot water, explain the process of diffusion. (A1) _____

_____

_____

CRITICAL THINKING ASSESSMENT

By answering yes or no, indicate which of the following provides an example of diffusion. (A1)

1. A perfume bottle is opened, and soon the odor can be smelled in all parts of the room. _____

2. A tea bag is dropped into a cup of hot water, and, without being stirred, all of the liquid becomes the color
   of tea leaves. _____

3. Water molecules move from a faucet through a garden hose when the faucet is turned on. _____

PART B: Assessments

Complete the following:

1. What was the change in the level of molasses in 10 minutes?_____   30 minutes? _____

2. How do you explain this change? (A3) _____

   _____

3. Using what you know about the process of osmosis, explain why celery and carrot pieces become more plump and crisp
   when placed overnight in a container of plain water. (A3) _____

   _____

CRITICAL THINKING ASSESSMENT

By answering yes or no, indicate which of the following involves osmosis. (A3)

1. A fresh potato is peeled, weighed, and soaked in a strong salt solution. The next day, it is discovered that the potato has
   lost weight. _____

2. Air molecules escape from a punctured tire as a result of high pressure inside. _____

3. Plant seeds soaked in water swell and become several times as large as before soaking. _____

PART C: Assessments

Complete the following:

1. In the spaces, sketch a few blood cells from each of the test tubes and indicate the magnification. Then, draw an arrow(s)
   indicating the movement of water (into the cell, out of the cell, or both equally). (A4)

Tube 1
(distilled water)

Tube 2
(0.9% NaCl)

Tube 3
(3.0% NaCl)

_____×   _____×   _____×

2. Based on your results, which tube contained a solution hypertonic to the blood cells? (A4) _____

Give the reason for your answer. (A4) _____

_____

3. Which tube contained a solution hypotonic to the blood cells? (A4) _____

Give the reason for your answer. (A4) _____

_____

4. Which tube contained a solution isotonic to the blood cells? (A4) _____

Give the reason for your answer. (A4) _____

_____

5. Observe the RBCs shown in figure 6.5. Select the solutions in which the cells were placed to illustrate the effects of tonicity on cells.

FIGURE 6.5 Three blood cells placed in three different solutions: distilled water, 0.9% NaCl, and 3% NaCl. Select the solutions in which the cells were placed, using the terms provided. (A4)

Terms:
Hypertonic
Hypotonic
Isotonic

(a) _____   (b) _____   (c) _____

CRITICAL THINKING ASSESSMENT

Explain the potential risk to a person who drinks unusually large quantities of plain water in a short period of time.
(A3) (A4)

_____

_____

_____

PART D: Assessments

Complete the following:

1. Which of the substances in the mixture you prepared passed through the filter paper into the filtrate? (A5) _____

_____

2. What evidence do you have for your answer to question 1? (A5) _____

_____

3. What force was responsible for the movement of substances through the filter paper? (A5) _____

4. What substances did not pass through the filter paper? (A5) _____

5. What factor prevented these substances from passing through? (A5) _____

   _____

6. During exercise, skeletal muscles increase their metabolic activity, and capillary blood flow increases. Would this increase *or* decrease filtration in these capillaries? _____ Explain your answer. (A5)

   _____

   _____

CRITICAL THINKING ASSESSMENT

By answering yes or no, indicate which of the following involves filtration. (A5)

1. Oxygen molecules move into a cell and carbon dioxide molecules leave a cell because of differences in the concentrations of these substances on either side of the plasma membrane. _____

2. Blood pressure forces water molecules and small dissolved solutes from the blood outward through the thin wall of a blood capillary. _____

3. Noninstant coffee is made using a coffeemaker. _____

Osmosis and Diffusion

LESSON 1

Varying Extracellular Concentration

The purpose of this exercise is to demonstrate how isotonic, hypotonic, and hypertonic solutions can affect red blood cells using a spectrophotometer to measure the color of different blood solutions. A solution consists of a solvent, which is usually water, and a solute, which are the molecules dissolved in the solvent. Molecules tend to move from areas of higher concentration to areas of lower concentration, which allows them to become evenly distributed throughout the solution. This is called diffusion. The concentration of solutes inside a blood cell and outside a blood cell is usually different, so a concentration gradient (flow of solutes) is established between the two areas across the cellular membrane. Water will move from the solution with a lower concentration of solutes to the one with the higher concentration of solutes. This is called osmosis.

In this lab you will examine the tonicity of sodium chloride solutions in relation to red blood cells. You will measure the color of ten different blood samples in different concentrations of sodium chloride with a spectrophotometer at 510 nm. Normally, this wavelength of light is reflected by the cell membranes. If the cells are placed in a hypotonic solution, they will burst, and less light will be reflected. This will cause the transmittance value from the spectrophotometer to increase. If the cells are placed in a hypertonic solution, they will shrink, and more light will be reflected, causing the transmittance value to decrease.

Before you begin, familiarize yourself with the following concepts (use your textbook as a reference):

- Movement across a membrane
- Osmotic pressure
- Crenation

State your hypothesis regarding the transmittance values related to the blood samples in different concentrations of sodium chloride.

_____

_____

_____

_____

1. Open Lesson 1, Osmosis and Diffusion: Varying Extracellular Concentration.
2. Read the objectives and introduction and take the pre-laboratory quiz.
3. After completing the pre-laboratory quiz, read through the wet lab.
4. The laboratory exercise will open when you click Continue after completing the wet lab (figure 6.6).

FIGURE 6.6   Opening screen for the laboratory exercise on Osmosis and Diffusion: Varying Extracellular Concentration.

5. Follow the instructions provided in the laboratory exercise and answer the following questions as you proceed.
6. Using the data from the Ph.I.L.S. journal, draw a graph of the NaCl concentration and % transmittance using the following axes. Use the table to record your different plot points. A6
7. On the preceding graph, indicate the areas that correspond to the blood cells being in each of the following solutions:
   (A) isotonic solution; (B) hypertonic solution; (C) hypotonic solution. A7 A8

8. Physiological saline (0.9%) is isotonic. What concentration (mM) would be equivalent to physiological saline? A9 _____
9. Describe the net movement of water (into the cell, out of the cell, no net movement) when cells are placed in a(n): A9
   a. isotonic solution _____
   b. hypotonic solution _____
   c. hypertonic solution _____

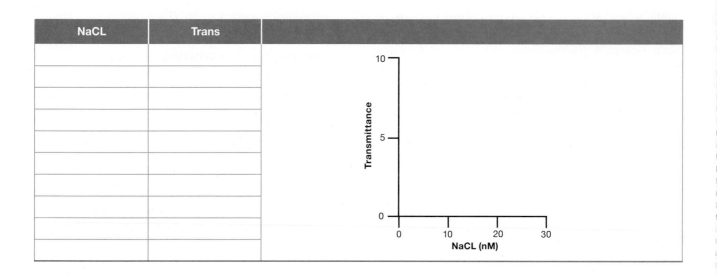

NaCL	Trans

CRITICAL THINKING ASSESSMENT

A patient is given an IV (intravenous) of deionized water (water with no solutes). If a large volume of fluid is administered, predict the effects. Choose one of the words in parentheses for each blank to complete the sentence. Plasma becomes

_____ (hypotonic/hypertonic) to the red blood cells.

_____ Water moves from the (plasma or red blood cells)

_____ to the (plasma or red blood cells)

_____ causing the red blood cells to (crenate or swell).

A person drinks ocean salt water (3.5% salt). If the person drinks a large volume of the salt water (the salt water is absorbed into the blood), the plasma becomes

_____ (hypotonic/hypertonic) to the red blood cells.

_____ Water moves from the (plasma or red blood cells)

_____ to the (plasma or red blood cells)

_____ causing the red blood cells to (crenate or swell).

What would you predict if a patient was administered an IV containing a physiological saline solution (0.9% sodium chloride)? _____

_____

Cell Cycle

Purpose of the Exercise

To review the phases in the cell cycle and to observe cells in various phases of their life cycles.

MATERIALS NEEDED

Models of animal cells during mitosis

Microscope slides of whitefish mitosis (blastula)

Compound light microscope

For Demonstration Activity:

Microscope slide of human chromosomes from leukocytes in mitosis

Oil immersion objective

Learning Outcomes AP|R

After completing this exercise, you should be able to:

O1 Describe the phases and structures of the cell cycle.

O2 Identify and sketch the phases (stages) in the life cycle of a particular cell.

O3 Arrange into a correct sequence a set of models or drawings of cells in various phases of their life cycles.

The O corresponds to the assessments A indicated in the Laboratory Assessment for this Exercise.

Pre-Lab

Carefully read the introductory material and examine the entire lab. Be familiar with the cell cycle from lecture or the textbook. Answer the pre-lab questions.

Pre-Lab Questions Select the correct answer for each of the following questions:

1. Which of the following is *not* part of interphase?
   a. cytokinesis
   b. first gap phase
   c. second gap phase
   d. synthesis phase

2. A cell in the stage of G-zero
   a. divides continuously.
   b. ceases cell division.
   c. exhibits cytokinesis.
   d. exists in a gap phase.

3. Which is the correct sequence of the M (mitotic) phase?
   a. telophase, prophase, metaphase, anaphase
   b. prophase, anaphase, metaphase, telophase
   c. prophase, telophase, metaphase, anaphase
   d. prophase, metaphase, anaphase, telophase

4. Which of the following events occurs in prophase?
   a. spindle disappears
   b. nuclear envelopes reassemble
   c. chromosomes condense
   d. sister chromatids separate

5. Which of the following events occurs during telophase?
   a. sister chromatids separate
   b. chromosomes align along equator
   c. cytokinesis
   d. spindle apparatus forms

6. Sister chromatids separate to opposite poles during
   a. telophase.
   b. anaphase.
   c. prophase.
   d. metaphase.

7. A karyotype is used to count centrioles and spindle fibers.
   a. True _____
   b. False _____

8. Daughter cells are formed from cell cycles.
   a. True _____
   b. False _____

The cell cycle consists of the series of changes a cell undergoes from the time it is formed until it divides. Interphase, mitotic phase, and differentiation are phases of a cell cycle. Typically, a newly formed diploid cell with 46 chromosomes grows to a certain size and then divides to form two new cells (*daughter cells*), each with 46 chromosomes.

Before the cell divides, it must synthesize biochemicals and other contents. This period of preparation is called *interphase.* The extensive period of interphase is divided into three phases. The S phase, when DNA synthesis occurs, is between two gap phases (G_1 and G_2), when cell growth occurs and cytoplasmic organelles duplicate. Interphase is followed by the *M (mitotic) phase,* which has two major portions: (1) division of the nucleus, called *mitosis,* and (2) division of the cell's cytoplasm, called *cytokinesis.* Mitosis includes four recognized phases: *prophase, metaphase, anaphase,* and *telophase.* Eventually, some specialized cells, such as skeletal muscle cells and most nerve cells, cease further cell division, but remain alive. A mature living cell that ceases cell division is in a state called G_0 (G-zero) for the remainder of its life span. This state can last from hours to decades.

A special type of cell division, called *meiosis,* occurs in the reproductive system to produce haploid gametes with 23 chromosomes. Meiosis is included in Laboratory Exercise 60.

PROCEDURE: Cell Cycle

A cell generally subsists for extended periods of time in interphase in which the metabolic processes within the nucleus and cytoplasm occur. Those processes were studied in Laboratory Exercise 5. The mitosis phase occurs for cellular division to enable growth and repair of cells. As you examine cells that have frequent cell division in this lab, you will find that cells in interphase are still the most frequently located. The slides selected for your study possess abundant mitosis phases so that you can more efficiently locate some of the interesting cellular structures that can be distinguished during mitosis. Keep in mind that the cell cycle is referred to by phases, not steps, as one phase gradually becomes the next phase. You will probably observe some phases that are debatable as to which phase is represented, but this is common. The figures selected for this lab for comparison purposes represent phases that are close to the mid-portion of the phase.

1. Study the various phases of the cell's life cycle represented in figures 7.1, 7.2, 7.3, and 7.4.
2. Using figure 7.5 as a guide, observe the animal mitosis models and review the major events in a cell's life cycle represented by each of them. Be sure you can arrange these models in correct sequence if their positions are changed. The acronym IPMAT can help you arrange the correct order of phases in the cell cycle. This includes interphase followed by the four phases of mitosis. Cytokinesis overlaps anaphase and telophase.
3. Complete Part A of Laboratory Assessment 7.

FIGURE 7.1 [AP|R] The cell cycle: interphase (G_1, S, and G_2) and M (mitotic) phase (mitosis and cytokinesis).

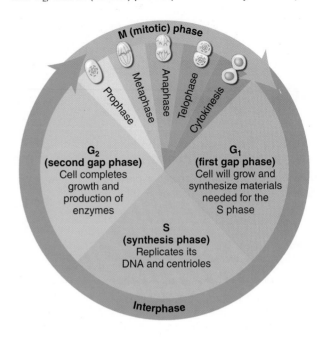

FIGURE 7.2 Cell in interphase (400×). The nucleoplasm contains a fine network of chromatin.

Plasma membrane
Cytoplasm
Nuclear envelope
Nucleoplasm

FIGURE 7.3 Cell in prophase (400×).

Nuclear envelope
Chromosomes
Centrosome
Plasma membrane

FIGURE 7.4 Structures in the dividing cell during anaphase.

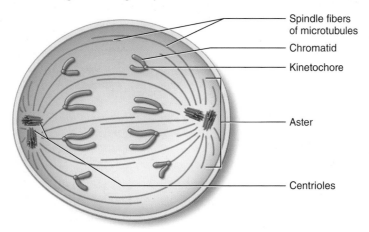

Spindle fibers of microtubules

Chromatid

Kinetochore

Aster

Centrioles

4. Obtain a slide of the whitefish mitosis (blastula).
   a. Examine the slide using the high-power objective of a microscope. The tissue on this slide was obtained from a developing embryo (blastula) of a fish, and many of the embryonic cells are undergoing mitosis. The chromosomes of these dividing cells are darkly stained.
   b. Search the tissue for cells in various phases of cell division. There are several sections on the slide. If you cannot locate different phases in one section, examine the cells of another section because the phases occur randomly on the slide. Frequently refer to figures 7.2, 7.3, and 7.5 as references for identifications.
   c. Each time you locate a cell in a different phase, sketch it in an appropriate circle in Part B of the laboratory assessment.
5. Complete Parts C and D of the laboratory assessment.

CRITICAL THINKING ACTIVITY

Which phase of the cell cycle was the most numerous in the blastula? _____

Explain your answer. _____

_____

_____

FIGURE 7.5 Mitosis drawings and photographs of the four phases and cytokinesis.

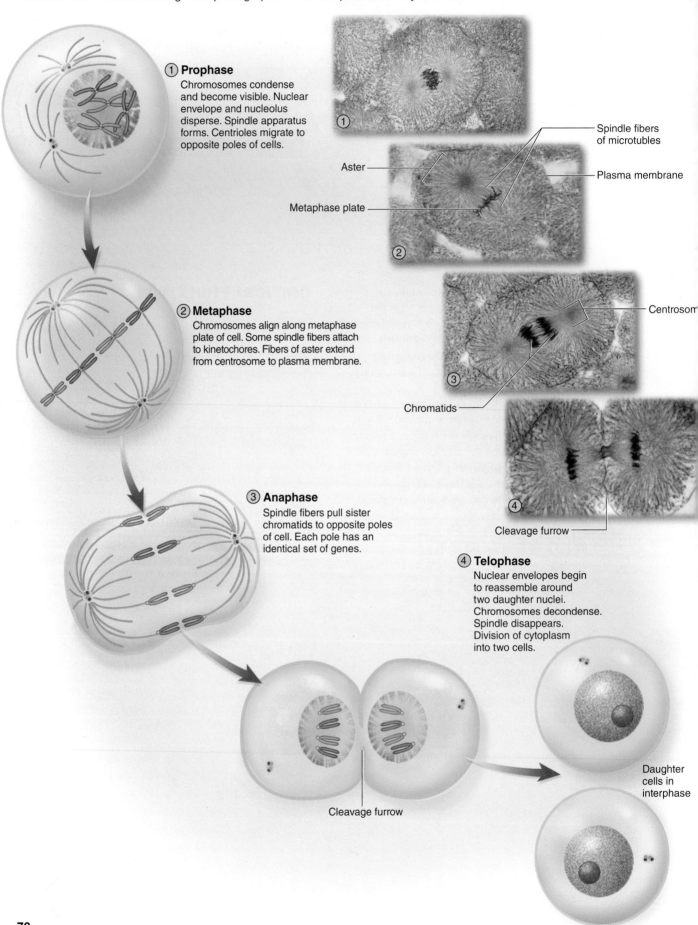

① **Prophase**
Chromosomes condense and become visible. Nuclear envelope and nucleolus disperse. Spindle apparatus forms. Centrioles migrate to opposite poles of cells.

② **Metaphase**
Chromosomes align along metaphase plate of cell. Some spindle fibers attach to kinetochores. Fibers of aster extend from centrosome to plasma membrane.

③ **Anaphase**
Spindle fibers pull sister chromatids to opposite poles of cell. Each pole has an identical set of genes.

④ **Telophase**
Nuclear envelopes begin to reassemble around two daughter nuclei. Chromosomes decondense. Spindle disappears. Division of cytoplasm into two cells.

Spindle fibers of microtubles

Aster

Plasma membrane

Metaphase plate

Centrosom

Chromatids

Cleavage furrow

Cleavage furrow

Daughter cells in interphase

Using the oil immersion objective of a microscope, see if you can locate some human chromosomes by examining a prepared slide of human chromosomes from leukocytes. The cells on this slide were cultured in a special medium and were stimulated to undergo mitosis. The mitotic process was arrested in metaphase by exposing the cells to a chemical called colchicine, and the cells were caused to swell osmotically. As a result of this treatment, the chromosomes were spread apart. A complement of human chromosomes should be visible when they are magnified about 1,000×. Each chromosome is double-stranded and consists of two chromatids joined by a common centromere (fig. 7.6).

FIGURE 7.6 AP|R (a) A complement of human chromosomes of a female (1,000×). (b) A karyotype can be constructed by arranging the homologous chromosome pairs together in a chart. A karyotype can aid in the diagnosis of genetic conditions and abnormalities. The completed karyotype indicates a normal male.

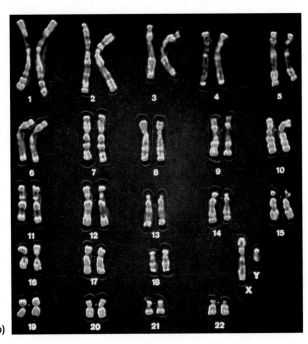

(a)

(b)

NOTES

Cell Cycle

Name _____

Date _____

Section _____

The Ⓐ corresponds to the indicated Learning Outcome(s) Ⓞ found at the beginning of the Laboratory Exercise.

PART A: Assessments

Complete the table by listing the major events of each phase in the box.

Phase		Major Events Occurring
Interphase	G_1, S, and G_2	
M (Mitotic) Phase	**Mitosis** Prophase	
	Metaphase	
	Anaphase	
	Telophase	
	Cytokinesis	

PART B: Assessments

Sketch an interphase cell and cells in different phases of mitosis to illustrate the whitefish cell's life cycle. Label the major cellular structures represented in the sketches and indicate cytokinesis locations. (The circles represent fields of view through the microscope.) A2

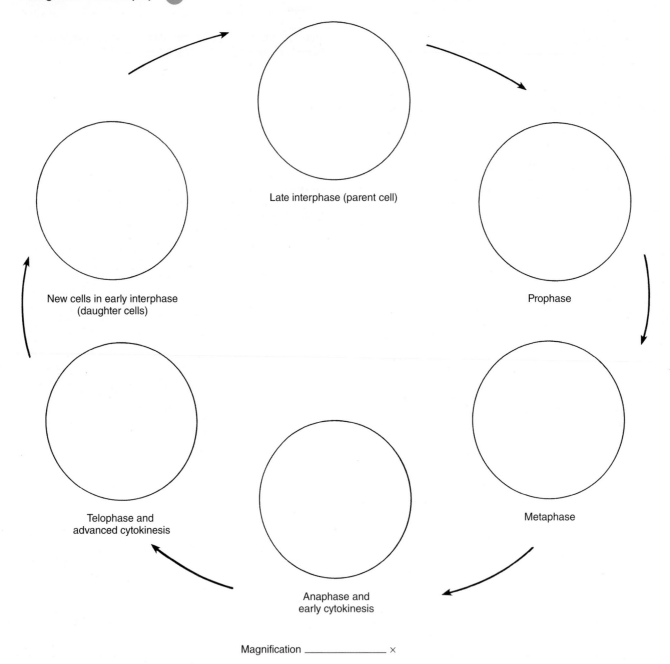

Late interphase (parent cell)

New cells in early interphase
(daughter cells)

Prophase

Telophase and
advanced cytokinesis

Metaphase

Anaphase and
early cytokinesis

Magnification _____ ×

PART C: Assessments

Complete the following:

1. In what ways are the new cells (daughter cells), which result from a cell cycle, similar? _____

_____

2. Distinguish between mitosis and cytokinesis. A1 _____

_____

PART D: Assessments

1. Identify the mitotic phase represented by each of the micrographs in figure 7.7 (a–d). A3

    a. _____   **c.** _____

    b. _____   **d.** _____

2. Identify the structures indicated by numbers in figure 7.7 by placing the correct numbers in the spaces provided.

    _____ Aster   _____ Metaphase plate

    _____ Centrosome   _____ Nuclear envelope

    _____ Chromatid   _____ Plasma membrane

    _____ Cleavage furrow   _____ Spindle fiber

FIGURE 7.7 Identify the mitotic phase and structures of the cell in each of these micrographs of the whitefish blastula (250×).

(a)

(b)

(c)

(d)

NOTES

Epithelial Tissues

1/26/17

Purpose of the Exercise

To review the characteristics of epithelial tissues and to observe examples.

MATERIALS NEEDED

Compound light microscope

Prepared slides of the following epithelial tissues:

    Simple squamous epithelium (lung)

    Simple cuboidal epithelium (kidney)

    Simple columnar epithelium (small intestine)

    Pseudostratified (ciliated) columnar epithelium (trachea)

    Stratified squamous epithelium (nonkeratinized; esophagus)

    Stratified squamous epithelium (keratinized; epidermis)

    Stratified cuboidal epithelium (salivary gland)

    Stratified columnar epithelium (urethra)

    Transitional epithelium (urinary bladder)

For Learning Extension Activity:

Colored pencils

Learning Outcomes AP|R

After completing this exercise, you should be able to:

O1 Sketch and label the characteristics of epithelial tissues that you were able to observe.

O2 Differentiate the special characteristics of each type of epithelial tissue.

O3 Indicate a location and function of each type of epithelial tissue.

O4 Inventory the general characteristics of epithelial tissues that you were able to observe.

The **O** corresponds to the assessments **A** indicated in the Laboratory Assessment for this Exercise.

Pre-Lab

Carefully read the introductory material and examine the entire lab. Be familiar with epithelial tissues from lecture or the textbook. Answer the pre-lab questions.

Pre-Lab Questions Select the correct answer for each of the following questions:

1. The tissues that cover external and internal body surfaces are
    a. epithelial.
    b. connective.
    c. muscle.
    d. nervous.

2. The cells with a flat shape are named
    a. stratified.
    b. cuboidal.
    c. columnar.
    d. squamous.

3. One edge of an epithelial cell attaches to the
    a. cilia.
    b. basement membrane.
    c. nucleus.
    d. cytoplasm.

4. If a structure cut has a circular appearance when observed, the sectional cut was a(n)
    a. oblique section.
    b. cross section.
    c. longitudinal section.
    d. lengthwise section.

5. Which of the following is *not* a function of epithelial tissues?
    a. protection
    b. secretion
    c. bind and support
    d. absorption

6. Which of the following epithelial tissues is more than one cell layer thick?
    a. transitional
    b. simple squamous
    c. pseudostratified
    d. simple cuboidal columnar

7. The study of tissues is called histology.
    a. True ✓
    b. False _____

8. The basement membrane is composed of cilia.
    a. True _____
    b. False ✓

A tissue is composed of a layer or group of cells similar in size, shape, and function. The study of tissues is called *histology.* Within the human body, there are four major types of tissues: (1) *epithelial,* which cover the body's external and internal surfaces and comprise most glands; (2) *connective,* which bind and support parts; (3) *muscle,* which make movement possible; and (4) *nervous,* which conduct impulses from one part of the body to another and help to control and coordinate body activities.

Epithelial tissues are tightly packed single (simple) to multiple (stratified) layers composed of cells that provide protective barriers. The underside of this tissue contains an acellular basement membrane layer of adhesive cellular secretions and collagen through which the epithelial cells anchor to an underlying connective tissue. A unique type of simple epithelium, pseudostratified columnar, appears to be multiple cells thick. However, this is a false appearance because all cells bind to the basement membrane. The cells readily divide and lack blood vessels (are avascular).

Epithelial tissues always have a free (apical) surface exposed to the outside or to an open space internally and a basal surface that attaches to the basement membrane. Blood vessels near the basement membrane allow epithelial cells to obtain their oxygen and nutrients. Many epithelial cell shapes, like squamous (flat), cuboidal, and columnar, exist and are used to name and identify the variations. The functions of epithelial tissues include protection, filtration, secretion, and absorption. Many of the prepared slides contain more than the tissue to be studied, so be certain that your view matches the correct tissue. Also be aware that stained colors of all tissues might vary.

The slides you observe might have representative tissue from more than one site, but the basic tissue structure will be similar to those represented in this laboratory exercise. Simple squamous epithelium is located in the air sacs (alveoli) of the lungs and inner linings of the heart and blood vessels. Simple cuboidal epithelium is situated in kidney tubules, thyroid gland, liver, and ducts of salivary glands. A nonciliated type of simple columnar epithelium is located in the linings of the uterus, stomach, and intestines, and a ciliated type lines the uterine tubes. A ciliated type of pseudostratified columnar epithelium is positioned in the linings of the upper respiratory tubes. A keratinized type of stratified squamous epithelium is found in the epidermis of the

skin, and a nonkeratinized type in the linings of the oral cavity, esophagus, vagina, and anal canal. Stratified cuboidal epithelium is represented in larger ducts of glands while stratified columnar epithelium lines the urethra of males. Transitional epithelium lines the urinary bladder, ureters, and part of the urethra. Various glands of the body are composed of primarily glandular epithelium. Studies of glands will be found in Laboratory Exercises 11, 39, and 40.

PROCEDURE: Epithelial Tissues

This procedure represents the first of three laboratory exercises in a study of histology. So that you may clearly observe the details of tissue sections, the sectioned tissue has been stained, and the colors you observe will depend upon the particular stains used for the preparation. The slice of the histological section is usually only one or two cells thick, which will allow light to be transmitted through the entire section. Use the fine adjustment knob on the microscope to view the depth of the structures within the tissue section. Refer to the figures and the table in the laboratory exercise often as you examine each tissue.

1. Use the microscope to observe the prepared slides of types of epithelial tissues. As you observe each tissue, look for its special distinguishing features such as cell size, shape, and arrangement. Compare your prepared slides of epithelial tissues to the micrographs in figure 8.1 and the characteristics of each specific tissue in table 8.1 on p. 81.
2. As you observe the tissues in figure 8.1 and the prepared slides, note characteristics epithelial tissues have in common.
3. As you observe each type of epithelial tissue, prepare a labeled sketch of a representative portion of the tissue in Part A of Laboratory Assessment 8. Add any personal notes near your sketches that might assist your recognition of the tissues for any future assessments.
4. Test your ability to recognize each type of epithelial tissue. To do this, have a laboratory partner select one of the prepared slides, cover its label, and focus the microscope on the tissue. Then see if you can correctly identify the tissue.
5. Review the introductory material and table 8.1.
6. Complete Parts B and C of the laboratory assessment.

(a) AP|R Simple squamous epithelium (side view) (from lung) (250×)

(b) Simple squamous epithelium (surface view) (250×)

(c) AP|R Simple cuboidal epithelium (from kidney) (165×)

(d) AP|R Simple columnar epithelium (nonciliated) (from intestine) (400×)

(e) AP|R Pseudostratified columnar epithelium with cilia (from trachea) (1,000×)

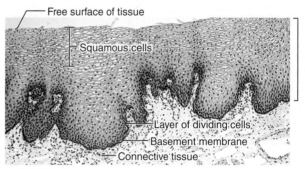

(f) AP|R Stratified squamous epithelium (nonkeratinized) (from esophagus) (100×)

FIGURE 8.1 *Continued.*

(g) AP|R Stratified squamous
epithelium (keratinized)
(from skin) (400×)

(h) AP|R Stratified cuboidal epithelium (from salivary gland) (600×)

(i) AP|R Stratified columnar epithelium (from male urethra) (230×)

(j) AP|R Transitional epithelium (unstretched) (from
urinary bladder) (675×)

(k) Transitional epithelium (stretched) (from urinary
bladder) (675×)

FIGURE 8.2 Three possible cuts of a banana: (a) cross section; (b) oblique section; (c) longitudinal section. Sections through an organ, as a body tube, frequently produce views similar to the cut banana.

(a)   (b)   (c)

LEARNING EXTENSION ACTIVITY

As you observe histology slides, be aware that tissues and some organs may have been sectioned in longitudinal (lengthwise cut), cross-sectional (cut across), or oblique (angular cut) ways. Observe figure 8.2 for how this can be demonstrated on cuts of a banana. The direction in which the tissue or organ was cut will result in a certain perspective when it is sectioned, just as the banana was cut three different ways. Often a tissue slide has more than one tissue represented and has blood vessels or other structures cut in various ways. Locate an example of a longitudinal section, cross section, and oblique section of some structure on one of your tissue slides.

Table 8.1 Epithelial Tissues, Descriptions, Functions, and Representative Locations

Tissue Type	Descriptions	Functions	Representative Locations
Simple squamous epithelium	Single thin layer; flattened cells	Filtration; diffusion; osmosis; secretion	Air sacs (alveoli) of lungs; walls of capillaries; linings of blood vessels and ventral body cavity
Simple cuboidal epithelium	Single layer; cube-shaped cells	Secretion; absorption	Surface of ovaries; linings of kidney tubules; linings of ducts of certain glands
Simple columnar epithelium	Single layer; elongated narrow cells; some ciliated	Protection; secretion; absorption	Linings of uterus, stomach, gallbladder, and intestines
Pseudostratified columnar epithelium	Single layer; elongated cells; some cells do not reach free surface; often ciliated	Protection; secretion; movement of mucus and substances	Linings of respiratory passages
Stratified squamous epithelium (nonkeratinized)	Many layers; surface cells flattened and remain alive	Protection; resists abrasion	Linings of oral cavity, esophagus, vagina, and anal canal
Stratified squamous epithelium (keratinized)	Many layers; surface cells dead and keratinized	Protection; resists abrasion; retards water loss	Epidermis of skin
Stratified cuboidal epithelium	2 to 3 layers; cube-shaped cells	Protection; secretion	Linings of larger ducts of mammary glands, sweat glands, salivary glands, and pancreas
Stratified columnar epithelium	Superficial layer of elongated cells; basal layers of cube-shaped cells	Protection; secretion	Part of the male urethra; parts of the pharynx
Transitional epithelium	Many layers; cube-shaped and elongated cells; thinner layers when stretched	Distensibility; protection	Linings of urinary bladder and ureters and part of urethra
Glandular epithelium	Unicellular or multicellular	Secretion	Salivary glands; sweat glands; endocrine glands

NOTES

Name _Julianne Klein_

Date _1/26/17_

Section _TTh 9:30-1220pm_

The (A) corresponds to the indicated Learning Outcome(s) (O) found at the beginning of the Laboratory Exercise.

Epithelial Tissues

PART A: Assessments

In the space that follows, sketch a few cells of each type of epithelium you observed. For each sketch, label the major characteristics, indicate the magnification used, write an example of a location in the body, and provide a function. Add any personal notes in any of the extra space within the box with your sketch that would assist your future ability to recognize the tissue. (A1) (A2) (A3)

Simple squamous epithelium (__4__ ×)
Location: _____
Function: _____

Simple cuboidal epithelium (__4__ ×)
Location: _____
Function: _____

Simple columnar epithelium (__4__ ×)
Location: _____
Function: _____

Pseudostratified columnar epithelium with cilia (__4__ ×)
Location: _____
Function: _____

1- simple cuboidal epithelium
2- simple cuboidal epithelium
3- pseudostratified columnar epithelium
4- transitional epithelium
6- transitional epithelium (31)

Nonkeratinized stratified squamous epithelium (__4__×)
Location: _____
Function: _____

Keratinized stratified squamous epithelium (__4__×)
Location: _____
Function: _____

Stratified cuboidal epithelium (__4__×)
Location: _____
Function: _____

Stratified columnar epithelium (__4__×)
Location: _____
Function: _____

Transitional epithelium (__4__×)
Location: _____
Function: _____

CRITICAL THINKING ASSESSMENT

As a result of your observations of epithelial tissues, which one(s) provide(s) the best protection? Explain your answer. **A3** _____

_____

_____

LEARNING EXTENSION ACTIVITY

Use colored pencils to differentiate various cellular structures in Part A. Select a different color for a nucleus, cytoplasm, plasma membrane, basement membrane, goblet cell, and cilia whenever visible.

PART B: Assessments

Match the tissues in column A with the characteristics in column B. Place the letter of your choice in the space provided. (Some answers may be used more than once.) **A2** **A3**

Column A

a. Simple columnar epithelium
b. Simple cuboidal epithelium
c. Simple squamous epithelium
d. Pseudostratified columnar epithelium
e. Stratified squamous epithelium
f. Stratified columnar epithelium
g. Stratified cuboidal epithelium
h. Transitional epithelium

Column B

h. h. g. **1.** Consists of several layers of cube-shaped, elongated, and irregular cells, allowing an expandable lining

d. **2.** Commonly possesses cilia that move dust and mucus out of the respiratory airways

c. e. **3.** Single layer of flattened cells

a. d. **4.** Single row of elongated cells, but some cells don't reach the free surface

c. **5.** Forms walls of capillaries and air sacs of lungs

f. **6.** Provides lining of urethra of males and parts of pharynx

c. **7.** Provides abrasion protection of skin epidermis and oral cavity

h. h. **8.** Forms inner lining of urinary bladder and ureters

b. **9.** Lines kidney tubules and ducts of salivary glands

a. **10.** Forms lining of stomach and intestines

g. **11.** Two or three layers of cube-shaped cells

e. **12.** Forms lining of oral cavity, anal canal, and vagina

PART C: Assessments

As you examined each specific epithelial tissue, what were the general characteristics they possessed? List and describe those characteristics that you were able to observe. **A4**

_____

_____

_____

_____

_____

_____

_____

85

IN order

epithelial - free surface

- sits on connective tissue (C.T.)
- cartilage
- muscle

*different staining
means new tissue *

Trachea

→ epithelial
→ C.T.
→ hyaline cartilage
→ muscle
→ C.T.

*nerves go
through C.T.

→ connective tissue
→ Lumen (squamos cells)

lumen

Shape of epithelial cells

1) squamos (nucleus same shape
as cell)
2) cuboidal (cube-like)
3) columnar (column-like)

Layering
- cuboidal no more than 2 layers
(glands + ducts)

transitional - urinary tract
(ureters, bladder + urethra)

Simple, stratified, pseudostratified

LABORATORY
EXERCISE 9

Connective Tissues

Purpose of the Exercise

To review the characteristics of connective tissues and to observe examples of the major types.

MATERIALS NEEDED

Compound light microscope
Prepared slides of the following:
Areolar connective tissue
Adipose tissue
Reticular connective tissue
Dense regular connective tissue
Dense irregular connective tissue
Elastic connective tissue
Hyaline cartilage
Fibrocartilage
Elastic cartilage
Bone (compact, ground, cross section)
Blood (human smear)

For Learning Extension Activity:
Colored pencils

Learning Outcomes AP|R

After completing this exercise, you should be able to:

01 Sketch and label the characteristics of connective tissues that you were able to observe.

02 Differentiate the special characteristics of each of the major types of connective tissue.

03 Indicate a location and function of each type of connective tissue.

04 Inventory the general characteristics of connective tissues that you were able to observe.

The O corresponds to the assessments A indicated in the Laboratory Assessment for this Exercise.

Pre-Lab

Carefully read the introductory material and examine the entire lab. Be familiar with connective tissues from lecture or the textbook. Answer the pre-lab questions.

Pre-Lab Questions Select the correct answer for each of the following questions:

1. Which of the following is *not* a function of connective tissues?
   a. fill spaces
   b. provide support
   c. bind structures together
   d. movement ✓

2. The most abundant fibers of connective tissues are
   a. reticular.
   b. elastic.
   c. collagen. ✓
   d. glycoprotein.

3. Which of the following is a loose connective tissue?
   a. fibrocartilage
   b. areolar ✓
   c. spongy bone
   d. elastic connective

4. Which connective tissue is liquid?
   a. blood ✓
   b. adipose
   c. reticular connective
   d. hyaline cartilage

5. The connective tissue that composes tendons and ligaments is
   a. dense regular connective. ✓
   b. elastic connective.
   c. elastic cartilage.
   d. dense irregular connective.

6. The connective tissue _____ contains abundant fibers of collagen and cells in a solid matrix.
   a. fibrocartilage
   b. adipose
   c. bone ✓
   d. areolar connective

7. Adipose, areolar, and reticular connective tissues are considered loose connective tissue types.
   a. True ✓
   b. False

87

special

8. Blood and lymph are considered loose connective tissue types.
   a. True ~~✓~~   b. False ✓
9. Cells secrete the fibers and a ground substance of a connective tissue.
   a. True ✓   b. False _____

Connective tissues contain a variety of cell types and occur in all regions of the body. They bind structures together, provide support and protection, fill spaces, store fat, and transport blood cells.

Connective tissue cells are not as close together as epithelial cells and are often widely scattered in an abundance of a noncellular *extracellular matrix*. This extracellular matrix varies in quantity depending upon the specific tissue type and is like a filler material between the cells. The cells produce and secrete two components of the extracellular matrix known as the *ground substance* and *fibers*. The ground substance varies from a liquid to semisolid to solid, and has functions in support and as a medium for substances to move between cells and blood vessels. Fibers, the other component of the extracellular matrix, consist of fibrous proteins including *collagen, reticular,* and *elastic* fibers.

Collagen fibers are thick threads, the most abundant of the three types, and are white in unstained tissues. Reticular fibers are fine threads of highly branched collagen with glycoprotein. Elastic fibers form complex networks of a highly branched protein called elastin, which is springlike and appears yellowish in unstained tissues. The fibers provide the binding properties within the tissue and between organs. You might compare connective tissue to making gelatin: the gelatin of various densities represents the ground substance, added fruit represents cells, and added strands represent the fibers.

Many of the prepared slides contain more than the tissue to be studied, so be certain that your view matches the correct tissue. Additional study of bone and blood will be found in Laboratory Exercises 12 and 41.

PROCEDURE: Connective Tissues AP|R

1. Use a microscope to observe the prepared slides of various connective tissues. As you observe each tissue, look for its special distinguishing features. Compare your prepared slides of connective tissues to the micrographs in figure 9.1 and the characteristics of each specific tissue in tables 9.1 and 9.2 on page 90.

FIGURE 9.1 AP|R Micrographs of connective tissues.

(a) Areolar connective (from fascia between muscles) (800×)

Elastic fiber
Collagen fiber
Fibroblast
Ground substance

(b) Adipose tissue (from hypodermis layer) (400×)

Plasma membrane
Cytoplasm
Nucleus
Fat-storage area
Adipocyte

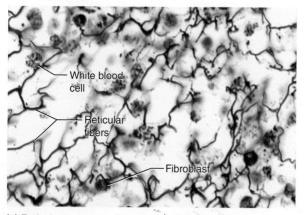

(c) Reticular connective (from spleen) (400×)

White blood cell
Reticular fibers
Fibroblast

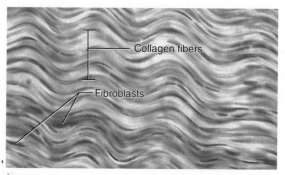

(d) Dense regular connective (from tendon) (400×)

Collagen fibers
Fibroblasts

FIGURE 9.1 AP|R *Continued.*

(e) Dense irregular connective (from dermis) (400×)

(f) Elastic connective (from artery wall) (600×)

(g) Hyaline cartilage (from trachea) (160×)

(h) Fibrocartilage (from intervertebral disc) (400×)

(i) Elastic cartilage (from ear) (400×)

(j) Compact bone (200×)

(k) Blood (400×)

6) Dense Regular CT
   Dense+Regular CT
7) Elastic CT
8) Blood (40x)
9) Blood (100x)
10) Compact Bone (10x)
11) Fibrocartilage (4x)

Table 9.1 Connective Tissues, Descriptions, and Functions

Tissue Type	Descriptions	Functions
Areolar connective	Cells in abundant fluid-gel matrix; loose arrangement of collagen and elastic fibers	Loosely binds organs; holds tissue fluids
Adipose	Cells in sparse fluid-gel matrix; closely packed cells	Protects; insulates; stores fat
Reticular connective	Cells in fluid-gel matrix; reticular fibers	Supportive framework of organs
Dense regular connective	Cells in fluid-gel matrix; parallel, wavy collagen fibers	Tightly binds body parts
Dense irregular connective	Cells in fluid-gel matrix; random collagen fibers	Sustains tissue tension; durable
Elastic connective	Cells in fluid-gel matrix; collagen fibers densely packed; branched elastic fibers	Provides elastic quality
Hyaline cartilage	Cells in firm solid-gel matrix; fine network of collagen fibers; appears glassy	Supports; protects; provides framework
Fibrocartilage	Cells in firm solid-gel matrix; abundant collagen fibers	Supports; protects; absorbs shock
Elastic cartilage	Cells in firm solid-gel matrix; weblike elastic fibers	Supports; protects; provides flexible framework
Bone	Cells in solid matrix; many collagen fibers	Supports; protects; provides framework; calcium storage
Blood	Cells and platelets in fluid matrix called plasma	Transports nutrients, wastes, and gases; defends against disease; clotting

Table 9.2 Types of Mature Connective Tissues and Representative Locations*

Loose Connective	Dense Connective	Cartilage	Bone	Liquid Connective
Areolar Connective Locations: around body organs; binds skin to deeper organs	*Dense Regular Connective* Locations: tendons; ligaments	*Hyaline Cartilage* Locations: nasal septum; larynx; costal cartilage; ends of long bones; fetal skeleton	*Compact Bone* Locations: bone shafts; beneath periosteum	*Blood* Locations: lumens of blood vessels; heart chambers
Adipose Locations: hypodermis (subcutaneous layer); around kidneys and heart; yellow bone marrow; breasts	*Dense Irregular Connective* Locations: dermis; heart valves; periosteum on bone	*Fibrocartilage* Locations: between vertebrae; between pubic bones; pads (meniscus) in knee	*Spongy (Cancellous) Bone* Locations: within ends of long bones; inside flat and irregular bones	*Lymph* Locations: lumens of lymphatic vessels
Reticular Connective Locations: spleen; thymus; lymph nodes; red bone marrow	*Elastic Connective* Locations: larger artery walls; vocal cords; some ligaments between vertebrae	*Elastic Cartilage* Locations: outer ear; epiglottis		

*This table represents a scheme to organize connective tissue relationships. Spongy bone and lymph tissues are not examined microscopically in this laboratory exercise.

2. As you observe the tissues in figure 9.1 and the prepared slides, note the characteristics connective tissues have in common.

3. As you observe each type of connective tissue, prepare a labeled sketch of a representative portion of the tissue in Part A of Laboratory Assessment 9. Add any personal notes near your sketches that might assist your recognition of the tissues for any future assessments.

4. Test your ability to recognize each of these connective tissues by having a laboratory partner select a slide, cover its label, and focus the microscope on this tissue. Then see if you correctly identify the tissue.

5. Complete Parts B and C of the laboratory assessment.

Name _____

Date _____

Section _____

The **A** corresponds to the indicated Learning Outcome(s) **O** found at the beginning of the Laboratory Exercise.

Connective Tissues

PART A: Assessments

In the space that follows, sketch a small section of each of the types of connective tissues you observed. For each sketch, label the major characteristics, indicate the magnification used, write an example of a location in the body, and provide a function. Add any personal notes in any of the extra space within the box with your sketch that would assist your future ability to recognize the tissue. **A1** **A2** **A3**

Areolar connective (_____×)
Location: _____
Function: _____

Adipose (_____×)
Location: _____
Function: _____

Reticular connective (_____×)
Location: _____
Function: _____

Dense regular connective (_____×)
Location: _____
Function: _____

Dense irregular connective (_____×)
Location: _____
Function: _____

Elastic connective (_____×)
Location: _____
Function: _____

Hyaline cartilage (_____×)
Location: _____
Function: _____

Fibrocartilage (_____×)
Location: _____
Function: _____

Elastic cartilage (_____×)
Location: _____
Function: _____

Compact bone (_____×)
Location: _____
Function: _____

```
┌─────────────────────────────────────┐
│                                     │
│                                     │
│                                     │
│                                     │
│                                     │
│                                     │
│                                     │
│                                     │
│ Blood (_____×)                      │
│ Location: _____ │
│ Function: _____ │
└─────────────────────────────────────┘
```

LEARNING EXTENSION ACTIVITY

Use colored pencils to differentiate various cellular structures in Part A. Select a different color for the cells, fibers, and ground substance whenever visible.

PART B: Assessments

Match the tissues in column A with the characteristics in column B. Place the letter of your choice in the space provided. (Some answers may be used more than once.) A2 A3

Column A	Column B
a. Adipose	_____ **1.** Forms framework of outer ear
b. Areolar connective	_____ **2.** Functions as heat insulator beneath skin
c. Blood	_____ **3.** Contains large amounts of fluid and transports wastes and gases
d. Bone (compact)	_____ **4.** Cells in a solid matrix arranged around central canal
e. Dense irregular connective	_____ **5.** Binds skin to underlying organs
f. Dense regular connective	_____ **6.** Main tissue of tendons and ligaments
g. Elastic cartilage	_____ **7.** Forms the flexible part of the nasal septum and ends of long bones
h. Elastic connective	_____ **8.** Pads between vertebrae that are shock absorbers
i. Fibrocartilage	_____ **9.** Main tissue of dermis
j. Hyaline cartilage	_____ **10.** Occurs in some ligament attachments between vertebrae and larger artery walls
k. Reticular connective	_____ **11.** Forms supporting tissue in walls of thymus and spleen
	_____ **12.** Cells in a fluid-gel matrix with parallel collagen fibers
	_____ **13.** Contains loose arrangement of elastic and collagen fibers
	_____ **14.** Transports nutrients and defends against disease

PART C: Assessments

As you examined each specific connective tissue, what were the general characteristics they possessed? List and describe those characteristics that you were able to observe. (A4)

_____

_____

_____

_____

_____

CRITICAL THINKING ASSESSMENT

Abdominal impact injuries in the region of the LUQ often involve the spleen. Explain the structural tissue characteristics within the spleen that make it so vulnerable to serious injury. (A3)

_____

_____

_____

_____

_____

LABORATORY
EXERCISE 10

Muscle and Nervous Tissues

Purpose of the Exercise

To observe muscle and nervous tissues and recognize the characteristics that make them unique.

MATERIALS NEEDED

Compound light microscope

Prepared slides of the following:

Skeletal muscle tissue

Smooth muscle tissue

Cardiac muscle tissue

Nervous tissue (spinal cord smear and/or cerebellum)

For Learning Extension Activity:

Colored pencils

Learning Outcomes AP|R

After completing this exercise, you should be able to:

01 Sketch and label the characteristics of the different muscle and nervous tissues that you were able to observe.

02 Differentiate the special characteristics of each type of muscle tissue and nervous tissue.

03 Indicate an example location and function of each type of muscle tissue and nervous tissue.

The **O** corresponds to the assessments **A** indicated in the Laboratory Assessment for this Exercise.

Pre-Lab

Carefully read the introductory material and examine the entire lab. Be familiar with muscle tissues and nervous tissue from lecture or the textbook. Answer the pre-lab questions.

Pre-Lab Questions Select the correct answer for each of the following questions:

1. Which muscle tissue is under conscious control?
   - (a.) skeletal
   - b. smooth
   - c. cardiac
   - d. stomach *3 something*

2. Which of the following organs lacks smooth muscle?
   - a. blood vessels
   - b. stomach
   - c. iris
   - (d.) heart

3. Which muscle tissue lacks striations?
   - a. cardiac
   - b. skeletal
   - (c.) smooth
   - d. voluntary

4. Which of the following **conduct** action potentials?
   - a. smooth muscle cells
   - (b.) neurons
   - c. neuroglia
   - d. skeletal muscle cells

5. Muscles of facial expression are
   - a. smooth muscles.
   - (b.) skeletal muscles.
   - c. involuntary muscles.
   - d. cardiac muscles.

6. Which muscle tissue is multinucleated?
   - (a.) skeletal
   - b. smooth
   - c. stomach
   - d. cardiac

7. Intercalated discs represent the junctions where heart muscle cells fit together. ✓
   - a. True ✓
   - b. False _____

8. Both cardiac and skeletal muscle cells are voluntary.
   - a. True _____
   - b. False ✓

   cardiac is involuntary

95

Muscle tissues are characterized by the presence of elongated cells, often referred to as muscle fibers, that can contract (shorten) to create movements. Many of our muscles are attached to the skeleton, but muscles are also components of many of our internal organs. During muscle contractions, considerable body heat is generated to help maintain our body temperature. Because more heat is generated than is needed to maintain our body temperature, much of the heat is dissipated from our bodies through the skin.

The three types of muscle tissues are *skeletal, smooth,* and *cardiac. Skeletal muscles* are usually under our conscious control and are considered voluntary. Although most skeletal muscles are attached to bones via tendons, the tongue, facial muscles, and voluntary sphincters do not attach directly to the bone. Functions of skeletal muscle include body movements, maintaining posture, breathing, speaking, controlling waste eliminations, and protection. Skeletal muscles also aid the movement of lymph and venous blood during muscle contractions. *Smooth muscle* is considered involuntary and is located in many visceral organs, the iris, blood vessels, respiratory tubes, and attached to hair follicles. Functions include the motions of visceral organs (peristalsis), controlling pupil size, blood flow, and airflow, and creating "goose bumps" if we are too cold or frightened. *Cardiac muscle* is located only in the heart wall. It is considered involuntary and functions to pump blood.

Nervous tissues occur in the brain, spinal cord, and peripheral nerves. The tissue consists of two cell types: *neurons* and *neuroglia.* Neurons, also called nerve cells, contain a cell body (soma) with the nucleus and most of the cytoplasm, and cellular processes that extend from the cell body. Cellular processes include one to many dendrites and a single axon (nerve fiber). Neurons are considered excitable cells because they can generate signals called action potentials (nerve impulses) along the neuron to another neuron or a muscle or gland. Neuroglia (glial cells) of various types are more abundant than neurons; they cannot generate or conduct nerve impulses, but they have important supportive and protective functions for neurons. Additional study of nervous tissue will be found in Laboratory Exercise 27.

PROCEDURE: Muscle and Nervous Tissues

1. Using the microscope, observe each of the types of muscle tissues on the prepared slides. Look for the special features of each type. Compare your prepared slides of muscle tissues to the micrographs in figure 10.1

FIGURE 10.1 AP|R Micrographs of muscle tissues.

(a) Skeletal muscle (from leg) (100×)

(b) Smooth muscle (from small intestine) (1,000×)

(c) Cardiac muscle (from heart) (250×)

and the characteristics of each specific tissue in tables 10.1 and 10.2.

2. As you observe each type of muscle tissue, prepare a labeled sketch of a representative portion of the tissue in Part A of Laboratory Assessment 10. Add any personal notes near your sketches that might assist your recognition of the tissues for future assessments.

3. Observe the prepared slide of nervous tissue and identify neurons (nerve cells), neuron cellular processes, and neuroglia (glial cells). Compare your prepared slide of nervous tissue to the micrograph in figure 10.2 and the characteristics in table 10.1.

4. Prepare a labeled sketch of nervous tissue along with any personal notes in Part A of the laboratory assessment.

5. Test your ability to recognize each of these muscle and nervous tissues by having your laboratory partner select a slide, cover its label, and focus the microscope on this tissue. Then see if you correctly identify the tissue.

6. Complete Part B of the laboratory assessment.

FIGURE 10.2 Micrograph of nervous tissue.

Nervous tissue (from spinal cord) (350×)

Table 10.1 Muscle and Nervous Tissues, Descriptions, Functions, and Representative Locations

Tissue Type	Descriptions	Functions	Representative Locations
Skeletal muscle	Long, threadlike cells; striated; many nuclei near plasma membrane	Voluntary movements of skeletal parts; facial expressions	Muscles usually attached to bones
Smooth muscle	Shorter fusiform (spindle-shaped) cells; single central nucleus	Involuntary movements of internal organs as peristalsis; control of blood vessel diameter	Walls of hollow internal organs
Cardiac muscle	Branched cells; striated; single nucleus (usually)	Heart contractions to pump blood; involuntary	Heart walls
Nervous tissue	Neurons with long cellular processes; neuroglia smaller and variable	Sensory reception and conduction of action potentials; neuroglia supportive	Brain, spinal cord, and peripheral nerves; glial cells in close proximity to neurons

Table 10.2 Muscle Tissue Characteristics

Characteristic	Skeletal Muscle	Smooth Muscle	Cardiac Muscle
Appearance of cells	Unbranched and relatively parallel	Short fusiform (spindle-shaped) cells tapering at ends	Branched and connected in complex networks
Striations (pattern of alternating dark and light bands across cells)	Present and obvious	Absent	Present but faint
Nucleus	Multinucleated	Uninucleated	Uninucleated (usually)
Intercalated discs (junction where cells fit together)	Absent	Absent	Present
Control	Voluntary (usually)	Involuntary	Involuntary

NOTES

Name _____

Date _____

Section _____

The (A) corresponds to the indicated Learning Outcome(s) (O) found at the beginning of the Laboratory Exercise.

Muscle and Nervous Tissues

PART A: Assessments

In the space that follows, sketch a few cells or fibers of each of the three types of muscle tissues and of nervous tissue as they appear through the microscope. For each sketch, label the major structures of the cells or fibers, indicate the magnification used, write an example of a location in the body, and provide a function. Add any personal notes in any of the extra space within the box with your sketch that would assist your future ability to recognize the tissue. (A1) (A2) (A3)

Skeletal muscle tissue (_____×) Location: _____ Function: _____	Smooth muscle tissue (_____×) Location: _____ Function: _____
Cardiac muscle tissue (_____×) Location: _____ Function: _____	Nervous tissue (_____×) Location: _____ Function: _____

LEARNING EXTENSION ACTIVITY

Use colored pencils to differentiate various cellular structures in Part A.

PART B: Assessments

Match the tissues in column A with the characteristics in column B. Place the letter of your choice in the space provided. (Some answers may be used more than once.) A2 A3

Column A

a. Cardiac muscle

b. Nervous tissue

c. Skeletal muscle

d. Smooth muscle

Column B

_____ **1.** Coordinates, regulates, and integrates body functions

_____ **2.** Contains intercalated discs

_____ **3.** Muscle that lacks striations

_____ **4.** Striated and involuntary

_____ **5.** Striated and voluntary

_____ **6.** Contains neurons and neuroglia

_____ **7.** Muscle attached to bones

_____ **8.** Muscle that composes heart

_____ **9.** Moves food through the digestive tract

_____ **10.** Conducts impulses along cellular processes

_____ **11.** Muscle under conscious control

_____ **12.** Muscle of blood vessels and iris

LABORATORY
EXERCISE 11

Integumentary System

Purpose of the Exercise

To observe the structures and tissues of the integumentary system and to review the functions of these parts.

MATERIALS NEEDED

Skin model

Hand magnifier or stereomicroscope

Forceps

Microscope slide and coverslip

Compound light microscope

Prepared microscope slide of human scalp or axilla

Prepared slide of dark (heavily pigmented) human skin

Prepared slide of thick skin (plantar or palmar)

For Learning Extension Activity:

Tattoo slide

Learning Outcomes AP|R

After completing this exercise, you should be able to:

O1 Locate and name the structures of the integumentary system.

O2 Describe the major functions of these structures.

O3 Distinguish the locations and tissues among epidermis, dermis, and the hypodermis.

O4 Identify and sketch the layers of the skin and associated structures observed on the prepared slide.

The **O** corresponds to the assessments **A** indicated in the Laboratory Assessment for this Exercise.

Pre-Lab

Carefully read the introductory material and examine the entire lab. Be familiar with skin layers and accessory structures of the skin from lecture or the textbook. Answer the pre-lab questions.

Pre-Lab Questions Select the correct answer for each of the following questions:

1. Which of the following is *not* a function of the integumentary system?
   - **a.** protection
   - **b.** excrete small amounts of waste
   - **c.** movement
   - **d.** aid in regulating body temperature

2. The two distinct skin layers are the
   - **a.** epidermis and dermis.
   - **b.** hypodermis and dermis.
   - **c.** hypodermis and epidermis.
   - **d.** dermis and hypodermis.

3. Apocrine sweat glands are located in _____ regions of the body.
   - **a.** forehead
   - **b.** axillary and genital
   - **c.** palmar
   - **d.** plantar

4. The hypodermis is composed of _____ tissues.
   - **a.** adipose connective and stratified squamous epithelial
   - **b.** areolar and dense irregular connective
   - **c.** stratified squamous epithelial and adipose connective
   - **d.** areolar and adipose connective

5. The _____ layer of the epidermis is only present in thick skin.
   - **a.** stratum corneum
   - **b.** stratum lucidum
   - **c.** stratum spinosum
   - **d.** stratum granulosum

6. Frequent cell division occurs in the _____ of the epidermis.
   - **a.** stratum corneum
   - **b.** stratum spinosum
   - **c.** stratum granulosum
   - **d.** stratum basale

7. The greatest concentration of melanin is in the dermis.

   a. True _____   **b.** False _____

8. Thick skin of the palms and soles contains five strata of the epidermis.

   a. True __✓__   **b.** False _____

The integumentary system includes the skin, hair, nails, and skin (cutaneous) glands. These structures provide a protective covering for deeper tissues, aid in regulating body temperature, retard water loss, house sensory receptors, synthesize various glandular chemicals, and excrete small quantities of wastes.

The skin consists of two distinct layers. The outer layer, the *epidermis,* consists of keratinized stratified squamous epithelium. The inner layer, the *dermis,* consists of a superficial papillary region of areolar connective tissue and a thicker and deeper reticular region of dense irregular connective tissue. Beneath the dermis is the *hypodermis* (subcutaneous layer; superficial fascia) composed of primarily adipose connective tissue and some areolar connective tissue. The hypodermis is not considered a true layer of the skin.

Accessory structures of the skin include *nails, hair follicles,* and *skin (cutaneous) glands.* The hair, which grows through a depression from the epidermis, possesses a *hair papilla* at the base of the hair. The papilla contains a network of capillaries that supply the nutrients for cell divisions required for hair growth within the hair bulb. As the cells of the hair are forced toward the surface of the body, they become keratinized and pigmented and die. Attached to the follicle is the *arrector pili muscle* that can pull the hair to a more upright position, causing goose bumps when experiencing cold temperatures or fear. A *sebaceous gland* secretes an oily sebum into the hair follicles, which keeps the hair and epidermal surface pliable and somewhat waterproof.

Sweat glands (*sudoriferous glands*) are distributed over most regions of the body and consist of two types of glands. The widespread *merocrine (eccrine) sweat glands* are most numerous on the palms, the soles of the feet, and the forehead. Their ducts open to the surface at a sweat pore. Their secretions increase during hot days, physical exercise, and stress; they serve a mini-excretory function and can help prevent our bodies from overheating. The *apocrine sweat glands* are most abundant in the axillary and genital regions. Apocrine sweat ducts open into the hair follicles and become active at puberty. Their secretions increase during stress and pain and have little influence on thermoregulation. Modified sweat glands of the integumentary system include *ceruminous glands* of the external auditory canal and *mammary glands* of the breasts.

FIGURE 11.1 AP|R Layers of skin and deeper hypodermis indicated on (a) an illustration and (b) cadaver skin.

PROCEDURE: Integumentary System

In this procedure you will use a skin model and make comparisons to the figures in the lab manual to locate the layers and accessory structures of the skin. Hair structures will be observed from your own body with different magnifications and then compared to the additional detail observed when viewing prepared slides. Several micrographs of different magnifications are provided showing vertical sections of the skin layers and accessory structures. Use a combination of all of the micrographs as you observe the skin slides available in your laboratory.

1. Use figures 11.1 and 11.2 to locate as many of these structures as possible on a skin model.
2. Use figure 11.3 as a guide to locate the specific epidermal layers (strata) on a skin model. Note the locations and descriptions from table 11.1.
3. Use the hand magnifier or stereomicroscope and proceed as follows:

   a. Observe the skin, hair, and nails on your hand.

   b. Compare the type and distribution of hairs on the front and back of your forearm.

FIGURE 11.2 Vertical section of the skin and hypodermis (subcutaneous layer).

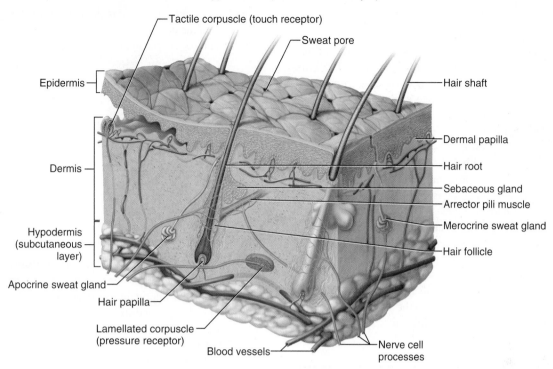

FIGURE 11.3 **AP|R** Epidermal layers in this section of thick skin from the fingertip (400×).

Table 11.1 Layers of the Epidermis

Layer	Location	Descriptions
Stratum corneum	Most superficial layer	Many layers of keratinized, dead epithelial cells; appear scaly and flattened; resists water loss, absorption, and abrasion
Stratum lucidum (only present in thick skin)	Between stratum corneum and stratum granulosum on soles and palms of thick skin	Cells appear clear; nuclei, organelles, and plasma membranes no longer visible
Stratum granulosum	Beneath the stratum corneum (or stratum lucidum of thick skin)	Three to five layers of flattened granular cells; contain shrunken fibers of keratin and shriveled nuclei
Stratum spinosum	Beneath the stratum granulosum	Many layers of cells with centrally located, large, oval nuclei; develop fibers of keratin; cells becoming flattened in superficial portion
Stratum basale	Deepest layer	A single row of cuboidal or columnar cells; layer also includes melanocytes; frequent cell division; some cells become parts of more superficial layers

4. Study the structure of a fingernail in figure 11.4. Use a hand magnifier or a stereomicroscope to observe various features of a fingernail. The *nail plate* is composed of dead epithelial cells derived from the stratum corneum. The exposed *free edge, nail body,* and the *nail root* under the skin make up the nail plate. The pinkish hue of most of the nail body results from underlying blood within dermal blood vessels, whereas the crescent-shaped *lunula* area appears whitish because of a thicker region of the *nail bed* that lies beneath the nail plate that obscures the color of blood. The nail bed is composed of epithelial cells derived from the stratum basale. On the proximal end of the nail bed is the *nail matrix* where cell division occurs, resulting in nail growth. The narrow *cuticle* is a region of dead skin.

5. Use low-power magnification of the compound light microscope and proceed as follows:
   a. Pull out a single hair with forceps and mount it on a microscope slide under a coverslip.
   b. Observe the root and shaft of the hair and note the scalelike parts that make up the shaft.

6. Complete Parts A and B of Laboratory Assessment 11.

FIGURE 11.4 AP|R Fingernail structures as seen from a surface view and a sagittal section.

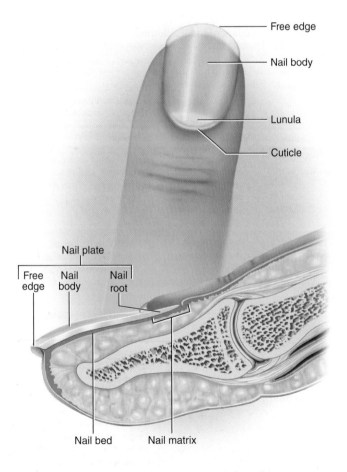

Free edge

Nail body

Lunula

Cuticle

Nail plate

Free edge | Nail body | Nail root

Nail bed   Nail matrix

CRITICAL THINKING ACTIVITY

Explain the advantage for melanin granules being located in the deep layers of the epidermis, and not in the dermis or deeper hypodermis (fig. 11.5c).

_____

_____

_____

_____

_____

_____

7. As vertical sections of human skin are observed, remember that the lenses of the microscope invert and reverse images. It is important to orient the position of the epidermis, dermis, and hypodermis layers using scan magnification before continuing with additional observations. Compare all of your skin observations to the various micrographs in figure 11.5. Use low-power magnification of the compound light microscope and proceed as follows:
   a. Observe the prepared slide of human scalp or axilla.
   b. Locate the epidermis, dermis, and hypodermis; a hair follicle; an arrector pili muscle; a sebaceous

LEARNING EXTENSION ACTIVITY

Observe a vertical section of human skin through a tattoo, using low-power magnification. Note the location of the dispersed ink granules within the upper portion of the dermis. From a thin vertical section of a tattoo, it is not possible to determine the figure or word of the entire tattoo as seen on the surface of the skin because only a small, thin segment of the image is in view at any one time. Compare this to the location of melanin granules found in dark (heavily pigmented) skin. Suggest reasons why a tattoo is permanent and a suntan is not.

_____

_____

_____

_____

_____

_____

FIGURE 11.5 Features of human skin are indicated in these variously stained micrographs: (a) and (b) various structures of epidermis and dermis; (c) epidermis of dark skin; (d) epidermis of light skin; (e) base of hair structures.

(a) AP|R (30×)

- Epidermis
- Papillary region of dermis
- Reticular region of dermis
- Sweat pore
- Hair follicle opening
- Dermal papilla
- Duct of merocrine sweat gland
- Dense irregular connective tissue
- Sebaceous glands
- Hair root
- Hair follicle
- Merocrine sweat gland

(b) AP|R (30×)

- Epidermis
- Dermis
- Hair follicle
- Arrector pili muscle
- Sebaceous gland

(c) (400×)

- Stratum corneum
- Stratum granulosum
- Stratum spinosum
- Stratum basale
- Melanin in cells of stratum basale
- Dermal papilla
- Basement membrane

(d) (400×)

- Stratum corneum of epidermis
- Dermal papilla
- Stratum basale cells of epidermis (no melanin)

(e) AP|R (36×)

- Hair follicle
- Dense irregular connective tissue of dermis
- Hair root
- Region of cell division
- Hair bulb
- Hair papilla
- Adipose cells of hypodermis

105

gland; and a sweat gland. Table 11.2 includes a summary of skin (cutaneous) glands, including locations and functions.

c. Focus on the epidermis with high power and locate the stratum corneum, stratum granulosum, stratum spinosum, and stratum basale. Note how the shapes of the cells in these layers differ as described in table 11.1.

d. Observe the projections of the papillary layer called the *dermal papillae,* located in the dermis directly beneath the epidermis. Larger projections found in the palms of the hands and soles of the feet help produce unique fingerprints (and toeprints).

e. Observe the dense irregular connective tissue that makes up the bulk of the dermis (reticular region).

f. Observe the adipose tissue that composes most of the hypodermis (subcutaneous layer) along with some areolar connective tissue.

8. Observe the prepared slide of dark (heavily pigmented) human skin with low-power magnification. Note that the pigment is most abundant in the deepest layers of the epidermis. Focus on this region with the high-power objective. The pigment-producing cells, or melanocytes, are located among the stratum basale cells. Some melanin is retained within some cells (keratinocytes) of the stratum spinosum as cells are forced closer to the surface of the skin. Differences in skin color are primarily due to the quantity of the pigment *melanin* produced by these cells. Exposure to ultraviolet (UV) rays of sunlight can increase the amount of melanin produced, causing a suntan. Melanin absorbs the UV radiation, which helps protect the nuclei of cells from the harmful effects of the sun.

9. Observe the prepared slide of thick skin from the palm of a hand or the sole of a foot. Locate the stratum lucidum, which is present only in thick skin. Locate the four other strata of thick skin and note the very thick stratum corneum (fig. 11.3).

10. Complete Part C of the laboratory assessment.

11. Using low-power magnification, locate a hair follicle sectioned longitudinally through its bulblike base. Also locate a sebaceous gland close to the follicle and find a sweat gland (fig. 11.5). Observe the detailed structure of these parts with high-power magnification.

12. Complete Parts D and E of the laboratory assessment.

Table 11.2 Skin (cutaneous) Glands

Skin (Cutaneous) Glands	Location	Functional Descriptions
Merocrine (eccrine) sweat glands	Most of body surface; ducts open onto body surface; abundant on forehead, palms, and soles of feet	Provide evaporative cooling during exercise and stress; they serve excretory function for water, salt, and nitrogen wastes
Apocrine sweat glands	Concentrations in axillary, genital areas; facial hair of males; ducts open into hair follicles	Secretions more viscous than merocrine secretions; more active after puberty; have little influence on thermoregulation; increased activity during pain and stress
Sebaceous glands	Associated with hair follicles	Secrete oily sebum into hair follicle; helps keep skin surface and hair pliable; more active after puberty
Ceruminous glands	External ear canal	Secrete cerumen (earwax); protective covering of external ear canal
Mammary glands	Within breasts	Produce breast milk during pregnancies; secrete milk during lactation into ducts that open on a nipple

Name _____

Date _____

Section _____

The Ⓐ corresponds to the indicated Learning Outcome(s) Ⓞ found at the beginning of the Laboratory Exercise.

Integumentary System

PART A: Assessments

1. Label the structures indicated in figure 11.6. Ⓐ1

FIGURE 11.6 Label the features of the skin.

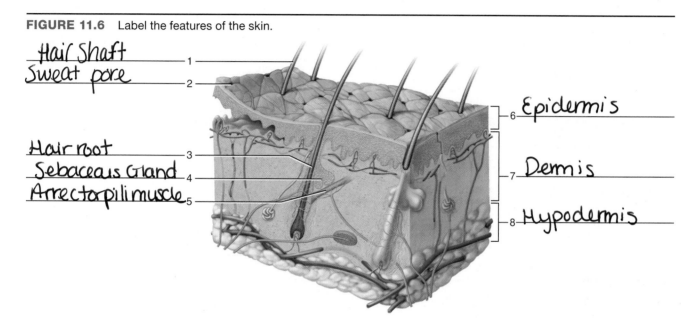

Hair Shaft — 1
Sweat pore — 2

Hair root — 3
Sebaceaus Gland — 4
Arrectarpili muscle — 5

6 — Epidermis
7 — Dermis
8 — Hypodermis

2. Match the structures in column A with the descriptions and functions in column B. Place the letter of your choice in the space provided. Ⓐ1 Ⓐ2

Column A	Column B
a. Apocrine sweat gland	_____ **1.** Portion of nail body near the cuticle
b. Arrector pili muscle	_____ **2.** Outermost layer of epidermis
c. Dermis	_____ **3.** Become active after puberty in axillary and groin regions
d. Epidermis	_____ **4.** Epidermal pigment
e. Hair follicle	_____ **5.** Inner layer of skin
f. Keratin	_____ **6.** Responds to elevated body temperature
g. Lunula	_____ **7.** General name of entire superficial layer of the skin
h. Melanin	_____ **8.** Gland that secretes an oily mixture called sebum
i. Merocrine sweat gland	_____ **9.** Tough protein of nails and hair
j. Sebaceous gland	_____ **10.** Cell division and deepest layer of epidermis
k. Stratum basale	_____ **11.** Tubelike part that contains the root of the hair
l. Stratum corneum	_____ **12.** Causes hair to stand on end and goose bumps to appear

PART B: Assessments

Complete the following:

1. How does the skin of your palm differ from that on the back (posterior) of your hand? **A1** _____

   _____

   _____

2. Describe the differences you observed in the type and distribution of hair on the front (anterior) and back (posterior) of your

   forearm. **A1** _____

   _____

3. Explain how a hair is formed. **A2** _____

   _____

4. What portion of the nail plate is not visible on the body surface when using a hand magnifier? **A1** _____

PART C: Assessments

Complete the following:

1. Complete the following chart, and then circle the layer that is found *beneath* the skin. **A3**

Layer	Location	Tissues	Distinguishing Characteristics
Epidermis			
Dermis			
Hypodermis			

2. How do the cells of stratum corneum and stratum basale differ? **A3** _____

   _____

   _____

3. State the specific location of melanin observed in dark skin. **A3** _____

   _____

4. Which is darker, the anterior forearm or the posterior forearm? _____ Explain why. **A3** _____

   _____

   _____

5. What special qualities, due to the presence of fibers, does the connective tissue of the dermis have? **A3** _____

   _____

6. How is the structure of the thick skin of the palms of the hands and soles of the feet properly suited for its function? **A4**

   _____

   _____

PART D: Assessments

Complete the following:

1. What part of the hair extends from the hair papilla to the body surface? (A1) _____

2. In which layer of skin are sebaceous glands found? (A1) _____

3. How are sebaceous glands associated with hair follicles, and what do they secrete? (A1) _____

_____

4. The ducts of apocrine sweat glands open into _____ . (A1)

5. Which type of sweat gland is most important in maintaining normal body temperature? (A1) _____

PART E: Assessments

Sketch a vertical section of human skin, using the scanning objective. Label the skin layers and a hair follicle, a sebaceous gland, and a sweat gland. (A4)

NOTES

LABORATORY EXERCISE 12

Bone Structure and Classification

Purpose of the Exercise

To review the way bones are classified and to examine the structure of a long bone and bone tissue.

MATERIALS NEEDED

Prepared microscope slide of ground compact bone

Human bone specimens, including long, short, flat, irregular, and sesamoid types

Human long bone, sectioned longitudinally

Fresh animal bones, sectioned longitudinally and transversely

Compound light microscope

Dissecting microscope

For Demonstration Activity:

Fresh chicken bones (radius and ulna from wings)

Vinegar or dilute hydrochloric acid

SAFETY

▶ Wear disposable gloves for handling fresh bones and for the demonstration of a bone soaked in vinegar or dilute hydrochloric acid.

▶ Wash your hands before leaving the laboratory.

Learning Outcomes AP|R

After completing this exercise, you should be able to:

O1 Classify bones into four groups based on their shapes and identify an example for each group. *Flat, short, Long, Irregular*

O2 Locate the major structures of a long bone. *D. Epiphysis, P. Epiphysis, Diaphysis*

O3 Distinguish between compact and spongy bone.

O4 Differentiate the structural and functional characteristics of bones.

The O corresponds to the assessments A indicated in the Laboratory Assessment for this Exercise.

Pre-Lab

Carefully read the introductory material and the entire lab. Be familiar with bone structure and bone tissue from lecture or the textbook. Answer the pre-lab questions.

Pre-Lab Questions Select the correct answer for each of the following questions:

1. Which of the following tissues is *not* part of a bone as an organ?
   - a. dense connective
   - b. cartilage
   - (c.) muscle
   - d. blood

2. The organic matter of living bone includes
   - a. calcium phosphate.
   - (b.) collagen fibers and cells.
   - c. calcium carbonate.
   - d. magnesium and fluoride.

3. The _____ is an example of a sesamoid bone.
   - a. vertebra
   - b. femur
   - c. carpal
   - (d.) patella

4. The epiphyseal plate represents the
   - a. ends of the epiphyses.
   - b. shaft between the epiphyses.
   - c. growth zone of hyaline cartilage.
   - d. membrane around a bone.

5. The central canal of a bone osteon contains
   - a. blood vessels and nerves.
   - b. osteocytes.
   - c. red bone marrow.
   - d. yellow bone marrow.

6. A _____ is an example of an irregular bone.
   - a. femur
   - b. carpal
   - c. rib
   - (d.) vertebra

111

7. A femur includes both compact and spongy bone tissues.
   a. True ✓____   b. False ____

8. Chicken bones, with both organic and inorganic components, possess the quality of tensile strength.
   a. True ____   b. False ____

9. Trabeculae are structural characteristics of compact bone.
   a. True ____   b. False ____

A bone represents an organ of the skeletal system. As such, it is composed of a variety of tissues, including bone tissue, cartilage, dense connective tissue, blood, and nervous tissue. Bones are not only alive but also multifunctional. They support and protect softer tissues, provide points of attachment for muscles, house blood-producing cells, and store inorganic salts.

Living bone is a combination of about one-third *organic matter* and two-thirds *inorganic matter.* The organic matter consists mostly of embedded cells and collagen fibers. The inorganic matter is mostly complex salt crystals, hydroxy-apatite, consisting of calcium phosphate. Lesser amounts of calcium carbonate and ions of magnesium, fluoride, and sodium become incorporated into the inorganic crystals.

Bones are classified according to their shapes as *long short, flat,* or *irregular.* Long bones are much longer than they are wide and have expanded ends. Short bones are somewhat cube shaped, with similar lengths and widths. *Sesamoid bones,* a special type of short bone, are small and embedded within a tendon near joints where compression often occurs. The patella is a sesamoid bone that is included in the total skeletal number of 206. Any additional sesamoid bones that may develop in compression areas of the hand or foot are not considered among the 206 bones of the adult skeleton. Flat bones have wide surfaces, but they are sometimes curved, such as those of the cranium. Irregular bones have numerous shapes and often have articulations with more than one other bone. Although bones of the skeleton vary greatly in size and shape, they have much in common structurally and functionally.

PROCEDURE: Bone Structure and Classification

During embryonic and fetal development, much of the supportive tissue is cartilage. Cartilage is retained in certain regions of the adult skeleton. Articular cartilage remains on the articulating surfaces of movable joints, costal cartilage connects the ribs to the sternum, and intervertebral discs are between the vertebrae. Hyaline cartilage composes the articular cartilage and the costal cartilage, while fibrocartilage is represented between the vertebrae.

Bones contain compact and spongy bone components. The compact bone structure contains cylinder-shaped units called osteons. Each osteon contains a central canal that includes blood vessels and nerves in living bone. The cells or osteocytes are located in concentric circles within small spaces known as lacunae. The cellular processes of the osteocytes pass through microscopic canaliculi, which allow for the transport of nutrient and waste substances between cells and the central canal. The extracellular matrix occupies most of the area of an osteon.

Spongy bone does not possess the typical osteon units of compact bone. The osteocytes of spongy bone are located within a lattice of bony plates known as trabeculae. Canaliculi allow for diffusion of substances between the cells and the marrow that is positioned between the trabeculae.

1. Observe the individual bone specimens and arrange them into groups, according to the following shapes and examples (figs. 12.1 and 12.2).

   long—femur; humerus; phalanges

   short—carpals; tarsals; sesamoid bones

   flat—ribs; scapula; sternum; most cranial bones

   irregular—vertebra; some facial bones such as sphenoid

CRITICAL THINKING ACTIVITY

Explain how bone cells (osteocytes) embedded in a solid ground substance obtain nutrients and eliminate wastes.

_____

_____

_____

_____

_____

FIGURE 12.1 Representative examples of bones classified by shape with a skeleton location icon for the bones illustrated.

(b) Tarsal bone of ankle, a short bone

(c) Parietal bone of skull, a flat bone

(d) Vertebra, an irregular bone

(a) Femur, a long bone

(e) Patella, a sesamoid bone

FIGURE 12.2 AP|R Radiograph (X ray) of a child's hand showing numerous epiphyseal plates. Only single epiphyseal plates develop in the hand and fingers.

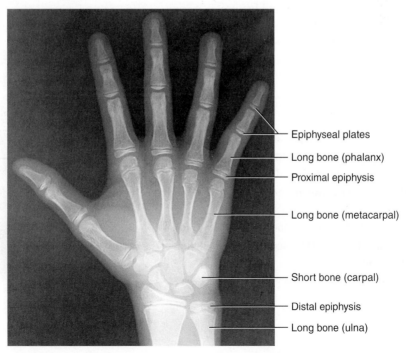

Epiphyseal plates

Long bone (phalanx)

Proximal epiphysis

Long bone (metacarpal)

Short bone (carpal)

Distal epiphysis

Long bone (ulna)

FIGURE 12.3 The major structures of a long bone (living femur) of an adult.

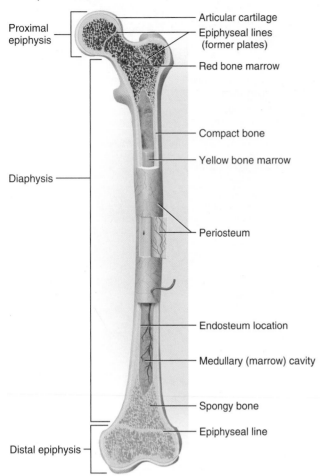

- Proximal epiphysis
- Diaphysis
- Distal epiphysis
- Articular cartilage
- Epiphyseal lines (former plates)
- Red bone marrow
- Compact bone
- Yellow bone marrow
- Periosteum
- Endosteum location
- Medullary (marrow) cavity
- Spongy bone
- Epiphyseal line

FIGURE 12.4 Representative locations of compact and spongy bone.

Trabeculae
Spongy bone
Compact bone

(a) Femur, cross section

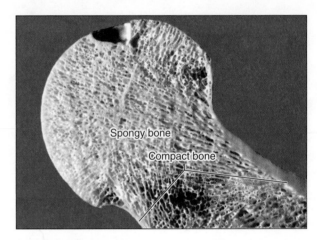

Spongy bone
Compact bone

(b) Femur, longitudinal section

Spongy bone
Compact bone

(c) Skull bone, sectioned

2. Become familiar with the structural features of a long bone indicated in figure 12.3.

3. Note representative compact and spongy bone locations (fig. 12.4).

4. Examine the microscopic structure of bone tissue by observing a prepared microscope slide of ground compact bone. Use figures 12.5 and 12.6 of bone tissue to locate the following features:

   osteon (Haversian system)—cylinder-shaped unit

   central canal (Haversian canal; osteonic canal)— contains blood vessels and nerves

   lamella—concentric ring of matrix around central canal

   lacuna—small chamber for an osteocyte

   bone extracellular matrix—collagen and calcium phosphate

   canaliculus—minute tube containing cellular process

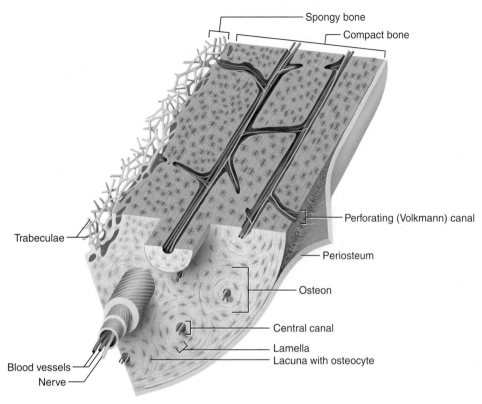

Spongy bone

Compact bone

Perforating (Volkmann) canal

Periosteum

Osteon

Central canal

Lamella

Lacuna with osteocyte

Trabeculae

Blood vessels

Nerve

FIGURE 12.6   **AP|R**   Micrograph of ground compact bone tissue (200×).

Osteon

Lamella

Bone extracellular matrix

Central canal

Lacuna (occupied by osteocyte in living bone)

Canaliculi

5. Complete Part A of Laboratory Assessment 12.
6. Examine the sectioned long bones and locate the following (figs. 12.2, 12.3, and 12.4):

epiphysis

proximal—nearest limb attachment to torso

distal—farthest from limb attachment to torso

epiphyseal plate—growth zone of hyaline cartilage

articular cartilage—on ends of epiphyses

diaphysis—shaft between epiphyses

periosteum—strong membrane around bone (except articular cartilage) of dense irregular connective tissue

compact (dense) bone—forms diaphysis and epiphyseal surfaces

spongy (cancellous) bone—within epiphyses

  trabeculae—a structural lattice of plates in spongy bone

medullary (marrow) cavity—hollow chamber

endosteum—thin membrane of reticular connective tissue that lines the medullary cavity

yellow bone marrow—occupies medullary cavity and stores adipose tissue

red bone marrow—occupies spongy bone in some epiphyses and flat bones and produces blood cells

7. Bone marrow is a soft tissue located within spaces of bones. The *red bone marrow* is very prevalent in a child and is a tissue that produces blood cells *(hemopoietic tissue)*. Much of the red bone marrow becomes a fatty *yellow bone marrow* in the adult. Figure 12.7 depicts the represented areas of an adult skeleton containing red and yellow bone marrow.

8. Use the dissecting microscope to observe the compact bone and spongy bone of the sectioned specimens. Also examine the marrow in the medullary cavity and the spaces within the spongy bone of the fresh specimen.

9. Complete Parts B and C of the laboratory assessment.

DEMONSTRATION ACTIVITY

Examine a fresh chicken bone and a chicken bone that has been soaked for several days in vinegar or overnight in dilute hydrochloric acid. Wear disposable gloves for handling these bones. This acid treatment removes the inorganic salts from the bone extracellular matrix. Rinse the bones in water and note the texture and flexibility of each (fig. 12.8a). The bone becomes soft and flexible without the support of the inorganic salts with calcium.

Examine the specimen of chicken bone that has been exposed to high temperature (baked at 121°C/250°F for 2 hours). This treatment removes the protein and other organic substances from the bone extracellular matrix (fig. 12.8b). The bone becomes brittle and fragile without the benefit of the collagen fibers. A living bone with a combination of the qualities of inorganic and organic substances possesses tensile strength. View the LabCam video on the chicken bone mentioned in the Pre-Lab.

FIGURE 12.7 Locations of bones containing red and yellow bone marrow in an adult skeleton. Bones in red contain red bone marrow; most bones in yellow contain yellow bone marrow.

FIGURE 12.8 Results of fresh chicken bone demonstration: (a) soaked in vinegar; (b) baked in oven.

(a)

(b)

Name _____

Date _____

Section _____

The (A) corresponds to the indicated Learning Outcome(s) (O) found at the beginning of the Laboratory Exercise.

Bone Structure and Classification

PART A: Assessments

Complete the following statements: (*Note:* Questions 1–6 pertain to bone classification by shape.)

1. A bone that has a wide surface is classified as a(an) ___flat___ bone. (A1)

2. The bones of the wrist are examples of ___short___ bones. (A1)

3. The bone of the thigh is an example of a(an) ___long___ bone. (A1)

4. Vertebrae are examples of ___irregular___ bones. (A1)

5. The patella (kneecap) is an example of a special type of short bone called a ___sesamoid___ bone. (A1)

6. The bones of the skull that form a protective covering for the brain are examples of ___flat___ bones. (A1)

7. Distinguish between the epiphysis and the diaphysis of a long bone. (A2) ___Epiphysis are the ends of the bone and diaphysis is the long shaft.___

8. Describe where cartilage is found on the surface of a long bone. (A2) ___Cartilage is found on ends of epiphysis of the bone.___

9. Describe where the periosteum is found on the surface of a long bone. (A2) ___Found on the diaphysis of the bone (outside.) (covering of the bone)___

PART B: Assessments

Complete the following:

1. Distinguish between the locations and tissues of the periosteum and those of the the the endosteum. (A2) _____

_____

_____

2. What structural differences did you note between compact bone and spongy bone? (A3) _____

_____

_____

_____

3. How are these structural differences related to the locations and functions of these two types of bone? (A3) _____

_____

_____

4. From your observations, how does the marrow in the medullary cavity compare with the marrow in the spaces of the spongy bone? (A4) _____

_____

_____

5. The humerus is the proximal bone of an upper limb; the femur is the proximal bone of a lower limb. In the adult skeleton only certain portions of these long bones retain functional red bone marrow. Describe the specific regions of these bones that retain blood cell production sites. (A4) _____

_____

_____

PART C: Assessments

Identify the structures indicated in figures 12.9, 12.10, and 12.11.

FIGURE 12.9   Label the structures of this long bone, using the terms provided. (A2) (A3)

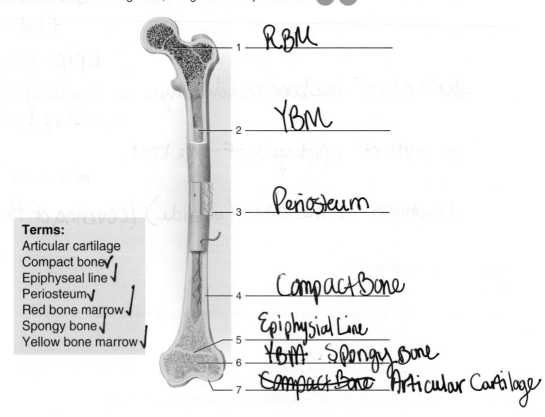

1 ___ RBM _____

2 ___ YBM _____

3 ___ Periosteum _____

Terms:
Articular cartilage
Compact bone ✓
Epiphyseal line ✓
Periosteum ✓
Red bone marrow ✓
Spongy bone ✓
Yellow bone marrow ✓

4 ___ Compact Bone _____

5 ___ Epiphysial Line _____

6 ___ ~~YBM~~ Spongy Bone _____

7 ___ ~~Compact Bone~~ Articular Cartilage _____

FIGURE 12.10 Identify the structures indicated in (a) the unsectioned long bone (fifth metatarsal) and (b) the partially sectioned long bone, using the terms provided. (A2) (A3)

Distal
Epiphysis — 1

Diaphysis — 2

Proximal
Epiphysis — 3

Terms:
Compact bone ✓
Diaphysis ✓
Epiphysis (distal) ✓
Epiphysis (proximal) ✓
Medullary cavity ✓
Spongy bone ✓

4 — Medullar cavity

Area not sectioned

5 — Compact Bone

6 — Spongy Bone

Distal

Proximal

(a)   (b)

FIGURE 12.11 Label the structures associated with bone, using the terms provided. (A3) (A4)

— 1 —

Terms:
Central canal
Compact bone
Osteon
Periosteum
Spongy bone
Trabeculae

Trabeculae — 2

Compact Bone — 3

4 — Osteon
5 — Central canal
6 — Periosteum

NOTES

LABORATORY EXERCISE 13

Organization of the Skeleton

Purpose of the Exercise

To explain the organization of the skeleton, introduce the major bones of the skeleton, and demonstrate the use of selected terms to describe bony structures.

MATERIALS NEEDED

Human skeleton, articulated

Human skeleton, disarticulated

For Demonstration Activity:
Radiographs (X rays) of skeletal structures
Stereomicroscope (dissecting microscope)

Learning Outcomes   AP|R

After completing this exercise, you should be able to:

O1 Locate and label the major bones of the human skeleton.

O2 Distinguish between the axial skeleton and the appendicular skeleton.

O3 Associate the terms used to describe skeletal structures and locate examples of such structures on the human skeleton.

The **O** corresponds to the assessments **A** indicated in the Laboratory Assessment for this Exercise.

Pre-Lab

Carefully read the introductory material and examine the entire lab. Be familiar with the axial and appendicular skeleton from lecture or the textbook. Answer the pre-lab questions.

Pre-Lab Questions Select the correct answer for each of the following questions:

1. The vertebral column does *not* include a
   a. rib.
   b. vertebra.
   c. sacrum.
   d. coccyx.

2. The _____ bone is part of the pectoral girdle.
   a. humerus
   b. sternum
   c. rib
   d. scapula

3. A _____ is a very large projection on a bone.
   a. fossa
   b. tubercle
   c. trochanter
   d. facet

4. A _____ is a shallow depression on a bone.
   a. head
   b. fossa
   c. condyle
   d. tuberosity

5. Bones that might form in the skull, but are *not* considered in the total number, are
   a. cartilaginous bones.
   b. sutural bones.
   c. middle ear bones.
   d. sesamoid bones.

6. The _____ is a bone in the upper limb.
   a. ulna
   b. scapula
   c. clavicle
   d. fibula

7. A _____ is a depression type of bone feature (bone marking).
   a. foramen
   b. crest
   c. sulcus
   d. tuberosity

The skeleton can be separated into two divisions: (1) the *axial skeleton*, which consists of the bones and cartilages of the head, neck, and trunk, and (2) the *appendicular skeleton*, which consists of the bones of the limbs and those that anchor the limbs to the axial skeleton. The bones that anchor the limbs include the pectoral and pelvic girdles.

Ossification is the formation of bone. This process occurs during fetal development and continues through childhood growth. Bone forms either from intramembranous origins or endochondral origins. During the growth of *intramembranous bones*, membrane-like connective tissue layers similar to the dermis develop in an area destined to become the flat bones of the skull. Eventually, bone-forming cells (osteoblasts) alter the membrane areas into bone tissue. In contrast, *endochondral bones* develop from hyaline cartilage and through the ossification process develop into most of the bones of the skeleton. In either type of ossification, both compact and spongy bone develops.

The number of bones in the adult skeleton is often reported to be 206. Men and women, although variations can exist, possess the same total bone number of 206. However, at birth the number of bones is closer to 275 as many ossification centers are still composed of cartilage, and some bones form during childhood. For example, each hip bone (coxal bone) includes an ilium, ischium, and pubis, and many long bones have three ossification centers separated by epiphyseal (growth) plates. The sternum, composed of a manubrium, body, and xiphoid process, becomes a single bone much later than when we reach our full height. Some people have additional bones not considered in the total number. *Sesamoid bones* other than the patellae may develop in the hand or the foot. Also, extra bones sometimes form in the skull within the sutures; these are called *sutural (wormian) bones*.

Special terminology is used to describe the features of a bone. The term used depends on whether the feature is a type of projection, articulation, depression, or opening. Many of these features can be noted when viewing radiographs. Some of the features can be *palpated* (touched) if they are located near the surface of the body.

PROCEDURE: Organization of the Skeleton

1. Study figure 13.1 and use it as a reference to examine the bones in the articulated human skeleton. As you locate the following bones, note the number of each in the skeleton. Palpate as many of the corresponding bones in your skeleton as possible.

axial skeleton
- ▪ skull
  - • cranium   (8)
  - • face   (14)
- ▪ middle ear bone   (6)
- ▪ hyoid bone—supports the tongue   (1)

- ▪ vertebral column
  - • vertebra   (24)
  - • sacrum   (1)
  - • coccyx   (1)
- ▪ thoracic cage
  - • rib   (24)
  - • sternum   (1)

appendicular skeleton
- ▪ pectoral girdle
  - • scapula   (2)
  - • clavicle   (2)
- ▪ upper limbs
  - • humerus   (2)
  - • radius   (2)
  - • ulna   (2)
  - • carpal   (16)
  - • metacarpal   (10)
  - • phalanx   (28)
- ▪ pelvic girdle
  - • hip bone (coxal bone; pelvic bone; innominate bone)   (2)
- ▪ lower limbs
  - • femur   (2)
  - • tibia   (2)
  - • fibula   (2)
  - • patella   (2)
  - • tarsal   (14)
  - • metatarsal   (10)
  - • phalanx   (28)

Total	**206 bones**

2. Complete Part A of Laboratory Assessment 13.
3. Bone features (bone markings) can be grouped together in a category of *projections, articulations, depressions,* or *openings*. Within each category more specific examples occur. The bones illustrated in figure 13.2 represent specific examples of locations of specific features in the human body. Locate each of the following features on the example bone from a disarticulated skeleton, noting the size, shape, and location in the human skeleton.

Projections: sites for tendon and ligament attachment

crest—ridgelike

epicondyle—superior to condyle

line (linea)—slightly raised ridge

process—prominent

protuberance—outgrowth

ramus—extension

spine—thornlike

trochanter—large

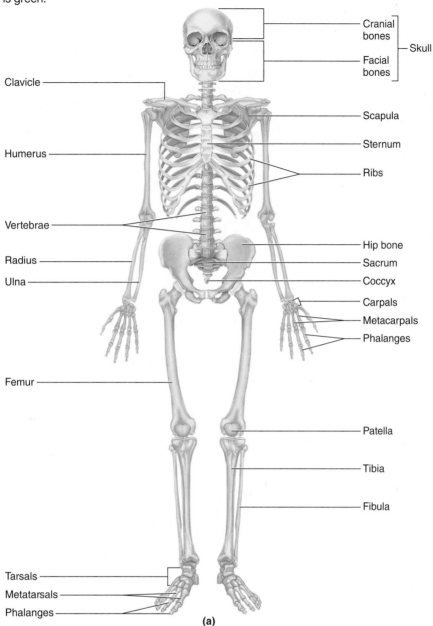

Cranial bones
Facial bones
— Skull
Clavicle
Scapula
Sternum
Humerus
Ribs
Vertebrae
Hip bone
Radius
Sacrum
Ulna
Coccyx
Carpals
Metacarpals
Phalanges
Femur
Patella
Tibia
Fibula
Tarsals
Metatarsals
Phalanges

(a)

tubercle—small knoblike
tuberosity—rough elevation

Articulations: **where bones connect at a joint or articulate with each other**
condyle—rounded process
facet—nearly flat
head—expanded end

Depressions: recessed areas in bones
alveolus—socket
fossa—shallow basin

fovea—tiny pit
notch—indentation on edge
sulcus—narrow groove

Openings: open spaces in bones
canal—tubular passage
fissure—slit
foramen—hole
meatus—tubelike opening
sinus—cavity

4. Complete Parts B, C, and D of the laboratory assessment.

FIGURE 13.1 AP|R *Continued.*

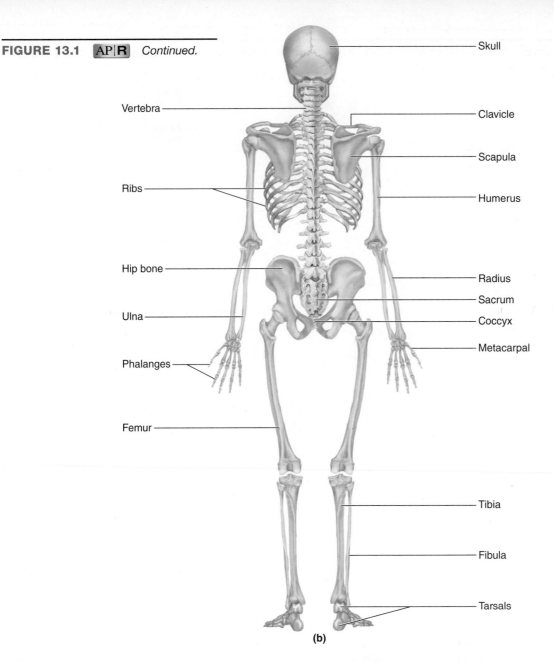

(b)

CRITICAL THINKING ACTIVITY

Locate and name the largest foramen in the skull.

_____

_____

Locate and name the largest foramen in the skeleton.

_____

_____

DEMONSTRATION ACTIVITY   AP|R

Images on radiographs (X rays) are produced by allowing X rays from an X-ray tube to pass through a body part and to expose photographic film positioned on the opposite side of the part. The image that appears on the film after it is developed reveals the presence of parts with different densities. Bone, for example, is very dense tissue and is a good absorber of X rays. Thus, bone generally appears light on the film. Air-filled spaces, on the other hand, absorb almost no X rays and appear as dark areas on the film. Liquids and soft tissues absorb intermediate quantities of X rays, so they usually appear in various shades of gray.

Examine the available radiographs of skeletal structures by holding each film in front of a light source. Identify as many of the bones and features as you can.

FIGURE 13.2 Representative examples of bone features (bone markings) on bones of the skeleton (a–h). The two largest foramina of the skeleton have complete labels.

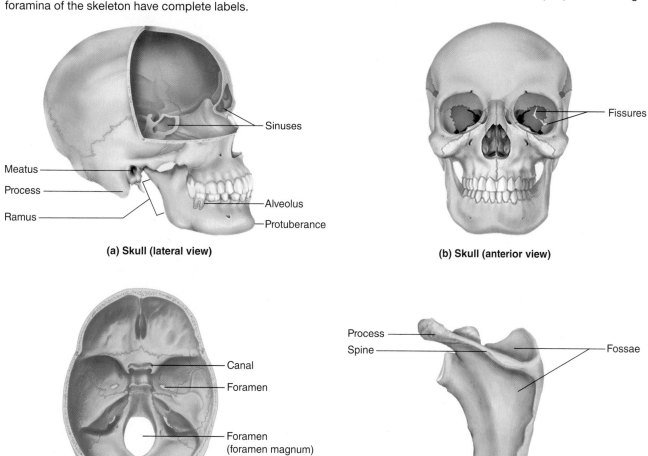

Sinuses

Meatus

Process

Ramus

Alveolus

Protuberance

(a) Skull (lateral view)

Fissures

(b) Skull (anterior view)

Canal

Foramen

Foramen
(foramen magnum)

Fossa

(c) Skull (cranial cavity, superior view)

Process

Spine

Fossae

(d) AP|R Scapula (posterior view)

Crest

Line

Notch

Spine

Notch

Tuberosity

Foramen (obturator foramen)

Tubercle

Ramus

(e) AP|R Hip bone (lateral view)

Foramen

Process

Facets

(f) AP|R Thoracic vertebra (superior view)

FIGURE 13.2 *Continued.*

Fovea
Head
Trochanters
Line
Epicondyles
Condyles

(g) AP|R Femur (posterior view)

Tubercles
Head
Sulcus
Tuberosity
Fossa
Condyle

(h) AP|R Humerus (anterior view)

Name _____

Date _____

Section _____

The Ⓐ corresponds to the indicated Learning Outcome(s) Ⓞ found at the beginning of the Laboratory Exercise.

Organization of the Skeleton

PART A: Assessments

Label the bones indicated in figure 13.3.

FIGURE 13.3 Label the major bones of the skeleton: (a) anterior view; (b) posterior view.

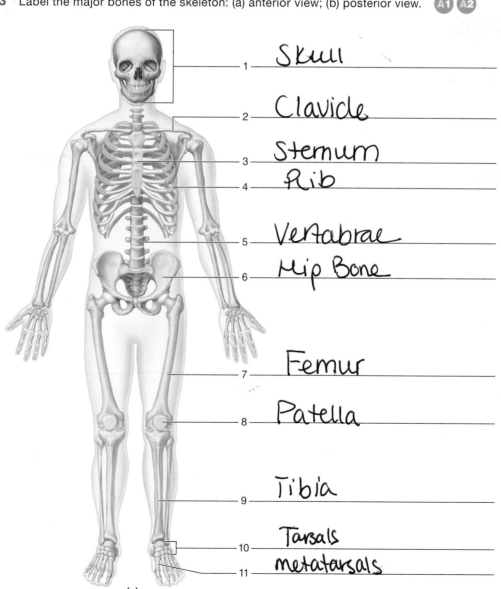

1. Skull
2. Clavicle
3. Sternum
4. Rib
5. Vertabrae
6. Hip Bone
7. Femur
8. Patella
9. Tibia
10. Tarsals
11. metatarsals

(a)

FIGURE 13.3 *Continued.*

Scapula 1
Humerus 2
Ulna 3

Phalanges 4

Femur 5

Tibia 6

7 Vertabrae

8 Radius
9 Sacrum

10 metacarpal

11 Fibula

(b)

PART B: Assessments

1. Identify the bones indicated in figure 13.4.

FIGURE 13.4 Identify the bones in this random arrangement, using the terms provided.

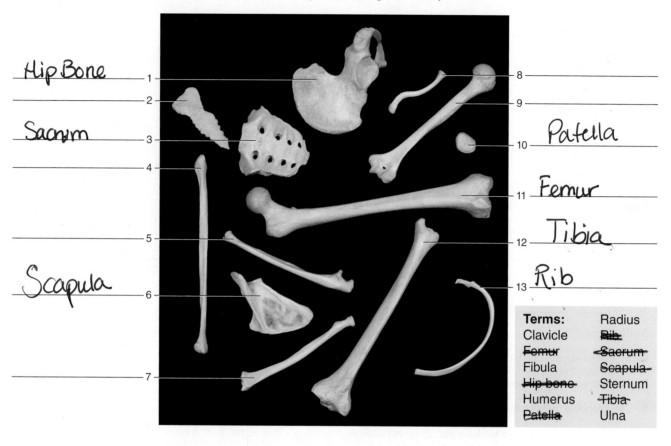

Hip Bone — 1

— 2

Sacrum — 3

— 4

— 5

Scapula — 6

— 7

8 —

9 —

Patella 10 —

Femur 11 —

Tibia 12 —

Rib 13 —

Terms: Radius
Clavicle   ~~Rib~~
~~Femur~~   ~~Sacrum~~
Fibula   ~~Scapula~~
~~Hip bone~~   Sternum
Humerus   ~~Tibia~~
~~Patella~~   Ulna

2. List any of the bones shown in figure 13.4 that are included as part of the axial skeleton.   _____

_____

PART C: Assessments

1. Match the terms in column A with the definitions in column B. Place the letter of your choice in the space provided.

Column A	Column B
a. Condyle	_____ 1. Small, nearly flat articular surface
b. Crest	_____ 2. Deep depression or shallow basin
c. Facet	_____ 3. Rounded process
d. Foramen	_____ 4. Opening or hole
e. Fossa	_____ 5. Projection extension
f. Line	_____ 6. Ridgelike projection
g. Ramus	_____ 7. Slightly raised ridge

2. Match the terms in column A with the definitions in column B. Place the letter of your choice in the space provided. Ⓐ3

Column A	Column B
a. Fovea	_____ **1.** Tubelike opening
b. Head	_____ **2.** Tiny pit or depression
c. Meatus	_____ **3.** Small, knoblike projection
d. Sinus	_____ **4.** Thornlike projection
e. Spine	_____ **5.** Rounded enlargement at end of bone
f. Trochanter	_____ **6.** Air-filled cavity within bone
g. Tubercle	_____ **7.** Relatively large process

PART D: Assessments

Complete the following statements:

1. The extra bones that sometimes develop between the flat bones of the skull are called_____ bones. Ⓐ1

2. Small bones occurring in some tendons in the hand or foot are called _____ bones. Ⓐ1

3. The cranium and facial bones compose the _____ . Ⓐ1

4. The _____ bone supports the tongue. Ⓐ1

5. The _____ at the inferior end of the sacrum is composed of several fused vertebrae. Ⓐ1

6. Most ribs are attached anteriorly to the ___Sternum_____ . Ⓐ1

7. The thoracic cage is composed of _____true_____ pairs of ribs. Ⓐ1

8. The scapulae and clavicles together form the ___pectoral_____ girdle. Ⓐ1

9. Which of the following bones is *not* part of the appendicular skeleton: clavicle, femur, scapula, sternum? _____ Ⓐ2

10. The wrist is composed of eight bones called ___carpals_____ . Ⓐ1

11. The hip bones (coxal bones) are attached posteriorly to the ___Vertabrae_____ . Ⓐ1

12. The ___patella_____ bone covers the anterior surface of the knee. Ⓐ1

13. The bones that articulate with the distal ends of the tibia and fibula are called ___tarsals_____ . Ⓐ1

14. All finger and toe bones are called ___Phalanges_____ . Ⓐ1

Skull

Purpose of the Exercise

To examine the structure of the human skull and to identify the bones and major features of the skull.

MATERIALS NEEDED

Human skull, articulated

Human skull, disarticulated

Human skull, sagittal section

Learning Outcomes AP|R

After completing this exercise, you should be able to:

01 Locate and label the bones of the skull and their major features.

02 Locate and label the major sutures of the cranium.

03 Locate the sinuses of the skull.

04 Distinguish the skull bones that comprise the orbit of an eye.

The O corresponds to the assessments A indicated in the Laboratory Assessment for this Exercise.

Pre-Lab

Carefully read the introductory material and the entire lab. Be familiar with the skull from lecture or the textbook. Answer the pre-lab questions.

Pre-Lab Questions Select the correct answer for each of the following questions:

1. The _____ is a skull bone that is *not* interlocked along sutures.
   - a. mandible
   - b. maxilla
   - c. temporal bone
   - d. occipital bone

2. Sinuses include all of the following functions *except*
   - a. warming and humidifying the air during breathing.
   - b. providing resonance to our voices.
   - c. making the skull lighter.
   - d. attachment sites for muscles.

3. The _____ bone serves for tongue and larynx muscle attachments.
   - a. mandible
   - b. maxilla
   - c. hyoid
   - d. temporal

4. Which of the following is a facial bone?
   - a. sphenoid bone
   - b. nasal bone
   - c. ethmoid bone
   - d. frontal bone

5. The _____ suture is located between the two parietal bones.
   - a. squamous
   - b. sagittal
   - c. coronal
   - d. lambdoid

6. The mastoid process, jugular foramen, and the external acoustic meatus are features of the _____ bone.
   - a. occipital
   - b. maxillary
   - c. temporal
   - d. ethmoid

7. Foramina, canals, and fissures serve as passageways for blood vessels and nerves.
   a. True _____ 　　　　　 b. False _____
8. The jugular foramen is larger than the foramen magnum.
   a. True _____ 　　　　　 b. False _____

A human skull consists of twenty-two bones that, except for the lower jaw, are firmly interlocked along sutures. Eight of these immovable bones make up the *braincase,* or *cranium,* and thirteen more immovable bones and the mandible form the *facial skeleton.*

Bones located within the temporal bone are the *auditory ossicles* (*middle ear bones*). They include the *malleus, incus,* and *stapes* that are examined in Laboratory Exercise 37. The stapes is considered the smallest bone in the human body among the total number of 206. The hyoid bone is a nonarticulating bone that is suspended from the temporal bones and serves for tongue and larynx muscle attachments. A fractured hyoid bone is sometimes used as an indication of strangulation as the cause of death.

After intramembranous ossification of the flat bones of the skull is complete, a fibrous type of joint remains between them. These joints are considered immovable and are known as *sutures.* The suture between the two parietal bones is the *sagittal suture.* The *lambdoid suture* is between the occipital and parietal bones. The *coronal suture* is bordered by the frontal bone and the parietal bones. The *squamous suture* is primarily along the superior border of the temporal bone with the parietal bone.

Sinus cavities are found in the frontal bone, ethmoid bone, sphenoid bone, and both maxillary bones. Because they are associated with the nasal passages, they are called *paranasal sinuses.* These air-filled sinuses are lined with a mucous membrane that is continuous with the nasal passages. The sinuses function to lighten the skull, provide resonance to our voices, and assist in warming and humidifying the air during breathing. An infection of the large maxillary sinuses is often misinterpreted to be a toothache.

PROCEDURE: Skull AP|R

1. Examine human articulated, disarticulated, and sectioned skulls available in the laboratory. Identify the bones and features illustrated in figures 14.1, 14.2, 14.3, 14.4, 14.5, and 14.6 on the actual skulls. Note that the figures are in full color, with the bones labeled in **boldface** and feature labels not in boldface. It is best to work with a partner or in small groups.
2. Examine the **cranial bones** of the articulated human skull and the sectioned skull. Also observe the corresponding disarticulated bones. Locate the following

FIGURE 14.1 AP|R Bones and features of the skull, anterior view. (**Boldface** indicates a bone name; not boldface indicates a feature.)

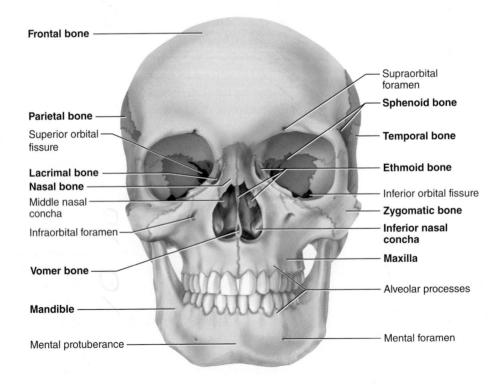

Frontal bone
Parietal bone
Superior orbital fissure
Lacrimal bone
Nasal bone
Middle nasal concha
Infraorbital foramen
Vomer bone
Mandible
Mental protuberance

Supraorbital foramen
Sphenoid bone
Temporal bone
Ethmoid bone
Inferior orbital fissure
Zygomatic bone
Inferior nasal concha
Maxilla
Alveolar processes
Mental foramen

FIGURE 14.2 **AP|R**   Bones and features of the skull, lateral view. (**Boldface** indicates a bone name; not boldface indicates a feature.)

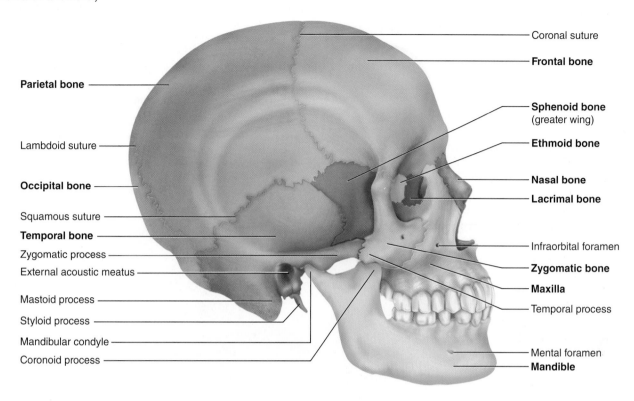

Coronal suture
Frontal bone
Parietal bone
Sphenoid bone (greater wing)
Ethmoid bone
Lambdoid suture
Nasal bone
Occipital bone
Lacrimal bone
Squamous suture
Temporal bone
Infraorbital foramen
Zygomatic process
Zygomatic bone
External acoustic meatus
Maxilla
Temporal process
Mastoid process
Styloid process
Mandibular condyle
Mental foramen
Coronoid process
Mandible

FIGURE 14.3 **AP|R**   Bones and features of the skull, inferior view. (**Boldface** indicates a bone name; not boldface indicates a feature.)

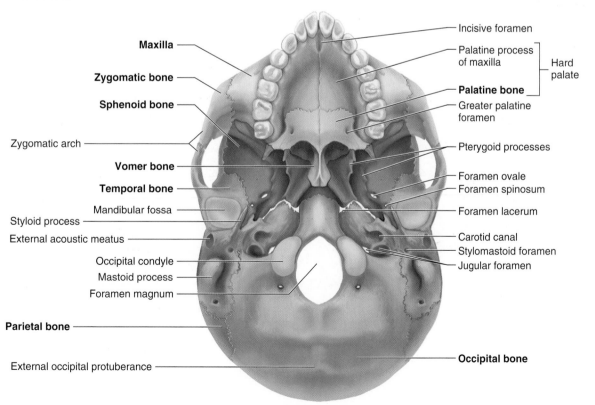

Incisive foramen
Maxilla
Palatine process of maxilla
Zygomatic bone
Hard palate
Sphenoid bone
Palatine bone
Greater palatine foramen
Zygomatic arch
Pterygoid processes
Vomer bone
Foramen ovale
Temporal bone
Foramen spinosum
Mandibular fossa
Foramen lacerum
Styloid process
External acoustic meatus
Carotid canal
Occipital condyle
Stylomastoid foramen
Mastoid process
Jugular foramen
Foramen magnum
Parietal bone
External occipital protuberance
Occipital bone

FIGURE 14.4 AP|R Bones and features of the floor of the cranial cavity, superior view. (**Boldface** indicates a bone name; not boldface indicates a feature.)

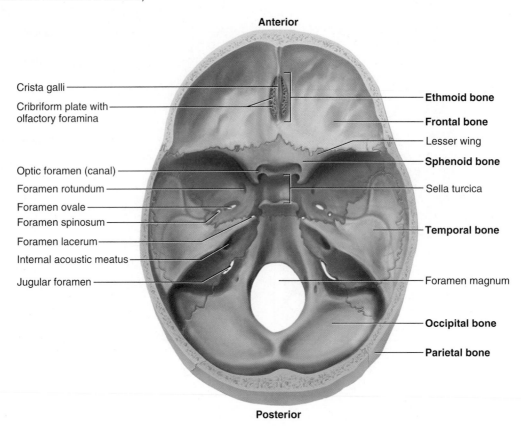

Anterior

Crista galli

Cribriform plate with olfactory foramina

Optic foramen (canal)

Foramen rotundum

Foramen ovale

Foramen spinosum

Foramen lacerum

Internal acoustic meatus

Jugular foramen

Ethmoid bone

Frontal bone

Lesser wing

Sphenoid bone

Sella turcica

Temporal bone

Foramen magnum

Occipital bone

Parietal bone

Posterior

FIGURE 14.5 AP|R Bones and features of the sagittal section of the skull. (**Boldface** indicates a bone name; not boldface indicates a feature.)

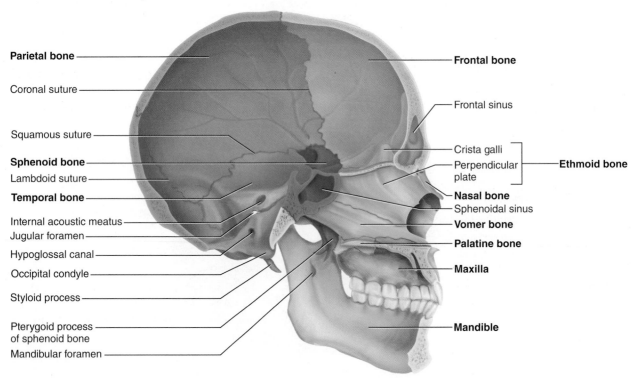

Parietal bone

Coronal suture

Squamous suture

Sphenoid bone

Lambdoid suture

Temporal bone

Internal acoustic meatus

Jugular foramen

Hypoglossal canal

Occipital condyle

Styloid process

Pterygoid process of sphenoid bone

Mandibular foramen

Frontal bone

Frontal sinus

Crista galli

Perpendicular plate

Ethmoid bone

Nasal bone

Sphenoidal sinus

Vomer bone

Palatine bone

Maxilla

Mandible

FIGURE 14.6   Bones of a disarticulated skull.

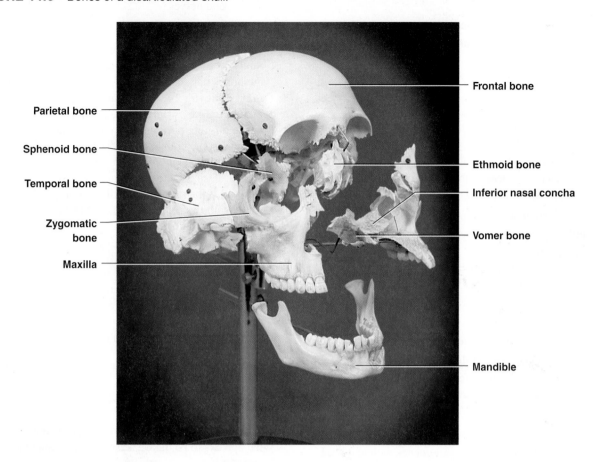

bones and features in the laboratory specimens and, at the same time, palpate as many of these bones and features in your skull as possible.

frontal bone   **(1)**

- supraorbital foramen (or notch)—transmits blood vessels and nerves
- frontal sinus—two present

parietal bone   **(2)**

- sagittal suture—where parietal bones meet
- coronal suture—where parietal bones meet frontal bone

occipital bone   **(1)**

- lambdoid suture—where parietal bones meet occipital bone
- external occipital protuberance—midline projection on posterior surface
- foramen magnum—where brainstem and spinal cord meet
- occipital condyle—articulates with first cervical vertebra

temporal bone   **(2)**

- squamous suture—where parietal bones meet temporal bone
- external acoustic meatus—opening of outer ear

- mandibular fossa—articulates with mandible
- mastoid process—attachment site for muscles
- styloid process—long, pointed projection
- carotid canal—transmits internal carotid artery
- jugular foramen—transmits internal jugular vein
- internal acoustic meatus—opening for cranial nerves
- zygomatic process—forms much of zygomatic arch

sphenoid bone   **(1)**

- sella turcica—saddle-shaped depression for pituitary gland
- greater and lesser wings—lateral extensions
- sphenoidal sinus—two present
- pterygoid process—inferior extensions for chewing muscle attachments

ethmoid bone   **(1)**

- cribriform plate—contains many olfactory foramina for olfactory nerves
- perpendicular plate—forms most of nasal septum
- superior nasal concha—small extensions in nasal cavity

- middle nasal concha—medium extensions in nasal cavity
- ethmoidal sinus—many present
- crista galli—attachment site of some brain membranes

3. Examine the **facial bones** of the articulated and sectioned skulls and the corresponding disarticulated bones. Locate the following:

maxilla (2)
- maxillary sinus—largest sinuses
- palatine process—forms anterior hard palate
- alveolar process—supports the teeth
- alveolar arch—formed by alveolar processes

palatine bone—L-shaped bones (2)

zygomatic bone (2)
- temporal process—extension that joins zygomatic process
- zygomatic arch—formed by zygomatic and temporal processes

lacrimal bone—thin and scalelike (2)

nasal bone—forms bridge of nose (2)

vomer bone—forms inferior part of nasal septum (1)

inferior nasal concha—largest of the conchae (2)

mandible (1)
- ramus—upward projection
- mandibular condyle—articulates with temporal bone
- coronoid process—attachment site for chewing muscles
- alveolar process—supports the teeth
- mandibular foramen—admits blood vessels and nerves for teeth

- mental foramen—passageway for blood vessels and nerves
- mental protuberance—chin projection

4. Complete Part A of Laboratory Assessment 14.
5. Study an anterior view of an orbit of the skull. Examine the bones that have a portion of the entire bone visible in an orbit (fig 14.7). A disarticulated skull, as shown in figure 14.6, would be helpful to distinguish skull bones forming portions of an orbit. The orbit of an eye includes some cranial and facial bones.
6. Examine any skulls that are sectioned to expose the paranasal sinuses. Sinuses are located in the frontal, ethmoid, sphenoid, and maxillary bones. Compare the skulls with figures 14.5, 14.6, and 14.8.
7. Some of the major passageways through the skull bones have specific names for the foramen, canal, fissure, or meatus. Reexamine figures 14.1 through 14.5 and the actual skulls and locate the passageways of the skull listed in table 14.1. The major structures transmitted through the specific passageway are included in the table.
8. Complete Parts B, C, and D of the laboratory assessment.

CRITICAL THINKING ACTIVITY

Examine the inside of the cranium on a sectioned skull. What appears to be the weakest area? Explain your answer.

_____

_____

_____

FIGURE 14.7 AP|R Anterior view of a left orbit. (**Boldface** indicates a bone name; not boldface indicates a feature.)

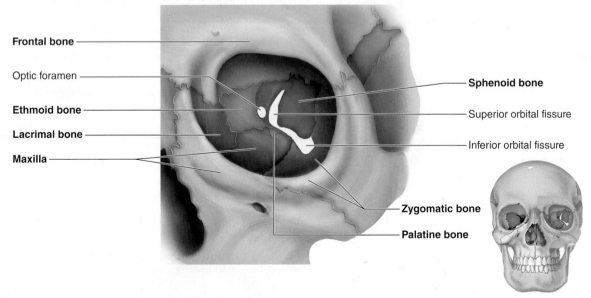

Frontal bone
Optic foramen
Ethmoid bone
Lacrimal bone
Maxilla
Sphenoid bone
Superior orbital fissure
Inferior orbital fissure
Zygomatic bone
Palatine bone

FIGURE 14.8 `AP|R` Paranasal sinuses are located in the frontal bone, ethmoid bone, sphenoid bone, and both maxillary bones.

Frontal sinus

Ethmoidal sinuses

Sphenoidal sinus

Maxillary sinus

Table 14.1 Major Passageways of the Skull

Passageway	Location Bone	Major Stuctures Transmitted Through
Carotid canal	Temporal bone	Internal carotid artery
Foramen lacerum	Between temporal, sphenoid, and occipital bones	Branch of pharyngeal artery (opening becomes mostly closed by connective tissues)
Foramen magnum	Occipital bone	Inferior part of brainstem connecting to spinal cord, vertebral arteries, and accessory nerve
Foramen ovale	Sphenoid bone	Mandibular division of trigeminal nerve
Foramen rotundum	Sphenoid bone	Maxillary division of trigeminal nerve
Foramen spinosum	Sphenoid bone	Middle meningeal blood vessels
Greater palatine foramen	Palatine bone	Palatine blood vessels and nerves
Hypoglossal canal	Occipital bone	Hypoglossal nerve
Incisive foramen	Maxilla	Nasopalatine nerves
Inferior orbital fissure	Between maxilla, sphenoid, and zygomatic bones	Infraorbital nerve and blood vessels
Infraorbital foramen	Maxilla	Infraorbital blood vessels and nerves
Internal acoustic meatus	Temporal bone	Branches of facial and vestibulocochlear nerves
Jugular foramen	Between temporal and occipital bones	Glossopharyngeal, vagus, and accessory nerves, and internal jugular vein
Mandibular foramen	Mandible	Inferior alveolar blood vessels and nerves
Mental foramen	Mandible	Mental nerve and blood vessels
Optic foramen (canal)	Sphenoid bone	Optic nerve and ophthalmic artery
Stylomastoid foramen	Temporal bone	Facial nerve and blood vessels
Superior orbital fissure	Sphenoid bone	Oculomotor, trochlear, and abducens nerves, and ophthalmic division of trigeminal nerve
Supraorbital foramen	Frontal bone	Supraorbital blood vessels and nerves

NOTES

Name _____

Date _____

Section _____

The Ⓐ corresponds to the indicated Learning Outcome(s) Ⓞ found at the beginning of the Laboratory Exercise.

Skull

PART A: Assessments

Identify the numbered bones and features of the skulls indicated in figures 14.9, 14.10, 14.11, and 14.12.

FIGURE 14.9 Identify the bones and features indicated on this anterior view of the skull, using the terms provided. (If the line lacks the word *bone*, label the particular feature.) Ⓐ1

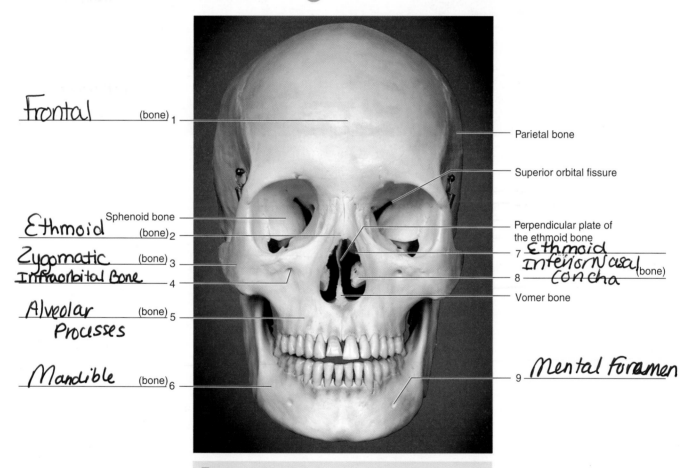

Frontal (bone) 1 — Parietal bone

— Superior orbital fissure

Sphenoid bone —
Ethmoid (bone) 2 — Perpendicular plate of the ethmoid bone

Zygomatic (bone) 3

Infraorbital Bone 4 — 7 Ethmoid

Alveolar (bone) 5 — 8 Inferior Nasal Concha (bone)

Processes — Vomer bone

Mandible (bone) 6 — 9 Mental Foramen

Terms:

Frontal bone	Mental foramen
Inferior nasal concha	Middle nasal concha
Infraorbital foramen	(of ethmoid bone)
Mandible	Nasal bone
Maxilla	Zygomatic bone

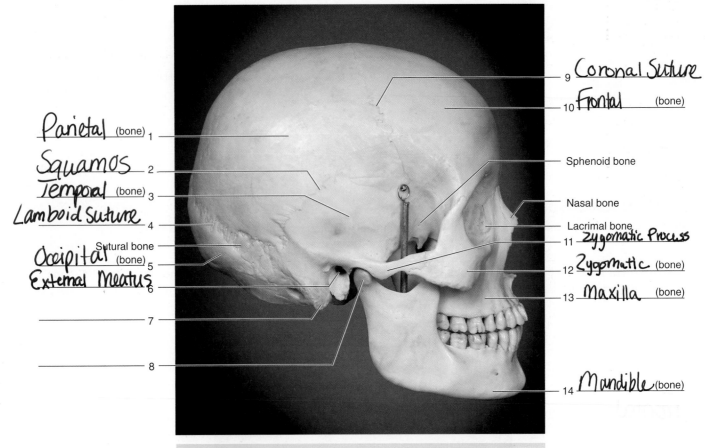

Parietal (bone) 1

Squamos 2

Temporal (bone) 3

Lamboid Suture 4

Occipital (bone) 5

External Meatus 6

7

8

Sutural bone

9 Coronal Suture

10 Frontal (bone)

Sphenoid bone

Nasal bone

Lacrimal bone

11 Zygomatic Process

12 Zygomatic (bone)

13 Maxilla (bone)

14 Mandible (bone)

Terms:

Coronal suture

External acoustic meatus

Frontal bone

Lambdoid suture

Mandible

Mandibular condyle

Mastoid process

Maxilla

Occipital bone

Parietal bone

Squamous suture

Temporal bone

Zygomatic bone

Zygomatic process

(of temporal bone)

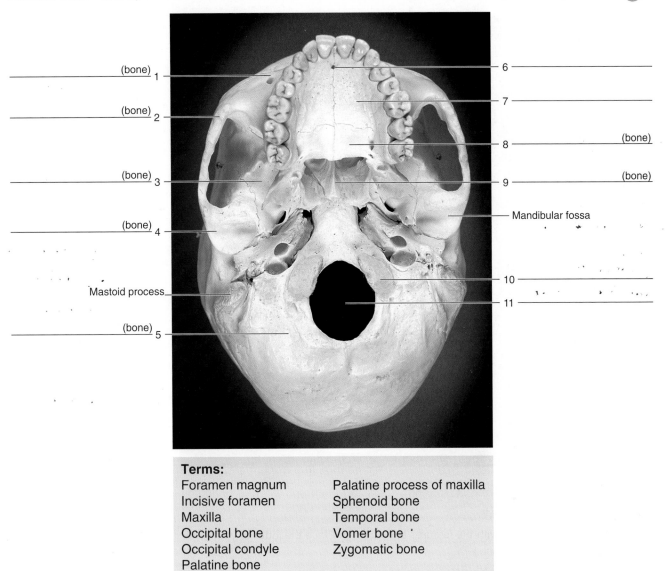

(bone) 1
(bone) 2
(bone) 3
(bone) 4
Mastoid process
(bone) 5

6
7
8 (bone)
9 (bone)
Mandibular fossa
10
11

Terms:

Foramen magnum	Palatine process of maxilla
Incisive foramen	Sphenoid bone
Maxilla	Temporal bone
Occipital bone	Vomer bone
Occipital condyle	Zygomatic bone
Palatine bone	

FIGURE 14.12 Identify the bones and features on this floor of the cranial cavity of a skull, using the terms provided.

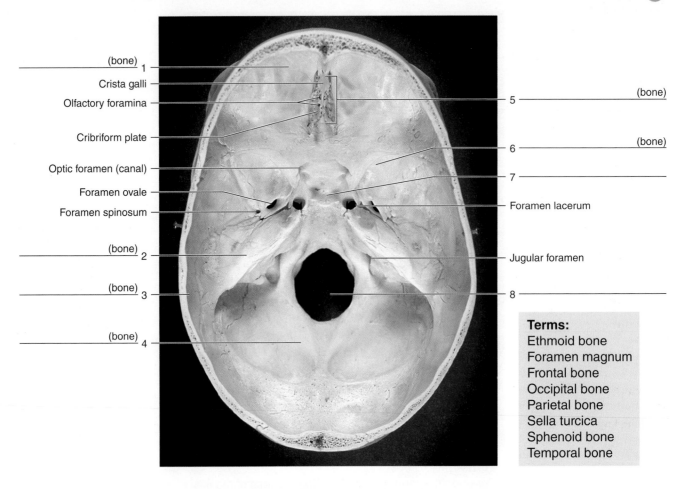

(bone) 1
Crista galli
Olfactory foramina
Cribriform plate
Optic foramen (canal)
Foramen ovale
Foramen spinosum
(bone) 2
(bone) 3
(bone) 4

(bone) 5
(bone) 6
7
Foramen lacerum
Jugular foramen
8

Terms:
Ethmoid bone
Foramen magnum
Frontal bone
Occipital bone
Parietal bone
Sella turcica
Sphenoid bone
Temporal bone

PART B: Assessments

Match the bones in column A with the features in column B. Place the letter of your choice in the space provided. (Some answers are used more than once.) A1

Column A	Column B
a. Ethmoid bone	_____ **1.** Forms sagittal, coronal, squamous, and lambdoid sutures
b. Frontal bone	_____ **2.** Cribriform plate
c. Occipital bone	_____ **3.** Crista galli
d. Parietal bone	_____ **4.** External acoustic meatus
e. Sphenoid bone	_____ **5.** Foramen magnum
f. Temporal bone	_____ **6.** Mandibular fossa
	_____ **7.** Mastoid process
	_____ **8.** Middle nasal concha
	_____ **9.** Occipital condyle
	_____ **10.** Sella turcica
	_____ **11.** Styloid process
	_____ **12.** Supraorbital foramen

PART C: Assessments

Complete the following statements:

1. The _____ suture joins the frontal bone to the parietal bones. (A2)

2. The parietal bones are firmly interlocked along the midline by the _____ suture. (A2)

3. The _____ suture joins the parietal bones to the occipital bone. (A2)

4. The temporal bones are joined to the parietal bones along the _____ sutures. (A2)

5. Name the three cranial bones that contain sinuses. (A3) _____

_____

6. Name a facial bone that contains a sinus. (A3) _____

7. Name six cranial bones that are visible on a lateral view of a skull. (A1) _____

_____

PART D: Assessments

Match the bones in column A with the characteristics in column B. Place the letter of your choice in the space provided. (A1) (A3) (A4)

Column A	Column B
a. Inferior nasal concha	_____ 1. Forms bridge of nose
b. Lacrimal bone	_____ 2. Only movable bone in the facial skeleton
c. Mandible	_____ 3. Contains coronoid process
d. Maxilla	_____ 4. Creates prominence of cheek inferior and lateral to the eye
e. Nasal bone	_____ 5. Contains sockets of upper teeth
f. Palatine bone	_____ 6. Forms inferior portion of nasal septum
g. Vomer bone	_____ 7. Forms anterior portion of zygomatic arch
h. Zygomatic bone	_____ 8. Scroll-shaped bone in nasal passage
	_____ 9. Forms anterior roof of mouth
	_____ 10. Contains mental foramen
	_____ 11. Forms posterior roof of mouth
	_____ 12. Small bone in medial wall of orbit
	_____ 13. Forms a very small portion of inferior orbit
	_____ 14. Contains large sinus inferior to orbit

CRITICAL THINKING ASSESSMENT

Assume that an orbit of a skull was damaged in an accident. Which bones of a skull might need to be considered for reconstruction in an orbital surgery? (A4) _____

_____

_____

Vertebral Column and Thoracic Cage

Purpose of the Exercise

To examine the vertebral column and the thoracic cage of the human skeleton and to identify the bones and major features of these parts.

MATERIALS NEEDED

Human skeleton, articulated

Samples of cervical, thoracic, and lumbar vertebrae

Human skeleton, disarticulated

Learning Outcomes AP|R

After completing this exercise, you should be able to:

01 Identify the structures and functions of the vertebral column.

02 Locate the features of a vertebra.

03 Distinguish the cervical, thoracic, and lumbar vertebrae and the sacrum and coccyx.

04 Identify the structures and functions of the thoracic cage.

05 Distinguish between true and false ribs.

The O corresponds to the assessments A indicated in the Laboratory Assessment for this Exercise.

Pre-Lab

Carefully read the introductory material and examine the entire lab. Be familiar with the vertebral column and the thoracic cage from lecture or the textbook. Answer the pre-lab questions.

Pre-Lab Questions Select the correct answer for each of the following questions:

1. The most superior bone of the vertebral column is the
   a. coccyx.   **(b.)** vertebra prominens.   **c.** axis.   **d.** atlas.

2. The vertebral column possesses
   a. four curvatures.   **(b)** three curvatures.
   c. one curvature.   **d.** no curvatures as it is straight.

3. Humans have _____ pairs of true ribs.
   a. two   **b.** five   ⊗ seven   **(d.)** twelve

4. The _____ ribs do *not* have costal cartilage attachments to the sternum.
   (a.) false   **b.** floating   **c.** true   **d.** superior

5. Humans possess _____ cervical vertebrae.
   a. twenty-six   **b.** twelve
   (c.) seven   ⊗ five

6. The superior end of the sacrum articulates with the
   a. coccyx.   **b.** femur.
   c. twelfth thoracic vertebra.   **d.** fifth lumbar vertebra.

7. The anterior (sternal) end of a rib articulates with a thoracic vertebra.
   a. True _____   **b.** False _____

8. All cervical, thoracic, and lumbar vertebrae possess a vertebral foramen.
   a. True _____   **b.** False _____

9. A feature of the second cervical vertebra is the dens.
   a. True _____   **b.** False _____

The vertebral column, consisting of twenty-six bones, extends from the skull to the pelvis and forms the vertical axis of the human skeleton. The *vertebral column* includes seven cervical vertebrae, twelve thoracic vertebrae, five lumbar vertebrae, one sacrum of five fused vertebrae, and one coccyx of usually four fused vertebrae. To help you remember the number of cervical, thoracic, and lumbar vertebrae from superior to inferior, consider this saying: breakfast at 7, lunch at 12, and dinner at 5. These vertebrae are separated from one another by cartilaginous *intervertebral discs* and are held together by ligaments.

The *thoracic cage* surrounds the thoracic and upper abdominal cavities. It includes the ribs, the thoracic vertebrae, the sternum, and the costal cartilages. The thoracic cage provides protection for the heart and lungs.

PROCEDURE A: Vertebral Column

The vertebral column extends from the first cervical vertebra adjacent to the skull to the inferior tip of the coccyx. The first cervical vertebra (C1) is also known as the *atlas* and has a posterior tubercle instead of a more pronounced spinous process. The second cervical vertebra (C2), known as the *axis,* has a superior projection, the dens (odontoid process) that serves as a pivot point for some rotational movements. The seventh cervical vertebra (C7) is often referred to as the *vertebra prominens* because the spinous process is elongated and easily palpated as a surface feature. The seven cervical vertebrae have the distinctive feature of transverse foramina for passageways of blood vessels serving the brain.

The twelve thoracic vertebrae have facets for the articulation sites of the twelve pairs of ribs. They are larger than the cervical vertebrae and have spinous processes that are rather long and have an inferior angle. The five lumbar vertebrae have the largest bodies, allowing better support and resistance to twisting of the trunk, and spinous processes that are rather short and blunt.

The five sacral vertebrae of a child fuse into a single bone by the age of about 26. The posterior ridge known as the *medial sacral crest* represents fused spinous processes. The usual four coccyx vertebrae fuse into a single bone by the age of about 30.

CRITICAL THINKING ACTIVITY

Note the four curvatures of the vertebral column. What functional advantages exist with curvatures for skeletal structure instead of a straight vertebral column?

_____

_____

_____

_____

_____

The four curvatures of the vertebral column develop either before or after birth. The *thoracic* and *sacral curvatures* (primary curvatures) form by the time of birth. The *cervical curvature* develops by the time a baby is able to hold the head erect and crawl, while the *lumbar curvature* forms by the time the child is able to walk. The cervical and lumbar curvatures represent the secondary curvatures. The four curvatures allow for flexibility and resiliency of the vertebral column and for it to function somewhat like a spring instead of a rigid rod.

1. Examine figure 15.1 and the vertebral column of the human skeleton. Locate the following bones and features. At the same time, locate as many of the corresponding bones and features in your skeleton as possible.

FIGURE 15.1 AP|R Bones and features of the vertebral column (right lateral view).

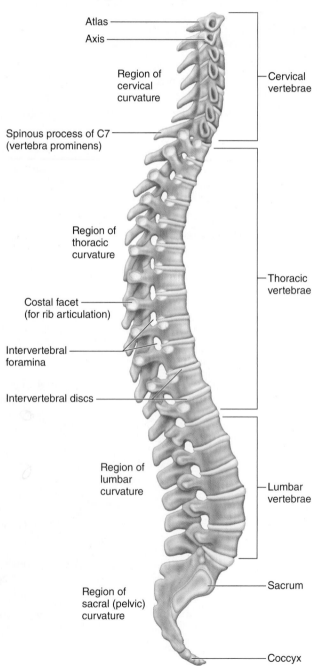

Atlas
Axis
Region of cervical curvature
Cervical vertebrae
Spinous process of C7 (vertebra prominens)
Region of thoracic curvature
Thoracic vertebrae
Costal facet (for rib articulation)
Intervertebral foramina
Intervertebral discs
Region of lumbar curvature
Lumbar vertebrae
Region of sacral (pelvic) curvature
Sacrum
Coccyx

cervical vertebrae (7)

- atlas (C1)
- axis (C2)
- vertebra prominens (C7)

thoracic vertebrae (12)

lumbar vertebrae (5)

sacrum (1)

coccyx (1)

intervertebral discs—fibrocartilage

vertebral canal—contains spinal cord

cervical curvature

thoracic curvature

lumbar curvature

sacral (pelvic) curvature

intervertebral foramina—passageway for spinal nerves

2. Compare the available samples of cervical, thoracic, and lumbar vertebrae along with figures 15.2 and 15.3. Note differences in size and shapes and locate the following features:

vertebral foramen—location of spinal cord

body—largest part of vertebra; main support portion

pedicles—form lateral area of vertebral foramen

laminae—thin plates forming posterior area of vertebral foramen

spinous process—posterior projection

transverse processes—lateral projections

facets—articulating surfaces

superior articular processes—superior projections

inferior articular processes—inferior projections

inferior vertebral notch—space for nerve passage

transverse foramina—only present on cervical vertebrae; passageway for blood vessels

---

FIGURE 15.2 AP|R The superior features of (a) the atlas, and the superior and right lateral features of (b) the axis. (The broken arrow indicates a transverse foramen.)

Number code:
1. Superior articular facet
2. Vertebral foramen
3. Dens (odontoid process)
4. Facet that articulates with occipital condyle
5. Transverse process
6. Transverse foramen
7. Spinous process
8. Body
9. Posterior tubercle

(a) Atlas (C1)

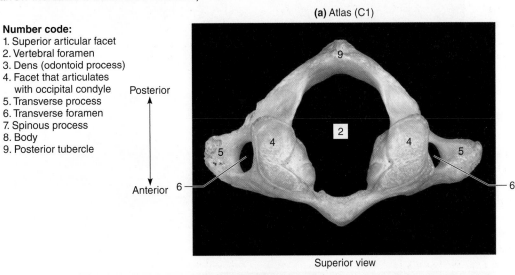

Superior view

(b) Axis (C2)

Superior view

Right lateral view

FIGURE 15.3 AP R The superior and right lateral features of the (a) cervical, (b) thoracic, and (c) lumbar vertebrae. (The broken arrow indicates a transverse foramen.)

Number code:
1. Superior articular process
2. Transverse process
3. Lamina
4. Spinous process
5. Pedicle
6. Body
7. Inferior vertebral notch
8. Vertebral foramen
9. Transverse foramen
10. Costal facets

Superior views   Right lateral views

(a) Cervical vertebra

Posterior

Anterior

(b) Thoracic vertebra

(c) Lumbar vertebra

Posterior ◄─────► Anterior

dens (odontoid process) of axis—superior process of C2 and is a pivot location at atlas

3. Examine the sacrum and coccyx along with figure 15.4. Locate the following features:

sacrum—formed typically by five fused vertebrae
- superior articular process—superior projection with a facet articulation site
- anterior sacral foramen— passageway for blood vessels and nerves
- posterior sacral foramen—passageway for blood vessels and nerves
- sacral promontory—anterior border of S1; important landmark for obstetricians
- sacral canal—portion of vetrebral canal
- median sacral crest—area of fused spinous processes
- sacral hiatus—inferior opening of vertebral canal

coccyx—formed by three to five fused vertebrae

4. Complete Parts A and B of Laboratory Assessment 15.

PROCEDURE B: Thoracic Cage

The twelve thoracic vertebrae are associated with the twelve pairs of ribs. The superior seven pairs of ribs are connected directly to the sternum with costal cartilage and are known as *true ribs*. The five inferior pairs, known as *false ribs,* either connect indirectly to the sternum with costal cartilage or do not connect to the sternum. Pairs eleven and twelve are called the *floating ribs* because they only connect with the thoracic vertebrae and not with the sternum.

The manubrium, body, and xiphoid process represent three regions that eventually fuse into a single flat bone, the sternum. The heart is located mainly beneath the body portion of the sternum. Any chest compressions administered during cardiopulmonary resuscitation should be over the body area of the sternum, not the xiphoid process region. Chest compressions over the xiphoid process region can force the xiphoid process deep into the liver or the inferior portion of the heart and cause a fatal hemorrhage to occur.

The thoracic cage protects organs in the thoracic and upper abdominal cavities. The thoracic cage also supports the pectoral girdle and upper limbs, functions in breathing, and serves for various muscle attachments.

1. Examine figures 15.5 and 15.6 and the thoracic cage of the human skeleton. Locate the following bones and features:

rib
- head—expanded end near thoracic vertebra
- tubercle—projection near thoracic vertebra
- neck—narrow region between head and tubercle
- shaft—main portion
- anterior (sternal) end—costal cartilage location
- facets—articulation surfaces
- true ribs—pairs 1–7
- false ribs—pairs 8–12; includes floating ribs
  - floating ribs—pairs 11–12

costal cartilages—hyaline cartilage

sternum
- jugular (suprasternal) notch—superior concave border of manubrium
- clavicular notch—articulation site of clavicle
- manubrium—superior part
- sternal angle—junction of manubrium and body at level of second rib pair
- body—largest, middle part
- xiphoid process—inferior part; remains cartilaginous until adulthood

2. Complete Parts C and D of the laboratory assessment.

FIGURE 15.4 The sacrum and coccyx.

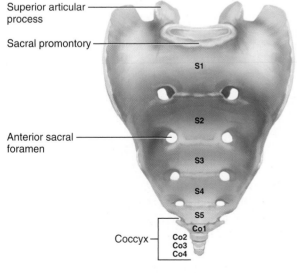

Superior articular process

Sacral promontory

S1

S2

Anterior sacral foramen

S3

S4

S5

Co1

Coccyx — Co2 Co3 Co4

(a) Anterior view

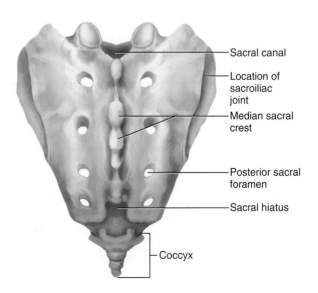

Sacral canal

Location of sacroiliac joint

Median sacral crest

Posterior sacral foramen

Sacral hiatus

Coccyx

(b) Posterior view

FIGURE 15.5 Superior view of a typical rib with the articulation sites with a thoracic vertebra.

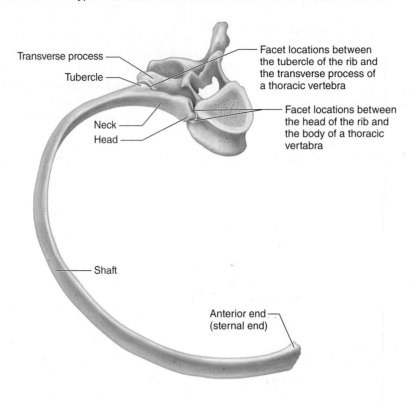

Transverse process

Tubercle

Neck

Head

Shaft

Anterior end (sternal end)

Facet locations between the tubercle of the rib and the transverse process of a thoracic vertebra

Facet locations between the head of the rib and the body of a thoracic vertabra

FIGURE 15.6 Bones and features of the thoracic cage (anterior view).

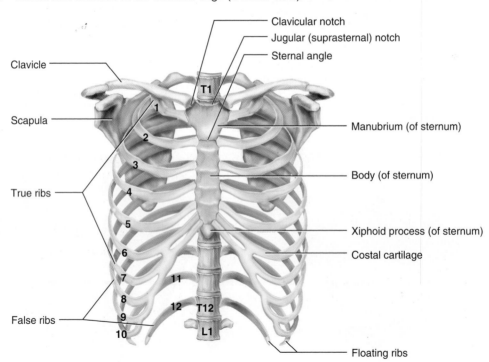

Clavicle

Scapula

True ribs

False ribs

Clavicular notch

Jugular (suprasternal) notch

Sternal angle

Manubrium (of sternum)

Body (of sternum)

Xiphoid process (of sternum)

Costal cartilage

Floating ribs

T1

T12

L1

1
2
3
4
5
6
7
8
9
10
11
12

Name _____

Date _____

Section _____

The Ⓐ corresponds to the indicated Learning Outcome(s) Ⓞ found at the beginning of the Laboratory Exercise.

Vertebral Column and Thoracic Cage

PART A: Assessments

Complete the following statements:

1. The vertebral column encloses and protects the _Vertebrae_. Ⓐ1

2. The vertebral column extends from the skull to the _coccyx_. Ⓐ1

3. The seventh cervical vertebra is called the _vertebra prominens_ and has an obvious spinous process surface feature that can be palpated. Ⓐ2

4. The _____ of the vertebrae support the weight of the head and trunk. Ⓐ2

5. The _____ separate adjacent vertebrae, and they soften the forces created by walking. Ⓐ1

6. The intervertebral foramina provide passageways for _____. Ⓐ1

7. Transverse foramina of _____ vertebrae serve as passageways for blood vessels leading to the brain. Ⓐ3

8. The first vertebra also is called the _atlas_. Ⓐ3

9. When the head is moved from side to side, the first vertebra pivots around the _axis_ of the second vertebra. Ⓐ2

10. The _lumbar_ vertebrae have the largest and strongest bodies. Ⓐ3

11. The typical number of vertebrae that fuse in the adult to form the sacrum is _____. Ⓐ1

PART B: Assessments

1. Based on your observations, compare typical cervical, thoracic, and lumbar vertebrae in relation to the characteristics indicated in the table. The table is partly completed. For your responses, consider characteristics such as size, shape, presence or absence, and unique features. Ⓐ2 Ⓐ3

Vertebra	Number	Size	Body	Spinous Process	Transverse Foramina
Cervical	7		smallest	C2 through C6 are forked (bifed)	
Thoracic		intermediate			
Lumbar					absent

2. Identify the bones and features in figures 15.7 and 15.8.

FIGURE 15.7 Label the bones and features of a lateral view of a vertebral column by placing the correct numbers in the spaces provided. A1 A2

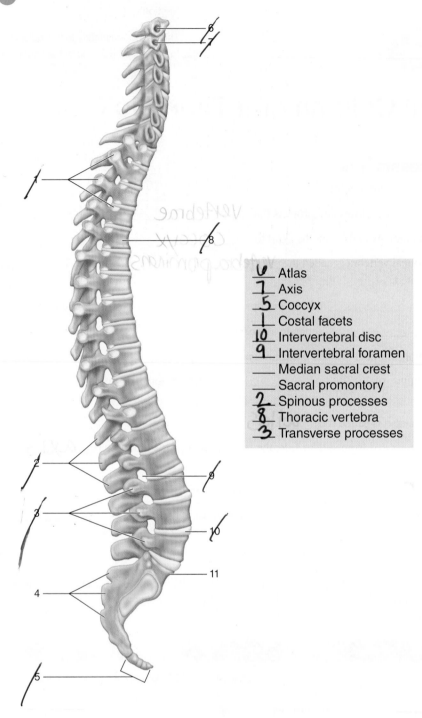

6	Atlas
7	Axis
5	Coccyx
1	Costal facets
10	Intervertebral disc
9	Intervertebral foramen
___	Median sacral crest
___	Sacral promontory
2	Spinous processes
8	Thoracic vertebra
3	Transverse processes

FIGURE 15.8 Identify the bones and features indicated in this radiograph of the neck (lateral view), using the terms provided. **A1** **A2**

Terms:
Atlas
Axis
Body
Intervertebral disc
Spinous process
Transverse process

Occipital bone

Posterior

Anterior

2 —— *Atlas* (bone)
3 —— *Axis* (bone)
—— Mandible
4 ——
5 ——
6 ——
1 ——

PART C: Assessments

Complete the following statements:

1. The manubrium, body, and xiphoid process form a bone called the _____. **A4**

2. The last two pairs of ribs that have no cartilaginous attachments to the sternum are sometimes called *Floating* ribs. **A5**

3. There are _____ pairs of true ribs. **A5**

4. Costal cartilages are composed of _____ tissue. **A4**

5. The manubrium articulates with the _____ on its superior border. **A4**

6. List three general functions of the thoracic cage. **A4** *Protection of lungs* _____
   _____

7. The sternal angle indicates the location of the _____ pair of ribs. **A4**

PART D: Assessments

Identify the bones and features indicated in figure 15.9.

FIGURE 15.9 Label the bones and features of the thoracic cage, using the terms provided. A4 A5

Terms:

Costal cartilage of false rib	Jugular notch
Costal cartilage of true rib	Sternal angle
Clavicular notch	Sternum
Floating rib	True rib

Pectoral Girdle and Upper Limb

Purpose of the Exercise

To examine the bones of the pectoral girdle and upper limb and to identify the major features of these bones.

MATERIALS NEEDED

Human skeleton, articulated

Human skeleton, disarticulated

Learning Outcomes AP|R

After completing this exercise, you should be able to:

O1 Locate and identify the bones of the pectoral girdle and their major features.

O2 Locate and identify the bones of the upper limb and their major features.

The **O** corresponds to the assessments **A** indicated in the Laboratory Assessment for this Exercise.

Pre-Lab

Carefully read the introductory material and examine the entire lab. Be familiar with the pectoral girdle and upper limb bones from lecture or the textbook. Answer the pre-lab questions.

Pre-Lab Questions Select the correct answer for each of the following questions:

1. The clavicle and the scapula form the
   a. pectoral girdle.
   b. pelvic girdle.
   c. upper limb.
   d. axial skeleton.

2. Anatomically, *arm* represents
   a. shoulder to fingers.
   b. shoulder to wrist.
   c. elbow to wrist.
   d. shoulder to elbow.

3. Which of the following is *not* part of the scapula?
   a. spine
   b. manubrium
   c. acromion
   d. supraspinous fossa

4. Which carpal is included in the proximal row?
   a. hamate
   b. capitate
   c. trapezium
   d. lunate

5. Which of the following is the most proximal part of the upper limb?
   a. styloid process of radius
   b. styloid process of ulna
   c. head of humerus
   d. medial epicondyle of humerus

6. Which of the following is the most distal feature of the humerus?
   a. anatomical neck
   b. deltoid tuberosity
   c. capitulum
   d. head

7. The capitate is one of the eight carpals in a wrist.
   a. True _____
   b. False _____

8. The clavicle articulates with the sternum and the humerus.
   a. True _____
   b. False _____

The pectoral girdle (shoulder girdle) consists of an anterior clavicle and a posterior scapula on each side of the body. The *pectoral girdle* represents an incomplete ring (girdle) of bones as the posterior scapulae do not meet each other, but muscles extend from their medial borders to the vertebral column. The two clavicles also do not meet as they articulate with the manubrium of the sternum. The pectoral girdle supports the upper limb and serves as the attachment point for various muscles that move the upper limb. This allows considerable flexibility of the shoulder. Relatively loose attachments of the pectoral girdle with the humerus allow a wide range of movements, but shoulder joint injuries are somewhat common. Additionally, the clavicle is a frequently broken bone when one reaches with an upper limb to break a fall.

Each upper limb includes a humerus in the arm, a radius and ulna in the forearm, and eight carpals, five metacarpals, and fourteen phalanges in the hand. (Anatomically, *arm* represents the region from shoulder to elbow, *forearm* is elbow to the wrist, and *hand* includes the wrist to the ends of digits.) These bones form the framework of the upper limb. They also function as parts of levers when muscles contract.

PROCEDURE A: Pectoral Girdle

The clavicle and scapula of the pectoral girdle provide for attachments of muscles of the neck and trunk. The clavicle is not a straight bone, but rather has two curves making it slightly S-shaped, and it serves as a brace bone to keep the upper limb to the side of the body. The bone is easily fractured because it is so close to the anterior surface and because the shoulder or upper limb commonly breaks a fall. Fortunately, bones have tensile strength, and when a force is exerted upon the clavicle from the lateral side of the body, the bone will bend to some extent at both curves until the threshold is reached, causing a fracture to occur.

The scapula has many tendon attachment sites for muscles of the neck, trunk, and upper limb. Because the scapula does not connect directly to the axial skeleton and because it has so many muscle attachments, there is a great deal of flexibility for shoulder movements. Unfortunately, this much flexibility makes the humerus vulnerable to dislocation from the scapula. This type of injury can occur during gymnastics maneuvers or when suddenly pulling or swinging a child by an upper limb.

1. Examine figures 16.1, 16.2, and 16.3 of the pectoral girdle. Locate the following features of the clavicle and

---

FIGURE 16.1   AP|R   Bones and features of the right shoulder and upper limb (anterior view).

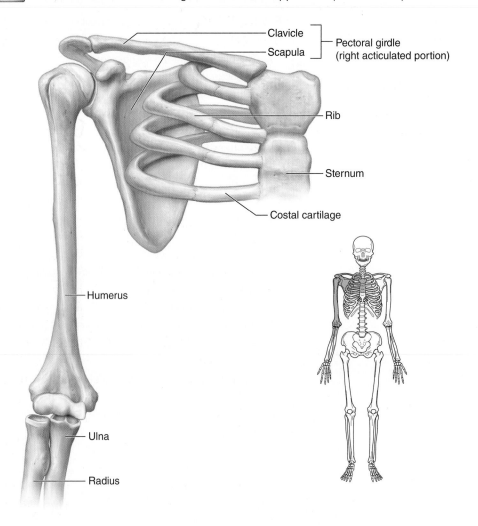

FIGURE 16.2 `AP|R` Right clavicle. The clavicle appears somewhat S-shaped when viewed from the superior and the inferior.

Acromial end

(a) Superior view

Sternal end

(b) Inferior view

the scapula. At the same time, locate as many of the corresponding surface bones and features of your own skeleton as possible.

clavicle

- sternal (medial) end—articulates with the manubrium of the sternum
- acromial (lateral) end—articulates with the acromion of the scapula

scapula

- spine
- acromion—lateral end of spine
- glenoid cavity—shallow socket; articulates with head of humerus
- coracoid process—beaklike projection
- borders
  - superior border
  - medial (vertebral) border
  - lateral (axillary) border

FIGURE 16.3 `AP|R` Right scapula (a) posterior surface (b) lateral view (c) anterior surface.

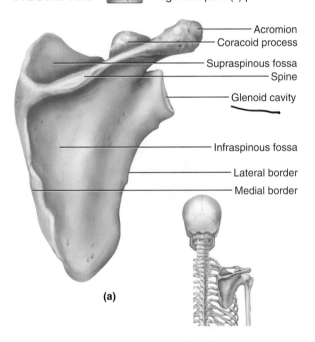

Acromion
Coracoid process
Supraspinous fossa
Spine
Glenoid cavity
Infraspinous fossa
Lateral border
Medial border

(a)

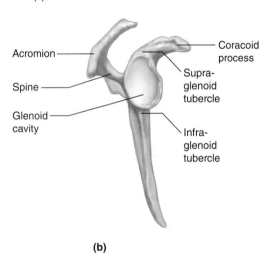

Acromion
Spine
Glenoid cavity
Coracoid process
Supra-glenoid tubercle
Infra-glenoid tubercle

(b)

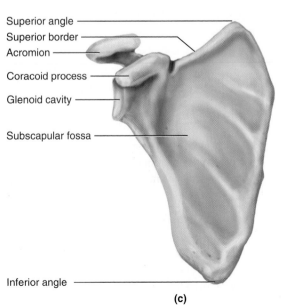

Superior angle
Superior border
Acromion
Coracoid process
Glenoid cavity
Subscapular fossa
Inferior angle

(c)

- fossae—shallow depressions
  - supraspinous fossa
  - infraspinous fossa
  - subscapular fossa
- angles
  - superior angle
  - inferior angle
- tubercles
  - supraglenoid tubercle
  - infraglenoid tubercle

2. Complete Parts A and B of Laboratory Assessment 16.

CRITICAL THINKING ACTIVITY

Why is the clavicle a bone that can easily fracture?

_____

_____

_____

_____

_____

_____

PROCEDURE B: Upper Limb

The humerus has sometimes been called the "funny bone" because of a tingling sensation (temporary pain) if it is bumped on the medial epicondyle where the ulnar nerve passes. The two bones of the forearm are nearly parallel when in anatomical position, with the radius positioned on the lateral side and the ulna on the medial side. All of the bones of the hand, including those of the wrist (carpals), palm (metacarpals), and digits (phalanges), are visible from an anterior view.

1. Examine figures 16.4, 16.5, and 16.6. Locate the following bones and features of the upper limb:

   humerus
   - proximal features
     - head—articulates with glenoid cavity
     - greater tubercle—on lateral side
     - lesser tubercle—on anterior side
     - anatomical neck—tapered region near head
     - surgical neck—common fracture site
     - intertubercular sulcus—furrow for tendon of biceps muscle
   - shaft
     - deltoid tuberosity
   - distal features
     - capitulum—lateral condyle; articulates with radius
     - trochlea—medial condyle; articulates with ulna

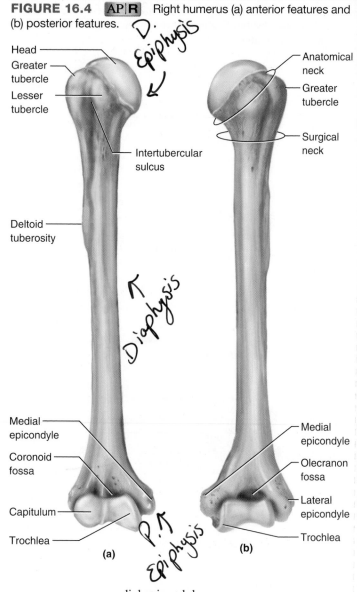

FIGURE 16.4 AP|R Right humerus (a) anterior features and (b) posterior features.

- medial epicondyle
- lateral epicondyle
- coronoid fossa—articulates with coronoid process of ulna
- olecranon fossa—articulates with olecranon process of ulna

radius—lateral bone of forearm
- head of radius—allows rotation at elbow
- radial tuberosity
- styloid process of radius
- ulnar notch of radius—articulation site with ulna

ulna—medial bone of forearm; longer than radius
- trochlear notch
- radial notch of ulna—articulation site with head of radius
- olecranon process
- coronoid process
- styloid process of ulna
- head of ulna—at distal end

FIGURE 16.5 AP|R   Anterior features of the right radius and ulna.

- Olecranon process
- Trochlear notch
- Coronoid process
- Radial notch of ulna
- Head of radius
- Radial tuberosity
- Radius
- Interosseous membrane
- Ulna
- Ulnar notch of radius
- Head of ulna
- Styloid process of ulna
- Styloid process of radius

carpal bones—positioned in two irregular rows
- proximal row (listed lateral to medial)
  - scaphoid   • triquetrum
  - lunate   • pisiform
- distal row (listed medial to lateral)
  - hamate   • trapezoid
  - capitate   • trapezium

The following mnemonic device will help you learn the eight carpals:

So Long Top Part
Here Comes The Thumb

The first letter of each word corresponds to the first letter of a carpal. This device arranges the carpals in order for the proximal, transverse row of four bones from lateral to medial, followed by the distal, transverse row from medial to lateral, which ends nearest the thumb. This arrangement assumes the hand is in the anatomical position.

metacarpals (I–V)

phalanges—located in digits (fingers)
- proximal phalanx
- middle phalanx—not present in first digit
- distal phalanx

2. Complete Parts C, D, and E of the laboratory assessment.

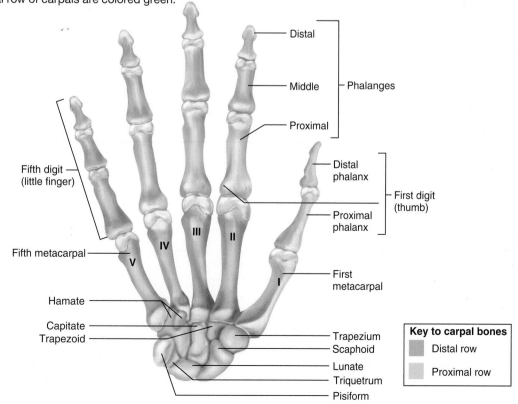

FIGURE 16.6 AP|R   Anterior (palmar) view of the bones of the right hand. The proximal row of carpals are colored yellow and the distal row of carpals are colored green.

- Distal
- Middle ⎱ Phalanges
- Proximal
- Distal phalanx
- Proximal phalanx ⎱ First digit (thumb)
- First metacarpal

- Fifth digit (little finger)
- Fifth metacarpal
- Hamate
- Capitate
- Trapezoid
- Trapezium
- Scaphoid
- Lunate
- Triquetrum
- Pisiform

V IV III II I

Key to carpal bones
- Distal row
- Proximal row

NOTES

ASSESSMENT
16

Name _____

Date _____

Section _____

The Ⓐ corresponds to the indicated Learning Outcome(s) Ⓞ found at the beginning of the Laboratory Exercise.

Pectoral Girdle and Upper Limb

PART A: Assessments

Complete the following statements:

1. The pectoral girdle is an incomplete ring because it is open in the back between the __Sternum__. Ⓐ1

2. The medial end of a clavicle articulates with the _____ of the sternum. Ⓐ1

3. The lateral end of a clavicle articulates with the _____ process of the scapula. Ⓐ1

4. The _____ is a bone that serves as a brace between the sternum and the scapula. Ⓐ1

5. The _____ divides the scapula into unequal portions. Ⓐ1

6. The lateral tip of the shoulder is the _____ of the scapula. Ⓐ1

7. Near the lateral end of the scapula, the _____ process of the scapula curves anteriorly and inferiorly from the clavicle. Ⓐ1

8. The glenoid cavity of the scapula articulates with the _____ of the humerus. Ⓐ1

PART B: Assessments

Label the structures indicated in figure 16.7.

FIGURE 16.7 AP|R Label the posterior surface of the right scapula, using the terms provided. Ⓐ1

1 — Acromion
2 — Coracoid Process
3 — Spine
4 — Infraspinous fossa
5 — Medial Border
6 — Inferior angle
7 — Glenoid Bone Cavity
8 — Lateral Border

Terms:
Acromion
Coracoid process
Glenoid cavity
Inferior angle
Infraspinous fossa
Lateral border
Medial border
Spine

PART C: Assessments

Match the bones in column A with the bones and features in column B. Place the letter of your choice in the space provided. A2

Column A	Column B
a. Carpals	_____ **1.** Capitate
b. Humerus	_____ **2.** Coronoid fossa
c. Metacarpals	_____ **3.** Deltoid tuberosity
d. Phalanges	_____ **4.** Greater tubercle
e. Radius	_____ **5.** Five palmar bones
f. Ulna	_____ **6.** Fourteen bones in digits
	_____ **7.** Intertubercular sulcus
	_____ **8.** Lunate
	_____ **9.** Olecranon fossa
	_____ **10.** Radial tuberosity
	_____ **11.** Trapezium
	_____ **12.** Trochlear notch

PART D: Assessments

Identify the bones and features indicated in the radiographs of figures 16.8, 16.9, and 16.10.

FIGURE 16.8 AP|R Identify the bones and features indicated on this radiograph of the right elbow (anterior view), using the terms provided. A2

1 _____ (bone)

2 _____

3 _____

4 _____

5 _____ (bone)

6 _____ (bone)

Terms:
Head of radius
Humerus
Medial epicondyle
Radius
Trochlea
Ulna

FIGURE 16.9 **AP|R** Identify the bones and features indicated on this radiograph of the anterior view of the right shoulder, using the terms provided. **A1** **A2**

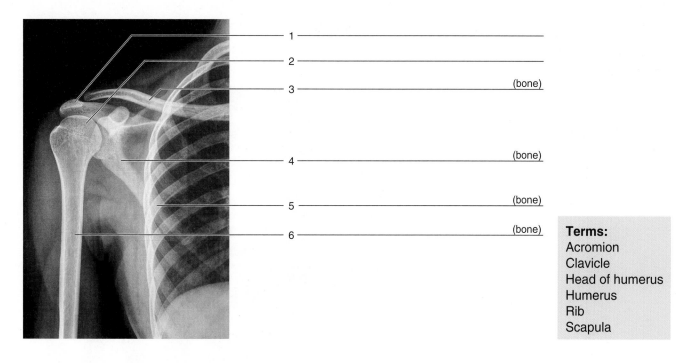

1 _____

2 _____

3 _____ (bone)

4 _____ (bone)

5 _____ (bone)

6 _____ (bone)

Terms:
Acromion
Clavicle
Head of humerus
Humerus
Rib
Scapula

FIGURE 16.10 **AP|R** Identify the bones indicated on this radiograph of the right hand (anterior view), using the terms provided. **A2**

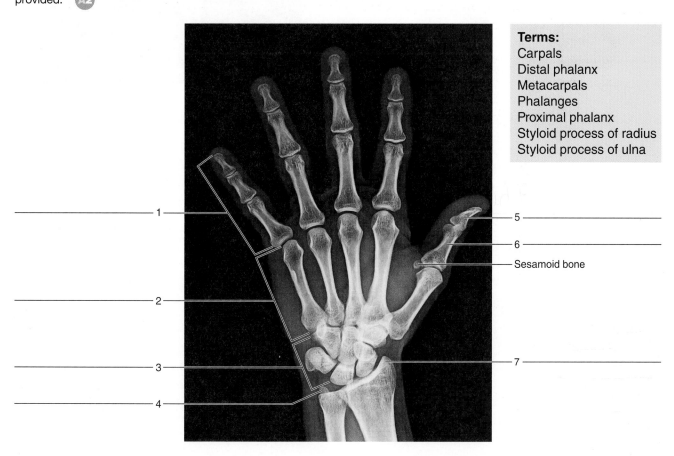

Terms:
Carpals
Distal phalanx
Metacarpals
Phalanges
Proximal phalanx
Styloid process of radius
Styloid process of ulna

1 _____

2 _____

3 _____

4 _____

5 _____

6 _____

Sesamoid bone

7 _____

PART E: Assessments

Identify the features of a humerus in figure 16.11 and the bones of the hand in figure 16.12.

FIGURE 16.11   Label the anterior features of a right humerus. (A2)

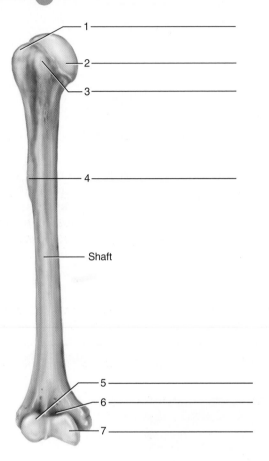

```
1 ——————————
2 ——————————
3 ——————————

4 ——————————


——— Shaft


5 ——————————
6 ——————————
7 ——————————
```

FIGURE 16.12   Complete the labeling of the bones numbered on this anterior view of the right hand by placing the correct numbers in the spaces provided. (A2)

_____	Capitate	___4___	Pisiform
_____	Distal phalanges	_____	Proximal phalanges
_____	Hamate	_____	Scaphoid
_____	Lunate	_____	Trapezium
_____	Metacarpals	_____	Trapezoid
_____	Middle phalanges	___3___	Triquetrum

LABORATORY
EXERCISE 17

Pelvic Girdle and Lower Limb 10/5/16

Purpose of the Exercise

To examine the bones of the pelvic girdle and lower limb, and to identify the major features of these bones.

MATERIALS NEEDED

Human skeleton, articulated
Human skeleton, disarticulated
Male and female pelves

Learning Outcomes AP|R

After completing this exercise, you should be able to:

O1 Locate and identify the bones of the pelvic girdle and their major features.

O2 Differentiate between a male and female pelvis.

O3 Locate and identify the bones of the lower limb and their major features.

The O corresponds to the assessments A indicated in the Laboratory Assessment for this Exercise.

Pre-Lab

Carefully read the introductory material and examine the entire lab. Be familiar with the pelvic girdle and the lower limb bones from lecture or the textbook. Answer the pre-lab questions.

Pre-Lab Questions Select the correct answer for each of the following questions:

1. The two hip bones articulate anteriorly at the
   a. acetabulum.   b. pubic arch.
   c. sacroiliac joint.   d. pubic symphysis.

2. Anatomically, *leg* refers to
   a. the lower limb.   b. hip to knee.
   c. knee to ankle.   d. hip to ankle.

3. The _____ is the largest portion of the hip bone.
   a. acetabulum   b. ilium
   c. ischium   d. pubis

4. The _____ is the lateral bone in the leg.
   a. tibia   b. fibula
   c. femur   d. patella

5. Which of the following bones is *not* a tarsal bone?
   a. metatarsal   b. talus
   c. calcaneus   d. cuboid

6. The ilium, ischium, and pubis are separate bones in a young child.
   a. True _____   b. False _____

7. Ischial spines, ischial tuberosities, and iliac crests are closer together in a pelvis of a female than in a pelvis of a male.
   a. True _____   b. False _____

8. Each digit of a foot has three phalanges.
   a. True _____   b. False _____

The pelvic girdle includes two hip bones that articulate with each other anteriorly at the pubic symphysis. Posteriorly, each hip bone articulates with a sacrum at a sacroiliac joint. Together, the pelvic girdle, sacrum, and coccyx constitute the pelvis. The pelvis, in turn, provides support for the trunk of the body and provides attachments for the lower limbs. The pelvis supports and protects the viscera in the pelvic region of the abdominopelvic cavity. The pelvic outlet, with boundaries that include the coccyx, inferior border of the pubic symphysis, and between the ischial tuberosities, is clinically important in females. The pelvic outlet must be large enough to successfully accommodate the fetal head during a vaginal delivery. Each acetabulum of a hip bone articulates with the head of the femur of a lower limb. The hip joint structures provide a more stable joint compared to a shoulder joint.

The bones of the lower limb form the framework of the thigh, leg, and foot. (Anatomically, *thigh* represents the region from hip to knee, *leg* is from knee to ankle, and *foot* includes the ankle to the end of the toes.) Each limb includes a femur in the thigh, a patella in the knee, a tibia and fibula in the leg, and seven tarsals, five metatarsals, and fourteen phalanges in the foot. These bones and large muscles provide weight-bearing support and locomotion and thus are considerably larger and possess more stable joints than those of an upper limb.

PROCEDURE A: Pelvic Girdle

The pelvic girdle supports the majority of the weight of the head, neck, and trunk. Each hip bone (coxal bone) originates in three separate ossification areas known as the ilium, ischium, and pubis. These are separate bones in a child but

fuse into an individual hip bone. All three parts of the hip bone fuse within the acetabulum, and the ischium and pubis also fuse along the inferior portion of the obturator foramen. The acetabulum is a well-formed deep socket that articulates with the head of the femur. The obturator foramen is the largest foramen in the skeleton; it serves as a passageway for blood vessels and nerves between the pelvic cavity and the thigh.

Several of the bone features (bone markings) of the pelvis can be palpated because they are located near the surface of the body. The median sacral crest is near the middle of the posterior surface of the pelvis, somewhat superior to the coccyx. The ilium is the largest portion of the hip bone, and its iliac crest can be palpated along the anterior and lateral portions. By following the anterior portion of the iliac crest, the anterior superior iliac spine can be felt very close to the surface.

1. Examine figures 17.1 and 17.2.
2. Observe the bones of the pelvic girdle and locate the following:

   hip bone (coxal bone; pelvic bone; innominate bone)
   - ilium
     - iliac crest
     - anterior superior iliac spine
     - anterior inferior iliac spine
     - posterior superior iliac spine
     - posterior inferior iliac spine
     - greater sciatic notch—portion in ilium
     - iliac fossa

FIGURE 17.1 AP|R   Bones of the pelvis (anterosuperior view).

Superior end of median sacral crest

Sacrum

Hip bone

Coccyx

Sacroiliac joint

Iliac crest

Iliac fossa

Sacrum

Pelvic inlet

Ischial spine

Pubic symphysis

Pubic arch

FIGURE 17.2 **AP|R** Right hip bone (a) lateral view (b) medial view. The three colors used enable the viewing of fusion locations of the ilium, ischium, and pubis of the adult skeleton.

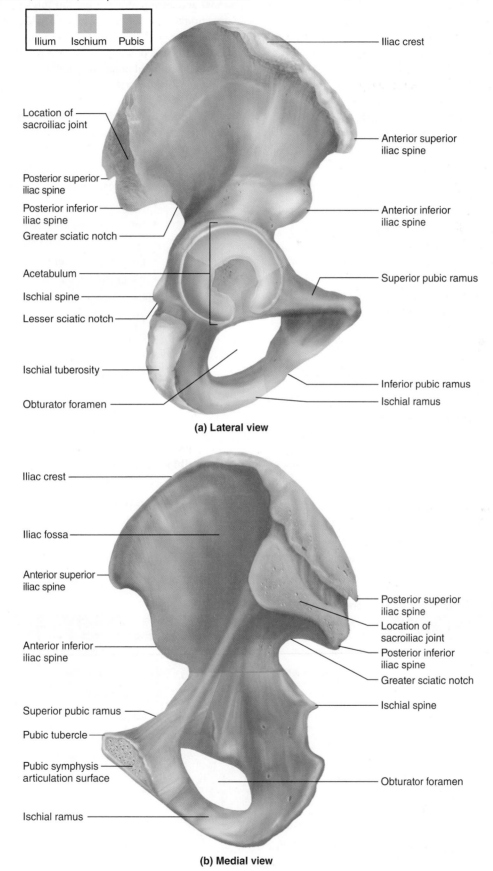

Ilium   Ischium   Pubis

(a) Lateral view

Iliac crest

Location of sacroiliac joint

Anterior superior iliac spine

Posterior superior iliac spine

Anterior inferior iliac spine

Posterior inferior iliac spine

Greater sciatic notch

Acetabulum

Superior pubic ramus

Ischial spine

Lesser sciatic notch

Ischial tuberosity

Obturator foramen

Inferior pubic ramus

Ischial ramus

(b) Medial view

Iliac crest

Iliac fossa

Anterior superior iliac spine

Posterior superior iliac spine

Location of sacroiliac joint

Anterior inferior iliac spine

Posterior inferior iliac spine

Greater sciatic notch

Superior pubic ramus

Ischial spine

Pubic tubercle

Pubic symphysis articulation surface

Obturator foramen

Ischial ramus

- ischium
  - ischial tuberosity—supports weight of body when seated
  - ischial spine
  - ischial ramus
  - lesser sciatic notch
  - greater sciatic notch—portion in ischium
- pubis
  - pubic symphysis—cartilaginous joint between pubic bones
  - pubic tubercle
  - superior pubic ramus
  - inferior pubic ramus
  - pubic arch—between pubic bones of pelvis
- acetabulum—formed by portions of ilium, ischium, and pubis
- obturator foramen—formed by portions of ischium and pubis

3. Observe the male pelvis and the female pelvis. Use table 17.1 as a guide as comparisons are made between the pelves of males and females.
4. Complete Part A of Laboratory Assessment 17.

PROCEDURE B: Lower Limb

The femur of the lower limb is the longest and strongest bone of the skeleton. The neck of the femur has somewhat of a lateral angle and is the weakest part of the bone and a common fracture site, especially if the person has some degree of osteoporosis. This fracture site is usually called a "broken hip"; however it is actually a broken femur in the region of the hip joint. Often this fracture is attributed to a fall, when many times the fracture occurs first, followed by the fall. The greater trochanter can be palpated along the proximal and lateral region of the thigh about one hand length below the iliac crest. The medial and lateral epicondyles can be palpated near the knee.

Large muscles are positioned on the anterior and posterior surfaces as well as the lateral and medial surfaces of the femur and the hip joint region. The thigh can be pulled in any direction depending upon which muscle group is contracting at that time. The large anterior muscles of the thigh (quadriceps femoris) possess a common patellar tendon with an enclosed sesamoid bone, the patella. The muscle attachment continues distally to the tibial tuberosity.

1. Examine figures 17.3, 17.4, and 17.5.
2. Observe the bones of the lower limb and locate each of the following:
   femur
   - proximal features
     - head
     - fovea capitis
     - neck
     - greater trochanter
     - lesser trochanter
   - shaft
     - gluteal tuberosity
     - linea aspera
   - distal features
     - lateral epicondyle
     - medial epicondyle
     - lateral condyle
     - medial condyle
   patella
   tibia—medial bone of leg
   - medial condyle
   - lateral condyle
   - tibial tuberosity
   - anterior border (crest; margin)
   - medial malleolus

Table 17.1 Differences Between Male and Female Pelves

Structure of Comparison	Male Pelvis AP|R	Female Pelvis AP|R
General structure	Heavier, thicker bones and processes	Lighter, thinner bones and processes
Sacrum	Narrower and longer	Wider and shorter
Coccyx	Less movable	More movable
Pelvic outlet	Smaller	Larger
Greater sciatic notch	Narrower	Wider
Obturator foramen	Round	Triangular to oval
Acetabula	Larger; closer together	Smaller; farther apart
Pubic arch	Usually 90° or less; more V-shaped	Usually greater than 90°
Ischial spines	Longer; closer together	Shorter; farther apart
Ischial tuberosities	Rougher; closer together	Smoother; farther apart
Iliac crests	Less flared; closer together	More flared; farther apart

FIGURE 17.3 **AP|R** The features of (a) the anterior surface and (b) the posterior surface of the right femur. The anterior and posterior views of the patella are included in the figure.

- Fovea capitis
- Head
- Neck
- Greater trochanter
- Lesser trochanter
- Gluteal tuberosity
- Linea aspera
- Shaft
- Medial epicondyle
- Lateral epicondyle
- Medial condyle
- Lateral condyle
- Patella
- Patella

(a) (b)

FIGURE 17.4 **AP|R** Features of the right tibia and fibula (anterior view).

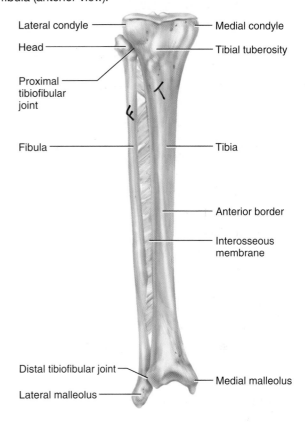

- Lateral condyle
- Head
- Proximal tibiofibular joint
- Fibula
- Medial condyle
- Tibial tuberosity
- Tibia
- Anterior border
- Interosseous membrane
- Distal tibiofibular joint
- Lateral malleolus
- Medial malleolus

FIGURE 17.5 **AP|R** Superior surface view of right foot. The proximal group of tarsals are colored yellow and the distal group of tarsals are colored green. 14 total

- Phalanges
  - Distal phalanx
  - Middle phalanx
  - Proximal phalanx
- Metatarsals
- Tarsals
  - Medial cuneiform
  - Intermediate cuneiform
  - Lateral cuneiform
  - Cuboid
  - Navicular
  - Talus (including articular surface for tibia)
  - Calcaneus

Key to tarsal bones
- Distal group
- Proximal group

fibula—lateral bone of leg
- ■ head
- ■ lateral malleolus

tarsal bones
- ■ talus
- ■ calcaneus
- ■ navicular
- ■ cuboid
- ■ lateral cuneiform
- ■ intermediate (middle) cuneiform
- ■ medial cuneiform

metatarsal bones

phalanges
- ■ proximal phalanx
- ■ middle phalanx—absent in first digit
- ■ distal phalanx

3. Complete Parts B, C, and D of the laboratory assessment.

CRITICAL THINKING ACTIVITY

Compare the size and depth of an acetabulum of a hip bone to the glenoid cavity of the scapula._____

_____

_____

_____

_____

How do these socket differences between the hip joint and the shoulder joint relate to strength and mobility?

_____

_____

_____

_____

_____

CRITICAL THINKING ACTIVITY

As a review of the entire skeleton, use the disarticulated skeleton and arrange all of the bones in relative position to re-create their normal positions. Work with a partner or in a small group and place the bones on the surface of a laboratory table. Check your results with the articulated skeletons.

Name _____

Date _10/5/16_____

Section _____

The Ⓐ corresponds to the indicated Learning Outcome(s) Ⓞ
found at the beginning of the Laboratory Exercise.

LABORATORY
ASSESSMENT
17

Pelvic Girdle and Lower Limb

PART A: Assessments

Complete the following statements:

1. The pelvic girdle consists of two _hip bones_ . Ⓐ1

2. The head of the femur articulates with the _acetabulum_ of the hip bone. Ⓐ1

3. The _ilium_ is the largest portion of the hip bone. Ⓐ1

4. The distance between the _____ represents the shortest diameter of the pelvic outlet. Ⓐ1

5. The pubic bones come together anteriorly to form a cartilaginous joint called the _~~sacroiliac joint~~_ . Ⓐ1
   pubic symphisis

6. The _iliac crest_ is the superior margin of the ilium that causes the prominence of the hip. Ⓐ1

7. When a person sits, the _tuberosity_ of the ischium supports the weight of the body. Ⓐ1

8. The angle formed by the pubic bones below the pubic symphysis is called the _pubic arch_ . Ⓐ1

9. The _obturator foramen_ is the largest foramen in the skeleton. Ⓐ1

10. The ilium joins the sacrum at the _sacroiliac_ joint. Ⓐ1

CRITICAL THINKING ASSESSMENT

Examine the male and female pelves. Look for major differences between them. Note especially the flare of the iliac bones, the angle of the pubic arch, the distance between the ischial spines and ischial tuberosities, and the curve and width of the sacrum. In what ways are the differences you observed related to the function of the female pelvis as a birth canal? Ⓐ2

_____

_____

_____

PART B: Assessments

Match the bones in column A with the features in column B. Place the letter of your choice in the space provided. Ⓐ3

Column A		Column B	
a. Femur	_e._ 1. Middle phalanx	_g._ 7. Tibial tuberosity
b. Fibula	_a._ 2. Lesser trochanter	_f._ 8. Talus
c. Metatarsals	_g._ 3. Medial malleolus	_a._ 9. Linea aspera
d. Patella	_a._ 4. Fovea capitis	_b._ 10. Lateral malleolus
e. Phalanges	_f._ 5. Calcaneus	_d._ 11. Sesamoid bone
f. Tarsals	_f._ 6. Lateral cuneiform	_c._ 12. Five bones that form the instep
g. Tibia		

PART C: Assessments

Identify the bones and features indicated in the radiographs of figures 17.6, 17.7, and 17.8.

FIGURE 17.6 AP|R Identify the bones and features indicated on this radiograph of the anterior view of the pelvic region, using the terms provided. A1 A3

Ilium _____ 1

Sacrum _____ 2

Pubic Symphysis _____ 3

Obturator Foramen _____ 4

5 _____ H.O.F.
6 _____ Pubis
7 _____ Ischium

Terms:
Head of femur ✓
Ilium ✓
Ischium ✓
Obturator foramen ✓
Pubic symphysis ✓
Pubis
Sacrum ✓

FIGURE 17.7 AP|R Identify the bones and features indicated in this radiograph of the right knee (anterior view), using the terms provided. A3

4 _____ Femur (bone)
5 _____ Patella (bone)
6 _____ Lateral condyle

medial Epicondyle _____ 1

H. O. Fibula _____ 2

7 _____ Tibia (bone)

fibula (bone) _____ 3

Terms:
Femur ✓
Fibula ✓
Head of fibula ✓
Lateral condyle ✓
Medial epicondyle ✓
Patella ✓
Tibia ✓

FIGURE 17.8 AP|R Identify the bones indicated in this radiograph of the right foot (medial view), using the terms provided. A3

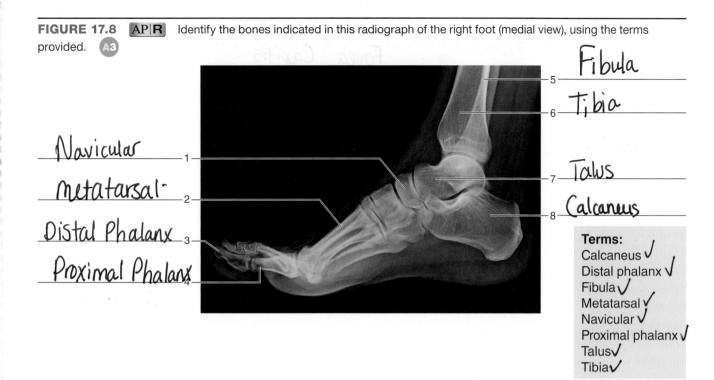

Navicular — 1
metatarsal — 2
Distal Phalanx — 3
Proximal Phalanx — 4

5 — Fibula
6 — Tibia
7 — Talus
8 — Calcaneus

Terms:
Calcaneus ✓
Distal phalanx ✓
Fibula ✓
Metatarsal ✓
Navicular ✓
Proximal phalanx ✓
Talus ✓
Tibia ✓

PART D: Assessments

Identify the bones of the foot in figure 17.9 and the features of a femur in figure 17.10.

FIGURE 17.9 Identify the bones indicated on this superior view of the right foot, using the terms provided. A3

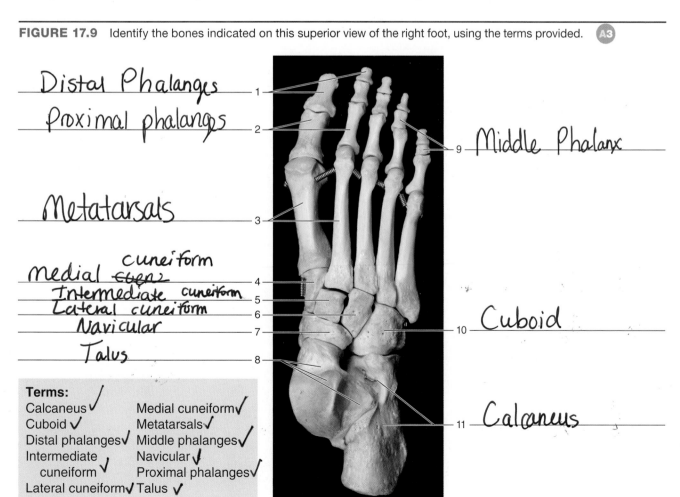

Distal Phalanges — 1
Proximal phalanges — 2

9 — Middle Phalanx

Metatarsals — 3

medial cuneiform — 4
Intermediate cuneiform — 5
Lateral cuneiform — 6
Navicular — 7
Talus — 8

10 — Cuboid

11 — Calcaneus

Terms:
Calcaneus ✓   Medial cuneiform ✓
Cuboid ✓   Metatarsals ✓
Distal phalanges ✓   Middle phalanges ✓
Intermediate   Navicular ✓
  cuneiform ✓   Proximal phalanges ✓
Lateral cuneiform ✓ Talus ✓

173

FIGURE 17.10 Label the anterior features of a right femur. A3

1 Fovea Capitis
2 Head
3 Neck
4 Greater Trochanter
5 Lesser trochanter

6 Shaft

7 Medial Epicondyle

LABORATORY
EXERCISE 18

Fetal Skeleton

Purpose of the Exercise

To examine and make comparisons between fetal skeletons and adult skeletons.

MATERIALS NEEDED

Human skull

Human adult skeleton

Human fetal skeleton

Meterstick

Metric ruler

Note: If a fetal skeleton is not available, use the figures represented in this laboratory exercise with the metersticks included in the photographs.

Learning Outcomes

After completing this exercise, you should be able to:

O1 Locate and describe the function of the fontanels of a fetal skull.

O2 Measure and compare the size of the cranial bones and the facial bones between a fetal skull and the adult skull.

O3 Compare the fetal frontal and temporal bones to adult examples.

O4 Measure and compare the total height and the upper and lower limb lengths for a fetal skeleton and an adult skeleton.

O5 Locate and describe six fetal bones or body regions that are not completely ossified but become single bones or groups of bones in the adult skeleton.

O6 Estimate the gestational age of a fetus from measured lengths.

O7 Distinguish intramembranous ossification bones from endochondral ossification bones.

The **O** corresponds to the assessments **A** indicated in the Laboratory Assessment for this Exercise.

Pre-Lab

Carefully read the introductory material and examine the entire lab. Be familiar with bone development and a fetal skull from lecture or the textbook. Answer the pre-lab questions.

Pre-Lab Questions Select the correct answer for each of the following questions:

1. Development of the embryo and fetus progresses from
   a. medial toward lateral.
   b. cephalic toward caudal.
   c. anterior toward posterior.
   d. inferior toward superior.

2. The last fontanel to close is the
   a. posterior.
   b. sphenoidal.
   c. mastoid.
   d. anterior.

3. Fontanels function for all of the following *except*
   a. formation of the vertebrae.
   b. compression of the cranium during vaginal delivery.
   c. additional cranial skull growth after birth.
   d. additional brain development after birth.

4. Measurements of height after 20 weeks of development are from
   a. crown-to-heel.
   b. crown-to-rump.
   c. head-to-toes.
   d. sitting height.

5. The anterior fontanel involves all of the following bones *except* the
   a. left parietal bone.
   b. right parietal bone.
   c. occipital bone.
   d. frontal bone.

175

A human adult skeleton consists of 206 bones. However, the skeleton of a newborn child may have nearly 275 bones due to the numerous *ossification centers* and *epiphyseal plates* present for many bones of the appendicular skeleton. Development of the embryo and the fetus progresses from cephalic toward caudal regions of the body. As a result, there is a noticeable difference between the proportions of the head, the entire body, and the upper and lower limbs between a fetus and an adult. Additionally, the bones most important for protection are among the early ossification sites.

The ossification process occurs during fetal development and continues through childhood growth. *Intramembranous ossification* forms the flat bones of the cranium, some facial bones, and portions of the mandible and clavicles. *Endochondral ossification* from hyaline cartilage forms most of the bones of the skeleton.

PROCEDURE A: Skulls

1. Examine the fetal skull from three views (figs. 18.1, 18.2, and 18.3). Areas of incomplete ossification of the flat bones of an infant's skull are called *fontanels* (fig. 18.1). Sometimes called "soft spots," fontanels allow for compression of the cranium during a vaginal delivery and for some additional brain development after birth. The posterior and lateral fontanels close during the first year after birth, but the anterior fontanel does not close until nearly two years of age. Locate the following fontanels:

   Mastoid fontanel (2)

   Sphenoidal fontanel (2)

   Posterior fontanel (1)

   Anterior fontanel (1)

2. Complete Part A of Laboratory Assessment 18.
3. Examine the anterior views of a fetal skull next to an adult skull (fig. 18.4). Use a meterstick and measure the height of the entire cranium (braincase) and the height of the face of the fetal skull (fig. 18.4a). The brain, which forms in early development, occupies the large cranial cavity. This gives a perspective of a large head. However, the small face is related to undersized maxillae, maxillary sinuses, mandible, lack of teeth eruptions, and nasal cavities. The facial portion of a fetal skull at birth represents about one-eighth of the entire skull,

FIGURE 18.1   Superior view of the fetal skull.

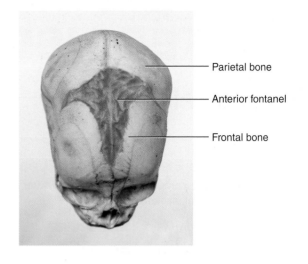

— Parietal bone
— Anterior fontanel
— Frontal bone

FIGURE 18.2   Lateral view of the fetal skull.

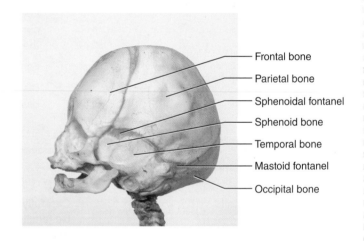

— Frontal bone
— Parietal bone
— Sphenoidal fontanel
— Sphenoid bone
— Temporal bone
— Mastoid fontanel
— Occipital bone

FIGURE 18.3   Posterior view of fetal skull.

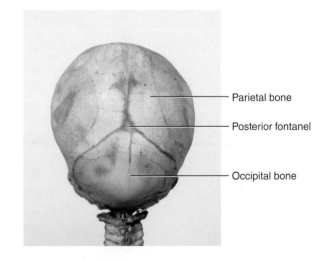

— Parietal bone
— Posterior fontanel
— Occipital bone

FIGURE 18.4 Anterior views of skulls next to metric scales: (a) fetal skull; (b) adult skull. The dotted lines indicate the division between the cranium and the face.

(a)

(b)

FIGURE 18.5 Anterior view of a fetal skeleton next to a metric scale. To estimate the gestational age of this fetus, measure the total height (crown-to-heel).

whereas the adult facial portion represents nearly one-half of the complete skull.

4. As growth of the fetus advances, the relative size of the cranium becomes smaller as the size of the face increases. Use a meterstick and measure the height of the cranium and face of an adult skull (fig. 18.4b).

5. Record your measurements in Part B of the laboratory assessment.

6. Make careful observations of the ossification of the frontal bone and the temporal bone.

7. Complete Part B of the laboratory assessment.

Table 18.1 Embryonic and Fetal Height (Length)

Gestational Age (Months)	Sitting Height (crown-to-rump)*	Total Height (crown-to-heel)*
1	0.6 cm	
2	3 cm	
3	7–9 cm	9–10 cm
4	12–14 cm	16–20 cm
5	18–19 cm	25–27 cm
6		30–32 cm
7		35–37 cm
8		40–45 cm
9		50–53 cm

*Measurements of height until about 20 weeks are more often from crown-to-rump; measurements after 20 weeks are more often from crown-to-heel.

PROCEDURE B: Entire Skeleton

1. Observe and then measure the total height (crown-to-heel) of the entire fetal skeleton (fig. 18.5) and an adult skeleton. Compare the height of the fetal skeleton with the data in table 18.1. Observe and then measure the total height (crown-to-heel) and the sitting height (crown-to-rump) of the human fetus shown in figure 18.6. Figure 18.7 illustrates the proportionate changes of bodies during development from an embryo to the adult. Record the measurements in Part C of the laboratory assessment.

2. Much of the development of the upper limb occurs during fetal development, especially during the second trimester. The rapid development of the lower limb is much greater after birth (figs. 18.5 and 18.7).

3. Observe and then measure the total length of the upper limb and the lower limb of the fetus (fig. 18.5). Record the measurements in Part C of the laboratory assessment.

4. Observe the adult skeleton and measure the total length of the upper limb and the lower limb. Record the measurements in Part C of the laboratory assessment.

5. Observe very carefully a hip bone, sternum, and sacrum of a fetal skeleton. Record your observations in Part C of the laboratory assessment.

FIGURE 18.6 Human fetus next to a metric scale. To estimate the gestational age of this young fetus, measure the total height (crown-to-heel) and the sitting height (crown-to-rump).

6. Make careful observations of the anterior knee, thoracic cage, wrist, and ankles of the fetal skeleton (fig. 18.5). Record your observations in Part C of the laboratory assessment.

7. Complete Parts C and D of the laboratory assessment.

FIGURE 18.7 During development, proportions of the head, entire body, and upper and lower limbs change considerably.

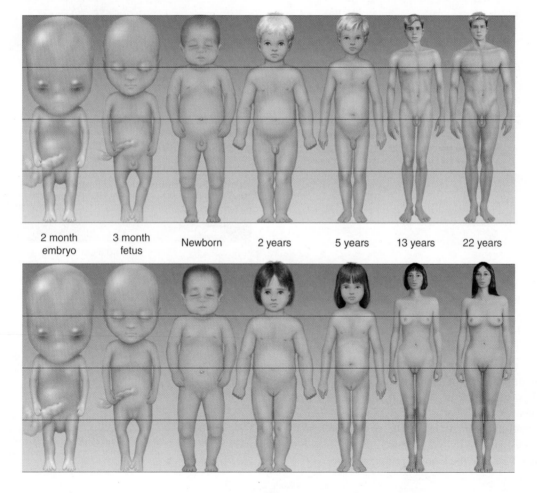

LABORATORY ASSESSMENT 18

Name _____

Date _____

Section _____

The (A) corresponds to the indicated Learning Outcome(s) (O) found at the beginning of the Laboratory Exercise.

Fetal Skeleton

PART A: Assessments

Complete the following:

1. List all of the bones that form part of the border with the mastoid fontanel. **A1**

_____

2. List all of the bones that form part of the border with the sphenoidal fontanel. **A1**

_____

3. List all of the bones that form part of the border with the posterior fontanel. **A1**

_____

4. List all of the bones that form part of the border with the anterior fontanel. **A1**

_____

5. Describe the function of fontanels. **A1**

_____

6. What is the final outcome of the fontanels? **A1**

_____

PART B: Assessments

1. Enter the measurements of the fetal and adult skulls in the table.

Skull	Total Height of Skull (cm)	Cranial Height (cm)	Facial Height (cm)
Fetus			
Adult			

2. Compare the frontal bone of the fetal skull with that of the adult skull. Describe some unusual aspects of the fetal frontal bone as compared to the adult structure.

_____

_____

_____

3. Compare the temporal bone of the fetal skull with that of the adult skull. Describe some unusual aspects of the fetal temporal bone as compared to the adult structure. (A3)

_____

_____

_____

4. Summarize the measured and observed relationships between a fetal skull and an adult skull. (A2)

_____

_____

_____

_____

CRITICAL THINKING ASSESSMENT

Why can it be difficult to match the identifications of baby pictures with those of adults? (A2)

_____

_____

_____

_____

PART C: Assessments

1. Enter the measurements of the fetal and adult skeletons in the table. (A4)

Skeleton	Total Height (cm)	Upper Limb Length (cm)	Lower Limb Length (cm)
Fetus			
Adult			

2. Summarize the relationship of total body height to limb lengths for the fetal skeleton and the adult skeleton. (A4)

_____

_____

_____

_____

3. Locate and describe the discernible ossification of the following fetal bones and body regions and compare them with the adult skeleton. (A5)

   a. Hip bone:

_____

_____

b. Sternum:

_____

_____

c. Sacrum:

_____

_____

d. Anterior knee:

_____

_____

e. Thoracic cage:

_____

_____

f. Wrist (carpals) and ankle (tarsals):

_____

_____

4. Estimate the age of fetal development of the fetal skeleton in figure 18.5 and the one that is represented at your school. Estimate the age of the human fetus shown in figure 18.6. Use the data provided in table 18.1 for your determination of the estimates. A6

    a. Fetal skeleton in figure 18.5: _____

    b. Fetal skeleton at your school: _____

    c. Fetus in figure 18.6: _____

PART D: Assessments

Identify the fetal skeleton structures in figures 18.8 and 18.9.

FIGURE 18.8 Label the fetal bones and fontanels by placing the correct numbers in the spaces provided.

____ Anterior fontanel
____ Frontal bone
____ Mandible
____ Mastoid fontanel
____ Parietal bone
____ Sphenoidal fontanel
____ Temporal bone

FIGURE 18.9 Distinguish the type of ossification formation by selecting the correct choice from the terms provided.

1 ——————————————————

2 ——————————————————

3 ——————————————————

4 ——————————————————

5 ——————————————————

Terms:
Endochondral ossification
Intramembranous ossification

Joint Structure and Movements

Purpose of the Exercise

To examine examples of the three types of structural and functional joints while exploring the major features of each joint. The types of movements produced at synovial joints will also be demonstrated.

MATERIALS NEEDED

Human skull

Human skeleton, articulated

Models of synovial joints (shoulder, elbow, hip, and knee)

For Demonstration Activities:

Fresh animal joint (knee joint preferred)

Radiographs of major joints

SAFETY ⚠

▶ Wear disposable gloves when handling the fresh animal joint.

▶ Wash your hands before leaving the laboratory.

Learning Outcomes AP|R

After completing this exercise, you should be able to:

01 Identify key features of fibrous, cartilaginous, and synovial joints.

02 Distinguish functional characteristics of synarthroses, amphiarthroses, and diarthroses.

03 Locate examples of each type of structural and functional joint.

04 Examine the structure and types of movements of the shoulder, elbow, hip, and knee joints.

05 Identify the types of movements that occur at synovial joints.

The ⓞ corresponds to the assessments Ⓐ indicated in the Laboratory Assessment for this Exercise.

Pre-Lab

Carefully read the introductory material and examine the entire lab. Be familiar with joint structures and movements from lecture or the textbook. Answer the pre-lab questions.

Pre-Lab Questions Select the correct answer for each of the following questions:

1. Fibrous joints are structural types containing
   - a. cartilage.
   - b. dense fibrous connective tissue.
   - c. synovial membranes.
   - d. synovial fluid.

2. Which synovial joint allows movements in all planes?
   - a. saddle
   - b. pivot
   - c. ball-and-socket
   - d. plane

3. Synarthosis pertains to functional joints that are
   - a. immovable.
   - b. slightly movable.
   - c. freely movable.
   - d. synovial types.

4. Movement away from the midline is called
   - a. rotation.
   - b. adduction.
   - c. extension.
   - d. abduction.

5. Movements that combine flexion, abduction, extension, and adduction are called
   - a. rotations.
   - b. excursions.
   - c. circumductions.
   - d. hyperextensions.

6. The synovial membrane and the fibrous capsule combine and form the
   - a. articular cartilage.
   - b. periosteum.
   - c. spongy bone.
   - d. joint (articular) capsule.

7. Rotation occurs at a pivot type of synovial joint.

    a. True ✓   **b.** False _____

8. Dorsiflexion and plantar flexion are characteristic movements of the hand.

    a. True _____   **b.** False ✓

                              Foot (plantar)

Joints are junctions between bones. Although they vary considerably in structure, they can be classified according to the type of tissue that binds the bones together. Thus, the three groups of structural joints can be identified as *fibrous joints, cartilaginous joints,* and *synovial joints.* Fibrous joints are filled with dense fibrous connective tissue, cartilaginous joints are filled with a type of cartilage, and synovial joints contain synovial fluid inside a joint cavity.

Joints can also be classified by the degree of functional movement allowed: *synarthroses* are immovable, *amphiarthroses* allow slight movement, and *diarthroses* allow free movement. Movements occurring at freely movable synovial joints are due to the contractions of skeletal muscles. In each case, the type of movement depends on the type of joint involved and the way in which the muscles are attached to the bones on either side of the joint.

PROCEDURE A: Types of Joints

Joints form where bones come together, or articulate with each other. Because there are variations in the types of connective tissues involved in the joint structure and variations in the degree of movement possible, two types of classifications are used. Structurally joints are fibrous, cartilaginous, or synovial; functionally joints are synarthrosis, amphiarthrosis, or diarthrosis. The joints are areas where bones are held together but accommodate various degrees of body movements. Some joints are only temporary; when complete ossification occurs, the entire joint becomes bone. Examples of these temporary joints include a right and left frontal bone becoming a single frontal bone, and the eventual complete ossification of the epiphyseal plates in the growth areas of the long bones.

1. Use table 19.1 as a reference and examine the human skull and articulated skeleton. Locate examples of the following types of structural and functional joints:

    fibrous joints

      ■ suture—synarthrosis

      ■ gomphosis—synarthrosis

      ■ syndesmosis—amphiarthrosis

Table 19.1 Classification of Joints

Structural Classification	Structural Features	Functional Classification and Movements	Location Examples
FIBROUS			
Suture	Dense fibrous connective tissue	Synarthrosis—immovable	Sutures of skull
Gomphosis	Periodontal ligament	Synarthrosis—immovable	Tooth sockets
Syndesmosis	Dense fibrous connective tissue	Amphiarthrosis—slightly movable	Tibiofibular joint
CARTILAGINOUS			
Synchondrosis	Hyaline cartilage	Synarthrosis—immovable	Epiphyseal plates
Symphysis	Fibrocartilage	Amphiarthrosis—slightly movable	Pubic symphysis; intervertebral discs
SYNOVIAL			
Plane (gliding)	All synovial joints contain articular cartilage, a synovial membrane, and a joint cavity filled with synovial fluid	Diarthrosis—uniaxial; sliding or twisting	Intercarpal; intertarsal; knee (between femur and patella)
Hinge		Diarthrosis—uniaxial; flexion and extension	Elbow; knee; interphalangeal
Pivot		Diarthrosis—uniaxial; rotation	Radioulnar at elbow; dens at atlas
Condylar (ellipsoid)		Diarthrosis—biaxial; flexion, extension, abduction, and adduction	Radiocarpal; metacarpophalangeal; knee (between femur and tibia when flexed)
Saddle		Diarthrosis—biaxial; variety of movements mainly in two planes	Base of thumb with trapezium
Ball-and-socket		Diarthrosis—multiaxial; movements in all planes	Hip; shoulder

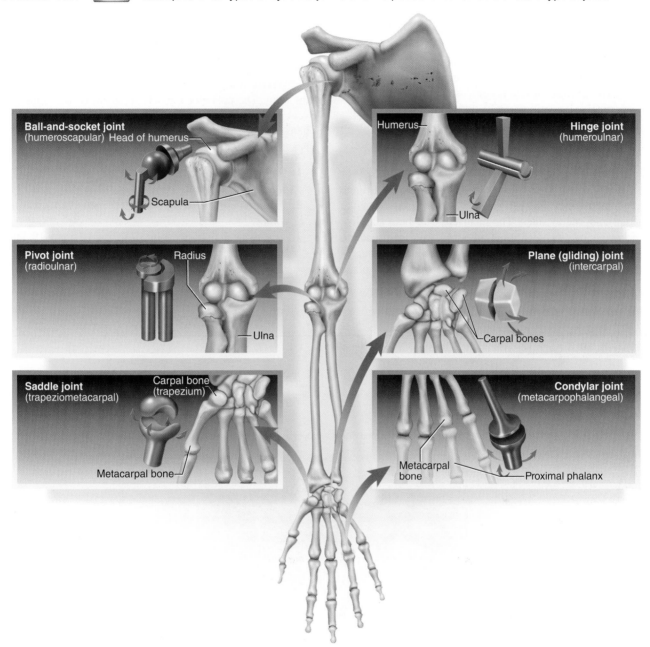

Ball-and-socket joint
(humeroscapular) Head of humerus
Scapula

Pivot joint
(radioulnar)
Radius
Ulna

Saddle joint
(trapeziometacarpal)
Carpal bone
(trapezium)
Metacarpal bone

Humerus
Hinge joint
(humeroulnar)
Ulna

Plane (gliding) joint
(intercarpal)
Carpal bones

Condylar joint
(metacarpophalangeal)
Metacarpal
bone
Proximal phalanx

cartilaginous joints

- synchondrosis—synarthrosis
- symphysis—amphiarthrosis

synovial joints—diarthrosis

2. Complete Part A of Laboratory Assessment 19.
3. Synovial, diarthrotic joints allow various degrees of freedom of movement. *Uniaxial* joints move in a single plane or axis; *biaxial* joints move in two planes or axes; *multiaxial* joints move in multiple planes or axes.

4. Using table 19.1 and figure 19.1 as references, locate examples of the following types of synovial joints in the skeleton. At the same time, examine the corresponding joints in the models and in your own body. Experiment with each joint to demonstrate the range of motion.

plane (gliding) joint	**hinge joint**
pivot joint	**condylar (ellipsoid) joint**
saddle joint	**ball-and-socket joint**

5. Complete Parts B and C of the laboratory assessment.

PROCEDURE B: Examples of Synovial Joints

Synovial joints have some characteristic structural features and functional characteristics. The structure of the synovial joint includes a joint (articular) capsule composed of two layers: an outer fibrous capsule and an inner synovial membrane. The fibrous capsule is composed of a dense connective tissue that is continuous with the periosteum of the bone, while the synovial membrane secretes a lubricating synovial fluid that fills the enclosed joint cavity. Articular (hyaline) cartilage covers the surface of the bone within the joint. Various ligaments and tendons also cross synovial joints.

In the knee, two crescent-shaped pieces of cartilage called menisci extend partway across the joint from the medial and lateral sides. The meniscus will help stabilize the knee joint, where a great amount of pressure occurs, and it also helps to absorb shock. Although the knee is basically a hinge joint, there are some condylar and plane components of the joint. The condylar component of the joint allows some limited rotation between the articulations of the rounded condyles of the femur and the tibia when the knee is partly in the flexed position. The plane joint between the patellar surface of the femur and the patella allows some sliding movements. The knee joint also consists of two cruciate ligaments that cross each other deep within the joint. The posterior cruciate ligament (PCL) helps prevent excessive flexion; the anterior cruciate ligament (ACL) helps prevent hyperextension of the knee joint. A torn ACL is a common type of knee injury.

1. Study the major features of a synovial joint in figure 19.2.
2. Examine models of the shoulder, elbow, hip, and knee joints.
3. Locate the major knee joint structures of figure 19.3 that are visible on the knee joint model.
4. Complete Part D of the laboratory assessment.

FIGURE 19.2 AP|R Basic structure of a synovial joint.

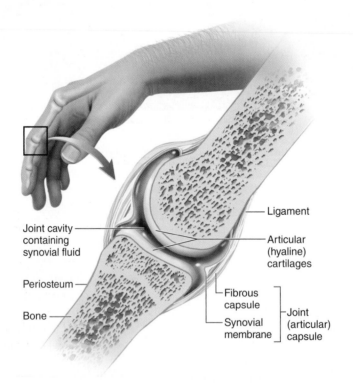

Joint cavity containing synovial fluid

Periosteum

Bone

Ligament

Articular (hyaline) cartilages

Fibrous capsule

Synovial membrane

Joint (articular) capsule

(a)

(c)

Femur

Patellar surface of femur

Medial condyle of femur

Joint cavity

Medial meniscus

Medial condyle of tibia

Tibial tuberosity

Patella (posterior surface)

Lateral condyle of femur

Anterior cruciate ligament

Lateral meniscus

Lateral chondyle of tibia

Patellar ligament (reflected)

Quadriceps tendon (reflected)

(b)

DEMONSTRATION ACTIVITY

Examine a longitudinal section of a fresh synovial animal knee joint. Locate the dense connective tissue that forms the fibrous capsule and the hyaline cartilage that forms the articular cartilage on the ends of the bones. Locate the synovial membrane on the inner surface of the joint capsule. Does the joint have any semilunar cartilages (menisci)? _____

PROCEDURE C: Joint Movements

1. Examine figures 19.4 and 19.5.
2. Most joints and body parts are extended and adducted when the body is positioned in anatomical position. Skeletal muscle action involves the more movable end (*insertion*) being pulled toward the more stationary end (*origin*). In the limbs, the origin is usually proximal to the insertion; in the trunk, the origin is usually medial to the insertion. Use these concepts as reference points as you move joints. Move various parts of your body to demonstrate the synovial joint movements described in tables 19.2 and 19.3.
3. Complete Part E of the laboratory assessment.

DEMONSTRATION ACTIVITY

Study the available radiographs of joints by holding the films in front of a light source. Identify the type of joint and the bones incorporated in the joint. Also identify other major visible features.

FIGURE 19.4 APR Examples of angular, rotational, and gliding movements of synovial joints (a–h).

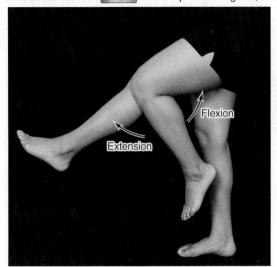

(a) Flexion and extension of the knee joint

(b) Flexion, extension, and hyper-extension of the wrist joint

(c) Flexion, extension, and hyperextension* of the shoulder joint (*Hyperextension of the shoulder and hip joints is movement past anatomical position.)

(d) Lateral flexion of the trunk (vertebral column)

(e) Abduction and adduction of the hip joint

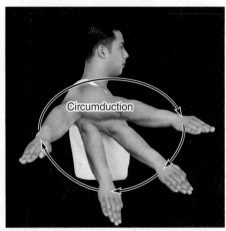

(f) Circumduction of the shoulder joint

(g) Medial and lateral rotation of the arm at the shoulder joint

(h) Gliding movements among the carpals of the wrist

FIGURE 19.5 AP|R Special movements of synovial joints (a–g).

(a) Supination and pronation of the forearm and hand involving movement at the radioulnar joint

(b) Elevation and depression of the shoulder (scapula)

(c) Dorsiflexion and plantar flexion of the foot at the ankle joint

(d) Inversion and eversion of the right foot at the ankle joint (The left foot is unchanged for comparison.)

(e) Protraction and retraction of the head

(f) Lateral and medial excursion of the mandible

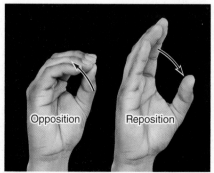

(g) Opposition and reposition of the thumb at the carpometacarpal (saddle) joint

Table 19.2 Angular, Rotational, and Gliding Movements

Movement	Description
Flexion	Decrease of an angle (usually in the sagittal plane)
Lateral flexion	Bending the trunk (vertebral column) to the side
Extension	Increase of an angle (usually in the sagittal plane)
Hyperextension	Extension beyond anatomical position
Abduction	Movement of a body part away from the midline (usually in the frontal plane)
Adduction	Movement of a body part toward the midline (usually in the frontal plane)
Circumduction	Circular movement (combines flexion, abduction, extension, and adduction)
Rotation	Movement of part around its long axis
Medial (internal)	Inward rotation
Lateral (external)	Outward rotation
Gliding	Back-and-forth and side-to-side sliding movements of plane joints

Table 19.3 Special Movements (pertain to specific joints)

Movement	Description
Supination	Rotation of forearm and palm of hand anteriorly or upward
Pronation	Rotation of forearm and palm of hand posteriorly or downward
Elevation	Movement of body part upward
Depression	Movement of body part downward
Dorsiflexion	Movement of ankle joint so dorsum (superior) of foot becomes closer to anterior surface of leg (as standing on heels)
Plantar flexion	Movement of ankle joint so the plantar surface of foot becomes closer to the posterior surface of leg (as standing on toes)
Inversion (called supination in some health professions)	Medial movement of sole of foot at ankle joint
Eversion (called pronation in some health professions)	Lateral movement of sole of foot at ankle joint
Protraction	Anterior movement in the transverse plane
Retraction	Posterior movement in the transverse plane

Excursion
   Lateral
   Medial
	Special movements of mandible when grinding food
	Movement of mandible laterally
	Movement of mandible medially

Opposition	Movement of the thumb to touch another finger of the same hand
Reposition	Return of thumb to anatomical position

CRITICAL THINKING ACTIVITY

Have your laboratory partner do some of the preceding movements and see if you can correctly identify the movements made.

CRITICAL THINKING ACTIVITY

Describe a body position that can exist when all major body parts are flexed._____

_____

_____

_____

Name _____

Date _____

Section __10/5/16_____

The (A) corresponds to the indicated Learning Outcome(s) (O)
found at the beginning of the Laboratory Exercise.

LABORATORY ASSESSMENT 19

Joint Structure and Movements

PART A: Assessments

Match the terms in column A with the descriptions in column B. Place the letter of your choice in the space provided.

(A1) (A2)

Column A		Column B
a. Gomphosis	**b**	**1.** Immovable joint between flat bones of the skull united by a thin layer of dense connective tissue
b. Suture		
c. Symphysis	**c.**	**2.** Fibrocartilage fills the slightly movable joint
d. Synchondrosis	**d.**	**3.** Temporary joint in which bones are united by bands of hyaline cartilage
e. Syndesmosis		
	e.	**4.** Slightly movable joint in which bones are united by interosseous membrane
	a.	**5.** Joint formed by union of tooth root in bony socket

PART B: Assessments

Identify the types of structural and functional joints numbered in figure 19.6. (A3)

Structural Classification	Functional Classification
1. a.) Fibrous	Suture (Synarthrosis)
2. b.) Cartilaginous	Fibrocartilage (Amphiarthrosis)
3. c.) Cartilaginous	Fibrocartilage (Amphiarthrosis)
4. d.) ~~Plane Joint (Gliding)~~	~~Intercarpal (sliding or twisting)~~
5. e.) ~~Pivot Joint~~ Fibrous	~~Ulna (rotation)~~ Sydesmoses
6. f.) Cartilaginous	Amphiarthrosis
7. g.) ~~condylar~~	~~Flexion, extension, hyperextension~~
8. h.) Synovial	Diarthrosis
9.	

FIGURE 19.6 **AP|R** Identify the types of structural and functional joints numbered in these illustrations (a–h).

(a)

(b)

(c)

1

2

3

Carpals

(d)

4

Humerus

5

Ulna

(e)

6

Atlas

Axis

(f)

Proximal phalanx

7

Metacarpal I

8

Trapezium

(g)

9

(h)

PART C: Assessments

Match the types of synovial joints in column A with the examples in column B. Place the letter of your choice in the space provided. **A3**

Column A

a. Ball-and-socket
b. Condylar (ellipsoid)
c. Hinge
d. Pivot
e. Plane (gliding)
f. Saddle

Column B

_**a.**_ 1. Hip joint
_____ 2. Metacarpal–phalanx
_**C.**_ 3. Proximal radius–ulna
_**C.**_ 4. Humerus–ulna of the elbow joint
_____ 5. Phalanx–phalanx
_**a.**_ 6. Shoulder joint
_**C.**_ 7. Knee joint between femur and tibia (main movement)
_____ 8. Carpal–metacarpal of the thumb
_____ 9. Carpal–carpal
_____ 10. Tarsal–tarsal

PART D: Assessments

Complete the missing components of the following table: **A4** **A5**

Name of Joint	Type of Joint	Bones Included	Types of Movement Possible
Shoulder joint		Humerus, scapula	
Elbow joint	Hinge, plane, pivot	tibia, fibula, ulna, humerus	Flexion and extension between humerus and ulna; twisting between radius and humerus; rotation between head of radius and ulna
Hip joint	Ball & socket	femur, pubic	Movements in all planes and rotation
Knee joint	Hinge (modified), condylar, plane	patella	

PART E: Assessments

Identify the types of joint movements numbered in figure 19.7. **A5**

1. _____ (of head)
2. _Elevation_ (of shoulder)
3. _Depression_ (of shoulder)
4. _Supination_
   _____ (of forearm and hand at radioulnar joint)
5. _Pronation_
   _____ (of forearm and hand at radioulnar joint)
6. _____ (of arm at shoulder)
7. _____ (of arm at shoulder)
8. _Flexion_ (of wrist)
9. _Hyperextension_ (of wrist)
10. _Abduction_ (of thigh at hip)
11. _Adduction_ (of thigh at hip)
12. _Rotation ? Circumduction_ (of lower limb at hip)
13. _Protraction_ (of chin/mandible)
14. _Retraction_ (of chin/mandible)
15. _flexion_ (of vertebral column)
16. _Hyperextension_ (of vertebral column)
17. _" "_ (of head and neck)
18. _"_ (of head and neck)

19. _____ (of arm at shoulder)	24. _Adduction_ (of thigh at hip)
20. _____ (of arm at shoulder)	25. _____ (of leg at knee)
21. _Flexion_ (of elbow)	26. _____ (of leg at knee)
22. _Extension_ (of elbow)	27. _Eversion_ (of foot at ankle)
23. _Abduction_ (of thigh at hip)	28. _Inversion_ (of foot at ankle)

FIGURE 19.7 Identify each of the types of movements numbered and illustrated: (a) anterior view; (b) lateral view of head; (c) lateral view. (Original concept of illustration drawn by Ross Martin.)

LABORATORY
EXERCISE 20

Skeletal Muscle Structure and Function

Purpose of the Exercise

To study the structure and function of skeletal muscles as cells and as organs.

MATERIALS NEEDED

Compound light microscope

Prepared microscope slide of skeletal muscle tissue (longitudinal section and cross section)

Human muscular model

Model of skeletal muscle fiber

For Demonstration Activity:

Fresh round beefsteak

Ph.I.L.S. 4.0, Lesson 5 Skeletal Muscle Function: Stimulus-Dependent Force Generation

Learning Outcomes AP|R

After completing this exercise, you should be able to:

01 Locate the structures of a skeletal muscle fiber (cell).

02 Describe how connective tissue is associated with muscle tissue within a skeletal muscle.

03 Distinguish between the origin and insertion of a muscle.

04 Describe and demonstrate the general actions of agonists, antagonists, synergists, and fixators.

05 Diagram and label a muscle twitch.

06 Demonstrate *threshold* (the minimum voltage required to observe the appearance of muscle contraction).

07 Illustrate *recruitment* (the stimulation of additional motor units in the muscle) by observing the change in amplitude of the contraction.

08 Determine maximum contraction (the stimulation of all motor units in the muscle) by observing no further increase in the amplitude (strength) of the contraction.

09 Measure amplitude of muscle contraction under varying strengths of stimulus voltage.

10 Integrate the concepts of recruitment and maximum stimulation as applied to the muscles of the body.

The **O** corresponds to the assessments **A** indicated in the Laboratory Assessment for this Exercise.

Pre-Lab

Carefully read the introductory material and examine the entire lab. Be familiar with skeletal muscle tissue and muscle structure and function from lecture or the textbook. Answer the pre-lab questions.

Pre-Lab Questions Select the correct answer for each of the following questions:

1. The outermost layer of connective tissue of a muscle is the
   - **a.** fascicle.
   - **b.** endomysium.
   - **c.** epimysium.
   - **d.** perimysium.

2. The thick myofibril filament of a sarcomere is composed of the protein
   - **a.** myosin.
   - **b.** actin.
   - **c.** titin.
   - **d.** sarcolemma.

3. The muscle primarily responsible for an action is the
   - **a.** synergist.
   - **b.** agonist.
   - **c.** antagonist.
   - **d.** origin.

4. The neuron and the muscle fibers it innervates is called the

a. neural impulse.
b. muscle fiber.
c. motor neuron.
d. motor unit.

5. The functional contractile unit of a muscle fiber (cell) is the

a. sarcoplasm.
b. sarcolemma.
c. sarcomere.
d. sarcoplasmic reticulum.

6. The action of a particular muscle is always the same.

a. True _____
b. False _____

7. A synergistic muscle contraction assists the agonist.

a. True _____
b. False _____

8. The plasma membrane of a muscle fiber (cell) is called the sarcolemma.

a. True _____
b. False _____

A **skeletal muscle represents** an organ of the muscular system and is composed of several types of tissues. These tissues include skeletal muscle tissue, nervous tissue, and various connective tissues. It is called skeletal muscle because the muscle is attached to bones via tendons.

Each skeletal muscle is encased within and permeated by connective tissue sheaths. The connective tissues surround and extend into the structure of a muscle and separate it into compartments. The entire muscle is encased by the *epimysium*. The *perimysium* covers bundles of cells (*fascicles*), within the muscle. The deepest connective tissue surrounds each individul muscle fiber (cell) as a thin *endomysium*. The connective tissues provide support and reinforcement during muscular contractions. The connective tissue often projects beyond the end of a muscle, providing an attachment to other muscles or to bones as a tendon. Some collagen fibers of the connective tissue are continuous with the tendon and the periosteum, making for a strong structural continuity.

Muscles are named according to their location, size, shape, action, attachments, number of origins, or the direction of the fibers. Examples of how muscles are named include: gluteus maximus (location and size); adductor longus (action and size); sternocleidomastoid (attachments); serratus anterior (shape and location); biceps brachii (two origins and location); and orbicularis oculi (direction of fibers and location).

Skeletal muscles are composed of many muscle fibers (cells). These muscle fibers are innervated by somatic motor neurons of the nervous system. Anatomically, each motor neuron is arranged to extend to a specific number of muscle fibers. (Note that each muscle fiber is innervated by only one motor neuron.) The motor neuron and the collection of muscle fibers it innervates is called a *motor unit*. When the motor neuron relays a neural impulse (action potential) to the muscle, all the muscle fibers that it innervates (i.e., the motor unit) will be stimulated to contract. If only a few motor units are stimulated, the muscle as a whole exerts little force. If a larger number of motor units are stimulated, the muscle will exert a greater force. Thus, to lift a heavy object like a suitcase requires the activation of more motor units in the biceps brachii than the lifting of a lighter object like a book.

PROCEDURE: Skeletal Muscle Structure and Function

1. Study a human muscular model and a skeletal muscle fiber model along with figure 20.1.
2. Examine the microscopic structure of a longitudinal section of skeletal muscle by observing a prepared microscope slide of this tissue. Use figure 20.2a of skeletal muscle tissue to locate the following features:

   skeletal muscle fiber (cell)

   nuclei

   striations (alternating light and dark)

FIGURE 20.1 AP|R Skeletal muscle structure from the gross anatomy to the microscopic arrangement. Note the distribution pattern of the epimysium, perimysium, and endomysium.

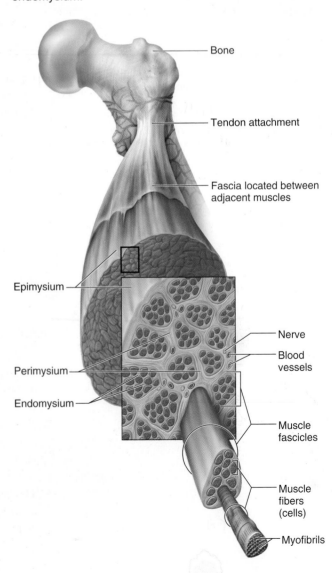

Bone

Tendon attachment

Fascia located between adjacent muscles

Epimysium

Perimysium

Endomysium

Nerve

Blood vessels

Muscle fascicles

Muscle fibers (cells)

Myofibrils

3. Skeletal muscle cells are stimulated by impulses (action potentials) over cellular processes (axons) of motor neurons (fig 20.2*a*). The *neuromuscular junction* is an area where the axon of a motor neuron terminates and synapses with a skeletal muscle fiber. The motor impulse results in a release of a chemical messenger, called a *neurotransmitter,* that allows the motor neuron to communicate with the muscle fiber (fig. 20.2*b*).

4. Study figures 20.1 and 20.3. Note the arrangement of muscle fibers in relation to the connective tissues of the

FIGURE 20.2 AP|R Skeletal muscle fibers with neuromuscular junctions: (a) micrograph of a longitudinal section (400×); (b) an illustration of one neuromuscular junction.

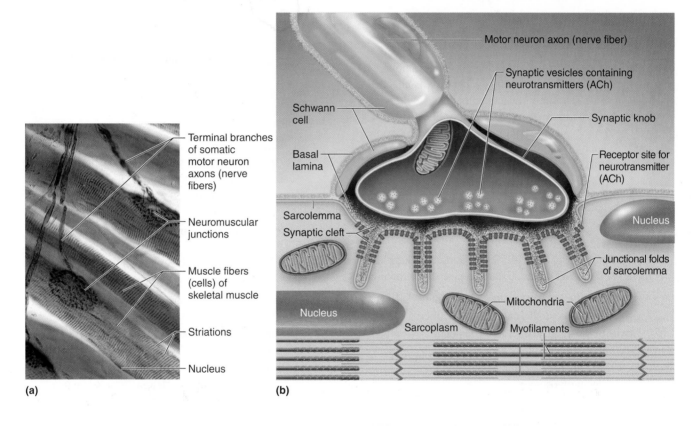

(a)

(b)

FIGURE 20.3 Micrograph of a cross section of a fascicle and associated connective tissues (800×).

epimysium, perimysium, and endomysium of an individual muscle. Examine a prepared slide of the microscopic structure of a cross section of skeletal muscle tissue along with figure 20.3 and locate the following features:

connective tissue sheaths

- epimysium—encases entire muscle
- perimysium—surrounds fascicles
- endomysium—covers individual fibers (cells)

fascicle—bundle of fibers within a muscle

muscle fiber

- nuclei—near plasma membrane (sarcolemma)
- myofibrils—protein filaments within a fiber

5. Examine the human torso model and locate examples of tendons and aponeuroses. An *origin tendon* is attached to a more fixed location, whereas the *insertion tendon* is attached to a more movable location. Sheets of connective tissue, called *aponeuroses,* also serve for some muscle attachments. An example of a large aponeurosis visible on the torso or a muscle chart is in the abdominal area. It appears as a broad white sheet of connective tissue that encloses or connects the abdominal muscles. Locate examples of cordlike tendons in your body. The large calcaneal tendon is easy to palpate in a posterior ankle just above the heel.

6. Using figures 20.4 and 20.5 as a guide, examine the model of the skeletal muscle fiber. Locate the following:

sarcolemma—plasma membrane of a muscle fiber

sarcoplasm—cytoplasm of a muscle fiber

myofibril—bundle of protein filaments (myofilaments)

- thick filament—composed of contractile protein myosin

- thin filament—composed mostly of contractile protein actin and some regulatory proteins troponin and tropomyosin
- elastic filament—composed of the springy protein titin

sarcoplasmic reticulum (SR)—smooth ER of muscle fiber

- terminal cisternae—extended ends of SR next to transverse tubules

transverse (T) tubules—inward extensions of sarcolemma through the muscle fiber

sarcomere—functional contractile unit within muscle fiber (fig. 20.5)

- A (anisotropic) band—dark band of thick filaments and some overlapping thin filaments

FIGURE 20.4 Structures of a segment of a muscle fiber (cell).

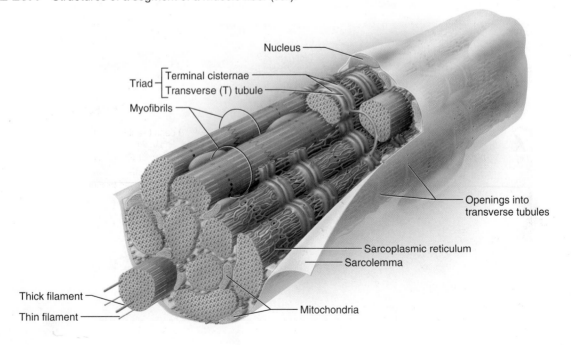

FIGURE 20.5 AP|R Sarcomere contractile unit of a relaxed muscle fiber: (a) micrograph (16,000×); (b) illustration of the repeating pattern of striations (bands).

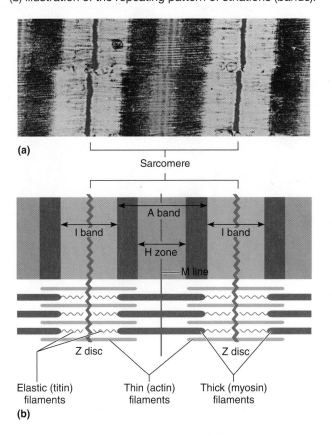

(a)

Sarcomere

A band

I band

I band

H zone

M line

Z disc

Z disc

Elastic (titin) filaments

Thin (actin) filaments

Thick (myosin) filaments

(b)

- I (isotropic) band—light band of thin filaments and bisecting Z disc
- H zone (band)—light band in middle of A band
- M line—dark line in middle of H zone
- Z disc (line)—thin and elastic filaments anchored at ends of sarcomere

7. Complete Parts A and B of Laboratory Assessment 20.
8. Study figure 20.6 and table 20.1.

Table 20.1   Various Roles of Muscles

Functional Category	Description
Agonist (prime mover)*	Muscle primarily responsible for the action (movement)
Antagonist	Muscle responsible for action in the opposite direction of an agonist or for resistance to an agonist
Synergist	Muscle that assists an agonist, often by supplementing the contraction force
Fixator	Special type of synergist muscle; muscle contraction will stabilize a joint so another contracting muscle exerts a force on something else

*The functional category agonist can be interchangeable with prime mover. Most actions (movements) involve several synergistic muscles participating in the force. The agonist responsible for the greatest amount of the force is often called the prime mover. An example would be the rotator cuff muscles (supraspinatus, infraspinatus, teres minor, and subscapularis) whose tendons act to stabilize the head of the humerus at the glenoid cavity of the scapula while the deltoid muscle lifts (abducts) the arm. These S.I.T.S. muscles are special synergists, called fixators, to the agonist (deltoid).

FIGURE 20.6   Antagonistic muscle pairs are located on opposite sides of the same body region as shown in an arm (posterior view). The muscle acting as the agonist depends upon the movement that occurs.

Scapula

Origin on scapula

Origin on humerus

Triceps brachii: long head

Triceps brachii: lateral head

(Action: extension of forearm at elbow)

Humerus

Insertion on ulna

Origins on scapula

Biceps brachii (Action: flexion of forearm at elbow)

Origin on humerus

Belly of biceps brachii

Brachialis (Action: flexion of forearm at elbow)

Tendon of biceps brachii

Insertion on ulna

Insertion on radius

Radius

Ulna

9. Locate the biceps brachii, brachialis, and triceps brachii and their origins and insertions in the human torso model and in your body.

10. Make various movements with your upper limb at the shoulder and elbow. For each movement, determine which muscles are functioning as *agonists* and as *antagonists*. When an agonist contracts (shortens) and a joint moves, its antagonist relaxes (lengthens). Antagonistic muscle pairs pull from opposite sides of the same body region. As an example, consider flexion of the forearm at the elbow versus extension of the forearm at the same elbow joint. The agonist and antagonist change, depending upon which movement is occurring. *Synergistic muscles* often supplement the contraction force of an agonist, or by acting as *fixators,* they might also stabilize nearby joints. **The role of a muscle as an agonist, antagonist, or synergist depends upon the movement under consideration, as their roles change.**

11. Complete Part C of the laboratory assessment.

Name _____

Date _____

Section _____

The (A) corresponds to the indicated Learning Outcome(s) (O) found at the beginning of the Laboratory Exercise.

LABORATORY ASSESSMENT 20

Skeletal Muscle Structure and Function

PART A: Assessments

Match the terms in column A with the definitions in column B. Place the letter of your choice in the space provided. (A1) (A2)

Column A	Column B
a. Endomysium	_____ **1.** Protein that comprises part of the thin filament with actin
b. Epimysium	_____ **2.** Cytoplasm of a muscle fiber
c. Fascia	_____ **3.** Connective tissue located between adjacent muscles
d. Fascicle	_____ **4.** Layer of connective tissue that separates a muscle into small bundles called fascicles
e. Myosin	
f. Perimysium	_____ **5.** Plasma membrane of a muscle fiber
g. Sarcolemma	_____ **6.** Layer of connective tissue that surrounds a skeletal muscle
h. Sarcomere	
i. Sarcoplasm	_____ **7.** Unit of alternating light and dark striations between Z discs (lines)
j. Sarcoplasmic reticulum	_____ **8.** Layer of connective tissue that surrounds an individual muscle fiber
k. Tendon	_____ **9.** Cellular organelle in muscle fiber corresponding to the endoplasmic reticulum
l. Troponin	
	_____ **10.** Cordlike part that attaches a muscle to a bone
	_____ **11.** Protein found within thick filament
	_____ **12.** Small bundle of muscle fibers within a muscle

PART B: Assessments

Provide the labels for the electron micrograph in figure 20.7.

FIGURE 20.7 Label this transmission electron micrograph (16,000×) of a relaxed sarcomere by placing the correct numbers in the spaces provided. (A1)

___ A band (dark)
___ H zone
___ I band (light)
___ M line
___ Sarcomere
___ Z disc

201

CRITICAL THINKING ASSESSMENT

When stimulated to contract, the sarcomeres shorten ⟶ myofibrils shorten ⟶ muscle fiber shortens ⟶ the muscle shortens. Complete the table by indicating the changes (if any) in two adjacent sarcomeres when they move from a relaxed state to a contracted state. **A1**

Contractile Unit	Changes that occur in relaxed ⟶ contracted sarcomeres
A band	
I band	
H zone	
Z discs	

PART C: Assessments

Complete the following statements:

1. The _____ of a muscle is attached to a more fixed location. **A3**

2. The _____ of a muscle is attached to a more movable location. **A3**

3. A muscle responsible for most of an action is called a(n) _____. **A4**

4. Assisting muscles of an agonist are called _____. **A4**

5. Antagonists are muscles that resist the actions of _____ and cause movement in the opposite direction. **A4**

6. When the forearm is extended at the elbow joint, the _____ muscle acts as the agonist. **A4**

7. When the brachialis acts as the agonist, the _____ muscle assists as a synergist. **A4**

8. A _____ is a synergistic muscle that will stabilize a joint when another contracting muscle exerts a force on something else. **A4**

LESSON 5 | Skeletal Muscle Function
Stimulus-Dependent Force Generation

The purpose of this exercise is to demonstrate how the twitch force generated in skeletal muscle is dependent on stimulus intensity. In this experiment, the gastrocnemius (calf) muscle is removed from a frog and attached to a recording device. As an electrical stimulus is applied at a range of voltages to the muscle, a chart of the timing and force (strength) of contraction is measured and graphed. If no contraction is seen when an electrically weak stimulus is applied, the stimulus is considered **subthreshold.** As the muscle is stimulated with an incremental increase in voltage, the first perceivable contraction means that **threshold** has been reached. Threshold is the minimum voltage required to generate an action potential in the muscle fiber and produce a contraction. In this Ph.I.L.S. lab, the stimulus (voltage) applied to the gastrocnemius muscle mimics the action potential generated by a motor neuron. You will change the voltage from 0 volts to 1.6 volts, while observing how increasing the stimulus intensity relates to an increase in skeletal muscle twitch force until a point of maximum contraction is achieved. Any voltage increase after maximum contraction will produce no further increase in force.

A stimulus at threshold or higher will trigger a skeletal **muscle twitch,** a phenomenon that does not occur *in vivo* because a single twitch by itself would not do any beneficial work. Normally, a succession of action potentials generated by the motor neuron would be delivered to the skeletal muscle. Because the gastrocnemius muscle in this experiment is being held at a fixed length, rather than changing its length as it contracts, the type of contraction is **isometric,** rather than isotonic. Studying an isometric skeletal muscle twitch *in vitro* provides some basic understanding of muscle physiology, allowing us to observe the distinct phases of a single contraction that include the latent period, contraction period, and relaxation period. The *latent period* is a delay between the onset of the stimulus and the onset of the contraction. During this time, the motor neuron action potential leads to the exposure of active sites on the actin filaments, creating an internal tension on the elastic components of the muscle. During the *contraction phase,* an external tension is created as the muscle contracts (shortens). The *relaxation phase* is seen when the level of intracellular Ca^{+2} falls and the crossbridge breaks, releasing the myosin from the actin.

A **motor unit** includes a somatic motor neuron and all of the skeletal muscle fibers that it activates (fig. 20.8). There are multiple motor units in a single muscle. The nervous system

FIGURE 20.8 AP|R   Motor unit.

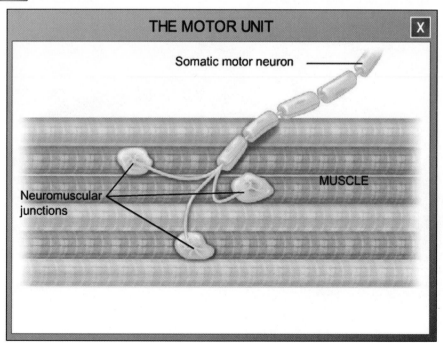

THE MOTOR UNIT

Somatic motor neuron

Neuromuscular junctions

MUSCLE

responds to the demands on a muscle by stimulating more motor units within that muscle to help it contract as a load increases; this is called *motor unit recruitment.* There are a limited number of motor units in a single muscle, so when all of the motor units are recruited, *maximum stimulation* has been reached. No matter how much will power you have, if all the motor units are recruited and the force generated is not moving the load, then you are out of luck.

Before you begin, familiarize yourself with the following concepts (use your textbook as a reference):

- Motor unit
- Neuromuscular junction
- Skeletal muscle twitch and its parts (latent period, contraction period, relaxation period)
- Difference between subthreshold, threshold, and maximal stimulus.
- Isometric contraction and isotonic contraction

State your hypothesis regarding how the isometric muscle twitch force of contraction is influenced by the changes in stimulus voltage.

_____

_____

_____

_____

_____

1. Open Lesson 5, Skeletal Muscle Function: Stimulus Dependent Force Generation.

2. Read the objectives and introduction and take the pre-laboratory quiz.

3. After completing the pre-laboratory quiz, read through the wet lab.

4. The laboratory exercise will open when you click Continue after completing the wet lab (fig. 20.9).

5. Follow the instructions provided in the laboratory exercise and answer the following questions as you proceed.

6. Record the data in Table 20.2 and complete the graph from the Ph.I.L.S. Journal of the muscle tension (g) at each increment of applied volts (V). (Compare to fig. 20.10 graph.)

7. Make notes of any pertinent observations here:

_____

_____

_____

8. What was the threshold voltage for stimulation of the frog's gastrocnemius muscle? A6 A9

_____

9. What voltage produced maximum contraction? A8 A9

_____

10. Create a graph of a muscle twitch at maximum contraction and label the y-axis, the x-axis, and the following phases: latent period, contraction period, and relaxation period. A5

FIGURE 20.9 Opening screen for the laboratory exercise on Skeletal Muscle Function: Stimulus-Dependent Force Generation.

FIGURE 20.10 Example: Line tracings showing the amount of tension produced by applying shocks of different voltages to the exposed frog gastrocnemius.

Table 20.2 Journal Data of Ph.I.L.S Lesson 5 **Skeletal Muscle Function:** *Stimulus-Dependent Force Generation*

Volts (V)	Muscle Tension (g)	
0.0		
0.1		
0.2		
0.3		
0.4		
0.5		
0.6		
0.7		
0.8		
0.9		
1.0		
1.1		
1.2		
1.3		
1.4		
1.5		
1.6		

11. What is occurring between the threshold and the maximum contraction? (A7) (A9)

_____

_____

_____

12. Explain the relationship of increasing voltage stimulation to the recruitment of motor units. (A6) (A7) (A8)

_____

_____

_____

CRITICAL THINKING ASSESSMENT

13. You are getting ready to leave the lab. You lift your pencil to put it in your book bag. Then, you lift your lab book and place it in your book bag. Which scenario resulted from the stimulation of more motor units? _____ Why? _____

_____

_____ (A10)

Which would require more neural signals being sent by the nervous system to the muscles of your arms? _____ Why? _____

_____

_____ (A10)

What concept covered in this lab (threshold, recruitment, or maximum contraction) explains why there are certain objects that are too heavy for you to lift? _____ Explain your reasoning.

_____

_____

_____ (A10)

14. Click the Post-Lab Quiz & Lab Report in the upper menu. Complete the 10 questions in the Post-Lab Quiz.

15. Read the conclusion on the computer screen.

16. Print the Lab Report for Skeletal Muscle Function: Stimulus-Dependent Force Generation.

Electromyography: BIOPAC© Exercise

Purpose of the Exercise

To measure and calculate fist clench strength from electromyograms obtained using the BIOPAC system and to listen to the EMG "sounds"; to measure and calculate the force generated during motor unit recruitment and to observe muscle fatigue.

MATERIALS NEEDED

Computer system (Mac OS X 10.5–10.8, or PC running Windows XP, Vista, 7, or 8)

BIOPAC Student Lab software version 4.0 or above

BIOPAC Acquisition Unit MP36, MP35, or MP45

BIOPAC electrode lead set (SS2L)

BIOPAC disposable vinyl electrodes (EL503), 6 electrodes per subject

BIOPAC electrode gel (GEL1) and abrasive pad (ELPAD), or alcohol wipes, or another type of skin cleanser

BIOPAC Headphones (OUT1/OUT1A for MP36/35 or 40 HP for MP45) optional (*Note:* OUT1A works only with the MP36)

BIOPAC Hand Dynamometer (SS25L or SS25LA), or Hand Clench Force Pump Bulb (SS56L) for Part 2. *Note:* If you are using the SS56L, the Clench Force Transducer Preference must be set before the calibration step.

SAFETY ⚠

▶ The BIOPAC electrode lead set is safe and easy to use and should be used only as described in the procedures section of the laboratory exercise.

▶ The electrode lead clips are color-coded. Make sure they are connected to the properly placed electrodes as demonstrated in figure 21.6.

▶ The vinyl electrodes are disposable and meant to be used only once. Each subject should use a new set.

Learning Outcomes

After completing this exercise, you should be able to:

O1 Record the maximum clench strength for dominant and nondominant hands.

O2 Examine, record, and correlate motor unit recruitment with increased power of skeletal muscle contraction.

O3 Listen to EMG "sounds" and correlate sound intensity with motor unit recruitment.

O4 Examine, record, and calculate motor unit recruitment with increased power of skeletal muscle contraction.

O5 Determine the maximum clench force between dominant and nondominant hands (and male and female, if possible).

O6 Record the force produced by clenched muscles during fatigue.

O7 Calculate the time it takes clenched muscles to fatigue.

The **O** corresponds to the assessments **A** indicated in the Laboratory Assessment for this Exercise.

Pre-Lab

Carefully read the introductory material and examine the entire lab. Be familiar with motor unit recruitment and muscle fatigue from lecture or the textbook.

This is the first of a series of laboratory exercises that will utilize the BIOPAC computer-based data acquisition and analysis system. This is a complete system consisting of hardware and software, which provides data similar to that obtained by a physiograph. Once the software package has been loaded onto your computer, each lesson will use a different set of hardware. Connections and how to use the instrumentation will be explained in each lesson.

The hardware includes the MP36/35 or MP45 Data Acquisition Unit, connection cables, transformers, transducers, electrodes and electrode cables, and other accessories. The electrodes and other accessories often contain sensors that pick up a signal and send it to the MP36/35 or MP45 Data Acquisition Unit. The software picks up electrical signals coming into the MP36/35 or MP45 Data Acquisition Unit, which are then displayed on the screen as a waveform. The software then guides the user through the lesson and manages data saving and review.

The BIOPAC computer-based data acquisition and analysis system works in a manner similar to other recording equipment in sensing electrical signals in the body. See figure 21.1 for the basic setup of the BIOPAC Student Lab System. The major advantage of a computerized system such as the BIOPAC system is that it has dynamic calibration. With other equipment such as a physiograph, the instructor or student needs to manually adjust the dials and settings to get a good signal for the recording. With the BIOPAC Student Lab, the adjustments are made automatically for each subject during the calibration procedure in each lesson.

Before you start any lesson, the software should already be installed on the computer, and you should read through the lesson completely. The BIOPAC Student Laboratory Manual that comes with the system has a full description of the lessons and some troubleshooting information. The lessons as described in this manual are shortened versions of these. *Note:* In the most recent upgrade of the BIOPAC Student Lab System, version 4.0, brief instructions for lab procedures are displayed in the Journal area below the data window. As you perform the lab, you may follow along with the brief instructions in the Journal, but you will have to refer to this lab manual for more detailed instructions. The instructions for data analysis are not provided in the Journal area, so you will have to refer to the lab manual for these. Figure 21.2 and table 21.1 have been included here as a quick reference guide to the BIOPAC Display Tools for Analysis as you proceed through the BIOPAC lessons.

In Laboratory Exercise 21, you will investigate the electrical properties of skeletal muscle using *electromyography.* This is a recording of skin-surface voltage that is produced by underlying skeletal muscle contraction. The voltage produced by contraction is weak, but it can be picked up by surface sensors to produce the recording, referred to as an *electromyogram (EMG).*

Skeletal muscle is one of three types of muscle tissue in the body. Skeletal muscle is mainly attached to the various bones of the skeleton; it helps provide a framework and support for the body, as well as the ability to move in response

FIGURE 21.1 The basic setup of the BIOPAC Student Lab System includes a computer, an MP36/35 Data Acquisition Unit, and various transducers.

Source: Courtesy of and © *BIOPAC* Systems, Inc. www.biopac.com

to willing an action to occur. (Cardiac muscle is found only in the heart, where it originates and spreads the heartbeats that keep us alive. Smooth muscle is located mostly in the tubular organs of our internal systems; it accomplishes many types of internal movement, such as the churning action of the stomach and the transport of urine from the kidney to the bladder.)

Each skeletal muscle in our body is composed of thousands of skeletal muscle fibers (cells). Skeletal muscle fibers are unique among the three types of muscle tissue in that they contract only in response to stimulation by a motor neuron. The response of a single skeletal muscle fiber to stimulation by its motor neuron is called a *twitch.* A twitch consists of a rapid period of tension development and contraction, followed by a decrease in tension and relaxation of the muscle.

Each motor neuron associated with skeletal muscle fibers innervates (stimulates, controls) a certain number of fibers, from fewer than 10 to thousands. A single motor neuron plus all of the muscle fibers it controls is called a *motor unit.* In a motor unit, whenever the motor neuron is activated, all of its muscle fibers are commanded to contract simultaneously. Therefore, in real life, entire motor units, rather than single muscle fibers, are contracting to accomplish body movements.

When a single muscle fiber is stimulated by its motor neuron to contract, then relaxes completely, and then contracts again, the subsequent contractions all produce the same amount of muscle tension (fig. 21.3a). However, if a muscle fiber is stimulated again, before it has time to relax completely, then the second twitch will generate more tension than the first. This is called *temporal (wave) summation,* and it is the body's way of increasing the force generated by muscles (fig. 21.3b). When a muscle fiber is stimulated to contract by several stimuli in rapid succession, and there is no opportunity for muscle relaxation between stimuli,

FIGURE 21.2 Basic setup and icons of the display window.

Source: Courtesy of and © *BIOPAC* Systems, Inc. www.biopac.com

Biopac Student Lab Lessons Display in analysis mode

Table 21.1 Display Tools for Analysis

Viewing Tool	Overview
Editing and Selection Tools	The **Editing and Selection tools** icon is located in the upper right corner of the display window.
	The **Arrow Tool** is a general-purpose cursor, used for selecting waveforms and scrolling through data.
I-Beam Tool	The **I-beam** icon is located in the upper right corner of the display window.
	The **I-beam Tool** is used to select an area for measurement. To activate it, click on it and move the mouse such that the cursor is positioned at the beginning of the region that you want to select. The cursor *must* be positioned within the data window. Hold the mouse button down and move (drag) the cursor to the second position. Release the mouse button to complete the selection. The selected area should remain darkened. Measurements will apply to the selected area.
Zoom Tool	The **Zoom** icon is in the upper right corner of the display window.
	Once selected, the **Zoom Tool** allows you to define an area of the waveform by click-hold-drag-releasing the mouse over the desired section. The horizontal (time) and vertical (amplitude) scales will expand to display just the selected section.
Zoom Back	Click on the **Display** menu to access this option.
	After you have used the **Zoom Tool, Zoom Back** will revert the display to the previous scale settings.
Horizontal (time) Scroll Bar	The **Horizontal Scroll Bar** is located below the time division scale. It only works when you are viewing just a portion of the waveform data.
	To move the display's time position, click on the scroll box arrows or hold the mouse button down on the scroll box and drag the box left (earlier) or right (later).
Vertical (amplitude) Scroll Bar	The **Vertical Scroll Bar** is located to the right of the amplitude (mV) division scale.
	To move the selected channel's vertical position, click on the scroll box or hold the mouse button down on the scroll box and drag the box up or down.
Autoscale waveforms	Click on the **Display** menu to access this option.
	Optimizes the vertical (amplitude) scale so that all the data will be shown on the screen.
Autoscale horizontal	Click on the **Display** menu to access this option.
	Compresses or expands the horizontal (time) scale so the entire recording will fit on the screen.
Grids	Grids provide a quick visual guide to determine amplitude and duration. To turn the grids display on or off, click on the **Show Grids** or **Hide Grids** button below the Menu Bar.
Adjust Baseline button	The **Adjust Baseline** button only appears in Laboratory Exercise 46, ECG.
	Allows you to position the waveform up or down in small increments so that the baseline can be set to exactly zero. This is not needed to get accurate amplitude measurements, but it may be desired before making a printout or when using grids. When the **Adjust Baseline** button is pressed, **Up** and **Down** buttons will appear. Simply click on these to move the waveform up or down.

FIGURE 21.3 Myograms of (a) isolated muscle twitches, (b) temporal summation, and (c) a tetanic contraction.

the tension increases until it reaches the maximum possible force for that muscle fiber, and then remains at that level until the stimulation stops; this type of sustained contraction is called *tetanus,* or a *tetanic contraction* (fig. 21.3c).

Although there are many motor units in a skeletal muscle, not all of them are stimulated during every muscle contraction. For example, if a person decides to pick up a 1 lb. weight, just a small number of motor units will be activated to contract. If, however, that person picks up a 10 lb. weight instead, many more motor units will be stimulated, in order to generate enough muscle tension to pick up the heavier object. An increase in the number of activated motor units in a skeletal muscle, in order to increase the tension in the muscle, is called *motor unit recruitment.* See figure 21.4 for an illustration of motor unit recruitment.

When skeletal muscles are at rest (not in the process of contracting), they maintain a slight amount of contraction, called muscle tone, or *tonus.* In resting skeletal muscles, motor units take turns contracting, so that the muscle is always ready to contract on short notice. This helps keep the motor units in good health, and in the case of the postural muscles in the back, it allows us to maintain good posture for many hours at a time.

Recruitment and tonus will be demonstrated in Part 1 of this laboratory exercise; as you increase the clench force of your fist, more motor units will be recruited to generate increasing amounts of muscle tension. You will be able to observe tonus in the short periods of time between fist clenches. The periods of tonus will show up in the BIOPAC data window as very low mV readings.

Strenuous exercise or work, which requires repeated or sustained maximal muscle contraction, sometimes results in *muscle fatigue,* a state in which muscle contraction decreases, even when the muscle is still being stimulated. Several factors appear to contribute to muscle fatigue. One of the main causes seems to be a buildup of lactic acid from the anaerobic respiration of glucose, the resulting decrease in blood pH in the vicinity, and the decreased response of the muscle at low pH. In other cases, muscle fatigue can be attributed to conduction failure, in which an accumulation of K^+ ions in the transverse tubules of skeletal muscle fibers eventually leads to an inability to produce subsequent action potentials. Part 2 of this lab exercise will demonstrate muscle fatigue. You will use a *hand dynamometer* or hand clench force pump bulb to generate a recording called a *dynagram.* See figure 21.5 for the appearance and proper grip positions of two different types of dynamometers and the pump bulb. Muscle fatigue will be induced by a sustained, maximum fist clench, and will appear on the dynagram as a steady decline in muscle tension.

PART 1: Electromyography: Standard and Integrated EMG

PROCEDURE A: Setup

1. With your computer turned **ON** and the BIOPAC MP36/35 unit turned **OFF,** plug the electrode lead set (SSL2) into Channel 1 and plug the headphones into the back of the unit if desired. If you are using the MP45, the USB cable must be connected, and the ready light must be ON. Plug the BIOPAC SS25 Hand Dynamometer or pump bulb into Channel 2 only if you are going to do Part 2 of this lab exercise. The dynamometer or pump bulb is not used in Part 1.

2. Turn on the MP36/35 Data Acquisition Unit.

3. Attach the electrodes to both forearms of the subject, as shown in figure 21.6. To help to ensure a good contact, make sure the area to be in contact with the electrodes is clean by wiping with the abrasive pad (ELPAD), skin cleanser, or an alcohol wipe. A small amount of BIOPAC electrode gel (GEL1) may also be used to make better contact between the sensor in the electrode and the skin. For optimal adhesion, the electrodes should be placed on the skin at least 5 minutes before the calibration procedure.

4. Attach the electrode lead set to the electrodes on the dominant arm (most often, the side of the writing hand), via the pinch connectors as shown in figure 21.6. Make sure the proper color is attached to the appropriate electrode.

FIGURE 21.4 Motor unit recruitment. (a) A small number of activated motor units only generates enough force to pick up a lightweight object. (b) Recruitment of more motor units has increased the force to the extent that a heavier object can be lifted.

(a)

(b)

FIGURE 21.5 These are the various types of hand dynamometers and/or hand clench force pump bulb that may be available in your lab. Notice the proper grip position for each type.

Source: Courtesy of and © *BIOPAC* Systems, Inc. www.biopac.com

SS25LA grip position   **SS25L grip position**

SS56L grip position

FIGURE 21.6 Electrode lead attachment.

Source: Courtesy of and © *BIOPAC* Systems, Inc. www.biopac.com

VIN- (White)

VIN+ (Red)

GND (Black)

5. Have the subject sit in a chair, with the dominant arm resting on his/her leg, electrodes facing up. The subject may hold a small rubber ball to clench during Part 1 of the exercise.
6. Start the BIOPAC Student Lab Program for Lesson 1, by clicking on L01-Electromyography (EMG) I.
7. Designate a unique filename to be used to save the subject's data.

PROCEDURE B: Calibration

1. Click on **Calibrate.**
2. Wait about 2 seconds, clench the fist as hard as possible, then release. Wait for the calibration to stop (this will take about 8 seconds).
3. The calibration recording should look similar to figure 21.7. If it does, click Continue. If it does not, click **Redo Calibration.** If the baseline is not at zero, the subject did not wait 2 seconds before clenching; in this case, you should redo the calibration.

PROCEDURE C: Recording

1. Fill out the subject profile in Laboratory Assessment 21. Click on **Record** and have the subject conduct a series of four clench-release-wait cycles, holding for 2 seconds during the clench and wait phases. During the first cycle the clench should be gentle; then increase the grip with each successive cycle so that the fourth clench is of maximum force.
2. Click **Suspend.** The recording should be similar to figure 21.8. If it is not, click **Redo** and repeat step 1.
3. Switch the pinch connectors of the electrode lead set to the nondominant arm. Click Continue, then click **Record,** and repeat from step 1 of Procedure C: Recording.
4. Click **Suspend** after the recording has finished. If the recording looks similar to figure 21.8, click **Stop.**
5. If you want to listen to the EMG signal, put on the headphones and click **Listen.** Have the subject experiment by changing the clench force as you watch the screen. When finished, click **Stop.** This data will not be saved for analysis.
6. Click **Done** to end the lesson.

PROCEDURE D: Data Analysis

1. You can either analyze the current file now, or save it and do the analysis later, using the **Review Saved Data** mode and selecting the correct file. Both options are available in the pop-up window that appears after clicking **Done** at the end of the lesson, under "Record or Analyze a BIOPAC Lesson."
2. Note the channel number designations: **CH 1** displays **EMG** (the actual voltage recording) and **CH 40** the **Integrated EMG** (which reflects the absolute intensity of the voltage).
3. Notice the measurement box settings: **CH 40** should be set on "mean," and the others should be set on "none."
4. After clicking the I-beam icon (this will activate the select function), select the first EMG cluster from the dominant arm, as shown in figure 21.9.

FIGURE 21.7 Calibration recording, part 1 (single clench).

Source: Courtesy of and © *BIOPAC* Systems, Inc. www.biopac.com

FIGURE 21.8 Sample recording of four clench-release-wait cycles of increasing strength.

Source: Courtesy of and © *BIOPAC* Systems, Inc. www.biopac.com

FIGURE 21.9 Proper selection of EMG data for one cluster.

Source: Courtesy of and © *BIOPAC* Systems, Inc. www.biopac.com

5. The values in the measurement boxes should be recorded in the table in Part IA of Laboratory Assessment 21.

6. Repeat for the other three clusters and record the data in Part IA of the laboratory assessment. Then record the data for all four clusters from the nondominant arm.

7. Complete Part I of the laboratory assessment.

8. If you are not going to complete Part 2, then exit the program by clicking **File,** and **Quit.**

PART 2: Electromyography: Motor Unit Recruitment and Fatigue
PROCEDURE A: Setup

If continuing with the same subject used in Part 1, move to setup step 5 below.

1. With your computer turned **ON** and the BIOPAC MP36/35 unit turned **OFF,** plug the electrode lead set (SSL2) into Channel 1 and BIOPAC Hand Dynamometer (SS25L or SS25LA), or BIOPAC Hand Clench Force Pump Bulb (SS56L) into Channel 2. If you are using the MP45, the USB cable must be connected, and the ready light must be ON. Plug the headphones into the back of the unit if desired.

2. Turn on the MP36/35 Data Acquisition Unit.

3. If necessary, attach the electrodes to both forearms of the subject, as shown in figure 21.6. To help to ensure a good contact, make sure the area to be in contact with the electrodes is clean by wiping with the abrasive pad (ELPAD) or alcohol wipe. A small amount of BIOPAC electrode gel (GEL1) may also be used to make better contact between the sensor in the electrode and the skin. For optimal adhesion, the electrodes should be placed on the skin at least 5 minutes before the calibration procedure.

4. Attach the electrode lead set to the electrodes on the dominant arm (most often, the side of the writing hand), via the pinch connectors, as shown in figure 21.6. Make

sure the proper color is attached to the appropriate electrode.

5. Have the subject sit in a chair, with the dominant arm resting on his/her leg, electrodes facing up.

6. Start the BIOPAC Student Lab Program for Electromyography II, lesson L02-EMG-2.

7. Designate a unique filename to be used to save the subject's data, then click **OK.**

8. Remember that if you are using the pump bulb, you will have to click on **File,** and **Lesson Preferences,** and set the preference for the SS56L before the calibration step.

PROCEDURE B: Calibration

1. Click **Calibrate.** Set the hand dynamometer or pump bulb on the table and click **OK.** This establishes a zero-force calibration. This portion of the calibration will run for only 1–2 seconds.

2. Grasp the hand dynamometer or pump bulb with the dominant hand when prompted. See figure 21.5 for proper grip positions. Note the position of your hand and try to replicate this during the recording phase.

3. Follow the instructions on the two successive pop-up windows and click **OK** when ready for each.

4. Wait 2 seconds, then clench the hand dynamometer or pump bulb as hard as possible, then release.

5. Wait for the calibration to stop. This will take about 8 seconds.

6. If the calibration recording resembles figure 21.10, click **Continue.** If it does not, click **Redo Calibration.**

7. Before proceeding to the recording procedure, the software will determine the force increments that you will use. The increments will be shown on the right side of the data window. Using the maximum force from the calibration of a hand dynamometer, the increments will be set as follows: for 0–25 kg, 5 kg increments will be used; for 25–50 kg, 10 kg increments will be used; for 50–75 kg, 15 kg increments will be used, and for

FIGURE 21.10 Sample calibration recording, part 2 (single clench).

Source: Courtesy of and © *BIOPAC* Systems, Inc. www.biopac.com

FIGURE 21.12 Sample recording of muscle fatigue.

Source: Courtesy of and © *BIOPAC* Systems, Inc. www.biopac.com

>75 kg, 20 kg increments will be used. If you are using a pump bulb, these increments will also be set by the software: for 0–5000 kgf/m², increments of 1000; for 5000–7500, increments of 1500; for 7500–10,000, increments of 2000; for 10,000–12,500, increments of 2,500; and for >12,500, increments of 3000 will be used.

PROCEDURE C: Recording

1. Click on **Record** and have the subject:
   - Clench-release-wait and repeat with increasing clench force by the increments determined in step 7 above until you reach your maximum clench force. (Increase your clench force by 1 grid line each time, as shown in figure 21.11.)
   - Each clench should be held for 2 seconds, and there should be a 2-second interval between clenches.
2. Click **Suspend.** The recording should resemble figure 21.11 for motor unit recruitment. If it does not, click **Redo** and repeat step 1.
3. Click **Continue,** and then **Record.**
4. Clench the hand dynamometer or pump bulb with your maximum force. Note the force and try to maintain it.
5. When the maximum clench force has decreased by 50%, click **Suspend.** The recording should resemble

figure 21.12 for muscle fatigue. If it does not, click **Redo** and repeat steps 3–5.
6. If correct, click **Continue.**
7. Switch the pinch connectors of the electrode lead set to the nondominant forearm. Grasp the hand dynamometer or pump bulb, and click **Record.** Proceed as with the dominant arm for recording recruitment and fatigue data.
8. When finished, click **Stop,** and then **Done.**

PROCEDURE D: Data Analysis

1. You can either analyze the current file or save it and do the analysis later using the **Review Saved Data** mode.
2. Note the channel number designations: **CH 1** displays EMG (hidden); **CH 40** displays the **Integrated EMG** (mV), and **CH 41** displays clench force.
3. Note the channel/measurement boxes as follows:

   CH 41: Mean; CH 40: Mean; CH 41: Value; and CH 40: DeltaT

4. Use the I-beam cursor to select an area on the plateau phase of the first clench of the dominant arm, as shown in figure 21.13. Record the measurements in Laboratory Assessment 21 in the table for Motor Unit Recruitment (Part II A1).

FIGURE 21.11 Sample recording of motor unit recruitment.

Source: Courtesy of and © *BIOPAC* Systems, Inc. www.biopac.com

FIGURE 21.13 Proper selection of the plateau area of the first clench.

Source: Courtesy of and © *BIOPAC* Systems, Inc. www.biopac.com

5. Repeat step 4 on the plateau for each successive clench.
6. Scroll to the second recording segment (for Fatigue).
7. Use the I-beam cursor to select a point of maximal clench force at the beginning of the muscle fatigue segment. Record the measurements in the laboratory assessment in the table for Fatigue (Part II A2).
8. Calculate 50% of the maximum clench force from step 7. Record the calculation in the laboratory assessment in the table for Fatigue (Part II A2).
9. Select the area from the point of 50% clench force to the maximum clench force by dragging the I-beam cursor. See figure 21.14 for proper selection of the area. Note the time to fatigue (CH 40/ΔT) and record in the laboratory assessment in the table for Fatigue (Part II A2).
10. Repeat from step 2 for the nondominant forearm.
11. Complete Part II of the laboratory assessment.
12. Exit the program by clicking **File** and **Quit.**

FIGURE 21.14 Proper selection of the area between 50% clench force and maximum clench force.

Source: Courtesy of and © *BIOPAC* Systems, Inc. www.biopac.com

NOTES

LABORATORY ASSESSMENT 21

Name _____

Date _____

Section _____

The corresponds to the indicated Learning Outcome(s) (O) found at the beginning of the Laboratory Exercise.

Electromyography: BIOPAC© Exercise

Subject Profile

Name _____   Height _____

Age _____   Weight _____

Gender: Male/Female   Dominant Forearm: right/left

I. Standard and Integrated EMG

PART A: Data and Calculations Assessments

1. EMG Measurements

Cluster #	DOMINANT FOREARM 40 Mean	NONDOMINANT FOREARM 40 Mean
1		
2		
3		
4		

Note: "Clusters" are the EMG bursts associated with each clench.

2. Use the mean measurement from the preceding table to compute the percentage increase in EMG activity recorded between the weakest clench and the strongest clench of the dominant arm.

Calculation:

Answer: _____ %

PART B: Assessments

Complete the following:

1. If you tested both forearms on the same subject, are the mean measurements for the right and left maximum grip EMG cluster the same? Which one suggests the greater grip strength? Explain. (A1)

_____

_____

_____

2. What factors in addition to gender contribute to observed differences in clench strength?

_____

_____

_____

3. Explain the source of signals detected by the EMG electrodes. A3

_____

_____

_____

4. What is meant by the term *motor unit recruitment?*

_____

_____

_____

5. Define electromyography.

_____

_____

_____

II. Motor Unit Recruitment and Fatigue
PART A: Data and Calculations Assessments

1. Increasing Clench Force (Recruitment) Data: A4 A5

Peak #	Assigned Force Increment SS25L/LA = Kg SS56L = kgf/m²	DOMINANT FOREARM		NONDOMINANT FOREARM	
		Force at Peak	Integrated EMG (mV)	Force at Peak	Integrated EMG (mV)
		41 Mean	40 Mean	41 Mean	40 Mean
1					
2					
3					
4					
5					
6					
7					
8					

2. Maximum Clench Force (Muscle Fatigue) Data:

DOMINANT FOREARM			NONDOMINANT FOREARM		
Maximum Clench Force	50% of Max Clench Force	Time* to Fatigue	Maximum Clench Force	50% of Max Clench Force	Time* to Fatigue
41 Value	calculate	40 Delta T	**41 Value**	calculate	40 Delta T

*Note: You do not need to indicate the Delta T (time to fatigue) polarity. The polarity of the Delta T measurement reflects the direction the I-beam cursor was dragged to select the data. Data selected left to right will have a positive ("+") polarity, while data selected right to left will have a negative ("–") polarity.

PART B: Assessments

Complete the following:

1. Compare the strength of the dominant forearm to that of the nondominant forearm. A5

_____

_____

_____

2. Is there a difference in the absolute values of force generated by males and females in your class? What might explain the difference? A5

_____

_____

_____

3. Is there a difference in fatigue times between the two forearms? Why might this happen?

_____

_____

_____

4. Define muscle fatigue.

_____

_____

_____

5. Define dynamometry.

_____

_____

_____

NOTES

Muscles of the Head and Neck

Purpose of the Exercise

To review the locations, actions, origins, and insertions of the muscles of the head and neck.

MATERIALS NEEDED

Human muscular model

Human skull

Human skeleton, articulated

For Learning Extension Activity:
Long rubber bands

Learning Outcomes AP|R

After completing this exercise, you should be able to:

O1 Locate and identify the muscles of facial expression, the muscles of mastication, the muscles that move the head and neck, and the muscles that move the hyoid bone and larynx.

O2 Describe and demonstrate the action of each of these muscles.

O3 Locate the origin and insertion of each of these muscles in a human skeleton and the musculature of the human torso model.

The **O** corresponds to the assessments **A** indicated in the Laboratory Assessment for this Exercise.

Pre-Lab

Carefully read the introductory material and examine the entire lab. Be familiar with the locations, actions, origins, and insertions of the muscles of the head and neck from lecture or the textbook. Answer the pre-lab questions.

Pre-Lab Questions Select the correct answer for each of the following questions:

1. Muscles of mastication are all inserted on the
   a. mandible.
   b. maxilla.
   c. tongue.
   d. teeth.

2. Muscles of mastication include the following *except* the
   a. lateral pterygoid.
   b. masseter.
   c. temporalis.
   d. platysma.

3. Facial expressions are so variable partly because many facial muscles are inserted
   a. on the mandible.
   b. in the skin.
   c. on the maxilla.
   d. on facial bones.

4. Which of the following muscles does *not* move the hyoid bone?
   a. sternocleidomastoid
   b. mylohyoid
   c. sternohyoid
   d. thyrohyoid

5. Which of the following is *not* a facial expression muscle?
   a. zygomaticus major
   b. buccinator
   c. orbicularis oris
   d. scalenes

6. Flexion of the head and neck is an action of the _____ muscle.
   a. semispinalis capitis
   b. splenius capitis
   c. sternocleidomastoid
   d. trapezius

7. Which of the following muscles is the most visible from an anterior view of a person in anatomical position?
   a. splenius capitis
   b. nasalis
   c. temporalis
   d. medial pterygoid

The skeletal muscles of the head include the muscles of facial expression. Human facial expressions are more variable than those of other mammals because many of the muscles have insertions in the skin, rather than on bones, allowing great versatility of movements and emotions.

The muscles of mastication (chewing) are all inserted on the mandible. Muscles for chewing allow the movements of depression and elevation of the mandible to open and close the jaw. However, jaw movements also involve protraction, retraction, lateral excursion, and medial excursion needed to bite off pieces of food and grind the food.

The muscles that move the head are located in the neck. The actions of flexion, extension, and hyperextension of the head and the cervical vertebral column occur when right and left paired muscles work simultaneously. Additional actions, such as lateral flexion and rotation, result from muscles on one side only contracting or alternating muscles contracting.

A group of neck muscles that have attachments on the hyoid bone and larynx assist in swallowing and speech. Some of the neck muscles that move the head and neck have their attachments in the thorax.

PROCEDURE: Muscles of the Head and Neck

The muscle lists and tables in this laboratory exercise reflect muscle groupings using a combination of related locations and functions. Study and master one group at a time as a separate task. Frequently refer to the illustrations, tables, skeletons, and models. Consider working with a partner or a small group to study and review.

1. Study figure 22.1 and table 22.1.
2. Locate the following muscles in the human muscular model and in your body whenever possible:

   muscles of facial expression
   - epicranius (occipitofrontalis)
     - frontal belly (frontalis)
     - occipital belly (occipitalis)
   - nasalis
   - orbicularis oculi
   - orbicularis oris
   - risorius
   - zygomaticus major
   - zygomaticus minor
   - buccinator
   - platysma
3. Demonstrate the actions of these muscles in your body.
4. Locate the origins and insertions of these muscles in the human skull and skeleton.
5. Study figures 22.1 and 22.2 and table 22.2.
6. Locate the following muscles in the human torso model and in your body whenever possible:

   muscles of mastication
   - masseter
   - temporalis
   - lateral pterygoid
   - medial pterygoid
7. Demonstrate the actions of these muscles in your body.

FIGURE 22.1 Muscles of facial expression and mastication (a) anterior view and (b) lateral view.

Table 22.1 Muscles of Facial Expression

Muscle	Origin	Insertion	Action
Epicranius	Occipital bone; galea aponeurotica	Skin of eyebrows; galea aponeurotica	Raises eyebrows; retracts scalp
Nasalis	Maxilla lateral to nose	Bridge of nose	Widens nostrils
Orbicularis oculi	Maxillary and frontal bones	Skin around eye	Closes eyes as in blinking
Orbicularis oris	Muscles near the mouth	Skin of lips	Closes lips; protrudes lips as for kissing
Risorius	Fascia near ear	Corner of mouth	Draws corner of mouth laterally
Zygomaticus major	Zygomatic bone	Corner of mouth	Raises corner of mouth as when smiling and laughing
Zygomaticus minor	Zygomatic bone	Corner of mouth	Raises corner of mouth as when smiling and laughing
Buccinator	Lateral surfaces of maxilla and mandible	Orbicularis oris	Compresses cheeks inward as when blowing air
Platysma	Fascia in upper chest	Lower border of mandible and skin at corner of mouth	Draws angle of mouth downward as when pouting or expressing horror

FIGURE 22.2 Deep muscles of mastication.

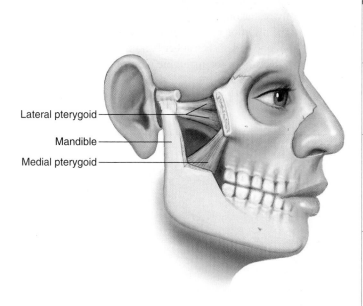

Lateral pterygoid
Mandible
Medial pterygoid

Table 22.2 Muscles of Mastication

Muscle	Origin	Insertion	Action
Masseter	Lower border of zygomatic arch	Lateral surface of mandible	Elevates mandible
Temporalis	Temporal bone	Coronoid process and anterior ramus of mandible	Elevates mandible
Lateral pterygoid	Sphenoid bone	Anterior surface of mandibular condyle	Depresses and protracts mandible and moves it from side to side as when grinding food
Medial pterygoid	Sphenoid, palatine, and maxillary bones	Medial surface of mandible	Elevates mandible and moves it from side to side as when grinding food

8. Locate the origins and insertions of these muscles in the human skull and skeleton.
9. Study figures 22.3, 22.4, and 22.5 and table 22.3.
10. Locate the following muscles in the human torso model and in your body whenever possible:

muscles that move head and neck

- sternocleidomastoid
- trapezius (superior part)
- scalenes (anterior, middle, and posterior)
- splenius capitis
- semispinalis capitis

11. Demonstrate the actions of these muscles in your body.
12. Locate the origins and insertions of these muscles in the human skull and skeleton.
13. Complete Part A of Laboratory Assessment 22.
14. Study figure 22.6 and table 22.4.

15. Locate the following muscles in the human torso model:

muscles that move hyoid bone and larynx

- suprahyoid muscles
  - digastric (2 parts)
  - stylohyoid
  - mylohyoid
- infrahyoid muscles
  - sternohyoid
  - omohyoid (2 parts)
  - sternothyroid
  - thyrohyoid

16. Demonstrate the actions of these muscles in your body.
17. Locate the origins and insertions of these muscles in the human skull and skeleton.
18. Complete Parts B, C, and D of the laboratory assessment.

LEARNING EXTENSION ACTIVITY

A long rubber band can be used to simulate muscle locations, origins, insertions, and actions on the human muscular model, the skeleton, or a laboratory partner. Hold one end of the rubber band firmly on the origin location of a muscle; then slightly stretch the rubber band and hold the other end on the insertion site. Allow the insertion end to slowly move toward the origin end to simulate the contraction and action of the muscle. For example, hold one end of the rubber band firmly on the zygomatic arch (the origin) and the other end on the mandible (the insertion); then move the mandible closer to the zygomatic arch to observe the action of the masseter muscle (elevate mandible).

FIGURE 22.3 Posterior muscles that move the head and neck.

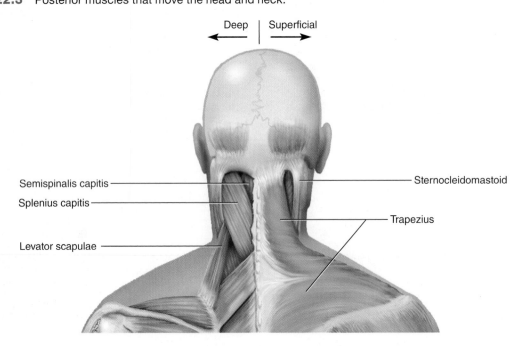

FIGURE 22.4 Muscles that move the head and neck, lateral view.

- Splenius capitis
- Sternocleidomastoid
- Levator scapulae
- Scalenes
- Trapezius
- Clavicle

FIGURE 22.5 Three scalene muscles located in the neck and upper thorax, anterior view. The scalenes assist in neck movements and respiration.

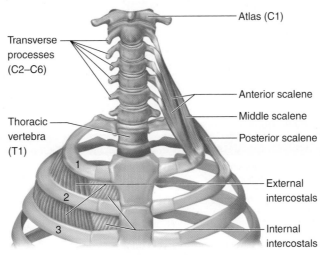

- Atlas (C1)
- Transverse processes (C2–C6)
- Thoracic vertebra (T1)
- Anterior scalene
- Middle scalene
- Posterior scalene
- External intercostals
- Internal intercostals

Table 22.3 Muscles that Move Head and Neck

Muscle	Origin	Insertion	Action
Sternocleidomastoid	Manubrium of sternum and medial clavicle	Mastoid process of temporal bone	Flexion of head and neck; rotation of head to left or right
Trapezius (superior part)	Occipital bone and spinous processes of C7 and several thoracic vertebrae	Clavicle and the spine and acromion of scapula	Extends head (as a synergist); (primary actions on scapula are covered in another lab)
Scalenes (anterior, middle, and posterior)	Transverse processes of C2–C6 cervical vertebrae	Ribs 1–2	Elevates ribs 1–2; flexion and rotation of neck when rib 1 is fixed
Splenius capitis	Spinous processes of C7–T6	Mastoid process and occipital bone	Extends head; rotates head
Semispinalis capitis	Processes of inferior cervical and superior thoracic vertebrae	Occipital bone	Extends head; rotates head

FIGURE 22.6 Muscles that move the hyoid bone and larynx assist in swallowing and speech and are grouped into suprahyoid and infrahyoid muscles.

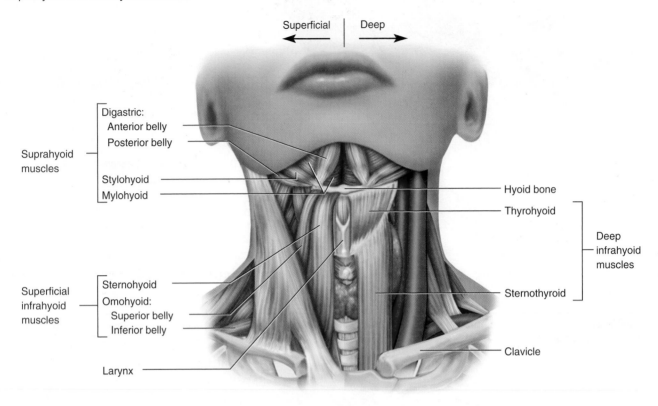

Table 22.4 Muscles that Move Hyoid Bone and Larynx

Muscle	Origin	Insertion	Action
Digastric (2 parts)	Inferior mandible (anterior belly) and mastoid process (posterior belly)	Hyoid bone	Opens mouth; depresses mandible; elevates hyoid bone
Stylohyoid	Styloid process of temporal bone	Hyoid bone	Retracts and elevates hyoid bone
Mylohyoid	Mandible	Hyoid bone	Elevates hyoid bone during swallowing
Sternohyoid	Manubrium and medial clavicle	Hyoid bone	Depresses hyoid bone
Omohyoid (2 parts)	Superior border of scapula	Hyoid bone	Depresses hyoid bone
Sternothyroid	Manubrium	Thyroid cartilage of larynx	Depresses larynx
Thyrohyoid	Thyroid cartilage of larynx	Hyoid bone	Depresses hyoid bone; elevates larynx

Name _____

Date _____

Section _____

The Ⓐ corresponds to the indicated Learning Outcome(s) Ⓞ found at the beginning of the Laboratory Exercise.

Muscles of the Head and Neck

PART A: Assessments

Identify the muscles indicated in the head and neck in figures 22.7 and 22.8.

FIGURE 22.7   Label the anterior muscles of the head. **A1**

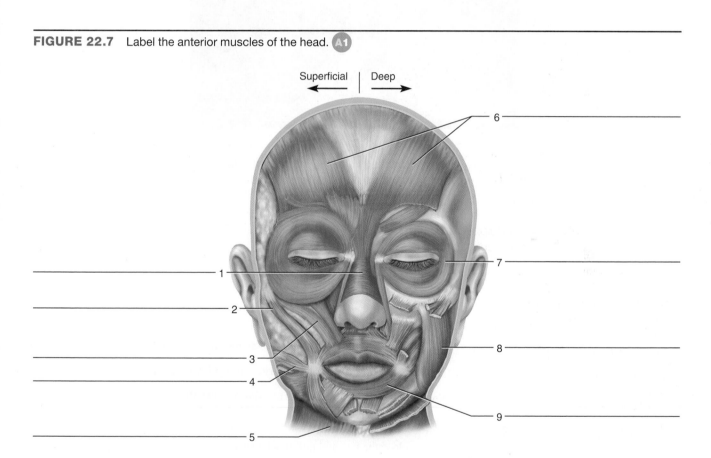

FIGURE 22.8 AP|R Identify the muscles of the head and neck in this lateral view of a cadaver, using the terms provided. A1

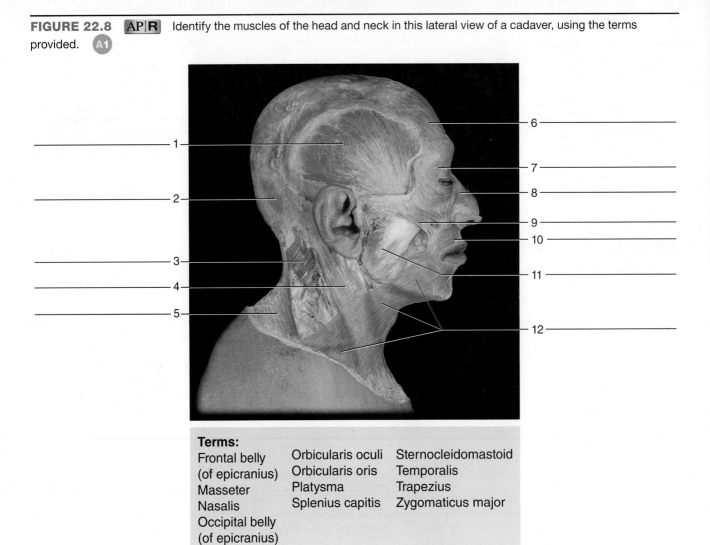

Terms:

Frontal belly (of epicranius)	Orbicularis oculi	Sternocleidomastoid
	Orbicularis oris	Temporalis
Masseter	Platysma	Trapezius
Nasalis	Splenius capitis	Zygomaticus major
Occipital belly (of epicranius)		

PART B: Assessments

Complete the following statements:

1. When the _____ contracts, the corner of the mouth is drawn upward and laterally when laughing. A2

2. The _____ acts to compress the wall of the cheeks when air is blown out of the mouth. A2

3. The _____ causes the lips to close and pucker during kissing, whistling, and speaking. A2

4. The temporalis acts to _____. A2

5. The _____ pterygoid can close the jaw and pull it sideways. A2

6. The _____ pterygoid can protrude the jaw, pull the jaw sideways, and open the mouth. A2

7. The _____ can close the eye, as in blinking. A2

8. The _____ can pull the head toward the chest. A2

9. The muscle used to pout and to express horror is the _____. A2

10. The muscle used to widen the nostrils is the _____. A2

11. The muscle used to elevate the hyoid bone during swallowing is the _____. A2

12. The _____ raises the eyebrows and moves the scalp. A2

PART C: Assessments

Name the muscle indicated by the following combinations of origin and insertion.

Origin	Insertion	Muscle
1. Manubrium of sternum	Thyroid cartilage of larynx	_____
2. Zygomatic arch	Lateral surface of mandible	_____
3. Sphenoid bone	Anterior surface below mandibular condyle	_____
4. Manubrium of sternum and medial clavicle	Mastoid process of temporal bone	_____
5. Lateral surfaces of mandible and maxilla	Orbicularis oris	_____
6. Fascia in upper chest	Lower border of mandible and skin around corner of mouth	_____
7. Temporal bone	Coronoid process and anterior ramus of mandible	_____
8. Spinous processes of cervical and thoracic vertebrae (C7–T6)	Mastoid process of temporal bone and occipital bone	_____
9. Styloid process of temporal bone	Hyoid bone	_____
10. Transverse processes of cervical vertebrae (C2–C6)	Ribs 1–2	_____
11. Frontal and maxillary bones	Skin around eye	_____
12. Medial clavicle and manubrium	Hyoid bone	_____

PART D: Assessments

CRITICAL THINKING ASSESSMENT

Identify the muscles of various facial expressions in the photographs of figure 22.9.

1. _____

2. _____

3. _____

4. _____

5. _____

FIGURE 22.9 Identify the muscles of facial expression being contracted in each of these photographs (a–c), using the terms provided. **A1** **A2**

(a)

(b)

(c)

Terms:
Epicranius (frontal belly)
Orbicularis oculi
Orbicularis oris
Platysma
Zygomaticus major

EXERCISE 23

Muscles of the Chest, Shoulder, and Upper Limb

Purpose of the Exercise

To review the locations, actions, origins, and insertions of the muscles in the chest, shoulder, and upper limb.

MATERIALS NEEDED

Human torso model

Human skeleton, articulated

Muscular models of the upper limb

For Learning Extension Activity:
Long rubber bands

Learning Outcomes AP|R

After completing this exercise, you should be able to:

O1 Locate and identify the muscles of the chest, shoulder, and upper limb.

O2 Describe and demonstrate the action of each of these muscles.

O3 Locate the origin and insertion of each of these muscles in a human skeleton and on the muscular models.

The **O** corresponds to the assessments **A** indicated in the Laboratory Assessment for this Exercise.

Pre-Lab

Carefully read the introductory material and examine the entire lab. Be familiar with the locations, actions, origins, and insertions of the muscles of the chest, shoulder, and upper limb from lecture or the textbook. Answer the pre-lab questions.

Pre-Lab Questions Select the correct answer for each of the following questions:

1. Chest and shoulder muscles that move the arm at the shoulder joint have insertions on the
   a. scapula.   **b.** clavicle and ribs.
   c. humerus.   **d.** radius or ulna.

2. Muscles located in the arm that move the forearm at the elbow joint have insertions on the
   a. scapula and clavicle.   **b.** humerus.
   c. radius or ulna.   **d.** carpals, metacarpals, and phalanges.

3. Rotator cuff muscles all have origins on the
   a. scapula.   **b.** clavicle.
   c. humerus.   **d.** radius.

4. Which of the following is *not* a rotator cuff muscle?
   a. supraspinatus   **b.** infraspinatus
   c. teres major   **d.** teres minor

5. The belly of the muscle is usually _____ the region pulled.
   a. located on   **b.** opposite
   c. distal to   **d.** proximal to

6. The belly of the flexor digitorum profundus muscle is located in the
  a. posterior superficial region of the forearm.
  b. anterior deep region of the forearm.
  c. anterior superficial region of the forearm.
  d. posterior superficial region of the arm.

7. There are more muscles that move the forearm than move the hand.
  a. True _____   **b.** False _____

8. The action of the brachialis muscle is flexion of the elbow joint.
  a. True _____   **b.** False _____

Many of the chest and shoulder muscles insert on the scapula and provide various movements of the pectoral girdle. Some chest and shoulder muscles crossing the shoulder joint have insertions on the humerus and are responsible for moving the arm in various ways at the shoulder joint. Several muscles of the chest are involved in respiration during inspiration and expiration. Muscles located in the arm having insertions on the radius and ulna are responsible for forearm movements at the elbow joint (flexion and extension), as well as rotation. Muscles located in the forearm with insertions on carpals, metacarpals, or phalanges are responsible for various movements of the hand.

The muscles of the upper limb vary from large sizes that provide powerful actions to numerous smaller muscles in the forearm and hand that allow finer motor skills. The human hand, with its opposable thumb, enables us to grasp objects and provides great dexterity for tasks such as writing, drawing, and playing musical instruments.

As you demonstrate the actions of muscles in an upper limb, use the standard anatomical regions to describe them. The *arm* refers to shoulder to elbow; the *forearm* refers to elbow to wrist; and the *hand* includes the wrist, palm, and fingers. Also remember that the belly of a muscle responsible for a joint movement is usually proximal to the region being pulled. If a muscle crosses two joints, the actions allowed are even more diverse.

PROCEDURE A: Muscles of the Chest and Shoulder

The muscle lists and tables in this laboratory exercise reflect muscle groupings using a combination of related locations and functions. Study and master one group at a time as a separate task. Frequently refer to the illustrations, tables, skeletons, and models. Consider working with a partner or a small group to study and review.

1. Study the major muscles of respiration in figures 23.1 and 23.2 and table 23.1. The major muscles of inspiration include the diaphragm and the external intercostals;

FIGURE 23.1 AP|R Major muscles of respiration: (a) lateral view of intercostals; (b) inferior view of diaphragm. (See figure 51.1 for anterior views of the major muscles of respiration as well as additional muscles that assist in respiration.)

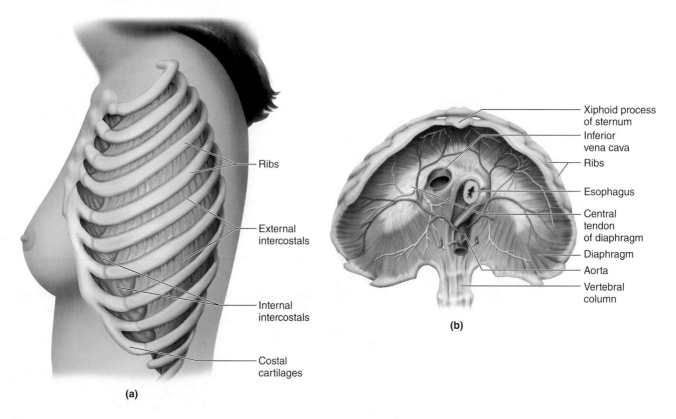

the internal intercostals are the primary muscles for forced expiration. Other muscles, such as the sternocleidomastoid, scalenes, pectoralis minor, rectus abdominis, and external oblique, serve as accessory muscles in respiration. Their actions in respiration are covered in more detail in Laboratory Exercise 51 as part of the respiratory system.

2. Locate the following muscles in the human torso model:

   diaphragm

   external intercostals

   internal intercostals

3. Demonstrate the actions of these muscles in your body.
4. Locate the origins and insertions of these muscles in the human skeleton.
5. Study figures 23.2 and 23.3 and table 23.2.

6. Locate the following muscles in the human torso model. Also locate in your body as many of the muscles as you can.

   muscles that move the pectoral girdle

   - anterior muscles
     - pectoralis minor
     - serratus anterior
   - posterior muscles
     - trapezius
     - rhomboid major
     - rhomboid minor
     - levator scapulae

7. Demonstrate the actions of these muscles in your body.
8. Locate the origins and insertions of these muscles in the human skeleton.
9. Study figures 23.2, 23.3, and 23.4, and table 23.3.

FIGURE 23.2 AP|R Anterior muscles of the chest, shoulder, and arm. Some regional muscles are also labeled.

Superficial | Deep

Trapezius

Internal intercostals

Deltoid

Pectoralis major

Biceps brachii

External oblique

Rectus abdominis

Brachioradialis

Flexor carpi radialis

External intercostals

Pectoralis minor

Coracobrachialis

Serratus anterior

Brachialis

Supinator

Flexor digitorum profundus

Flexor pollicis longus

Table 23.1 Major Muscles of Respiration

Muscle	Origin	Insertion	Action
Diaphragm	Costal cartilages and ribs 7–12, xiphoid process, and lumbar vertebrae	Central tendon of diaphragm	Flattens during contraction, which expands thoracic cavity and compresses abdominal viscera during inspiration
External intercostals	Inferior border of superior rib	Superior border of inferior rib	Elevates ribs, which expands thoracic cavity during inspiration
Internal intercostals	Superior border of inferior rib	Inferior border of superior rib	Depresses ribs, which compresses thoracic cavity during forced expiration

FIGURE 23.3 AP|R Posterior muscles of the shoulder, back, and arm.

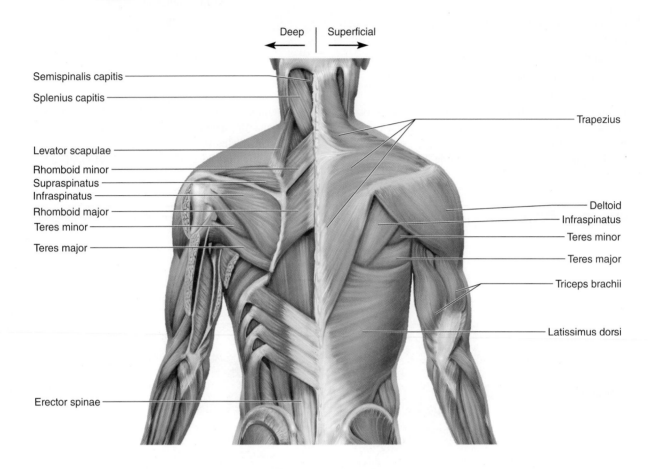

Deep | Superficial

Semispinalis capitis

Splenius capitis

Levator scapulae

Rhomboid minor

Supraspinatus

Infraspinatus

Rhomboid major

Teres minor

Teres major

Erector spinae

Trapezius

Deltoid

Infraspinatus

Teres minor

Teres major

Triceps brachii

Latissimus dorsi

Table 23.2   Muscles that Move the Pectoral Girdle

Muscle	Origin	Insertion	Action
Pectoralis minor	Sternal ends of ribs 3–5	Coracoid process of scapula	Pulls scapula forward and downward; raises ribs
Serratus anterior	Lateral surfaces of most ribs	Medial border of scapula	Pulls scapula anteriorly (protracts scapula)
Trapezius	Occipital bone and spinous processes of C7 and upper thoracic vertebrae	Clavicle and the spine and acromion of scapula	Rotates scapula; various fibers raise scapula, pull scapula medially, or pull scapula and shoulder downward
Rhomboid major	Spinous processes of thoracic vertebrae (T2–T5)	Medial border of scapula	Retracts and elevates scapula
Rhomboid minor	Spinous processes of cervical vertebra C7 and thoracic vertebra T1	Medial border of scapula	Retracts and elevates scapula
Levator scapulae	Transverse processes of cervical vertebrae (C1–C4)	Medial border of scapula	Elevates scapula

FIGURE 23.4 AP|R Rotator cuff muscles: (a) anterior view; (b) posterior view; (c) lateral view. Some regional bones, features, and muscles are also labeled.

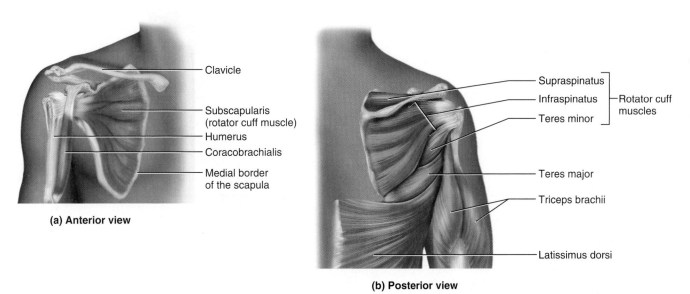

(a) Anterior view

Clavicle
Subscapularis (rotator cuff muscle)
Humerus
Coracobrachialis
Medial border of the scapula

(b) Posterior view

Supraspinatus
Infraspinatus — Rotator cuff muscles
Teres minor
Teres major
Triceps brachii
Latissimus dorsi

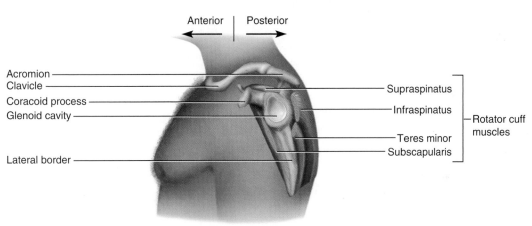

(c) Lateral view

Anterior | Posterior

Acromion
Clavicle
Coracoid process
Glenoid cavity
Lateral border
Supraspinatus
Infraspinatus — Rotator cuff muscles
Teres minor
Subscapularis

Table 23.3 Muscles that Move the Arm

Muscle	Origin	Insertion	Action
Pectoralis major	Clavicle, sternum, and costal cartilages of ribs 1–7	Intertubercular sulcus of humerus	Flexes, adducts, and rotates arm medially
Latissimus dorsi	Spinous processes of lumbar and lower thoracic vertebrae, iliac crest, and lower ribs	Intertubercular sulcus of humerus	Extends, adducts, and rotates the arm medially
Deltoid	Acromion and spine of the scapula and the clavicle	Deltoid tuberosity of humerus	Abducts, extends, and flexes arm
Teres major	Lateral border of scapula	Intertubercular sulcus of humerus	Extends, adducts, and rotates arm medially
Coracobrachialis	Coracoid process of scapula	Shaft of humerus	Flexes and adducts the arm
Supraspinatus	Supraspinous fossa of scapula	Greater tubercle of humerus	Abducts the arm
Infraspinatus	Infraspinous fossa of scapula	Greater tubercle of humerus	Rotates arm laterally
Teres minor	Lateral border of scapula	Greater tubercle of humerus	Rotates arm laterally
Subscapularis	Subscapular fossa of scapula	Lesser tubercle of humerus	Rotates arm medially

10. Locate the following muscles in the human torso model and models of the upper limb. Also locate in your body as many of the muscles as you can.

    muscles that move the arm
- origins on axial skeleton
  - pectoralis major
  - latissimus dorsi
- origins on scapula
  - deltoid
  - teres major
  - coracobrachialis
- rotator cuff (SITS) muscles (origins also on scapula)—SITS is an acronym that corresponds to the first letter of each muscle in the group
  - supraspinatus
  - infraspinatus
  - teres minor
  - subscapularis

11. Demonstrate the actions of these muscles in your body.
12. Locate the origins and insertions of these muscles in the human skeleton.
13. Complete Part A of Laboratory Assessment 23.

PROCEDURE B: Muscles of the Upper Limb

1. Study figures 23.2, 23.3, 23.5, and 23.6 and table 23.4.

FIGURE 23.5 AP|R The (a) superficial, (b) intermediate, and (c) deep anterior forearm muscles.

Pronator teres
Brachioradialis
Flexor carpi radialis
Palmaris longus
Flexor carpi ulnaris
Pronator quadratus
Palmar aponeurosis

(a) Anterior muscles: superficial

Flexor digitorum superficialis

(b) Anterior muscles: intermediate

Supinator
Flexor digitorum profundus
Flexor pollicis longus
Pronator quadratus

(c) Anterior muscles: deep

2. Locate the following muscles in the human torso model and models of the upper limb. Also locate in your body as many of the muscles as you can.

muscles that move the forearm

- muscle bellies in arm
  - biceps brachii
  - brachialis
  - triceps brachii

- muscle bellies in forearm
  - brachioradialis
  - anconeus
  - supinator
  - pronator teres
  - pronator quadratus

3. Demonstrate the actions of these muscles in your body.
4. Locate the origins and insertions of these muscles in the human skeleton.

FIGURE 23.6 **AP|R** The (a) superficial and (b) deep posterior forearm muscles.

(a) Posterior muscles: superficial

(b) Posterior muscles: deep

Table 23.4 Muscles that Move the Forearm

Muscle	Origin	Insertion	Action
Biceps brachii	Coracoid process (short head) and tubercle above glenoid cavity of scapula (long head)	Radial tuberosity	Flexes elbow and supinates forearm and hand
Brachialis	Distal anterior shaft of humerus	Coronoid process of ulna	Flexes elbow (agonist)
Triceps brachii	Tubercle below glenoid cavity and lateral and medial surfaces of humerus	Olecranon process of ulna	Extends elbow
Brachioradialis	Distal lateral end of humerus	Lateral surface of radius near styloid process	Flexes elbow
Anconeus	Lateral epicondyle of humerus	Olecranon process of ulna	Extends elbow
Supinator	Lateral epicondyle of humerus and ulna distal to radial notch	Lateral surface of proximal radius	Supinates forearm and hand
Pronator teres	Medial epicondyle of humerus and coronoid process of ulna	Lateral surface of radius	Pronates forearm and hand
Pronator quadratus	Anterior distal end of ulna	Anterior distal end of radius	Pronates forearm and hand

Table 23.5   Muscles that Move the Hand

Muscle	Origin	Insertion	Action
Palmaris longus (sometimes absent)	Medial epicondyle of humerus	Palmar aponeurosis	Flexes wrist
Flexor carpi radialis	Medial epicondyle of humerus	Base of metacarpals II–III	Flexes wrist and abducts hand
Flexor carpi ulnaris	Medial epicondyle of humerus and olecranon process of ulna	Carpals (pisiform and hamate) and metacarpal V	Flexes wrist and adducts hand
Flexor digitorum superficialis	Medial epicondyle of humerus, coronoid process of ulna, and radius	Middle phalanges of fingers 2–5	Flexes fingers and wrist
Flexor digitorum profundus	Anterior ulna and interosseous membrane	Distal phalanges of fingers 2–5	Flexes fingers 2–5
Flexor pollicis longus	Anterior radius and interosseous membrane	Distal phalanx of thumb	Flexes thumb
Extensor carpi radialis longus	Lateral supracondylar ridge of humerus	Base of metacarpal II	Extends wrist and abducts hand
Extensor carpi radialis brevis	Lateral epicondyle of humerus	Base of metacarpal III	Extends wrist and abducts hand
Extensor digitorum	Lateral epicondyle of humerus	Posterior phalanges of fingers 2–5	Extends fingers
Extensor digiti minimi	Lateral epicondyle of humerus	Proximal phalanx of finger 5	Extends finger 5 (little finger)
Extensor carpi ulnaris	Lateral epicondyle of humerus and posterior ulna	Base of metacarpal V	Extends wrist and adducts hand
Abductor pollicis longus	Posterior radius and ulna and interosseous membrane	Metacarpal I and trapezium	Abducts thumb
Extensor pollicis longus	Posterior ulna and interosseous membrane	Distal phalanx of thumb	Extends thumb
Extensor pollicis brevis	Posterior radius and interosseous membrane	Proximal phalanx of thumb	Extends thumb
Extensor indicis	Posterior ulna and interosseous membrane	Middle and distal phalanges of finger 2	Extends finger 2 (index finger)

5. Study figures 23.5 and 23.6 and table 23.5.
6. Locate the following muscles in the human torso model and models of the upper limb. Also locate in your body as many of the muscles as you can.

   muscles that move the hand
   - anterior superficial flexor muscles
     - palmaris longus (absent sometimes)
     - flexor carpi radialis
     - flexor carpi ulnaris
     - flexor digitorum superficialis (intermediate depth)
   - anterior deep flexor muscles
     - flexor digitorum profundus
     - flexor pollicis longus
   - posterior superficial extensor muscles
     - extensor carpi radialis longus
     - extensor carpi radialis brevis
     - extensor digitorum
     - extensor digiti minimi
     - extensor carpi ulnaris
   - posterior deep extensor muscles
     - abductor pollicis longus
     - extensor pollicis longus
     - extensor pollicis brevis
     - extensor indicis

7. Demonstrate the actions of these muscles in your body.
8. Locate the origins and insertions of these muscles in the human skeleton.
9. Complete Parts B, C, D, and E of the laboratory assessment.

LEARNING EXTENSION ACTIVITY

A long rubber band can be used to simulate muscle locations, origins, insertions, and actions on muscular models, the skeleton, or a laboratory partner. Hold one end of the rubber band firmly on the origin location of a muscle; then slightly stretch the rubber band and hold the other end on the insertion site. Allow the insertion end to slowly move toward the origin end to simulate the contraction and action of the muscle.

Name _____

Date _____

Section _____

The Ⓐ corresponds to the indicated Learning Outcome(s) Ⓞ found at the beginning of the Laboratory Exercise.

Muscles of the Chest, Shoulder, and Upper Limb

PART A: Assessments

Identify the chest, shoulder, and respiratory muscles shown in figures 23.7, 23.8, and 23.9.

FIGURE 23.7   AP|R   Label the deep anterior muscles of the chest.

Clavicle
Coracoid process
1
2
3
4
Xiphoid process

FIGURE 23.8 AP|R Label the muscles of respiration of the posterior body wall (anterior view). The heart and lungs are removed. A1

Esophagus

Aorta

Inferior vena cava

1

2

3

FIGURE 23.9 AP|R Identify the posterior muscles of the left shoulder of a cadaver, using the terms provided. The deltoid and trapezius muscles have been removed. A1

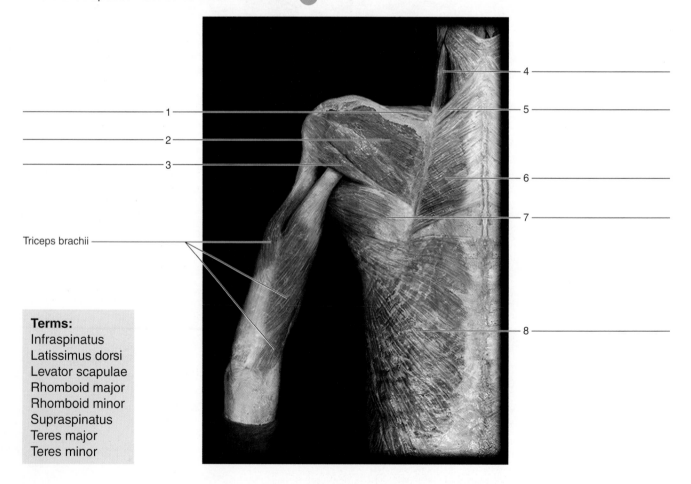

1

2

3

4

5

6

7

8

Triceps brachii

Terms:
Infraspinatus
Latissimus dorsi
Levator scapulae
Rhomboid major
Rhomboid minor
Supraspinatus
Teres major
Teres minor

PART B: Assessments

Identify the muscles indicated in the chest, shoulder, arm, and forearm in figures 23.10 and 23.11.

FIGURE 23.10 AP|R Label the anterior muscles of the chest, shoulder, and upper limb. A1

Superficial | Deep

Trapezius

1
2
3
4
5
6

7
8
9

FIGURE 23.11 AP|R Identify the anterior muscles of the right forearm of a cadaver, using the terms provided. A1

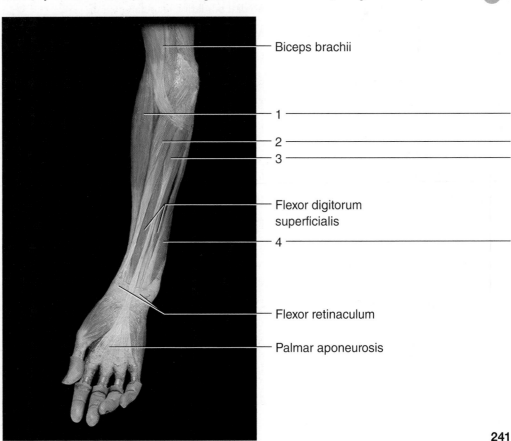

Biceps brachii

1
2
3

Flexor digitorum superficialis

4

Flexor retinaculum

Palmar aponeurosis

Terms:
Brachioradialis
Flexor carpi radialis
Flexor carpi ulnaris
Palmaris longus

PART C: Assessments

Match the muscles in column A with the actions in column B. Place the letter of your choice in the space provided. **A2**

Column A	Column B
a. Brachialis	_____ **1.** Abducts, extends, and flexes arm
b. Coracobrachialis	_____ **2.** Pulls arm anteriorly (flexion) and across chest (adduction) and rotates arm medially
c. Deltoid	
d. Extensor carpi ulnaris	_____ **3.** Flexes wrist and adducts hand
e. Flexor carpi ulnaris	_____ **4.** Raises and adducts (retracts) scapula
f. Infraspinatus	_____ **5.** Raises ribs in forceful inhalation or pulls scapula forward and downward
g. Pectoralis major	
h. Pectoralis minor	_____ **6.** Used to thrust shoulder anteriorly (protraction), as when pushing something
i. Rhomboid major	_____ **7.** Flexes the forearm at the elbow
j. Serratus anterior	_____ **8.** Flexes and adducts arm at the shoulder along with pectoralis major
k. Teres major	
l. Triceps brachii	_____ **9.** Extends the forearm at the elbow
	_____ **10.** Extends, adducts, and rotates arm medially
	_____ **11.** Extends and adducts hand at the wrist
	_____ **12.** Rotates arm laterally

PART D: Assessments

Name the muscle indicated by the following combinations of origin and insertion. **A3**

Origin	Insertion	Muscle
1. Spinous processes of thoracic vertebrae	Medial border of scapula	_____
2. Lateral surfaces of most ribs	Medial border of scapula	_____
3. Sternal ends of ribs 3–5	Coracoid process of scapula	_____
4. Coracoid process of scapula	Shaft of humerus	_____
5. Lateral border of scapula	Intertubercular sulcus of humerus	_____
6. Subscapular fossa of scapula	Lesser tubercle of humerus	_____
7. Lateral border of scapula	Greater tubercle of humerus	_____
8. Distal anterior shaft of humerus	Coronoid process of ulna	_____
9. Medial epicondyle of humerus	Palmar aponeurosis	_____
10. Anterior radius and interosseous membrane	Distal phalanx of thumb	_____
11. Posterior ulna and interosseous membrane	Middle and distal phalanges of finger 2	_____
12. Clavicle and spine and acromion of scapula	Deltoid tuberosity of humerus	_____
13. Ribs 7–12 and xiphoid process	Central tendon	_____
14. Superior border of inferior rib	Inferior border of superior rib	_____

PART E: Assessments

CRITICAL THINKING ASSESSMENT

Identify the muscles indicated in figure 23.12.

FIGURE 23.12 Label these muscles that appear as body surface features in these photographs (a, b, and c) by placing the correct numbers in the spaces provided. A1

____ Biceps brachii
____ Deltoid
____ External oblique
____ Pectoralis major
____ Rectus abdominis
____ Serratus anterior
____ Sternocleidomastoid
____ Trapezius

(a)

____ Biceps brachii
____ Deltoid
____ Infraspinatus
____ Latissimus dorsi
____ Trapezius
____ Triceps brachii

(b)

(continued)

FIGURE 23.12 *Continued.*

____ Biceps brachii
____ Brachioradialis
____ Deltoid
____ Pectoralis major
____ Serratus anterior
____ Trapezius
____ Triceps brachii

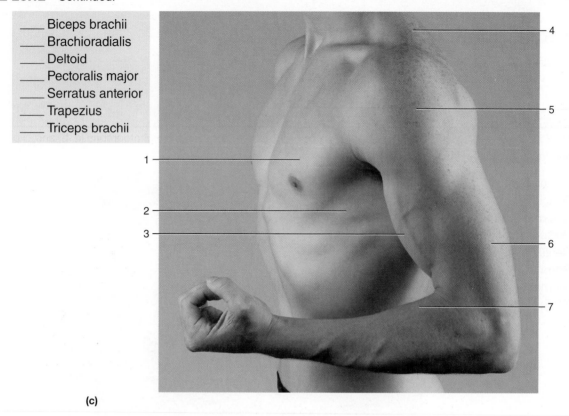

(c)

LABORATORY
EXERCISE 24

Muscles of the Vertebral Column, Abdominal Wall, and Pelvic Floor

Purpose of the Exercise

To review the actions, origins, and insertions of the muscles of the vertebral column, abdominal wall, and pelvic floor.

MATERIALS NEEDED

Human torso model with musculature

Human skeleton, articulated

Muscular models of male and female pelves

Learning Outcomes AP|R

After completing this exercise, you should be able to:

O1 Locate and identify the muscles of the vertebral column, abdominal wall, and pelvic floor.

O2 Describe the action of each of these muscles.

O3 Locate the origin and insertion of each of these muscles in a human skeleton or on muscular models.

The **O** corresponds to the assessments **A** indicated in the Laboratory Assessment for this Exercise.

Pre-Lab

Carefully read the introductory material and examine the entire lab. Be familiar with the locations, actions, origins, and insertions of the muscles of the vertebral column, abdominal wall, and the pelvic floor from lecture or the textbook. Answer the pre-lab questions.

Pre-Lab Questions Select the correct answer for each of the following questions:

1. Which of the following muscles is *not* part of the erector spinae group?
   a. iliocostalis   b. longissimus
   c. spinalis   d. quadratus lumborum

2. The abdominal wall muscles possess the following functions *except*
   a. extend the spine.
   b. compress the abdominal visceral organs.
   c. assist in forceful exhalation.
   d. help maintain posture.

3. The abdominal muscle nearest the midline is the
   a. external oblique.   b. internal oblique.
   c. rectus abdominis.   d. transversus abdominis.

4. Which of the following muscles would be most important in relation to a vaginal delivery?
   a. bulbospongiosus   b. quadratus lumborum
   c. spinalis   d. erector spinae

5. The four muscles of the abdominal wall
   a. contain muscle fibers in the same direction.
   b. contain muscle fibers in four different directions.
   c. are responsible for trunk extension.
   d. fully overlap each other.

245

6. The iliocostalis represents the lateral muscle group of the erector spinae.
   a. True _____   b. False _____
7. The pelvic floor muscles are all arranged in a single superficial layer.
   a. True _____   b. False _____

The deep muscles of the back extend the vertebral column. Because the muscles have many origins, insertions, and subgroups, the muscles overlap each other. The deep back muscles can extend the spine when contracting as a group but also help to maintain posture and normal spine curvatures.

The anterior and lateral walls of the abdomen contain broad, flattened muscles arranged in layers. These muscles connect the rib cage and vertebral column to the pelvic girdle. The abdominal wall muscles compress the abdominal visceral organs, help maintain posture, assist in forceful exhalation, and contribute to trunk flexion and waist rotation.

The muscles in the region of the perineum form the pelvic floor (outlet) and can be divided into two triangles: a urogenital triangle and an anal triangle. The anterior urogenital triangle contains superficial and deeper muscle layers associated with the external genitalia. The posterior anal triangle contains the pelvic diaphragm and the anal sphincter. The pelvic floor is penetrated by the urethra, vagina, and anus in a female; thus, pelvic floor muscles are important in obstetrics.

PROCEDURE A: Muscles of the Vertebral Column and Abdominal Wall

The muscles that move the vertebral column are located deep in the back from the pelvic girdle to the skull. The erector spinae group includes three major columns of muscles. Each group is composed of several muscles (table 24.1). For the purpose of this exercise, the entire erector spinae group is considered to have the common action of extending the vertebral column when they contract bilaterally; if they contract unilaterally, lateral flexion of the trunk occurs. A familiar lifting injury, known as a muscle strain (pulled muscle), can occur especially in the lumbar region of the erector spinae group. The three muscles of the semispinalis group have actions similar to the erector spinae group. The quadratus lumborum is another major muscle of the inferior vertebral column and functions in extension and lateral flexion of the vertebral column.

The splenius capitis and the semispinalis capitis muscles, located in the posterior neck extending from the vertebral column to the skull, are large muscles associated with the vertebral column and act in extensions and rotations of the head and neck. Smaller muscles of the vertebral column, including the serratus posterior and multifidus, assist in either breathing, extension, or rotation functions.

The abdominal wall is composed of four muscles situated as broad, flat muscle layers. Their muscle fibers are oriented in different directions much like plywood. This provides additional strength to support and compress the abdominal viscera. The muscle layers attach to structures such as the linea alba, inguinal ligament, or aponeuroses, since the abdominal region lacks skeletal attachments for these muscles.

1. Study figure 24.1 and table 24.1.
2. Locate the following muscles in the human torso model:
   erector spinae group
   - iliocostalis group (lateral group)
   - longissimus group (intermediate group)
   - spinalis group (medial group)

   semispinalis group

   quadratus lumborum
3. Demonstrate the actions of these muscles in your body.
4. Locate the origins and insertions of these muscles in the human skeleton.
5. Study figures 24.2 and 24.3, and table 24.2.
6. Locate the following muscles in the human torso model:
   external oblique

   internal oblique

   transversus abdominis (transverse abdominal)

   rectus abdominis
7. Demonstrate the actions of these muscles in your body.

Table 24.1 Muscles of the Vertebral Column

Muscle	Origin	Insertion	Action
Erector spinae (three columns of muscles: iliocostalis group, longissimus group, and spinalis group)	Various sites including cervical, thoracic, and lumbar vertebrae, ribs, and sacrum	Various sites including occipital bone, mastoid process, cervical and thoracic vertebrae	Maintains posture; extends vertebral column; lateral flexion when unilateral contractions occur
Semispinalis group (semispinalis capitis, cervicis, and thoracis)	Transverse processes of cervical and thoracic vertebrae	Spinous processes of cervical and thoracic vertebrae and occipital bone	Extends head and vertebral column; lateral flexion of neck and vertebral column when unilateral contractions occur
Quadratus lumborum	Iliac crest	Transverse processes of L1–L4 and rib 12	Extends vertebral column; lateral flexion when unilateral contractions occur

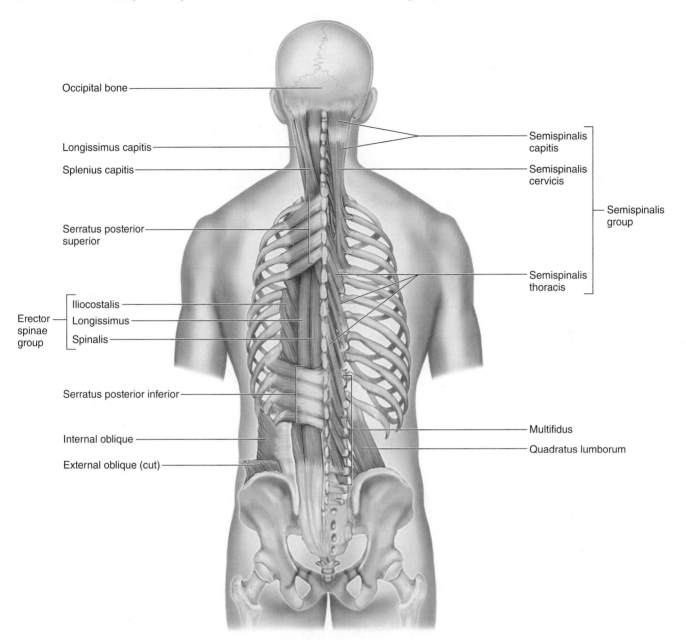

Occipital bone

Longissimus capitis

Splenius capitis

Serratus posterior superior

Erector spinae group
- Iliocostalis
- Longissimus
- Spinalis

Serratus posterior inferior

Internal oblique

External oblique (cut)

Semispinalis capitis

Semispinalis cervicis

Semispinalis group

Semispinalis thoracis

Multifidus

Quadratus lumborum

FIGURE 24.2 APR Muscles of the abdominal wall.

External oblique

Tendinous intersection

Rectus sheath

Umbilicus

Linea alba

Aponeurosis

Inguinal ligament

Rectus abdominis

Transversus abdominis

Internal oblique

FIGURE 24.3 Transverse view of the midsection of the abdominal wall.

Posterior

Anterior

Linea alba

Parietal peritoneum

Aponeuroses

Subcutaneous fat

Skin

Rectus sheath

Transversus abdominis

Internal oblique

External oblique

Rectus abdominis

Table 24.2 Muscles of the Abdominal Wall

Muscle	Origin	Insertion	Action
External oblique	Anterior surfaces of ribs 5–12	Anterior iliac crest and linea alba by aponeurosis	Tenses abdominal wall; compresses abdominal contents; aids in trunk rotation
Internal oblique	Iliac crest and inguinal ligament	Cartilages of lower ribs, linea alba, and crest of pubis	Same as external oblique
Transversus abdominis	Costal cartilages of lower ribs, iliac crest, and inguinal ligament	Linea alba and pubic crest	Compresses abdominal contents
Rectus abdominis	Pubic crest and pubic symphysis	Xiphoid process of sternum and costal cartilages of ribs 5–7	Flexes vertebral column; compresses abdominal contents

8. Locate the origins and insertions of these muscles in the human skeleton.
9. Complete Parts A and B of Laboratory Assessment 24.

PROCEDURE B: Muscles of the Pelvic Floor

The diamond-shaped region between the lower limbs is the *perineum*. It is located between the ischial tuberosities on the lateral sides, the anterior pubic symphysis, and the posterior coccyx. The perineum possesses an anterior *urogenital triangle* and a posterior *anal triangle*. The urogenital triangle includes the external genitalia, the urogenital openings, and associated muscles. The anal triangle includes the anus as the digestive opening and associated muscles. The pelvic floor is composed of three layers of muscles. The term *diaphragm* is used in conjunction with some of these muscle layers, forming a type of partition or wall, and referred to as a urogenital diaphragm and a pelvic diaphragm.

1. Study figure 24.4 and table 24.3.
2. Locate the following muscles in the models of the male and female pelves:

   levator ani

   coccygeus

   superficial transverse perineal muscle

   bulbospongiosus

   ischiocavernosus

3. Locate the origins and insertions of these muscles in the human skeleton.
4. Complete Part C of the laboratory assessment.

Table 24.3 Muscles of the Pelvic Floor

Muscle	Origin	Insertion	Action
Levator ani	Pubis and ischial spine	Coccyx	Supports pelvic viscera and provides sphincterlike action in anal canal and vagina
Coccygeus	Ischial spine	Sacrum and coccyx	Supports pelvic viscera
Superficial transverse perineal muscle	Ischial tuberosity	Central tendon of perineum	Supports pelvic viscera
Bulbospongiosus	Central tendon of perineum	Males: Ensheaths root of penis Females: Ensheaths root of clitoris	Males: Assists emptying of urethra and assists in erection of penis Females: Constricts vagina and assists in erection of clitoris
Ischiocavernosus	Ischial tuberosity	Pubic symphysis near base of penis or clitoris	Males: Erects penis Females: Erects clitoris

FIGURE 24.4 Muscles of the pelvic floor: (a) superficial perineal muscles of male and female; (b) urogenital diaphragm (next deeper perineal muscles) of male and female; (c) pelvic diaphragm (deepest layer) of female.

Male

Female

Penis

Ischiocavernosus — Clitoris

Perineal raphe — Urethra

Bulbospongiosus — Urogenital triangle

Superficial transverse perineal muscle — Vagina

Levator ani — Anus

Gluteus maximus — Anal triangle

(a) Superficial region of perineum, inferior view

Urethra

Pubic symphysis — Vagina

External urethral sphincter — Anus

Deep transverse perineal muscle

Central tendon

External anal sphincter

(b) Urogenital diaphragm of perineum, inferior view

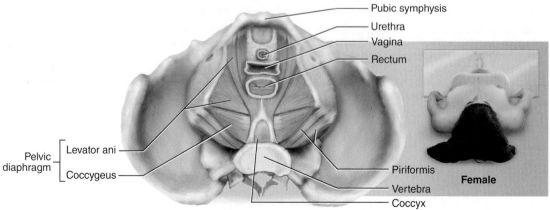

Pubic symphysis

Urethra

Vagina

Rectum

Pelvic diaphragm { Levator ani

Coccygeus }

Piriformis

Female

Vertebra

Coccyx

(c) AP|R Female pelvic diaphragm, superior view

Name _____

Date _____

Section _____

The Ⓐ corresponds to the indicated Learning Outcome(s) Ⓞ found at the beginning of the Laboratory Exercise.

Muscles of the Vertebral Column, Abdominal Wall, and Pelvic Floor

PART A: Assessments

Identify the muscles indicated in figures 24.5 and 24.6.

FIGURE 24.5 AP|R Label the deep back muscles of a cadaver, using the terms provided. The trapezius, latissimus dorsi, rhomboids, and serratus posterior have been removed. Ⓐ1

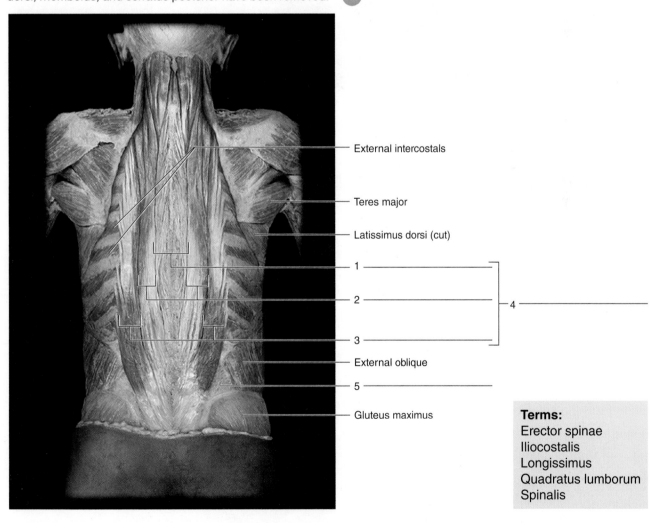

— External intercostals

— Teres major

— Latissimus dorsi (cut)

1 _____ ⌐
2 _____ │ 4 _____
3 _____ ⌐

— External oblique

5 _____

— Gluteus maximus

Terms:
Erector spinae
Iliocostalis
Longissimus
Quadratus lumborum
Spinalis

FIGURE 24.6 AP|R Label the muscles of the abdominal wall.

Pectoralis minor

Serratus anterior

Rectus sheath

1

2

Inguinal ligament

Internal intercostals

External intercostals

3

4

CRITICAL THINKING ACTIVITY

List the abdominal muscles from superficial to deep for an appendectomy incision. A1

_____

_____

_____

_____

PART B: Assessments

Complete the following statements:

1. A band of tough connective tissue in the midline of the anterior abdominal wall called the _____ serves as a muscle attachment. **A1**

2. The _____ muscle spans from the costal cartilages and xiphoid process to the pubic bones. **A3**

3. The _____ forms the third layer (deepest layer) of the abdominal wall muscles. **A1**

4. The action of the external oblique muscle is to _____. **A2**

5. The action of the rectus abdominis is to _____. **A2**

6. The iliocostalis, longissimus, and spinalis muscles together form the _____ group. **A1**

7. The _____ has the origin on the iliac crest and the insertion on rib 12 and L1–L4 vertebrae. **A3**

8. The origin of the _____ is on the iliac crest and inguinal ligament. **A3**

9. The insertion of the _____ is on the xiphoid process and costal cartilages. **A3**

10. Name four muscles that compress the abdominal contents. **A2**

    _____

    _____

    _____

    _____

11. The actions of the _____ group of vertebral column muscles are similar to the erector spinae group and the quadratus lumborum. **A2**

12. Name the three muscles included in the semispinalis group. **A1**

    _____

    _____

    _____

PART C: Assessments

Complete the following statements:

1. The levator ani and coccygeus together form the _____ diaphragm. (A1)

2. The posterior portion of the perineum is called the _____ triangle. (A1)

3. The _____ ensheaths the root of the penis. (A1)

4. In females, the bulbospongiosus acts to constrict the _____. (A2)

5. In the female, the pelvic floor (outlet) is penetrated by the urethra, vagina, and _____. (A1)

6. In the female, the _____ muscles are separated by the vagina, urethra, and anal canal. (A1)

7. The insertion of the levator ani is on the _____. (A3)

8. The coccygeus muscle has an origin on the _____. (A3)

9. The superficial transverse perineal muscle acts to support the _____. (A2)

10. Name five muscles of the pelvic floor that males and females both have in common. (A1) (There are more than five correct answers.)

    _____

    _____

    _____

    _____

    _____

LABORATORY EXERCISE 25

Muscles of the Hip and Lower Limb

Purpose of the Exercise

To review the actions, origins, and insertions of the muscles that move the thigh, leg, and foot.

MATERIALS NEEDED

Human muscular model

Human skeleton, articulated

Muscular models of the lower limb

For Learning Extension Activity:
Long rubber bands

Learning Outcomes AP|R

After completing this exercise, you should be able to:

O1 Locate and identify the muscles that move the thigh, leg, and foot.

O2 Describe and demonstrate the actions of each of these muscles.

O3 Locate the origin and insertion of each of these muscles in a human skeleton and on muscular models.

The **O** corresponds to the assessments **A** indicated in the Laboratory Assessment for this Exercise.

Pre-Lab

Carefully read the introductory material and examine the entire lab. Be familiar with the locations, actions, origins, and insertions of the muscles of the hip and lower limb from lecture or the textbook. Answer the pre-lab questions.

Pre-Lab Questions Select the correct answer for each of the following questions:

1. Muscles that move the thigh at the hip joint have origins on the
   a. pelvis.   b. femur.
   c. tibia.   d. patella.

2. Anterior thigh muscles serve as prime movers of
   a. flexion of the leg at the knee joint.
   b. abduction of the thigh at the hip joint.
   c. adduction of the thigh at the hip joint.
   d. extension of the leg at the knee joint.

3. The muscles of the lower limb _____ than those of the upper limb.
   a. are more numerous   b. are larger
   c. conduct actions with   d. are less powerful
      more precision

4. Which of the following muscles is *not* part of the quadriceps group?
   a. rectus femoris   b. vastus lateralis
   c. semitendinosus   d. vastus medialis

5. Which of the following muscles is *not* part of the hamstring group?
   a. rectus femoris   b. biceps femoris
   c. semitendinosus   d. semimembranosus

6. Muscles located in the lateral leg include actions of
   a. extension of the knee joint.   b. eversion of the foot.
   c. flexion of the toes.   d. extension of the toes.

7. Contractions of the sartorius muscle involve movements across two different joints.

   True _____   False _____

8. The insertions of the four quadriceps femoris muscles are on the femur.

   True _____   False _____

The muscles that move the thigh at the hip joint have their origins on the pelvis and insertions usually on the femur. Those muscles attached on the anterior pelvis act to flex the thigh at the hip joint; those attached on the posterior pelvis act to extend the thigh at the hip joint; those attached on the lateral side of the pelvis act as abductors and rotators of the thigh at the hip joint; and those attached on the medial pelvis act as adductors of the thigh at the hip joint.

The muscles that move the leg at the knee joint have their origins on the femur or the hip bone. The anterior thigh muscles have their insertions on the proximal tibia and serve as prime movers of extension of the leg at the knee joint. The posterior thigh muscles have their insertions on the tibia or fibula and act as prime movers of flexion of the leg at the knee joint.

The muscles that move the foot have their origins on the tibia, fibula, or femur. The anterior leg muscles have actions of dorsiflexion and extension of the toes; the posterior leg muscles have actions of plantar flexion and flexion of the toes; the lateral leg muscles have actions of eversion of the foot.

In contrast to the upper limb muscles, those in the lower limb tend to be much larger and are involved in powerful contractions for walking or running as well as isometric contractions in standing. Although the lower limb has more stability than the upper limb, it has a more limited range of motion. As compared to the forearm, the leg has fewer muscles because the foot is not as involved in actions that require precision movements as the hand.

As you demonstrate the actions of muscles in the lower limb, refer to standard anatomical regions of the limb. The *thigh* refers to hip to knee; the *leg* refers to knee to ankle; and the *foot* refers to ankle, metatarsal area, and toes. Several muscles in the hip and lower limb cross more than one joint and involve movements of both joints. Examples of muscles crossing more than one joint include the sartorius and gastrocnemius.

PROCEDURE: Muscles of the Hip and Lower Limb

1. Study figures 25.1, 25.2, 25.3, and 25.4, and table 25.1.
2. Locate the following muscles in the human torso model and in the lower limb models. Also locate as many of them as possible in your body.

   muscles that move the thigh/hip joint
   - ■ anterior hip muscles
     - • iliopsoas group
       - psoas major
       - iliacus

FIGURE 25.1   Muscles of the anterior right hip and thigh.

Psoas major

Iliacus

Anterior superior iliac spine

Tensor fasciae latae

Medial adductor muscles

Iliotibial tract (band)

Rectus femoris

Sartorius

Vastus lateralis

Vastus medialis

Quadriceps femoris tendon

Patellar ligament

- ■ posterior and lateral hip muscles
  - • gluteus maximus
  - • gluteus medius
  - • gluteus minimus
  - • tensor fasciae latae
  - • piriformis
- ■ medial adductor muscles
  - • pectineus
  - • adductor longus
  - • adductor magnus
  - • adductor brevis
  - • gracilis

3. Demonstrate the actions of these muscles in your body.

FIGURE 25.2 Selected individual muscles of the anterior right hip and medial thigh.

Iliopsoas
Psoas major
Iliacus
Pectineus
Adductor brevis
Adductor longus
Adductor magnus
Gracilis

FIGURE 25.3 Muscles of the posterior right hip and thigh.

Gluteus medius
Gluteus maximus
Gracilis
Adductor magnus
Long head
Vastus lateralis
Semitendinosus
Iliotibial tract
Biceps femoris
Short head
Semimembranosus
Gastrocnemius

FIGURE 25.4 Superficial and deep posterior hip muscles.

Table 25.1 Muscles that Move the Thigh

Muscle	Origin	Insertion	Action
Psoas major	Intervertebral discs, bodies, and transverse processes of T12–L5	Lesser trochanter of femur	Flexes thigh at hip; flexes trunk at hip (when thigh is fixed)
Iliacus	Iliac fossa of ilium	Lesser trochanter of femur	Same as psoas major
Gluteus maximus	Sacrum, coccyx, and posterior iliac crest	Gluteal tuberosity of femur and iliotibial tract	Extends thigh at hip
Gluteus medius	Lateral surface of ilium	Greater trochanter of femur	Abducts and medially rotates thigh
Gluteus minimus	Lateral surface of ilium	Greater trochanter of femur	Same as gluteus medius
Tensor fasciae latae	Iliac crest and anterior superior iliac spine	Iliotibial tract	Abducts and medially rotates thigh
Piriformis	Anterior surface of sacrum	Greater trochanter of femur	Abducts and laterally rotates thigh
Pectineus	Pubis	Femur distal to lesser trochanter	Adducts and flexes thigh
Adductor longus	Pubis near pubic symphysis	Linea aspera of femur	Adducts and flexes thigh
Adductor magnus	Pubis and ischial tuberosity	Linea aspera of femur	Adducts thigh; posterior portion extends thigh, and anterior portion flexes thigh
Adductor brevis	Pubis	Linea aspera of femur	Adducts thigh
Gracilis	Pubis	Medial surface of tibia	Adducts thigh and flexes leg at knee

4. Locate the origins and insertions of these muscles in the human skeleton.
5. Study figures 25.1, 25.3, and 25.5 and table 25.2.
6. Locate the following muscles in the human torso model and in the lower limb models. Also locate as many of them as possible in your body.

   muscles that move the leg/knee joint
   - anterior thigh muscles
     - sartorius
     - quadriceps femoris group
       rectus femoris
       vastus lateralis
       vastus medialis
       vastus intermedius
   - posterior thigh muscles
     - hamstring group
       biceps femoris
       semitendinosus
       semimembranosus

7. Demonstrate the actions of these muscles in your body.
8. Locate the origins and insertions of these muscles in the human skeleton.
9. Complete Part A of Laboratory Assessment 25.
10. Study figures 25.6, 25.7, and 25.8, and table 25.3.

FIGURE 25.5 Selected quadriceps femoris muscles of the anterior right thigh. The rectus femoris removed to expose the deeper muscles and tendons of the quadriceps femoris group.

Vastus intermedius

Vastus lateralis

Vastus medialis

Quadriceps femoris tendon

Patella

Patellar ligament

Tibial tuberosity

Table 25.2 Muscles that Move the Leg

Muscle	Origin	Insertion	Action
Sartorius	Anterior superior iliac spine	Proximal medial surface of tibia	Flexes, abducts, and laterally rotates thigh at hip; flexes leg at knee
Quadriceps Femoris Group			
Rectus femoris	Anterior inferior iliac spine and superior margin of acetabulum	Patella by common quadriceps tendon, which continues as patellar ligament to tibial tuberosity	Extends leg at knee; flexes thigh at hip
Vastus lateralis	Greater trochanter and linea aspera of femur		Extends leg at knee
Vastus medialis	Linea aspera of femur		Extends leg at knee
Vastus intermedius	Anterior and lateral surfaces of femur		Extends leg at knee
Hamstring Group			
Biceps femoris	Ischial tuberosity (long head) and linea aspera of femur (short head)	Head of fibula and lateral condyle of tibia	Flexes leg at knee; rotates leg laterally; extends thigh
Semitendinosus	Ischial tuberosity	Proximal medial surface of tibia	Flexes leg at knee; rotates leg medially; extends thigh
Semimembranosus	Ischial tuberosity	Posterior medial condyle of tibia	Flexes leg at knee; rotates leg medially; extends thigh

FIGURE 25.6 Muscles of the anterior right leg (a and b).

(a) Superficial view

(b) Selected individual muscles

FIGURE 25.7 Muscles of the posterior right leg (a and b).

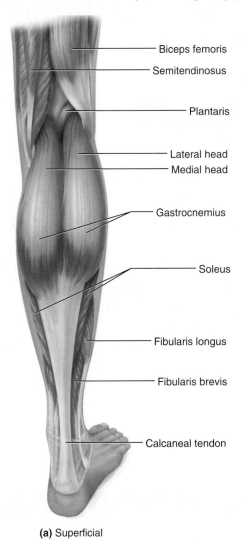

Biceps femoris
Semitendinosus
Plantaris
Lateral head
Medial head
Gastrocnemius
Soleus
Fibularis longus
Fibularis brevis
Calcaneal tendon

(a) Superficial

Plantaris (cut)
Gastrocnemius (cut)
Soleus (cut)
Fibularis longus
Tibialis posterior
Flexor digitorum longus
Flexor hallucis longus
Fibularis brevis

(b) Deep

11. Locate the following muscles in the human torso model and in the lower limb models. Also locate as many of them as possible in your body.

muscles that move the foot
- anterior leg muscles
  - tibialis anterior
  - extensor digitorum longus
  - extensor hallucis longus
  - fibularis (peroneus) tertius
- posterior leg muscles
  - superficial group
    gastrocnemius
    soleus
    plantaris
  - deep group
    tibialis posterior
    flexor digitorum longus
    flexor hallucis longus

- lateral leg muscles
  - fibularis (peroneus) longus
  - fibularis (peroneus) brevis

12. Demonstrate the actions of these muscles in your body.
13. Locate the origins and insertions of these muscles in the human skeleton.
14. Complete Parts B, C, and D of the laboratory assessment.

LEARNING EXTENSION ACTIVITY

A long rubber band can be used to simulate muscle locations, origins, insertions, and actions on muscular models, the skeleton, or a laboratory partner. Hold one end of the rubber band firmly on the origin location of a muscle; then slightly stretch the rubber band and hold the other end on the insertion site. Allow the insertion end to slowly move toward the origin end to simulate the contraction and action of the muscle.

FIGURE 25.8   Muscles of the lateral right leg.

Biceps femoris

Vastus lateralis

Head of fibula

Gastrocnemius

Soleus

Tibialis anterior

Fibularis longus

Extensor digitorum longus

Fibularis brevis

Extensor retinacula

Fibularis tertius

Calcaneal tendon

Metatarsal V

Fibular retinaculum

Table 25.3   **Muscles that Move the Foot**

Muscle	Origin	Insertion	Action
Tibialis anterior	Lateral condyle and proximal tibia	Medial cuneiform and metatarsal I	Dorsiflexion and inversion of foot
Extensor digitorum longus	Lateral condyle of tibia and anterior surface of fibula	Dorsal surfaces of second and third phalanges of toes 2–5	Extends toes 2–5 and dorsiflexion
Extensor hallucis longus	Anterior surface of fibula	Distal phalanx of the great toe	Extends great toe and dorsiflexion
Fibularis tertius	Anterior distal surface of fibula	Dorsal surface of metatarsal V	Dorsiflexion and eversion of foot
Gastrocnemius	Lateral and medial condyles of femur	Calcaneus via calcaneal tendon	Plantar flexion of foot and flexes knee
Soleus	Head and shaft of fibula and posterior surface of tibia	Calcaneus via calcaneal tendon	Plantar flexion of foot
Plantaris	Superior to lateral condyle of femur	Calcaneus	Plantar flexion of foot and flexes knee
Tibialis posterior	Posterior tibia and fibula	Tarsals (several) and metatarsals II–IV	Plantar flexion and inversion of foot
Flexor digitorum longus	Posterior surface of tibia	Distal phalanges of toes 2–5	Flexes toes 2–5 and plantar flexion and inversion of foot
Flexor hallucis longus	Posterior distal fibula	Distal phalanx of great toe	Flexes great toe and plantar flexes foot
Fibularis longus	Head and shaft of fibula and lateral condyle of tibia	Medial cuneiform and metatarsal I	Plantar flexion and eversion of foot; also supports arch
Fibularis brevis	Distal fibula	Metatarsal V	Plantar flexion and eversion of foot

LABORATORY ASSESSMENT
25

Name _____

Date _____

Section _____

The Ⓐ corresponds to the indicated Learning Outcome(s) Ⓞ found at the beginning of the Laboratory Exercise.

Muscles of the Hip and Lower Limb

PART A: Assessments

Identify the muscles indicated in the hip and thigh in figures 25.9 and 25.10.

FIGURE 25.9 Label the right anterior muscles of the hip and thigh. Ⓐ1

1 _____
2 _____
3 _____
4 _____
5 _____
6 _____
7 _____
8 _____
9 _____
Quadriceps femoris tendon

Patellar ligament

FIGURE 25.10 AP|R Label the right posterior hip and thigh muscles of a cadaver, using the terms provided. The gluteus maximus has been removed. Ⓐ1

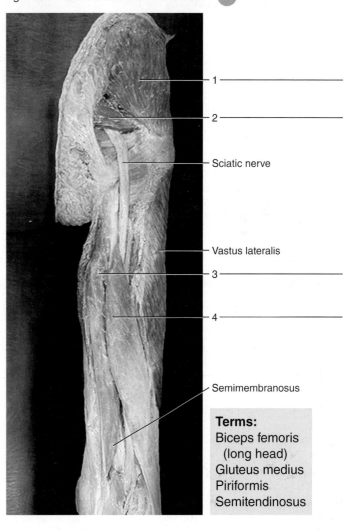

1 _____
2 _____
Sciatic nerve

Vastus lateralis
3 _____
4 _____

Semimembranosus

Terms:
Biceps femoris
  (long head)
Gluteus medius
Piriformis
Semitendinosus

PART B: Assessments

Match the muscles in column A with the actions in column B. Place the letter of your choice in the space provided.

Column A

a. Biceps femoris

b. Fibularis (peroneus) longus

c. Gluteus medius

d. Gracilis

e. Psoas major and iliacus

f. Quadriceps femoris group

g. Sartorius

h. Tibialis anterior

i. Tibialis posterior

Column B

1. Adducts thigh and flexes knee

2. Plantar flexion and eversion of foot

3. Flexes thigh at the hip

4. Abducts and flexes thigh and rotates it laterally; flexes leg at knee

5. Abducts thigh and rotates it medially

6. Plantar flexion and inversion of foot

7. Flexes leg at the knee and laterally rotates leg

8. Extends leg at the knee

9. Dorsiflexion and inversion of foot

PART C: Assessments

Name the muscle indicated by the following combinations of origin and insertion.

Origin	Insertion	Muscle
1. Lateral surface of ilium	Greater trochanter of femur	_____
2. Anterior superior iliac spine	Medial surface of proximal tibia	_____
3. Lateral and medial condyles of femur	Calcaneus	_____
4. Iliac crest and anterior superior iliac spine	Iliotibial tract	_____
5. Greater trochanter and linea aspera of femur	Patella to tibial tuberosity	_____
6. Ischial tuberosity	Medial surface of proximal tibia	_____
7. Linea aspera of femur	Patella to tibial tuberosity	_____
8. Posterior surface of tibia	Distal phalanges of four lateral toes	_____
9. Lateral condyle and proximal tibia	Medial cuneiform and first metatarsal	_____
10. Ischial tuberosity and pubis	Linea aspera of femur	_____
11. Anterior sacrum	Greater trochanter of femur	_____

PART D: Assessments

Identify the muscles in the leg in figure 25.11.

FIGURE 25.11 Label the right leg muscles of a cadaver, lateral view, using the terms provided.

Terms:
Extensor digitorum longus
Fibularis longus
Gastrocnemius
Soleus
Tibialis anterior

CRITICAL THINKING ASSESSMENT

Identify the muscles indicated in figure 25.12.

FIGURE 25.12 Label these muscles that appear as lower limb surface features in these photographs (a and b), by placing the correct numbers in the spaces provided. **A1**

Medial   Lateral

(a) Left thigh, anterior view

____ Rectus femoris
____ Vastus lateralis
____ Vastus medialis

(b) Right lower limb, medial view

____ Gastrocnemius
____ Sartorius
____ Soleus
____ Tibialis anterior
____ Vastus medialis

266

Surface Anatomy

Purpose of the Exercise

To examine the surface features of the human body and the terms used to describe them.

MATERIALS NEEDED

Small round stickers
Colored pencils (black and red)

Learning Outcomes

After completing this exercise you should be able to:

O1 Locate and identify major surface features of the human body.

O2 Arrange surface features by body region.

O3 Distinguish between surface features that are bony landmarks and soft tissue.

The **O** corresponds to the assessments **A** indicated in the Laboratory Assessment for this Exercise.

Pre-Lab

Carefully read the introductory material and examine the entire lab. Be familiar with body regions, the skeletal system, and the muscular system from lecture or the textbook. Answer the pre-lab questions.

Pre-Lab Questions Select the correct answer for each of the following questions:

1. The technique of examining surface features with hands and fingers is called
   a. exploration.   b. touching.
   c. palpation.   d. feeling.

2. Soft tissues include the following *except*
   a. bone.   b. muscle.
   c. ligament.   d. skin.

3. Which of the following is a bony feature of the upper torso?
   a. pectoralis major   b. anterior axillary fold
   c. jugular notch   d. deltoid

4. Which of the following is a soft tissue feature of the lower limb?
   a. tibial tuberosity   b. head of the fibula
   c. medial malleolus   d. vastus lateralis

5. Palpation enables us to determine all of the following characteristics of a structure *except*
   a. location.   b. function.
   c. size.   d. texture.

6. The laryngeal prominence can be palpated in the anterior neck.
   a. True _____   b. False _____

7. The linea alba is a feature of the posterior torso (trunk).
   a. True _____   b. False _____

External landmarks, called surface anatomy, located on the human body provide an opportunity to examine surface features and to help locate other internal structures. This exercise will focus on bony landmarks and superficial soft tissue structures, primarily related to the skeletal and muscular systems. The technique used for the examination of surface features is called *palpation* and is accomplished by touching with the hands or fingers. This enables us to determine the location, size, and texture of the underlying anatomical structure. Some of the surface features can be visible without the need of palpation, especially on body builders and very thin individuals. If the individual involved has considerable subcutaneous adipose tissue, additional pressure might be necessary to palpate some of the structures.

These surface features will help us to locate additional surface and deeper structures during the study of the organ systems. Surface features will help us to determine the pulse locations of superficial arteries, to assess injection sites, to describe pain and injury sites, to insert examination tubes, and to listen to heart, lung, and bowel sounds with a stethoscope. For those who choose careers in emergency medical services, nursing, medicine, physical education, physical therapy, chiropractic, and massage therapy, surface anatomy is especially valuable.

Many clues to a person's overall health can be revealed from the body surface. The general nature of the surface texture and coloration can be observed and examined quickly. Palpation of various surface features can provide additional clues to possible ailments. With the use of different pressures during palpations, clues to the depth of organs and possible abnormalities become more evident. An external examination is an important aspect of a physical exam. The usefulness of this laboratory exercise will become even more apparent as you continue the study of anatomy and physiology and related careers.

PROCEDURE: Surface Anatomy

1. Review the earlier laboratory manual exercises and figures on body regions, skeletal system, and muscular system.
2. Examine the surface features labeled in figures 26.1 through 26.8. Features in **boldface** are *bony features.* Other labeled features (not boldface) are *soft tissue features.* A bony feature is a part of a bone and the skeletal system. Soft tissue features are composed of tissue other than bone, such as a muscle, tendon, ligament, cartilage, or skin. Palpate the labeled surface features to distinguish the texture and degree of firmness of the bony features and the soft tissue features. Most of the surface features included in this laboratory exercise can be palpated on your own body.

FIGURE 26.1 AP|R   Surface features of the posterior shoulder and torso. (**Boldface** indicates bony features; not boldface indicates soft tissue features.)

Spine of scapula
C7
T1 and superior angle of scapula
Medial border of scapula
T7 and inferior angle of scapula
Latissimus dorsi
T12 and rib 12
L4 and iliac crest

Upper trapezius
Middle trapezius
Erector spinae
Inferior angle of scapula
Lower trapezius
Iliac crest
Posterior superior iliac spine
Sacrum

FIGURE 26.2 **AP|R** Surface features of (a) the anterior head and neck, (b) the lateral head and neck, and (c) the posterior head and neck. (**Boldface** indicates bony features; not boldface indicates soft tissue features.)

Frontalis

Zygomatic arch

Mandible

Hyoid bone

Laryngeal prominence ("Adam's apple") of thyroid cartilage

(a)

Zygomatic arch

External acoustic meatus

Temporomandibular joint

Sternocleidomastoid

Trapezius (upper)

(b)

External occipital protuberance

Mastoid process

Occipital ridge

Trapezius (upper)

Vertebra prominens (C7)

Acromion

Deltoid (posterior)

Medial border of scapula

(c)

FIGURE 26.3 AP|R Anterior surface features of (a) upper torso and (b) lower torso. (**Boldface** indicates bony features; not boldface indicates soft tissue features.)

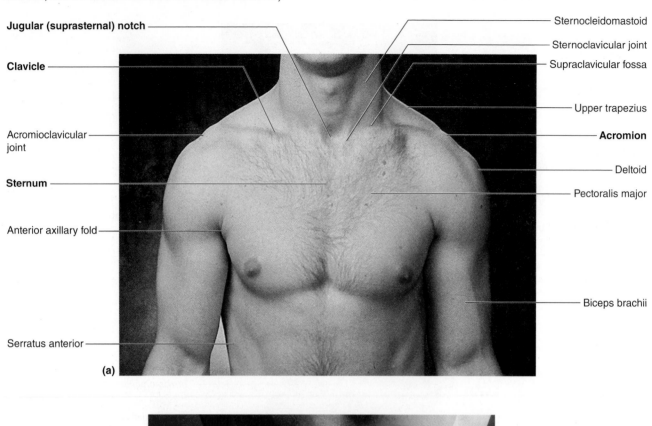

Jugular (suprasternal) notch

Clavicle

Acromioclavicular joint

Sternum

Anterior axillary fold

Serratus anterior

(a)

Sternocleidomastoid

Sternoclavicular joint

Supraclavicular fossa

Upper trapezius

Acromion

Deltoid

Pectoralis major

Biceps brachii

Sternum

Xiphoid process

Rectus abdominis

Umbilicus

Iliac crest

(b)

Axilla

Serratus anterior

Linea alba

External oblique

Anterior superior iliac spine (ASIS)

FIGURE 26.4 AP|R Surface features of the lateral shoulder and upper limb. (**Boldface** indicates bony features; not boldface indicates soft tissue features.)

Acromion

Acromioclavicular joint

Greater tubercle of humerus

Anterior deltoid

Middle deltoid

Posterior deltoid

Biceps brachii

Triceps brachii

Olecranon process

Kyphotic curve of spine

Posterior axillary fold

Lordotic curve of spine

FIGURE 26.5 AP|R Surface features of the upper limb (a) anterior view and (b) posterior view. (**Boldface** indicates bony features; not boldface indicates soft tissue features.)

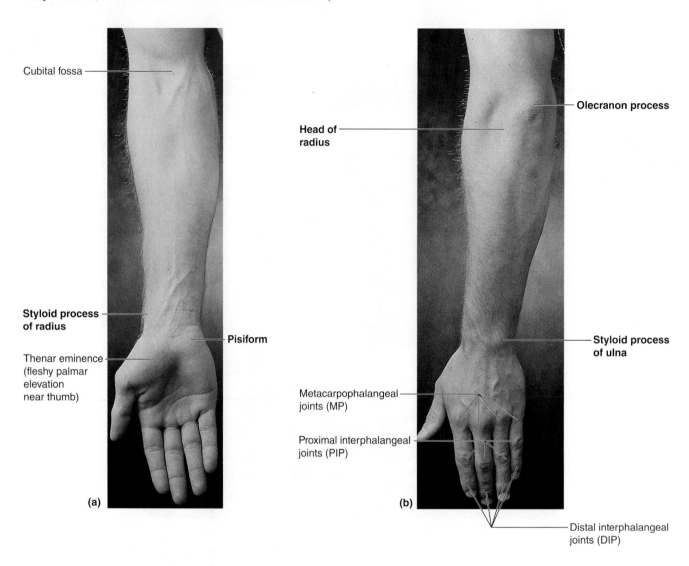

Cubital fossa

Head of radius

Olecranon process

Styloid process of radius

Thenar eminence (fleshy palmar elevation near thumb)

Pisiform

Styloid process of ulna

Metacarpophalangeal joints (MP)

Proximal interphalangeal joints (PIP)

Distal interphalangeal joints (DIP)

(a)

(b)

FIGURE 26.6 [AP|R] Surface features of the anterior view of the body. (**Boldface** indicates bony features; not boldface indicates soft tissue features.)

Anterior superior iliac spine

Greater trochanter of femur

Vastus lateralis

Vastus medialis

Patella

Head of fibula

Tibial tuberosity

Tibialis anterior

Gastrocnemius

Tibia

Medial malleolus

Lateral malleolus

FIGURE 26.7 [AP|R] Surface features of the plantar surface of the foot. (**Boldface** indicates bony features; not boldface indicates soft tissue features.)

Transverse arch

Head of metatarsal I

Medial longitudinal arch

FIGURE 26.8 AP|R Surface features of the (a) posterior lower limb, (b) lateral lower limb, and (c) medial lower limb. (**Boldface** indicates bony features; not boldface indicates soft tissue features.)

Popliteal fossa

Gastrocnemius

Calcaneal tendon

Lateral malleolus

Calcaneus

(a)

Iliotibial tract

Patella

Patella

Head of fibula

Gastrocnemius

Soleus

Calcaneal tendon

Medial malleolus

Lateral malleolus

(b)

(c)

3. Complete Parts A and B of Laboratory Assessment 26.
4. Use figures 26.1 through 26.8 to determine which items listed in table 26.1 are bony features and which items are soft tissue features. Examine the list of bony and soft tissue surface features in the first column. If the feature listed is a bony surface feature, place an "X" in the second column; if the feature listed is a soft tissue feature, place an "X" in the third column.
5. Complete Part C of the laboratory assessment.

Table 26.1 Representative Bony and Soft Tissue Surface Features

(*Note:* After indicating your "X" in the appropriate column, use the extra space in columns 2 and 3 to add personal descriptions of the nature of the bony feature or the soft tissue feature palpation.)

Bony and Soft Tissues	Bony Features	Soft Tissue Features
Biceps brachii		
Calcaneal tendon		
Deltoid		
Gastrocnemius		
Greater trochanter		
Head of fibula		
Iliac crest		
Iliotibial tract		
Lateral malleolus		
Mandible		
Mastoid process		
Medial border of scapula		
Rectus abdominis		
Sacrum		
Serratus anterior		
Sternocleidomastoid		
Sternum		
Thenar eminence		
Tibialis anterior		
Zygomatic arch		

NOTES

26

Name _____

Date _____

Section _____

The Ⓐ corresponds to the indicated Learning Outcome(s) Ⓞ found at the beginning of the Laboratory Exercise.

Surface Anatomy

PART A: Assessments

Match the body regions in column A with the surface features in column B. Place the letter of your choice in the space provided. Ⓐ1 Ⓐ2

Column A	Column B
a. Head	_____ **1.** Umbilicus
b. Trunk (torso, including shoulder and hip)	_____ **2.** Medial malleolus
c. Upper limb	_____ **3.** Iliac crest
d. Lower limb	_____ **4.** Transverse arch
	_____ **5.** Spine of scapula
	_____ **6.** External occipital protuberance
	_____ **7.** Cubital fossa
	_____ **8.** Sternum
	_____ **9.** Olecranon process
	_____ **10.** Zygomatic arch
	_____ **11.** Mastoid process
	_____ **12.** Thenar eminence
	_____ **13.** Popliteal fossa
	_____ **14.** Metacarpophalangeal joints

Label figure 26.9 with the surface features provided.

FIGURE 26.9 Using the terms provided, label the surface features of the (a) anterior view of the body and (b) posterior view of the body. (**Boldface** indicates bony features; not boldface indicates soft tissue features.) **A1**

(a)

Terms:

Acromion	Thenar eminence
Biceps brachii	**Tibia**
Clavicle	**Tibial tuberosity**
Greater trochanter	Trapezius (upper)
Head of fibula	Vastus lateralis
Iliac crest	**Xiphoid process**
Pectoralis major	**Zygomatic arch**
Rectus abdominis	

FIGURE 26.9 *Continued.*

— Ischial tuberosity

(b)

Terms:

C7	**Iliac crest**	**Olecranon process**
Calcaneal tendon	Iliotibial tract	Popliteal fossa
Distal interphalangeal joint (DIP)	**Inferior angle of scapula**	**Sacrum**
Erector spinae	**Mastoid process**	Soleus
External occipital protuberance	**Medial border of scapula**	**Styloid process of ulna**

PART C: Assessments

Using table 26.1, indicate the locations of the surface features with an "X" on figure 26.10. Use a black "X" for the bony features and a red "X" for the soft tissue features.

FIGURE 26.10   Indicate bony surface features with a black "X" and soft tissue surface features with a red "X" using the results from table 26.1, on the (a) anterior and (b) posterior diagrams of the body.   A3

Terms from table 26.1:
Biceps brachii
Calcaneal tendon
Deltoid
Gastrocnemius
Greater trochanter
Head of fibula
Iliac crest
Iliotibial tract
Lateral malleolus
Mandible
Mastoid process
Medial border of scapula
Rectus abdominis
Sacrum
Serratus anterior
Sternocleidomastoid
Sternum
Thenar eminence
Tibialis anterior
Zygomatic arch

(a)

(b)

LABORATORY EXERCISE 27

Nervous Tissue and Nerves

Purpose of the Exercise

To review the characteristics of nervous tissue and to observe neurons, neuroglia, and various features of a nerve.

MATERIALS NEEDED

Compound light microscope

Prepared microscope slides of the following:

Spinal cord (smear)

Dorsal root ganglion (section)

Neuroglia (astrocytes)

Peripheral nerve (cross section and longitudinal section)

Neuron model

For Learning Extension Activity:

Prepared microscope slide of Purkinje cells from cerebellum

Learning Outcomes AP|R

After completing this exercise, you should be able to:

O1 Describe the location and characteristics of nervous tissue.

O2 Distinguish structural and functional characteristics between neurons and neuroglia.

O3 Identify and sketch the major structures of a neuron and a nerve.

The **O** corresponds to the assessments **A** indicated in the Laboratory Assessment for this Exercise.

Pre-Lab

Carefully read the introductory material and examine the entire lab. Be familiar with the structures and functions of nervous tissue and nerves from lecture or the textbook. Answer the pre-lab questions.

Pre-Lab Questions Select the correct answer for each of the following questions:

1. The cell body of a neuron contains the
   a. nucleus.
   b. dendrites.
   c. axon.
   d. neuroglia.

2. A multipolar neuron contains
   a. one dendrite and many axons.
   b. many dendrites and one axon.
   c. one dendrite and one axon.
   d. a single process with the dendrite and axon.

3. Neuroglia that produce myelin insulation in the CNS are
   a. microglia.   b. astrocytes.
   c. ependymal cells.   d. oligodendrocytes.

4. The PNS contains
   a. 12 pairs of cranial nerves only.
   b. 31 pairs of spinal nerves only.
   c. 12 pairs of cranial nerves and 31 pairs of spinal nerves.
   d. 43 pairs of spinal nerves.

5. Schwann cells
   a. are only in the brain.
   b. are only in the spinal cord.
   c. are throughout the CNS.
   d. have a myelin sheath and neurilemma.

6. A _____ neuron is the most common structural neuron in the brain and spinal cord.
   a. multipolar   b. tripolar
   c. bipolar   d. unipolar

7. Sensory neurons conduct impulses from the spinal cord to a muscle or a gland.
   a. True _____   b. False _____

8. Astrocytes have contacts between blood vessels and neurons in the CNS.
   a. True _____   b. False _____

Nervous tissue, which occurs in the brain, spinal cord, and peripheral nerves, contains *neurons* (nerve cells) and *neuroglia* (neuroglial cells; glial cells). Neurons are irritable (excitable) and easily respond to stimuli, resulting in the conduction of an *action potential (impulse; nerve impulse; nerve signal)*. A neuron contains a cell body with a nucleus and most of the cytoplasm, and elongated cell processes (typically, several dendrites and a single axon) along which impulse conductions occur. Neurons can be classified according to structural variations of their cell processes: *multipolar* with many dendrites and one axon (nerve fiber), *bipolar* with one dendrite and one axon, and *unipolar* with a single process where the dendrite leads directly into the axon. Functional classifications are used according to the direction of the impulse conduction: *sensory (afferent) neurons* conduct impulses from receptors toward the central nervous system (CNS), *interneurons (association neurons)* conduct impulses within the CNS, and *motor (efferent) neurons* conduct impulses away from the CNS to effectors (muscles or glands).

Neuroglia (supportive cells) are located in close association with neurons. Four types of neuroglia are found in the CNS: *astrocytes, microglia, oligodendrocytes,* and *ependymal cells.* Astrocytes are numerous, and their branches support neurons and blood vessels. Microglia can phagocytize microorganisms and nerve tissue debris. Oligodendrocytes produce the myelin insulation in the CNS. Ependymal cells line the brain ventricles and secrete CSF (cerebrospinal fluid), and the cilia on their apical surfaces aid the circulation of the CSF.

Two types of neuroglia are located in the peripheral nervous system (PNS): *Schwann cells* and *satellite cells.* Schwann cells surround nerve fibers numerous times. The wrappings nearest the nerve fiber represent the myelin sheath and serve to insulate and increase the impulse speed, while the last wrapping, called the neurilemma, contains most of the cytoplasm and the nucleus and functions in nerve fiber regeneration in PNS neurons. Satellite cells surround and support the cell body regions, called ganglia, of peripheral neurons.

The PNS contains 12 pairs of *cranial nerves* that arise from the brain and 31 pairs of *spinal nerves* that arise from the spinal cord, all containing parallel axons of neurons and neuroglia representing the nervous tissue components. Most nerves are mixed in that they contain both sensory and motor neurons, but some contain only sensory or motor components. The nerves represent an organ structure as they also contain small blood vessels and fibrous connective tissue. The fibrous connective tissue around each nerve fiber and the Schwann cell is the *endoneurium;* a bundle of nerve fibers, representing a *fascicle,* is surrounded by a *perineurium;* and the entire nerve is surrounded by an *epineurium.* The fibrous connective components of a nerve provide protection and stretching during body movements.

PROCEDURE A: Nervous Tissue AP|R

In this procedure you will be asked to identify the locations and characteristics of structures in nervous tissue. Two tables of nervous tissue include the structural and functional characteristics of neurons, and neuroglia within the central nervous system (CNS) and the peripheral nervous system (PNS). Diagrams of neurons and neuroglia are provided as sources of comparison for your microscopic observations using the compound microscope. When you make sketches in the laboratory assessment, include all the observed structures with labels.

1. Examine tables 27.1 and 27.2 and figures 27.1, 27.2, and 27.3.

Table 27.1   Structural and Functional Types of Neurons

CLASSIFIED BY STRUCTURE		
Type	**Structural Characteristics**	**Location**
Multipolar neuron	Cell body with one axon and multiple dendrites *multiple char.*	Most common type of neuron in the brain and spinal cord
Bipolar neuron	Cell body with one axon and one dendrite *2 char.*	In receptor parts of the eyes, nose, and ears; rare type
Unipolar neuron	Cell body with a single process that divides into two branches and functions as an axon; only the receptor ends of the peripheral (distal) process function as dendrites	Most sensory neurons
CLASSIFIED BY FUNCTION		
Type	**Functional Characteristics**	**Structural Characteristics**
Sensory (afferent) neuron	Conducts impulses from receptors in peripheral body parts into the brain or spinal cord	Most unipolar; some bipolar
Interneuron	Transmits impulses between neurons in the brain and spinal cord	Multipolar
Motor (efferent) neuron	Conducts impulses from the brain or spinal cord out to muscles or glands (effectors)	Multipolar

Table 27.2 Types of Neuroglia in the CNS and PNS

Type	Structural Characteristics	Functions
CNS (BRAIN AND SPINAL CORD)		
Astrocytes	Star-shaped cells contacting neurons and blood vessels; most abundant type	Structural support; formation of scar tissue to replace damaged neurons; transport of substances between blood vessels and neurons; communicate with one another and with neurons; remove excess ions and neurotransmitters *blood brain barrier*
Microglia	Small cells with slender cellular processes	Structural support; phagocytosis of microorganisms and damaged tissue
Oligodendrocytes	Shaped like astrocytes, but with fewer cellular processes; processes wrap around axons	Form myelin sheaths in the brain and spinal cord
Ependyma	Cuboidal cells lining cavities of the brain and spinal cord	Secrete and assist in circulation of cerebrospinal fluid (CSF)
PNS (PERIPHERAL NERVES)		
Schwann cells	Cells that wrap tightly around the axons of peripheral neurons	Form myelin sheaths in PNS and speed neurotransmission
Satellite cells	Flattened cells that surround cell bodies of neurons in ganglia	Support and protect ganglia

(handwritten note: Similar — bracketing Oligodendrocytes and Ependyma)

FIGURE 27.1 Structural types of neurons (a) multipolar neuron, (b) bipolar neuron, and (c) unipolar neuron.

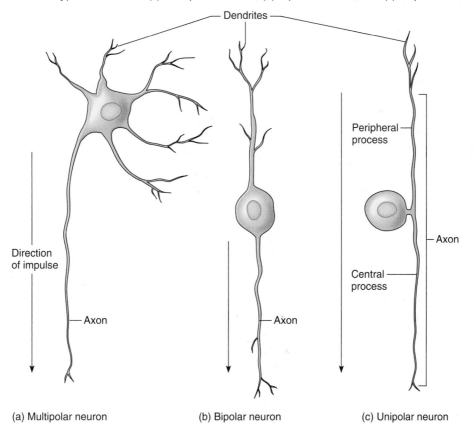

(a) Multipolar neuron (b) Bipolar neuron (c) Unipolar neuron

FIGURE 27.2 AP|R Diagram of a multipolar motor neuron.

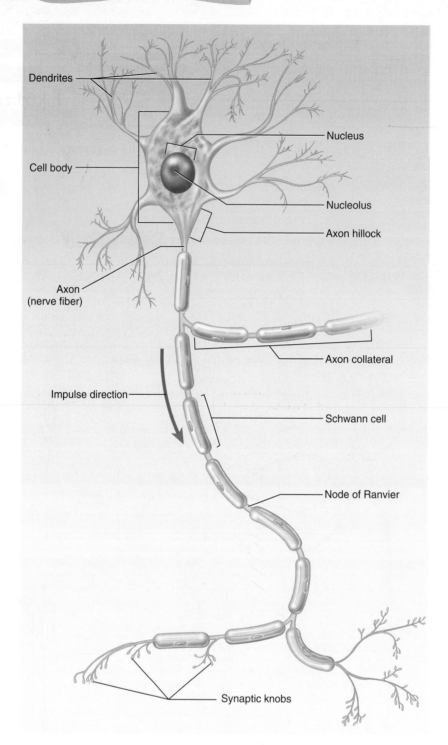

Dendrites

Nucleus

Cell body

Nucleolus

Axon hillock

Axon
(nerve fiber)

Axon collateral

Impulse direction

Schwann cell

Node of Ranvier

Synaptic knobs

FIGURE 27.3 AP|R Diagram of a cross section of a myelinated axon (nerve fiber) of a spinal nerve.

— Schwann cell nucleus

— Myelin sheath

— Neurofibrils

— Axon (nerve fiber)

— Neurilemma

2. Complete Parts A and B of Laboratory Assessment 27.
3. Obtain a prepared microscope slide of a spinal cord smear. Using low-power magnification, search the slide and locate the relatively large, deeply stained cell bodies of multipolar motor neurons.
4. Observe a single multipolar motor neuron, using high-power magnification (fig. 27.4), and note the following features:

cell body (soma)
- nucleus
- nucleolus
- Nissl bodies (chromatophilic substance)—a type of rough ER in neurons
- neurofibrils—threadlike structures extending into axon
- axon hillock—origin region of axon

dendrites—conduct impulses toward cell body

axon (nerve fiber)—conducts impulse away from cell body

Compare the slide to the neuron model and to figure 27.4. Also note small, darkly stained nuclei of neuroglia around the motor neuron.

FIGURE 27.4 AP|R Micrograph of a multipolar neuron and neuroglia from a spinal cord smear (350×).

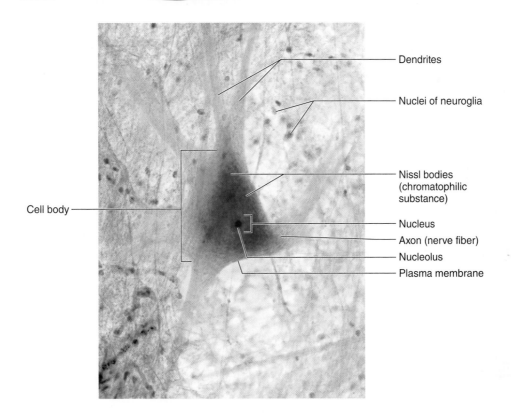

Dendrites

Nuclei of neuroglia

Nissl bodies (chromatophilic substance)

Nucleus

Axon (nerve fiber)

Nucleolus

Plasma membrane

Cell body

5. Sketch and label a motor (efferent) neuron in the space provided in Part C of the laboratory assessment.
6. Obtain a prepared microscope slide of a dorsal root ganglion. Search the slide and locate a cluster of sensory neuron cell bodies. You also may note bundles of nerve fibers passing among groups of neuron cell bodies (fig. 27.5).

FIGURE 27.5 Micrograph of a dorsal root ganglion containig cell bodies of sensory (afferent) neurons (100×).

Sensory neuron cell bodies

Nerve fibers (axons)

7. Sketch and label a sensory (afferent) neuron cell body in the space provided in Part C of the laboratory assessment.
8. Examine figure 27.6 and table 27.2.
9. Obtain a prepared microscope slide of neuroglia. Search the slide and locate some darkly stained astrocytes with numerous long, slender processes (fig. 27.7).
10. Sketch a neuroglia in the space provided in Part C of the laboratory assessment.

LEARNING EXTENSION ACTIVITY

Obtain a prepared microscope slide of Purkinje cells (fig. 27.8). To locate these neurons, search the slide for large, flask-shaped cell bodies. Each cell body has one or two large, thick dendrites that give rise to extensive branching networks of dendrites. These large cells are located in a particular region of the brain (cerebellar cortex).

PROCEDURE B: Nerves

In this procedure you will compare an illustration of a peripheral spinal nerve with actual microscopic views of nerves. A nerve is an organ of the nervous system and contains various tissues in combination with the nervous tissue. The nerve contains the nerve fibers (axons) of the neurons and the Schwann cells representing some neuroglia in the PNS.

1. Study figure 27.9 of a cross section of a spinal nerve. Examine the fibrous connective tissue pattern including the epineurium around the entire nerve, the perineurium around fascicles, and the endoneurium around an

FIGURE 27.6 Types of neuroglia in the CNS.

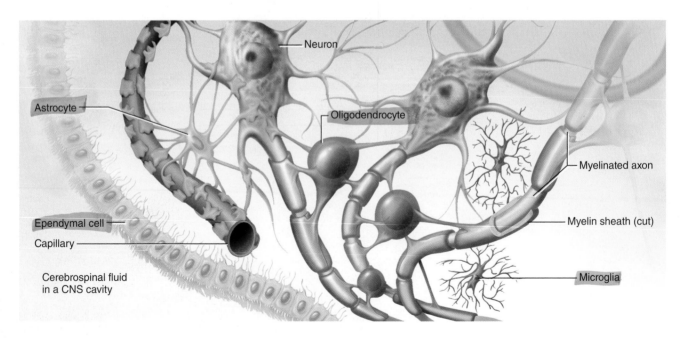

Neuron

Astrocyte

Oligodendrocyte

Myelinated axon

Myelin sheath (cut)

Ependymal cell

Capillary

Microglia

Cerebrospinal fluid in a CNS cavity

FIGURE 27.7 Micrograph of astrocytes (1,000×).

Blood vessel

Astrocyte

FIGURE 27.8 Large multipolar Purkinje cell from cerebellum of the brain (400×).

Dendrites with extensive branching

Cell body

Axon

FIGURE 27.9 AP|R Diagram of a cross section of a peripheral spinal nerve.

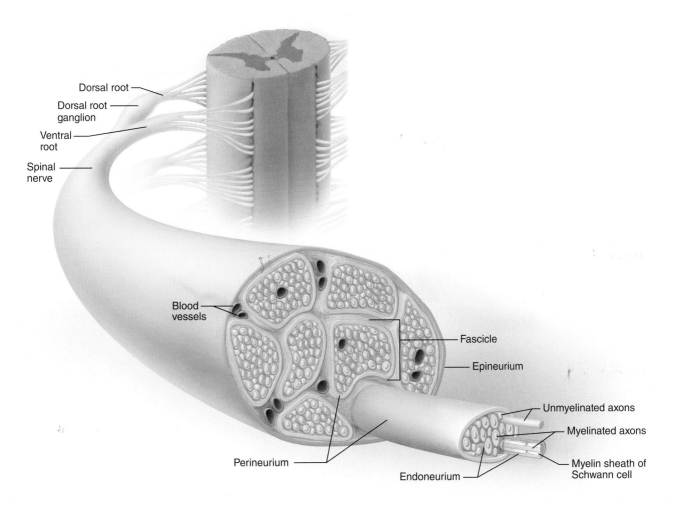

Dorsal root

Dorsal root ganglion

Ventral root

Spinal nerve

Blood vessels

Fascicle

Epineurium

Unmyelinated axons

Myelinated axons

Myelin sheath of Schwann cell

Perineurium

Endoneurium

FIGURE 27.10 Cross section of a bundle of axons within a nerve (400×).

- Connective tissue (perineurium)
- Connective tissue (endoneurium)
- Axon (nerve fiber)
- Myelin sheath of Schwann cell
- Neurilemma of Schwann cell

FIGURE 27.11 [AP|R] Longitudinal section of a nerve (2,000×).

- Node of Ranvier
- Axon (nerve fiber)
- Neurilemma of Schwann cell
- Myelin sheath of Schwann cell

individual axon and Schwann cell. Reexamine figure 27.3 to note additional cross-sectional detail of the Schwann cell layers.

2. Obtain a prepared microscope slide of a nerve. Locate the cross section of the nerve and note the many round nerve fibers inside. Also note the dense layer of connective tissue (perineurium) that encircles a fascicle of nerve fibers and holds them together in a bundle. The individual nerve fibers are surrounded by a layer of more delicate connective tissue (endoneurium) (fig. 27.10).

3. Using high-power magnification, observe a single nerve fiber and note the following features:

   axon

   myelin sheath of Schwann cell around the axon (most of the myelin may have been dissolved and lost during the slide preparation)

   neurilemma of Schwann cell

4. Sketch and label a nerve fiber with Schwann cell (cross section) in the space provided in Part D of the laboratory assessment.

5. Locate the longitudinal section of the nerve on the slide (fig. 27.11). Note the following:

   axon

   myelin sheath of Schwann cell

   neurilemma of Schwann cell

   node of Ranvier—narrow gap between the Schwann cells (figs. 27.2 and 27.11)

6. Sketch and label a nerve fiber with Schwann cell (longitudinal section) in the space provided in Part D of the laboratory assessment.

Name _____

Date _____

Section _____

The Ⓐ corresponds to the indicated Learning Outcome(s) Ⓞ found at the beginning of the Laboratory Exercise.

Nervous Tissue and Nerves

PART A: Assessments

Match the terms in column A with the descriptions in column B. Place the letter of your choice in the space provided.

Column A		Column B
a. Astrocyte	f. ▨	1. Sheath of Schwann cell containing cytoplasm and nucleus that encloses myelin
b. Axon	h.	2. Corresponds to rough endoplasmic reticulum in other cells
c. Collateral	g.	3. Network of threadlike structures within cell body and extending into axon
d. Dendrite	c.	4. Substance of Schwann cell composed of lipoprotein that insulates axons and increases impulse speed
e. Myelin		
f. Neurilemma	d.	5. Neuron process with many branches that conducts an action potential (impulse) toward the cell body
g. Neurofibrils	e. ▨	6. Branch of an axon
h. Nissl bodies (chromatophilic substance)	a.	7. Star-shaped neuroglia between neurons and blood vessels
i. Unipolar neuron	b.	8. Nerve fiber arising from a slight elevation of the cell body that conducts an action potential (impulse) away from the cell body
	i.	9. Possesses a single process from the cell body

PART B: Assessments

Match the terms in column A with the descriptions in column B. Place the letter of your choice in the space provided.

Column A		Column B
a. Effector	d.	1. Transmits impulse from sensory to motor neuron within central nervous system
b. Ependymal cell	f.	2. Transmits impulse out of the brain or spinal cord to effectors (muscles and glands)
c. Ganglion		
d. Interneuron (association neuron)	h.	3. Transmits impulse into brain or spinal cord from receptors
e. Microglia	g.	4. Myelin-forming neuroglia in brain and spinal cord
f. Motor (efferent) neuron	e.	5. Phagocytic neuroglia
g. Oligodendrocyte	a.	6. Structure capable of responding to motor impulse
h. Sensory (afferent) neuron	c.	7. Specialized mass of neuron cell bodies outside the brain or spinal cord
	b.	8. Cells that line cavities of the brain and secrete cerebrospinal fluid

PART C: Assessments

In the space that follows, sketch the indicated cells. Label any of the cellular structures observed, and indicate the magnification of each sketch. Add notes in the extra space within the box with the sketch that would assist in your understanding of the structures and functions. **A2** **A3**

Motor neuron (_____×)	Sensory neuron cell body (_____×)

Neuroglia (_____×)

PART D: Assessments

In the space that follows, sketch the indicated view of a nerve fiber (axon). Label any structures observed, and indicate the magnification of each sketch. Add notes in the extra space within the box with the sketch that would assist in your understanding of the structures and functions. **A2** **A3**

Nerve fiber cross section with Schwann cell (_____×)	Nerve fiber longitudinal section with Schwann cell (_____×)

EXERCISE 28

Meninges, Spinal Cord, and Spinal Nerves

Purpose of the Exercise

To review the characteristics of the meninges, spinal cord, spinal nerves, and nerve plexuses and to observe the major anatomical features of these structures.

MATERIALS NEEDED

Spinal cord model with meninges

Vertebral column model with spinal nerves

Compound light microscope

Prepared microscope slide of a spinal cord cross section with spinal nerve roots

Limb models with plexuses and nerves

For Demonstration Activity:

Preserved spinal cord with meninges intact

Learning Outcomes AP|R

After completing this exercise, you should be able to:

01 Describe the arrangement of the spinal cord meninges and the structure of each.

02 Identify the major anatomical features and functions of the spinal cord.

03 Locate the distribution and functions of the spinal nerves and plexuses.

The ⓞ corresponds to the assessments Ⓐ indicated in the Laboratory Assessment for this Exercise.

Pre-Lab

Carefully read the introductory material and examine the entire lab. Be familiar with the structures and functions of the spinal cord, spinal nerves, and meninges from lecture or the textbook. Answer the pre-lab questions.

Pre-Lab Questions Select the correct answer for each of the following questions:

1. The _____ is the most superficial membrane of the meninges.
   - a. subarachnoid space
   - b. pia mater
   - c. arachnoid mater
   - d. dura mater

2. The inferior end of the adult spinal cord terminates
   - a. inferior to L1.
   - b. inferior to L5.
   - c. in the sacrum.
   - d. in the coccyx.

3. The dorsal root of spinal nerves contains
   - a. interneurons.
   - b. sensory neurons.
   - c. motor neurons.
   - d. sensory and motor neurons.

4. Which of the following is *not* part of the gray matter of the spinal cord?
   - a. gray commissure
   - b. posterior horn
   - c. lateral funiculus
   - d. lateral horn

5. The central canal of the spinal cord is located within the
   - a. white matter.
   - b. epidural space.
   - c. gray commissure.
   - d. subarachnoid space.

6. The brachial plexus is formed from components of spinal nerves C5–T1.
   - a. True _____
   - b. False _____

7. There are eight pairs of cervical spinal nerves.
   - a. True _____
   - b. False _____

8. The major ascending (sensory) and descending (motor) tracts compose the gray matter of the spinal cord.
   - a. True _____
   - b. False _____

Three membranes, or *meninges,* surround the entire CNS. They consist of three protective layers of fibrous connective tissue membranes located between the bones of the skull and vertebral column and the soft tissues of the central nervous system. The most superficial layer, the *dura mater,* is a tough membrane with an epidural space containing blood vessels, loose connective tissue, and adipose tissue between the membrane and the vertebrae. A weblike *arachnoid mater* adheres to the inside of the dura mater. The *subarachnoid space,* located beneath the arachnoid mater, contains *cerebrospinal fluid* (*CSF*) and serves as a protective cushion for the spinal cord and brain. The delicate innermost membrane, the *pia mater,* adheres to the surface of the spinal cord and brain. The *denticulate ligaments,* extending from the pia mater to the dura mater, anchor the spinal cord. An inferior extension of the pia mater, the *filum terminale,* anchors the spinal cord to the coccyx.

The spinal cord is a column of nerve fibers and supporting glial cells that extends from the foramen magnum through the vertebral canal to about the level of L1. Together with the brain, it makes up the central nervous system. In the cervical and lumbar regions of the spinal cord, enlargements give rise to spinal nerves to the upper and lower limbs, respectively. The terminal end of the spinal cord, the *conus medullaris,* is located just inferior to lumbar vertebra L1 in the adult.

There are 31 pairs of spinal nerves attached to the spinal cord. At close proximity to the cord, the spinal nerve has two branches: a *dorsal* (*posterior*) *root* containing *sensory neurons* and dorsal root ganglion, and a *ventral* (*anterior*) *root* containing *motor neurons.* The spinal nerves emerge through the nearby intervertebral foramina; however, most lumbar and sacral nerves extend inferiorly through the vertebral canal as the *cauda equina* (resembling a horse's tail) to emerge in their respective regions of the vertebral column.

The spinal cord has a central region, the *gray matter,* with paired *posterior, lateral,* and *anterior horns* connected by a *gray commissure* containing the *central canal.* The gray matter is a processing center for spinal reflexes and synaptic integration. The *white matter* of the spinal cord, represented by paired *posterior, lateral,* and *anterior funiculi* (*columns*), contains *ascending* (*sensory*) and *descending* (*motor*) *tracts.* At various levels of the spinal cord, many of the tracts cross over (decussate) to the opposite side of the spinal cord or the brainstem.

PROCEDURE A: Meninges

Three membranes, or meninges, surround the entire CNS. The superficial, tough dura mater of the spinal cord meninges has an epidural space between the membrane and the vertebrae. A subarachnoid space exists between the middle arachnoid mater and the pia mater that contains cerebrospinal fluid (CSF). The delicate pia mater closely adheres to the surface of the brain and spinal cord. Collectively, the meninges enclose and provide physical protective layers around the delicate brain and spinal cord. The CSF provides an additional protective cushion from sudden jolts to the head or the back. The cranial meninges, that are more directly associated with the brain and skull, are covered in Laboratory Exercise 30.

1. Study figures 28.1, 28.2, and 28.3.
2. Observe the spinal cord model with meninges and locate the following features from superficial to deep:

   epidural space—between vertebra and dura mater; contains connective tissues and blood vessels; site where epidural anesthesia is administered.

   dura mater (dural sheath)—most superficial of the three meninges

---

FIGURE 28.1   The meninges, associated near the spinal cord and spinal nerves, anterior view.

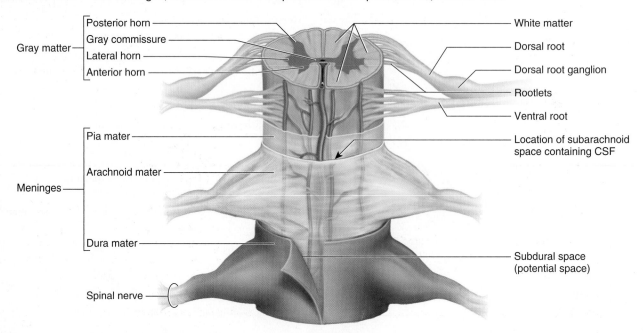

FIGURE 28.2 Posterior view of a segment of the cervical spinal cord. The three meninges surround the spinal cord. Rootlets of dorsal and ventral spinal nerves enter the spinal cord. Posterior portions of the cervical vertebrae have been removed to expose all the labeled structures.

FIGURE 28.3 Features of the spinal cord cross section and surrounding structures.

subdural space—a narrow potential space with a thin film of fluid

arachnoid mater (membrane)—thin, weblike middle layer of meninges

subarachnoid space—contains protective CSF

denticulate ligament—extensions of pia mater to dura mater to suspend and anchor the spinal cord

pia mater—adheres closely to spinal cord

3. Complete Part A of Laboratory Assessment 28.

PROCEDURE B: Structure of the Spinal Cord

1. Study figures 28.1, 28.3, and 28.4, along with the cross section model of the spinal cord.

2. Obtain a prepared microscope slide of a spinal cord cross section. Use the scan and low power of the microscope to locate the following features:

   posterior median sulcus
   anterior median fissure

FIGURE 28.4 Cross section of the spinal cord showing ascending (sensory) tracts (pathways) on one side and the descending (motor) tracts (pathways) on the other side. Ascending and descending tracts are present on both sides (bilateral) of the spinal cord in the white matter.

Ascending tracts
Descending tracts

Posterior median sulcus
Gray matter
Dorsal ramus
Ventral ramus
Anterior median fissure

Posterior funiculus
Central canal
Dorsal root containing sensory neurons
Dorsal root ganglion
Lateral funiculus
Ventral root containing motor neurons
Anterior funiculus

central canal

gray matter

- gray commissure
- posterior (dorsal) horn
- lateral horn
- anterior (ventral) horn

white matter

- posterior (dorsal) funiculus (column)
- lateral funiculus (column)
- anterior (ventral) funiculus (column)

roots of spinal nerve

- dorsal (posterior) roots—contain sensory neurons
- dorsal (posterior) root ganglia—contain cell bodies of sensory neurons
- ventral (anterior) roots—contain motor neurons

3. Observe the model of the vertebral column with spinal nerves along with figure 28.5. Note the length of the spinal cord from the foramen magnum to the conus medullaris. Note also the 31 pairs of spinal nerves, the cauda equina, and the filum terminale.
4. Complete Part B of the laboratory assessment.

PROCEDURE C: Spinal Nerves and Nerve Plexuses

There are 31 pairs of spinal nerves that emerge between bones of the vertebral column. Cervical spinal nerve pair C1 emerges between the foramen magnum of the occipital bone and the atlas. The other cervical nerves emerge inferior to the 7 cervical vertebrae, resulting in 8 pairs of cervical nerves. The rest of the spinal nerves emerge inferiorly to the corresponding vertebrae, resulting in 12 pairs of thoracic nerves, 5 pairs of lumbar nerves, 5 pairs of sacral nerves, and 1 coccygeal pair. Recall that the single sacrum of the adult is a result of fusions of 5 sacral vertebrae. Because the inferior end of the spinal cord of

the adult terminates just inferior to L1, many of the spinal nerves give rise to a bundle of nerve roots, the cauda equina, extending inferiorly through the remainder of the lumbar vertebrae and the sacrum (fig. 28.5).

Each spinal nerve possesses a dorsal root and a ventral root adjacent to the spinal cord. The dorsal root contains the sensory neurons; the ventral root contains the motor neurons. The resulting main spinal nerve, formed from a fusion of the dorsal and ventral roots, is referred to as a mixed nerve because sensory and motor signals (action potentials) exist within the same nerve. A short distance from the spinal cord, some of the spinal nerves merge (anastomose) and form a weblike plexus. Major anastomoses include the cervical, brachial, lumbar, and sacral nerve plexuses, and a small coccygeal plexus. Major nerves branching from each plexus include: phrenic nerve of the cervical plexus; axillary, musculoskeleton, median, ulnar, and radial nerves of the brachial plexus; femoral nerve of the lumbar plexus; and sciatic nerve of the sacral plexus (table 28.1).

1. Examine figures 28.2 and 28.5.
2. Observe a vertebral column model with spinal nerves extending from the intervertebral foramina and limb models with plexuses and nerves. Compare the emerging locations of the spinal nerves with the names of the vertebrae.
3. The *dorsal* and *ventral roots* have multiple *rootlets* attached to the spinal cord (figs. 28.1, 28.2, and 28.3). All of the 31 pairs of spinal nerves are mixed, containing sensory and motor nerve fibers. The actual spinal nerve is very short as it divides after emerging from its foramen, where it branches into a *dorsal (posterior) ramus,* a *ventral (anterior) ramus,* and a *meningeal branch.* The tiny meningeal branch reenters the vertebral canal and innervates the meninges and vertebrae, while the relatively large ventral ramus has communicating rami that connect to the autonomic fibers of the *sympathetic trunk ganglion (sympathetic chain ganglion)* (fig. 28.6).

FIGURE 28.5 Posterior view of the origins and categories of the 31 pairs of spinal nerves on the right. The spinal cord begins at the foramen magnum and ends at the conus medullaris. The plexuses formed by various spinal nerves are illustrated on the left.

Vertebra C1 (atlas)
Cervical plexus (C1–C5)
Brachial plexus (C5–T1)
Vertebra T1
Lumbar enlargement region
Vertebra L1
Lumbar plexus (L1–L4)
Sacral plexus (L4–S4)
Coccygeal plexus (S4–Co1)
Sciatic nerve

Foramen magnum
C1
C2
C3
C4
C5
C6
C7
C8
Cervical nerves (C1–C8)
Cervical enlargement region
T1
T2
T3
T4
T5
T6
T7
T8
T9
T10
T11
T12
Thoracic nerves (T1–T12)
L1
L2
L3
L4
L5
Conus medullaris
Lumbar nerves (L1–L5)
Cauda equina
S1
S2
S3
S4
S5
Sacral nerves (S1–S5)
Filum terminale
Coccygeal nerve (Co1)

Table 28.1   Nerve Plexuses

Plexus	Spinal Nerves Contributing Fibers	Most Significant Nerves of Plexus	Area Served	Comments
Cervical (figs. 28.7 and 28.8)	C1–C5	Phrenic (C3–C5)	Neck, shoulder, diaphragm, and portions of head	Some do not include C5 as part of cervical plexus; only a few fibers of C5 contribute to the phrenic nerve; the phrenic nerve (C3–C5) is essential in breathing
Brachial (figs. 28.9 and 28.10)	C5–C8 and T1	Musculocutaneous, axillary, radial, median, ulnar	Shoulder, arm, forearm, and hand	Small contribution of fibers from C4 and T2; a very complex network of nerves
Lumbar (fig. 28.11)	T12 (minor contributions) and L1–L4	Femoral, obturator	Inferior abdominopelvic and anterior thigh	The lumbar and sacral plexuses have connecting fibers and are sometimes called the lumbosacral plexus; minor contributions from T12
Sacral (fig. 28.12)	L4–L5 and S1–S4	Sciatic (common fibular and tibial) (figs. 28.11 and 28.12)	Buttock, posterior thigh, leg, and foot	The sciatic nerve has fibers from five spinal nerves (L4–L5 and S1–S3); the large sciatic nerve is located deep to the gluteus maximus and posterior thigh and is vulnerable to injuries and sciatica pain
Coccygeal (fig. 28.12)	S4–S5 and Co1	Pudendal	Pelvic floor	Tiny plexus; fibers from sacral plexus contribute to pudendal nerve

FIGURE 28.6 Transverse view of the spinal cord and a pair of spinal nerves with its branches associated with a vertebra.

4. In many areas along the spinal cord, the ventral (anterior) rami form *nerve plexuses* with complex networks of branches and fibers, comparable to an electrical junction box, primarily serving the limbs (table 28.1 and fig. 28.5). In some areas, some fibers intermingle with an adjacent plexus. The nerve plexuses include: cervical plexus (figs. 28.7 and 28.8); brachial plexus (figs. 28.9 and 28.10); lumbar plexus (fig. 28.11); and sacral and coccygeal plexuses (fig. 28.12).

5. The brachial plexus, considered the most complex, can be palpated just lateral to the sternocleidomastoid muscle and superior to the clavicle (fig. 28.10). The ventral rami of C5–T1 represent the five *roots* of the brachial plexus. The five roots unite into three *trunks;* each trunk divides into *anterior* and *posterior divisions.* The anterior and posterior divisions give rise to three bundles called *lateral, medial,* and *posterior cords.* Arising from the three cords in the axillary region, the five most significant nerves of the brachial plexus emerge (figs. 28.9 and 28.10). The brachial plexus has a characteristic **W** or **M** shape when viewed in a careful dissection of a cadaver (fig. 28.10).

6. The dorsal root near the spinal cord possesses sensory neurons that conduct nerve impulses from peripheral regions of the body. This sensory input correlates to specific areas of the skin. A specific area of the skin that a particular cutaneous branch of a spinal nerve innervates is called a *dermatome.* Figure 28.13 represents a *dermatome map* of anterior and posterior cutaneous regions of the body innervated by specific spinal nerves. Testing various dermatomes by light pinpricks or pokes can be used to diagnose possible spinal cord injuries and spinal nerve damage. The specific spinal nerve damaged may not be precisely identified because there is some overlap of neurons with close neighboring dermatomes.

7. Complete Parts C and D of the laboratory assessment.

FIGURE 28.7 The cervical plexus (C1–C5). See table 28.1 and figure 28.8, the cervical plexus of the cadaver.

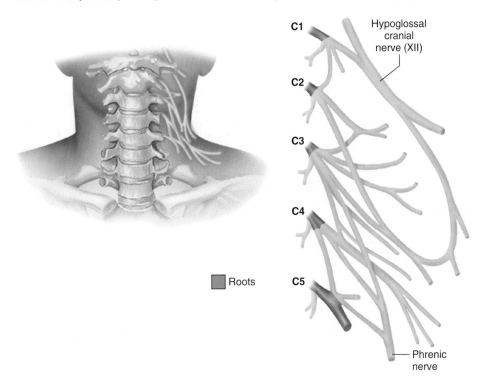

FIGURE 28.8 The cervical plexus of a cadaver, anterior view. The phrenic nerve, containing some fibers from C3, C4, and C5 innervates the diaphragm and is essential in breathing. See table 28.1.

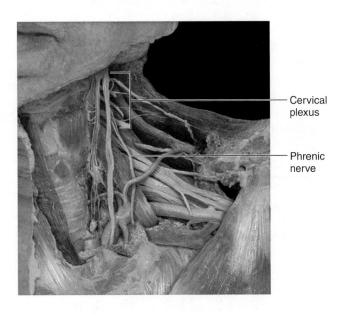

FIGURE 28.9 The brachial plexus (C5–C8 and T1). See table 28.1 and figure 28.10, the brachial plexus of the cadaver.

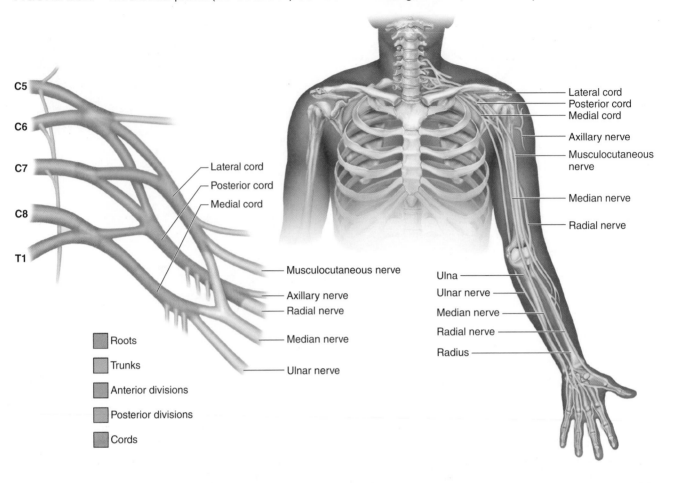

C5
C6
C7
C8
T1

Lateral cord
Posterior cord
Medial cord

Musculocutaneous nerve
Axillary nerve
Radial nerve
Median nerve
Ulnar nerve

Roots
Trunks
Anterior divisions
Posterior divisions
Cords

Lateral cord
Posterior cord
Medial cord
Axillary nerve
Musculocutaneous nerve
Median nerve
Radial nerve

Ulna
Ulnar nerve
Median nerve
Radial nerve
Radius

FIGURE 28.10 The brachial plexus of the left shoulder of the cadaver. The five most significant nerves of the plexus are exposed in this anterior view. See table 28.1.

Lateral cord
Posterior cord
Musculocutaneous nerve
Axillary nerve
Medial cord
Radial nerve
Median nerve
Ulnar nerve

298

FIGURE 28.11 The lumbar plexus (T12 and L1–L4). See table 28.1.

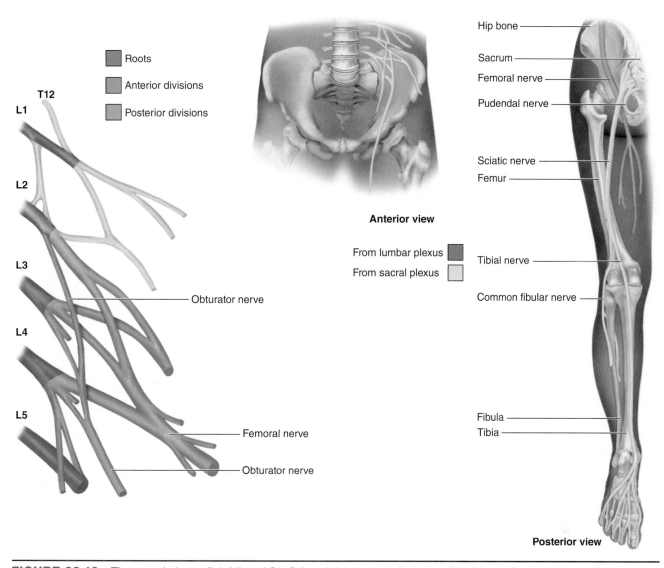

Roots
Anterior divisions
Posterior divisions

T12
L1
L2
L3
L4
L5

Obturator nerve
Femoral nerve
Obturator nerve

From lumbar plexus
From sacral plexus

Anterior view

Hip bone
Sacrum
Femoral nerve
Pudendal nerve
Sciatic nerve
Femur
Tibial nerve
Common fibular nerve
Fibula
Tibia

Posterior view

FIGURE 28.12 The sacral plexus (L4–L5 and S1–S4) and the coccygeal plexus (S4–S5 and Co1). See table 28.1 and the posterior view of figure 28.11.

L4
L5
S1
S2
S3
S4
S5
Co1

Roots
Anterior divisions
Posterior divisions

Common fibular nerve
Tibial nerve
Sciatic nerve
Pudendal nerve

FIGURE 28.13 Dermatome map of the body surface (a) anterior; (b) posterior. Spinal nerve C1 does not innervate the skin, but all of the other spinal nerves possess sensory branches to a zone indicated on the dermatome map.

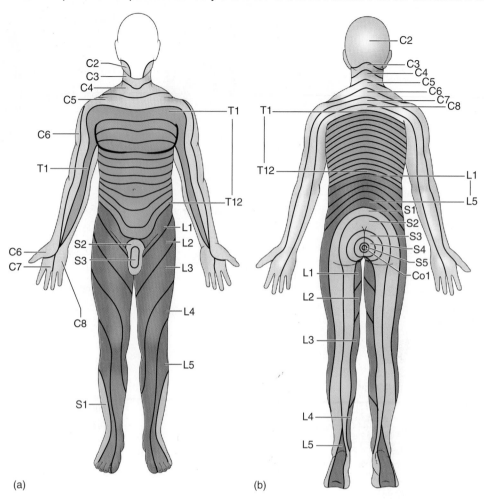

(a)

(b)

DEMONSTRATION ACTIVITY

Observe the preserved section of a spinal cord with meninges intact. Note the appearance of the central gray matter and the surrounding white matter. Note the heavy covering of dura mater, firmly attached to the cord on each side by a set of ligaments (denticulate ligaments) originating in the pia mater. The intermediate layer of meninges, the arachnoid mater, is devoid of blood vessels, but in a live human being, the space beneath this layer contains cerebrospinal fluid. The pia mater, closely attached to the surface of the spinal cord, contains many blood vessels.

Name _____

Date _____

Section _____

The ⒶE corresponds to the indicated Learning Outcome(s) ⓄE
found at the beginning of the Laboratory Exercise.

Meninges, Spinal Cord, and Spinal Nerves

PART A: Assessments

Match the terms in column A with the descriptions in column B. Place the letter of your choice in the space provided.

Column A	Column B
a. Arachnoid mater	b. 1. Connections from pia mater to dura mater that anchor the spinal cord
b. Denticulate ligaments	e. ea 2. Inferior continuation of pia mater to the coccyx
c. Dura mater	c. 3. Outermost layer of meninges
d. Epidural space	f. 4. Follows irregular contours of spinal cord surface
e. Filum terminale	g. 5. Contains a protective cushion of cerebrospinal fluid
f. Pia mater	a. 6. Thin, weblike middle membrane
g. Subarachnoid space	d. 7. Separates dura mater from bone of vertebra
h. Subdural space	h. 8. Potential narrow space with a thin film of fluid

PART B: Assessments

Identify the features indicated in the spinal cord cross section of figure 28.14.

FIGURE 28.14 Micrograph of a spinal cord cross section with spinal nerve roots (7.5×). Label the features by placing the correct numbers in the spaces provided. (A2) (A3)

Posterior/Dorsal

Dura mater

Anterior/Ventral

_8_ Anterior median fissure	_2_ Dorsal root of spinal nerve	_4_ Ventral root of spinal nerve
_5_ Central canal	_1_ Gray matter	_7_ White matter
_3_ Dorsal root ganglion	_6_ Posterior median sulcus	

PART C: Assessments

Complete the following statements:

1. The spinal cord gives rise to 31 pairs of _Spinal nerves_. (A2) (A3)

2. The bulge in the spinal cord that gives off nerves to the upper limbs is called the _____ enlargement. (A2)

3. The bulge in the spinal cord that gives off nerves to the lower limbs is called the _____ enlargement. (A2)

4. The _____ is a groove that extends the length of the spinal cord posteriorly. (A2)

5. In a spinal cord cross section, the posterior _funiculus_ of the gray matter resemble the upper wings of a butterfly. (A2)

6. The axons of motor neurons are found in the _____ roots of spinal nerves. (A3)

7. The _central canal_ connects the gray matter on the left and right sides of the spinal cord. (A2)

8. The _central canal_ in the gray commissure of the spinal cord contains cerebrospinal fluid and is continuous with the cavities of the brain. (A2)

9. The white matter of the spinal cord is divided into anterior, lateral, and posterior _funiculi_ that contain ascending and descending tracts. (A2)

10. There are _____ 8 _____ ~~(1 above, 1 below C_1)~~ pairs of cervical spinal nerves. (A3)

11. There are _____ 5 _____ pairs of sacral spinal nerves. (A3)

12. Cervical spinal nerve pair C1 originates between the foramen magnum of the occipital bone and the _____ atlas _____. (A3)

13. Spinal nerves L4 through S4 form a _____ sacral _____ plexus. (A3)

14. The gray matter of the spinal cord is divided into anterior, lateral, and posterior _____ ~~dorsal sections~~ horns _____. (A2)

15. The spinal cord ends inferior to L1 in a tapered point called the conus medullaris (A2)

PART D: Assessments

Complete the following:

1. To block chronic pain in a patient, sometimes doctors will sever the dorsal root of a spinal nerve. Provide a rationale for such a procedure. (A2) (A3) _____

_____

_____

_____

2. Describe what the result would be if the ventral root of a spinal nerve were severed? (A2) (A3) _____

_____

_____

_____

3. Injuries to the vertebral column and spinal cord are always of concern. Explain why injuries, especially to cervical regions, are of the greatest concern. Include specific cervical areas as part of your answer. (A2) (A3) _____

_____

_____

_____

CRITICAL THINKING ASSESSMENT

Why are intramuscular injections given in the gluteus medius muscle rather than in the larger gluteus maximus muscle in the buttock region?

_____

_____

_____

_____

Reflex Arc and Somatic Reflexes

Purpose of the Exercise

To review the components of reflex arcs and reflexive behavior. Several common somatic reflexes will also be demonstrated.

MATERIALS NEEDED

Rubber percussion hammer
Reflex arc model

Learning Outcomes

After completing this exercise, you should be able to:

O1 Demonstrate and record results of stretch reflexes that occur under normal laboratory conditions and from mental distractions.

O2 Describe the components of a reflex arc.

O3 Analyze the components and patterns of stretch reflexes.

The **O** corresponds to the assessments **A** indicated in the Laboratory Assessment for this Exercise.

Pre-Lab

Carefully read the introductory material and examine the entire lab. Be familiar with reflexes from lecture or the textbook. Answer the pre-lab questions.

Pre-Lab Questions Select the correct answer for each of the following questions:

1. The impulse over a motor neuron will lead to
   a. an interneuron.   b. the spinal cord.
   c. a receptor.   d. an effector.

2. Stretch reflex receptors are called
   a. effectors.   b. muscle spindles.
   c. interneurons.   d. motor neurons.

3. Stretch reflexes include all of the following *except* the _____ reflex.
   a. withdrawal   b. patellar
   c. calcaneal   d. biceps

4. A withdrawal reflex could occur from
   a. striking the patellar ligament.
   b. striking the calcaneal tendon.
   c. striking the triceps tendon.
   d. touching a hot object.

5. The calcaneal reflex response is
   a. pain interpretation.   b. flexion of the leg at the knee joint.
   c. plantar flexion of the foot.   d. a separation of toes.

6. The quadriceps femoris is the effector muscle of the patellar reflex.
   a. True _____   b. False _____

7. The dorsal roots of spinal nerves contain the axons of the motor neurons.
   a. True _____   b. False _____

8. The normal patellar reflex response involves extension of the leg at the knee joint.
   a. True _____   b. False _____

A reflex arc represents the simplest type of nerve pathway found in the nervous system. This pathway begins with a *receptor* at the dendrite end of a *sensory (afferent) neuron*. The sensory neuron leads into the central nervous system (*integration center*) and may communicate with one or more *interneurons*. Some of these interneurons, in turn, communicate with *motor (efferent) neurons,* whose axons (nerve fibers) lead outward to *effectors*. Thus, when a sensory receptor is stimulated by a change occurring inside or outside the body, impulses may pass through a *reflex arc,* and, as a result, effectors may respond. Such an automatic, subconscious response is called a *reflex*.

Skeletal muscles are involved in *somatic reflexes*. Somatic reflexes include *stretch reflexes* and *withdrawal reflexes*. Some reflexes involve a single synapse (monosynaptic) between a sensory and motor neuron; more complex arcs or circuits involve two or more synapses (polysynaptic) and some interneurons. A reflex arc includes five basic components:

1. **Receptor**—location of the stimulus at some site in the PNS
2. **Sensory neuron**—conducts impulse from receptor to the CNS (brain or spinal cord)
3. **Integration center**—the brain or spinal cord where synapses occur with some interneurons
4. **Motor neuron**—conducts impulse from the integration center to the effector
5. **Effector**—a muscle or a gland that responds with a contraction or a secretion

Skeletal muscles function as the effectors for the somatic stretch reflexes included in this laboratory exercise. Withdrawal reflex examples are described and illustrated in this laboratory exercise, but they will not be demonstrated for safety reasons.

Smooth muscle, cardiac muscle, and glands are the effectors of *autonomic reflexes*. Some examples of autonomic reflexes are included in Laboratory Exercise 36.

A stretch reflex involves a single synapse (monosynaptic) between a sensory and a motor neuron within the gray matter of the spinal cord. A branch of a sensory fiber is also involved in an inhibitory pathway with an interneuron and motor neuron to an antagonist (fig. 29.1). Examples of stretch reflexes include the patellar, calcaneal, biceps, triceps, and plantar reflexes. Other more complex withdrawal reflexes involve interneurons in combination with sensory and motor neurons; thus they are polysynaptic. Examples of withdrawal reflexes include responses to touching hot objects or stepping on sharp objects (fig. 29.2).

Reflexes demonstrated in this laboratory exercise are stretch reflexes. When a muscle is stretched by a tap over

---

FIGURE 29.1 AP|R Diagram and pathway of a stretch (monosynaptic) reflex arc, and the reciprocal inhibition of its antagonistic muscle. The patellar reflex represents a specific example of a stretch reflex.

① Tap on patellar ligament excites nerve endings of muscle spindle in quadriceps femoris.

② Sensory impules travel to spinal cord and synapse with motor neuron and interneuron.

③ Sensory neuron stimulates motor neuron in spinal cord of a monosynaptic reflex arc.

④ Motor impulses stimulate quadriceps (effector muscle) to contract, producing knee-jerk response.

⑤ At same time, a branch of the sensory nerve fiber stimulates an interneuron in spinal cord.

⑥ That neuron inhibits motor neuron that supplies hamstring muscles.

⑦ Hamstring contraction is inhibited (reciprocal inhibition), so hamstrings (knee flexors) do not antagonize quadriceps (knee extensor).

⊕ Excitatory synapse
⊖ Inhibitory synapse

its tendon, stretch receptors (proprioceptors) called *muscle spindles* are stretched within the muscle, which initiates an impulse over a reflex arc. A sensory neuron conducts an impulse from the muscle spindle into the gray matter of the spinal cord, where it synapses with a motor neuron, which conducts the impulse to the effector muscle. The stretched muscle responds by contracting to resist or reverse further stretching. These stretch reflexes are important to maintaining proper posture, balance, and movements. For example, if one is trying to stand up straight and starts to sway, resulting in a potential loss of balance, the stretched muscle will trigger a reflex arc, causing the stretched muscle to contract to correct the stance. Observations of many of these reflexes in clinical tests on patients may indicate damage to a level of the spinal cord or peripheral nerves of the particular reflex arc. Indications of certain diseases can be diagnosed if a reflex is absent, hypoactive (diminished), or hyperactive (exaggerated).

PROCEDURE: Reflex Arc and Somatic Reflexes

1. Study figure 29.1 as an example of a stretch reflex. Compare the stretch reflex with the withdrawal reflex shown in figure 29.2. Examine a model of a reflex arc.
2. Work with a laboratory partner to demonstrate each of the reflexes listed. (See fig. 29.3*a–e* also.) *It is important that muscles involved in the reflexes be totally relaxed to observe proper responses.* If a person is trying too

FIGURE 29.2 Polysynaptic withdrawal reflex along with crossed-extensor reflex on opposite limb. Note that when the agonist muscle contracts, the antagonist muscle is inhibited. For safety reasons, withdrawal reflexes are not tested in this laboratory exercise.

⊕ Excitatory synapse
⊖ Inhibitory synapse

② Sensory neuron activates multiple interneurons

③ Motor neurons to flexor excited on same limb

⑤ Opposite side motor neurons to extensor excited

Extensor muscle (antagonist) inhibited

④ Flexor muscle contracts

Flexor muscle (antagonist) inhibited

⑥ Opposite limb extensor contracts to support weight and balance

① Stepping on glass stimulates pain receptors in right foot

Withdrawal of right leg (withdrawl reflex)

Extension of left leg (crossed-extensor reflex)

hard to experience the reflex or is trying to suppress the reflex, assign a multitasking activity while the stimulus with the rubber percussion hammer occurs. For example, assign a physical task with upper limbs along with a complex mental activity during the patellar reflex demonstration. After each demonstration, record your observations in the table provided in Part A of Laboratory Assessment 29.

a. *Patellar (knee-jerk) reflex.* Have your laboratory partner sit on a table (or sturdy chair) with legs relaxed and hanging freely over the edge without touching the floor. Gently strike your partner's patellar ligament (just below the patella) with the blunt side of a rubber percussion hammer (fig. 29.3a). The reflex arc involves the femoral nerve and the spinal cord. The normal response is a moderate extension of the leg at the knee joint.

b. *Calcaneal (ankle-jerk) reflex.* Have your partner kneel on a chair with back toward you and with feet slightly dorsiflexed over the edge and relaxed. Gently strike the calcaneal tendon (just above its insertion on the calcaneus) with the blunt side of the rubber hammer (fig. 29.3b). The reflex arc involves the tibial nerve and the spinal cord. The normal response is plantar flexion of the foot.

c. *Biceps (biceps-jerk) reflex.* Have your partner place a bare arm bent about 90° at the elbow on the table. Press your thumb on the inside of the elbow over the tendon of the biceps brachii, and gently strike your thumb with the rubber hammer (fig. 29.3c). The reflex arc involves the musculocutaneous nerve and the spinal cord. Watch the biceps brachii for a response. The response might be a slight twitch of the muscle or flexion of the forearm at the elbow joint.

d. *Triceps (triceps-jerk) reflex.* Have your partner lie supine with an upper limb bent about 90° across the abdomen. Gently strike the tendon of the triceps brachii near its insertion just proximal to the olecranon process at the tip of the elbow (fig. 29.3d). The reflex arc involves the radial nerve and the spinal cord. Watch the triceps brachii for a response. The response might be a slight twitch of the muscle or extension of the forearm at the elbow joint.

e. *Plantar reflex.* Have your partner remove a shoe and sock and lie supine with the lateral surface of the foot resting on the table. Draw the metal tip of the rubber hammer, applying firm pressure, over the sole from the heel to the base of the large toe (fig. 29.3e). The normal response is flexion (curling) of the toes and plantar flexion of the foot. If the toes spread apart and dorsiflexion of the great toe occurs, the reflex is the abnormal *Babinski reflex* response (normal in infants until the nerve fibers have complete myelinization). If the Babinski reflex occurs later in life, it may indicate damage to the corticospinal tract of the CNS.

FIGURE 29.3 Demonstrate each of the following reflexes: (a) patellar reflex; (b) calcaneal reflex; (c) biceps reflex; (d) triceps reflex; and (e) plantar reflex.

(a) Patellar reflex

(b) Calcaneal reflex

3. Various factors such as mental distractions and fatigue might influence the degree of the reflex reaction. To test the mental distraction factor, repeat each of the reflex demonstrations while someone is purposely distracting the subject. To test the fatigue factor, have your partner exercise and then repeat all of the reflex demonstrations again. Record your observations in the table in Part A of the laboratory assessment.

4. Complete Part B of the laboratory assessment.

FIGURE 29.3 *Continued.*

(c) Biceps reflex

(d) Triceps reflex

(e) Plantar reflex

NOTES

Name _____

Date _____

Section _____

The (A) corresponds to the indicated Learning Outcome(s) (O) found at the beginning of the Laboratory Exercise.

Reflex Arc and Somatic Reflexes

PART A: Assessments

Complete the following table:

Reflex Tested	Response Observed	Effector Muscle Involved	Response During Mental Distraction or Fatigue (no change, more, or less)
Patellar	moderate extension of the leg below knee joint		
Calcaneal	Plantar flexion of the foot		
Biceps	Slight twitch of the muscle		
Triceps	Slight twitch of muscle		
Plantar	Curling of the toes		

What conclusions can you make when comparing results of somatic reflexes under typical laboratory conditions to the results under conditions of mental distraction or fatigue? (A1) _____

_____

_____

PART B: Assessments

Complete the following statements:

1. A withdrawal reflex employs _____ in conjunction with sensory and motor neurons. **A2**

2. Interneurons in a withdrawal reflex are located in the _____. **A2**

3. A reflex arc begins with the stimulation of a _____ at the dendrite end of a sensory neuron. **A2**

4. Effectors of somatic reflex arcs are _____ muscles. **A2**

5. A monosynaptic reflex arc involves a synapse between a _____ neuron and a _____ neuron. **A2**

6. The effector muscle of the patellar reflex is the _____. **A2**

7. The sensory stretch receptors (muscle spindles) of the patellar reflex are located in the _____ muscle. **A2**

8. The dorsal root of a spinal nerve contains the _____ neurons. **A2**

9. The normal plantar reflex results in _____ of toes. **A2**

10. Stroking the sole of the foot in infants results in dorsiflexion and toes that spread apart, called the _____ reflex. **A2**

11. In a stretch reflex arc, when the agonist muscle contracts, reciprocal inhibition occurs in its _____ muscle. **A2**

12. Synapses, whether excitatory or inhibitory, occur in the _____ matter of the spinal cord. **A2**

13. Effectors of autonomic reflexes include glands, smooth muscle, and _____. **A2**

14. List the major events that occur in the patellar reflex, from the striking of the patellar ligament to the resulting response. **A2** **A3**

   _____

   _____

   _____

CRITICAL THINKING ASSESSMENT

What characteristics do the reflexes you demonstrated have in common? **A3**

_____

LABORATORY EXERCISE 30

Brain and Cranial Nerves

Purpose of the Exercise

To review the structural and functional characteristics of the human brain and cranial nerves.

MATERIALS NEEDED

Dissectible model of the human brain
Human brain ventricles model
Human brain model with cranial nerves
Preserved human brain
Anatomical charts of the human brain

For testing cranial nerves:
Aromatic substances
Snellen eye chart
Pen flashlight
Sugar
Salt
Unsweetened lemon juice
Tuning fork

Learning Outcomes AP|R

After completing this exercise, you should be able to:

O1 Identify the locations and functions of the cranial meninges.

O2 Locate and describe the structural and functional regions of the brain.

O3 Identify the twelve pairs of cranial nerves.

O4 Differentiate the functions of each cranial nerve.

O5 Distinguish the cranial nerve possibly damaged from an observed impaired function.

The **O** corresponds to the assessments **A** indicated in the Laboratory Assessment for this Exercise.

Pre-Lab

Carefully read the introductory material and examine the entire lab. Be familiar with the brain and cranial nerves from lecture or the textbook. Answer the pre-lab questions.

Pre-Lab Questions Select the correct answer for each of the following questions:

1. Each hemisphere of the cerebrum regulates
   a. motor functions on the opposite side of the body.
   b. motor functions on the same side of the body.
   c. only functions within the brain.
   d. only functions within the spinal cord.

2. There are _____ pairs of cranial nerves.
   a. 2   b. 12
   c. 31   d. 43

3. Which of the following is *not* part of the brainstem?
   a. midbrain   b. pons
   c. medulla oblongata   d. thalamus

4. The _____ is the deep lobe of the cerebrum.
   a. temporal   b. insula
   c. parietal   d. occipital

5. The _____ separates the precentral and postcentral gyrus.
   a. lateral sulcus   c. central sulcus
   b. parieto-occipital sulcus   d. longitudinal fissure

6. The cranial meninges include the dura mater, arachnoid mater, and pia mater.
   a. True ___✓___   b. False _____

313

7. Ventricles of the brain contain cerebrospinal fluid.

    a. True _____   **b.** False _____

8. Functions of the cerebellum include reasoning, memory, and regulation of body temperature.

    a. True _____   **b.** False _____

The brain, the largest and most complex part of the nervous system, contains nerve centers associated with sensory perceptions and integration. It issues motor commands to skeletal muscles and carries on higher mental activities. The brain coordinates muscular movements and regulates the functions of internal organs. The cranial meninges surround the brain and are continuous with the spinal cord meninges.

The cerebral cortex, comprised of gray matter with billions of interneurons, represents areas for conscious awareness and decision-making processes. Sensory areas receive information from various receptors, association areas interpret sensory input, and motor areas involve planning and controlling muscle movements. All of these functional regions are influenced and integrated together in making complex decisions. Each hemisphere primarily interprets sensory and regulates motor functions on the opposite (contralateral) side of the body.

Twelve pairs of cranial nerves arise from the ventral surface of the brain and are designated by number and name. The cranial nerves are part of the PNS; most arise from the brainstem region of the brain. Although most of these nerves conduct both sensory and motor impulses, some contain only sensory fibers associated with special sense organs. Others are primarily composed of motor fibers and are involved with the activities of muscles and glands.

PROCEDURE A: Cranial Meninges

The cranial meninges, as well as the spinal cord meninges, include the *dura mater, arachnoid mater,* and *pia mater.* The thick, opaque, outermost layer, the dura mater, adheres to the inside of the skull. The weblike, semitransparent, middle layer, the arachnoid mater, spans the fissures of the brain. A *subarachnoid space* containing protective *cerebrospinal fluid (CSF)* is between the arachnoid mater and the innermost pia mater. The thin, delicate pia mater adheres tightly to the surfaces of the ridges (*gyri*) and grooves (*sulci*) on the surface of the brain (fig. 30.1).

The superficial dura mater is a single meningeal layer around the spinal cord, but it becomes a double layer in the cranial cavity due to the fusion of the meningeal layer with the periosteal layer of the skull (fig. 30.1). The two layers of the dura mater separate in some cranial areas, forming *dural sinuses.* The blood-filled dural sinuses are modified pathways for collecting venous blood of the brain. The large dural sinus, the *superior sagittal sinus,* is shown in figure 30.1.

In certain areas, the meningeal layer of the dura mater forms *dural septa,* membranous partitions separating various parts of the brain. These partitions function to support and stabilize the brain within the cranial cavity. The largest partition, the *falx cerebri,* extends into the longitudinal fissure separating the cerebral hemispheres (figs. 30.1 and 30.2). Other dural septa include one that separates the cerebellum from the cerebrum (*tentorium cerebelli*) and a partial septum separating the inferior cerebellar hemispheres (*falx cerebelli*).

The cranial subarachnoid space contains cerebrospinal fluid as the protective cushion of the brain. The periosteal layer of the dura mater adheres tightly to the inside of the skull, and therefore there is no cranial epidural space as seen

FIGURE 30.1 Cranial meninges associated with the brain.

FIGURE 30.2 `AP|R` Superior view of the surface of the brain within the skull of a cadaver. A portion of the dura mater, the outer meningeal layer, forms the falx cerebri within the longitudinal fissure.

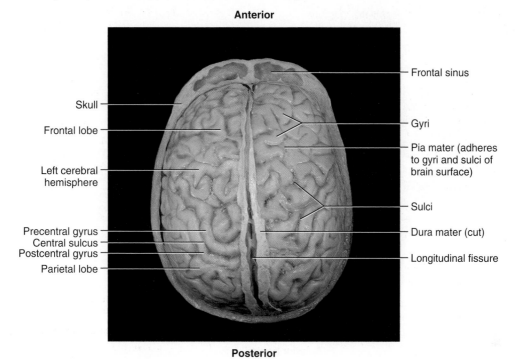

Anterior

Skull

Frontal lobe

Left cerebral
hemisphere

Precentral gyrus
Central sulcus
Postcentral gyrus
Parietal lobe

Frontal sinus

Gyri

Pia mater (adheres
to gyri and sulci of
brain surface)

Sulci

Dura mater (cut)

Longitudinal fissure

Posterior

in the spinal cord. An epidural space and the subdural space are potential spaces in the cranial cavity. Head injuries can result in accumulations of blood in the epidural or subdural potential spaces, referred to as an epidural hematoma or a subdural hematoma. Such head injuries can result in compression on brain tissue and possible death.

The study of the brain will include several regional categories: the ventricles, external surface features, cerebral hemispheres, diencephalon, brainstem, and cerebellum. During each regional study, examine available anatomical charts, dissectible models, and a human brain.

1. Examine figures 30.3 and 30.4 illustrating the ventricles of the brain. The four ventricles contain a clear cerebrospinal fluid (CSF) secreted by blood capillaries named choroid plexuses. The CSF flows from the lateral ventricles into the third ventricle and the fourth ventricle and then into the central canal of the spinal cord and through pores into the subarachnoid space. Locate each of the following features using a human brain ventricles model and dissectible brain models:

   ventricles
   - lateral ventricles—the largest ventricles; one located in each cerebral hemisphere
   - third ventricle—located within the diencephalon inferior to the corpus callosum
   - fourth ventricle—located between the pons and the cerebellum

2. Examine figures 30.2 and 30.5 for a study of the external surface features of the brain. Locate the following features using dissectible models and the human brain:

gyri—elevated surface ridges
- precentral gyrus
- postcentral gyrus

sulci—shallow grooves
- central sulcus—divides the frontal from the parietal lobe
- lateral sulcus (fissure)—divides the temporal from the parietal lobe
- parieto-occipital sulcus—divides the occipital from the parietal lobe

fissures—deep grooves
- longitudinal fissure—separates the cerebral hemispheres; contains falx cerebri, a dural septum
- transverse fissure—separates the cerebrum from the cerebellum; contains tentorium cerebelli, a dural septum

lobes of cerebrum—names associated with bones of the cranium
- frontal lobe
- parietal lobe
- temporal lobe
- occipital lobe
- insula (insular lobe)—deep within cerebrum; not visible on surface

3. Examine figures 30.2, 30.4, 30.5, and 30.6 and table 30.1 during the study of the cerebral hemispheres. The cerebrum is the largest part of the brain and has two hemispheres connected by the corpus callosum. The lobes of the cerebrum were included in the surface features

315

FIGURE 30.3 AP|R Four ventricles of the brain and their connections: (a) right lateral view and (b) anterior view.

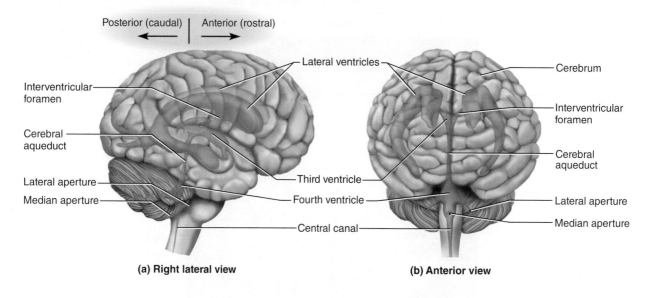

Posterior (caudal) | Anterior (rostral)

Lateral ventricles

Interventricular foramen

Cerebral aqueduct

Lateral aperture

Median aperture

Third ventricle

Fourth ventricle

Central canal

(a) Right lateral view

Cerebrum

Interventricular foramen

Cerebral aqueduct

Lateral aperture

Median aperture

(b) Anterior view

FIGURE 30.4 AP|R Transverse section of the human brain showing some ventricles, surface features, and internal structures of the cerebrum (superior view). This section exposes the anterior and posterior horns of both lateral ventricles.

Anterior (rostral)

Longitudinal fissure

Cerebral cortex (gray matter)

Cerebral white matter

Lateral ventricle (left)

Corpus callosum (anterior part)

Basal nuclei of cerebrum

Thalamus of diencephalon

Sulcus

Gyrus

Third ventricle in diencephalon

Choroid plexus

Corpus callosum (posterior part)

Longitudinal fissure

Posterior (caudal)

FIGURE 30.5 `AP|R` Lobes of the cerebrum. Retractors are used to expose the deep insula.

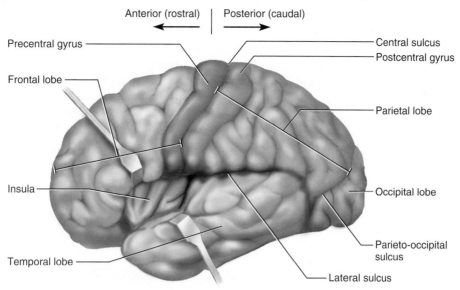

Anterior (rostral) | Posterior (caudal)

Precentral gyrus

Central sulcus
Postcentral gyrus

Frontal lobe

Parietal lobe

Insula

Occipital lobe

Temporal lobe

Parieto-occipital sulcus

Lateral sulcus

FIGURE 30.6 `AP|R` Diagram of a sagittal (median) section of the brain.

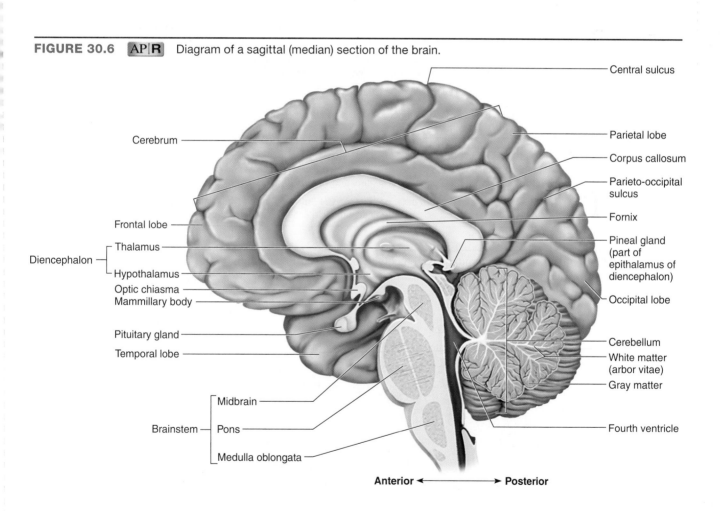

Central sulcus

Cerebrum

Parietal lobe

Corpus callosum

Parieto-occipital sulcus

Fornix

Frontal lobe

Diencephalon — Thalamus

Pineal gland (part of epithalamus of diencephalon)

Hypothalamus

Optic chiasma
Mammillary body

Occipital lobe

Pituitary gland

Temporal lobe

Cerebellum
White matter (arbor vitae)
Gray matter

Midbrain

Brainstem — Pons

Fourth ventricle

Medulla oblongata

Anterior ◄──────► Posterior

because they are also visible on the surface, except the deep insula. The functions associated with these lobes will be included in step 4. Locate the following features using dissectible models and a human brain:

cerebral cortex—thin surface layer of gray matter containing very little myelin; area of conscious awareness and processing information

cerebral white matter—largest portion of the cerebrum containing myelinated nerve fibers; transmits impulses between cerebral areas and lower brain centers

basal nuclei—masses of gray matter deep within the white matter; sometimes called basal ganglia as a clinical term; relay motor impulses from the cerebral cortex to the brainstem and spinal cord; the main structures include the caudate nucleus, putamen, and globus pallidus

4. Examine figures 30.5 and 30.7 to compare the structural lobes of the cerebrum with the functional regions of the lobes. Functional areas are not visible as distinct parts on an actual human brain. Examine the labeled areas in figure 30.7 that represent the following functional regions of the cerebrum:

sensory areas

- primary somatosensory cortex—receives information from skin receptors and proprioceptors
- somatosensory association cortex—integrates sensory information from primary cortex

Table 30.1 **AP|R** Major Regions and Functions of the Brain

Region	Functions
1. Cerebrum	Controls higher brain functions, including sensory perception, storing memory, reasoning, and determining intelligence; initiates voluntary muscle movements
2. Diencephalon	
a. Thalamus	Relay station for sensory impulses ascending from other parts of the nervous system to the cerebral cortex
b. Hypothalamus	Helps maintain homeostasis by regulating visceral activities and by linking the nervous and endocrine systems; regulates body temperature, sleep cycles, emotions, and autonomic nervous system control
3. Brainstem	
a. Midbrain	Contains visual reflex centers that move the eyes and head and auditory reflex centers; contains fiber tracts that connect the pons with the cerebrum; contains aqueduct between third and fourth ventricles where CSF flows
b. Pons	Relays impulses between higher and lower brain regions; helps regulate rate and depth of breathing
c. Medulla oblongata	Conducts ascending and descending impulses between the brain and spinal cord; contains cardiac, vasomotor, and respiratory control centers and various nonvital reflex control centers
4. Cerebellum	Processes information from other parts of the CNS by nerve tracts; integrates sensory information concerning the position of body parts; and coordinates muscle activities and maintains posture

FIGURE 30.7 Some structural and functional areas of the left cerebral hemisphere. (*Note:* These areas are not visible as distinct parts of the brain.)

- Wernicke's area—processes spoken and written language
- visual areas—consists of a primary and association (interpretation) area for vision
- auditory areas—consist of a primary and association area for hearing
- olfactory association area—interpretation of odors
- gustatory cortex—perceptions of taste

motor areas

- primary motor cortex—controls skeletal muscles
- motor association (premotor) area—planning body movements
- Broca's area—planning speech movements

5. Examine figures 30.6 and 30.8 and table 30.1 during the study of the diencephalon, brainstem, and cerebellum. Locate the following features using anatomical charts, dissectible models, and a human brain:

diencephalon

- thalamus—largest portion
- hypothalamus—inferior portion
- optic chiasma (chiasm)—optic nerves meet

- mammillary bodies—pair of small humps
- pineal gland—formed from epithalamus; secretes hormone melatonin

brainstem

- midbrain—superior region of brainstem
  - cerebral peduncles—connect pons to cerebrum
  - corpora quadrigemina—four bulges (two superior colliculi [visual reflexes] and two inferior colliculi [auditory reflexes])
  - cerebral aqueduct—cerebrospinal fluid flows from third to fourth ventricle
- pons—bulge on underside of brainstem
- medulla oblongata—inferior region of brainstem

cerebellum—cauliflower-like appearance

- right and left hemispheres
- vermis—connects the two hemispheres
- cerebellar cortex—gray matter portion
- arbor vitae—deeper branching pattern of white matter

6. Complete Parts A, B, and C of Laboratory Assessment 30.

FIGURE 30.8 Cerebellum and brainstem (a) median section and (b) superior view of cerebellum.

(a) Median section **AP|R**

(b) Superior view

PROCEDURE B: Cranial Nerves

1. The cranial nerves are part of the PNS. Examine figures 30.9, 30.10, 30.11, and 30.12 and table 30.2.
2. Observe the brain model with cranial nerves and preserved specimen of the human brain, and locate as many of the following cranial nerves as possible as you differentiate their associated functions:

olfactory nerves (I)
optic nerves (II)
oculomotor nerves (III)
trochlear nerves (IV)
trigeminal nerves (V)
abducens nerves (VI)
facial nerves (VII)
vestibulocochlear nerves (VIII)
glossopharyngeal nerves (IX)
vagus nerves (X)
accessory nerves (XI)
hypoglossal nerves (XII)

The following mnemonic device will help you learn the twelve pairs of cranial nerves in the proper order:

Old **Op**ie **oc**casionally **tr**ies **trig**onometry, and **f**eels **v**ery **glo**omy, **vagu**e, and **hypo**active.[1]

3. Examine table 30.3, which contains symptoms of a particular cranial nerve damaged, and the basic clinical laboratory tests to detect the damaged nerve.
4. With a laboratory partner, conduct each of the twelve clinical tests of the cranial nerves indicated in the last column of table 30.3. Summarize your results of the basic clinical tests in Part F of the laboratory assessment.
5. Complete Parts D, E, and F of the laboratory assessment.

[1]From *HAPS-Educator,* Winter 2002. An official publication of the Human Anatomy & Physiology Society (HAPS).

FIGURE 30.9 AP|R   Photograph of the cranial nerves attached to the base of the human brain.

Table 30.2 The Cranial Nerves and Functions

Number and Name		Type	Function
I	Olfactory	Sensory	Sensory impulses associated with smell
II	Optic	Sensory	Sensory impulses associated with vision
III	Oculomotor	Primarily motor*	Motor impulses to superior, inferior, and medial rectus and inferior oblique muscles that move the eyes, adjust the amount of light entering the eyes, focus the lenses, and raise the eyelids
IV	Trochlear	Primarily motor*	Motor impulses to superior oblique muscles that move the eyes
V	Trigeminal	Mixed	
	Ophthalmic division		Sensory impulses from the surface of the eyes, tear glands, scalp, forehead, and upper eyelids
	Maxillary division		Sensory impulses from the upper teeth, upper gum, upper lip, lining of the palate, and skin of the face
	Mandibular division		Sensory impulses from the scalp, skin of the jaw, lower teeth, lower gum, and lower lip Motor impulses to muscles of mastication and to muscles in the floor of the mouth
VI	Abducens	Primarily motor*	Motor impulses to lateral rectus muscles that move the eyes laterally
VII	Facial	Mixed	Sensory impulses associated with taste receptors of the anterior tongue Motor impulses to muscles of facial expression, tear glands, and salivary glands
VIII	Vestibulocochlear	Sensory	
	Vestibular branch		Sensory impulses associated with equilibrium
	Cochlear branch		Sensory impulses associated with hearing
IX	Glossopharyngeal	Mixed	Sensory impulses from the pharynx, tonsils, posterior tongue, and carotid arteries Motor impulses to salivary glands and to muscles of the pharynx used in swallowing
X	Vagus	Mixed	Somatic motor impulses to muscles associated with speech and swallowing; autonomic motor impulses to the viscera of the thorax and abdomen Sensory impulses from the pharynx, larynx, esophagus, and viscera of the thorax and abdomen
XI	Accessory	Primarily motor*	
	Cranial branch		Motor impulses to muscles of the soft palate, pharynx, and larynx
	Spinal branch		Motor impulses to muscles of the neck and shoulder
XII	Hypoglossal	Primarily motor*	Motor impulses to muscles that move the tongue

*These nerves contain a small number of sensory impulses from proprioceptors.

Table 30.3 Clinical Tests to Detect Possible Cranial Nerve Damage

Cranial Nerve	Symptoms of Nerve Damage	Basic Clinical Tests to Detect Damaged Cranial Nerve
Olfactory I (fig. 30.10)	Impaired smell or loss of smell	Sniff and determine if aromatic substances such as clove oil, perfume, and coffee are recognized
Optic II	Impaired vision or loss of vision	Read the time on the wall clock; identify the letters on a Snellen eye chart
Oculomotor III	Drooping eyelid; impaired eye movements; dilated pupil	Follow a finger without moving head; use a pen flashlight to gently shine light into the eye to check if pupils constrict
Trochlear IV	Impaired eye movements	Follow a finger and check if eye can be rotated
Trigeminal V (fig. 30.11)	Loss of sensations of touch, hot, cold, and pain in facial areas; impaired jaw movements	Touch the face in areas corresponding to the branches of the nerve to see if touch is detected; open and close mouth and move jaw various directions
Abducens VI	Loss of lateral eye movements	Follow a finger to test for lateral eye movements
Facial VII (fig. 30.12)	Distorted taste sensations on anterior region of tongue; dry eye and mouth; impaired facial muscle expressions	Place grains of sugar (or salt) on tip of tongue for detection of sweet (or salty) sensation; test ability to smile and whistle
Vestibulocochlear VIII	Loss of balance; loss of hearing	Try to walk a straight line to check balance (vestibular branch); use a tuning fork to test hearing ability (cochlear branch)
Glossopharyngeal IX	Dry mouth; impaired swallowing; impaired taste sensation on posterior tongue	Check for ability to swallow and cough; check posterior tongue with unsweetened lemon juice for detection of sour sensation
Vagus X	Impaired swallowing; difficulty in speaking and experiencing hoarseness	Check for swallowing impairments (cranial nerves IX and X innervate pharynx muscles); check for monotone and hoarseness in voice
Accessory XI	Impaired shoulder and rotating head movements	Examine ability to "shrug" shoulders and turn head against some resistance
Hypoglossal XII	Deviations or inability to stick out tongue; speech and swallowing difficulties	Check for ability to protrude and retract tongue without any deviations in position

FIGURE 30.10　AP|R　Olfactory cranial nerve.

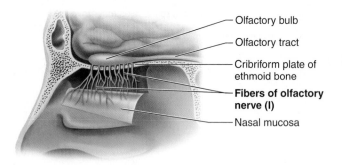

Olfactory bulb

Olfactory tract

Cribriform plate of ethmoid bone

Fibers of olfactory nerve (I)

Nasal mucosa

FIGURE 30.11 AP|R Trigeminal cranial nerve and its three major branches.

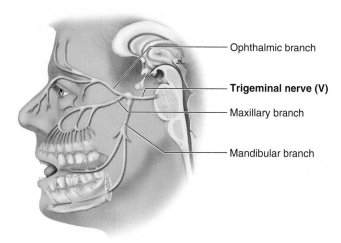

Ophthalmic branch

Trigeminal nerve (V)

Maxillary branch

Mandibular branch

FIGURE 30.12 AP|R Facial cranial nerve: (a) illustration of facial nerve and its five major branches; (b) photograph to help remember the facial areas served by the five branches.

Facial nerve (VII)

Temporal

Zygomatic

Buccal

Mandibular

Cervical

(a)

Temporal

Zygomatic

Buccal

Mandibular

Cervical

(b)

NOTES

Name _____

Date _____

Section _____

The Ⓐ corresponds to the indicated Learning Outcome(s) Ⓞ found at the beginning of the Laboratory Exercise.

Brain and Cranial Nerves

PART A: Assessments

Match the terms in column A with the descriptions in column B. Place the letter of your choice in the space provided.

A1 A2 A3

Column A	Column B
a. Central sulcus	_____ **1.** Structure formed by the crossing-over of the optic nerves
b. Cerebral cortex	_____ **2.** Part of diencephalon that forms lower walls and floor of third ventricle
c. Corpus callosum	
d. Gyrus	_____ **3.** Cone-shaped gland in the upper posterior portion of diencephalon
e. Hypothalamus	_____ **4.** Connects cerebral hemispheres
f. Insula	_____ **5.** Ridge on surface of cerebrum with pia mater closely attached
g. Medulla oblongata	_____ **6.** Separates frontal and parietal lobes
h. Midbrain	_____ **7.** Part of brainstem between diencephalon and pons
i. Optic chiasma	_____ **8.** Rounded bulge on underside of brainstem
j. Pineal gland	_____ **9.** Part of brainstem continuous with the spinal cord
k. Pons	_____ **10.** Internal brain chamber filled with CSF
l. Ventricle	_____ **11.** Cerebral lobe located deep within lateral sulcus
	_____ **12.** Thin layer of gray matter on surface of cerebrum

PART B: Assessments

Complete the following statements:

1. The _____ and the _____ together compose the dura mater in the cranial cavity. A1

2. The subarachnoid space contains a protective _____. A1

3. The _____, a dural septum, is located within the longitudinal fissure between the cerebral hemispheres. A1

4. The superior sagittal sinus collects and contains _____. A1

5. The cerebral cortex contains the _____ matter. A2

6. Grooves on the surface of the brain are sulci; ridges on the surface are _____. A2

7. The auditory areas of the brain are part of the _____ lobe. A2

8. The vision areas of the brain are part of the _____ lobe. A2

9. The left cerebral hemisphere primarily controls the _____ side of the body. A2

10. The brainstem includes the pons, the midbrain, and the _____. A2

11. The delicate _____ membrane is located on the surface of the brain. **A1**

12. The _____ fissure separates the two cerebral hemispheres. **A2**

13. The primary motor cortex is located within the _____ gyrus. **A2**

14. Arbor vitae and vermis are components of the _____. **A2**

15. The _____ ventricle is located between the pons and the cerebellum. **A2**

16. The _____ connects the two hemispheres of the cerebellum. **A2**

PART C: Assessments

Identify the features indicated in the median section of the right half of the human brain in figure 30.13.

FIGURE 30.13 **AP|R** Label the features on this median section of the right half of the human brain by placing the correct numbers in the spaces provided. **A2**

____ Brainstem
____ Cerebellum
____ Cerebrum
____ Corpus callosum
____ Diencephalon
____ Hypothalamus
____ Medulla oblongata
____ Midbrain
____ Pons
____ Thalamus

PART D: Assessments

Identify the cranial nerves that arise from the base of the brain in figure 30.14.

FIGURE 30.14 **AP|R** Complete the labeling of the twelve pairs of cranial nerves as viewed from the base of the brain. The Roman numerals indicated are also often used to reference a cranial nerve. **A3**

1 _____ (I)
(tract and bulb from)

2 _____ (II)

3 _____ (III)

4 Trochlear nerve _____ (IV)

5 _____ (V)

6 _____ (VI)

7 Facial nerve _____ (VII)

8 _____ (VIII)

9 _____ (IX)

10 _____ (X)

11 _____ (XI)

12 _____ (XII)

Spinal nerves

PART E: Assessments

Match the cranial nerves in column A with the associated functions in column B. Place the letter of your choice in the space provided. **A4**

Column A	Column B
a. Abducens	_____ **1.** Regulates thoracic and abdominal viscera
b. Accessory	_____ **2.** Equilibrium and hearing
c. Facial	_____ **3.** Stimulates superior oblique muscle of eye
d. Glossopharyngeal	_____ **4.** Sensory impulses from teeth and face
e. Hypoglossal	_____ **5.** Adjusts light entering eyes and eyelid opening
f. Oculomotor	_____ **6.** Smell
g. Olfactory	_____ **7.** Controls neck and shoulder movements
h. Optic	_____ **8.** Controls tongue movements
i. Trigeminal	_____ **9.** Vision
j. Trochlear	_____ **10.** Stimulates lateral rectus muscle of eye
k. Vagus	_____ **11.** Sensory from anterior tongue and controls salivation and secretion of tears
l. Vestibulocochlear	_____ **12.** Sensory from posterior tongue and controls salivation and swallowing

PART F: Assessments

1. Summarize the results of the basic clinical tests performed to detect any possible damage to a particular cranial nerve. (A4) (A5)

   Name the cranial nerves that indicated normal functional results based upon the clinical tests performed. _____
   _____
   _____
   _____

   Name any of the cranial nerves that had indications of impaired functional results based upon the clinical tests performed, and describe the specific impaired function observed. _____
   _____
   _____
   _____

2. Place the number and name of the cranial nerve best associated with each of the following normal abilities or impaired functions. All twelve cranial nerves are represented for the answers. (A4) (A5)

   _____ **1.** Experiences motion sickness and seems intoxicated

   _____ **2.** Unable to rotate eyeball inferolaterally

   _____ **3.** Able to depress shoulder joint, but unable to elevate shoulder

   _____ **4.** Identifies tissues using the compound microscope

   _____ **5.** Experiences difficulty with tongue movements when talking and swallowing

   _____ **6.** Abducts the eyeballs

   _____ **7.** Detects the odor of burning fall leaves

   _____ **8.** Detects taste sensation of a medicine placed on the back of the tongue

   _____ **9.** Moves jaw from side to side and chews food

   _____ **10.** One pupil of an eye has a different shape and size

   _____ **11.** Hoarseness experienced in voice and difficulty speaking

   _____ **12.** Experiences facial muscle paralysis and sagging corner of mouth (Bell palsy)

LABORATORY EXERCISE 31A

Reaction Time: BIOPAC© Exercise

Purpose of the Exercise

To measure reaction time in response to random-interval and fixed-interval stimuli using the BIOPAC Student Lab system, to observe differences in reaction time between the dominant and nondominant hands, and to conduct a brief statistical analysis of the reaction times of a group of subjects.

MATERIALS NEEDED

Computer system (Mac OS X 10.5–10.8), or PC running Windows XP, Vista, 7, or 8

BIOPAC Student Lab software, version 4.0 or above

BIOPAC Data Acquisition Unit MP36, MP35, or MP45

BIOPAC Hand Switch (SS10L)

BIOPAC Headphones (OUT-1 or OUT-1A)
Note: OUT-1A is compatible only with the MP36

SAFETY

▶ The BIOPAC system is safe and easy to use. There are no special safety considerations in this lab exercise, but subjects should wash their hands before and after holding the hand switch, to reduce the chances of spreading infection.

Learning Outcomes

After completing this exercise, you should be able to:

01 Explain the effects of learning on reaction time in response to random-interval and fixed-interval stimuli.

02 Compare the reaction times between the dominant and nondominant hands.

03 Explain individual differences in reaction time.

04 Analyze the reaction times of a group of subjects through a series of statistical calculations of the group mean, variance, and standard deviation.

The ○ corresponds to the assessments Ⓐ indicated in the Laboratory Assessment for this Exercise.

Pre-Lab

Carefully read the introductory material and examine the entire lab. Be familiar with the structure and function of synapses, neuromuscular junctions, and neural pathways from lecture or the textbook.

329

In order to maintain homeostasis, our bodies have to be able to respond to changes in our internal and external environments. Some of these responses are involuntary (automatic, not under conscious control), while others are voluntary (willed to occur, under conscious control). For example, if a person walks out of a warm house on a cold morning, the cold temperature will be the stimulus for both involuntary and voluntary responses. Involuntary (reflex) responses include the redistribution of blood flow, with more blood being sent toward the core of the body, inactivation of the sweat glands, and perhaps shivering. Voluntary reactions to the same stimulus may include zipping up your jacket, putting on warmer clothes, or simply going back into the house.

The neural pathways for voluntary and involuntary responses are similar, but not identical. Figure 31A.1 illustrates the basic pathway for a neural response to a stimulus.

Both voluntary and involuntary responses require stimulation of specific receptors, the transmission of nerve impulses over a series of neurons (one or more in the central nervous system), and the activation of effectors that carry out the response. However, involuntary reflex responses occur more quickly than voluntary reactions; they involve fewer neurons and do not require input from the cerebral cortex. Therefore, the reaction time for involuntary reflex responses is shorter than that of voluntary reactions.

FIGURE 31A.1 A general neural pathway for a voluntary or involuntary response to a stimulus.

The neural pathway leading to an involuntary reflex response is called a reflex arc. Figure 31A.2 shows the components of a spinal reflex arc:

1. *Sensory receptor,* which detects a specific type of change or stimulus
2. *Sensory (afferent) neuron,* which transmits nerve impulses from a receptor into the central nervous system (CNS), in this case, the spinal cord
3. *Interneuron (association neuron),* which integrates and coordinates the information received from sensory neurons (present in some types of reflex arcs, but not others; the processing is done in the CNS, whether or not an interneuron is present)
4. *Motor (efferent) neuron,* which transmits a motor command from an interneuron (or sensory neuron, in some cases) to an effector
5. *Effector,* a muscle or gland that carries out the reflex response, according to the instructions from a motor neuron

FIGURE 31A.2 The components of a spinal reflex arc (involuntary).

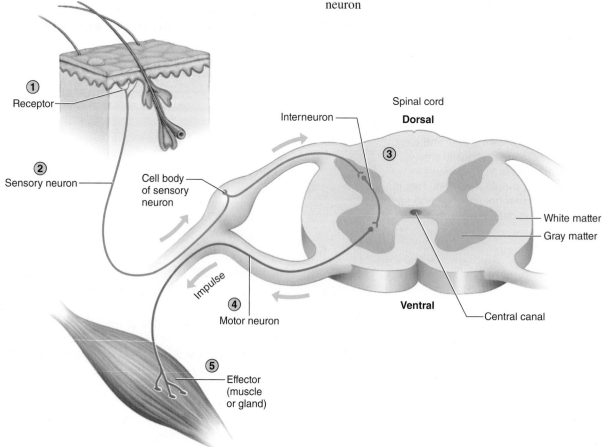

The response to laying a hand on the surface of a hot stove is conducted through a reflex arc such as the one shown in figure 31A.2. Since it causes a person to withdraw a limb from a painful stimulus, it is called a withdrawal reflex. When a hand is placed on the hot stove, several types of receptors are stimulated: warm receptors, pain receptors, touch receptors, and pressure receptors, but the pain receptors elicit the reflex response. Each receptor is connected to a sensory neuron, which transmits sensory information to an interneuron in the spinal cord. The interneuron coordinates the information from the various pain receptors and formulates a motor command, which it sends to a motor neuron. The motor neuron then sends the command to the effectors of the reflex, which in this case are flexor muscles of the arm. As a result of contraction of these muscles, the hand is lifted away from the painful stimulus, reducing or eliminating tissue damage. Although the brain is made aware of the reflex response, it is not a necessary component of the reflex arc.

In a voluntary reaction, however, a person chooses to respond in a certain way to a particular stimulus. This laboratory exercise is an example of a voluntary reaction to a stimulus. In order to measure reaction time, subjects are asked to hold a hand switch, and press a button in response to hearing a click. Since this type of reaction requires a conscious choice and involvement of the brain, its neural pathway is more complex than the reflex arc of an involuntary response. The neural pathway for the voluntary reaction to an auditory stimulus, which is the basis for this lab exercise, is shown in the following flowchart: Auditory stimulus (click) → sensory (hearing) receptors in inner ear → sensory neurons in vestibulocochlear nerve → interneurons in brainstem → interneurons in thalamus → interneurons in auditory cortex of cerebrum → descending pathways to spinal cord → motor neurons to proper muscles to move the thumb → pressing of hand switch.

In this lab exercise, you will test your reaction time by listening for clicks through headphones, and pressing a button on a hand switch as soon as possible after hearing each click. Reaction times will be tested in response to random-interval and fixed-interval stimuli, and compared between your dominant and nondominant hands. You will also gather data from other students in your class and observe individual differences, and you will determine data variability through calculation of variance and standard deviation. The BIOPAC program will record the reaction times and print them in the Journal area below the data window. You will then be guided through the statistical calculations.

PROCEDURE A: Setup

1. Before starting the lab exercise, you should designate a **Subject** to perform the experiment, a **Director,** who will inform the subject of each step in the experiment, and a **Recorder,** who will run the computer.
2. With your computer turned **ON** and the BIOPAC MP35/36 unit turned **OFF,** plug the hand switch (SSL10) into Channel 1 and the headphones (OUT-1

or OUT-1A) into the back of the Data Acquisition Unit. If you are using the MP45, the USB cable must be connected, and the ready light must be on. (Note: The OUT-1A headphones can only be used with the MP36.)
3. Turn on the MP36/35.
4. Start the BIOPAC Student Lab program, and click on **L11-Reaction Time I;** then click **OK.**
5. Designate a filename for the subject's data, and click **OK.**

PROCEDURE B: Calibration

1. Have the subject get ready for the calibration recording by sitting in a chair in a relaxed position, with eyes closed. The subject should put the headphones on and hold the hand switch upright in the dominant hand, with the thumb ready to press the button on top of the hand switch.
2. Click **Calibrate.** The calibration recording will last for 8 seconds and will stop automatically. About 4–5 seconds into the calibration, the subject will hear a click through the headphones. The subject should firmly press the button on the hand switch as soon as possible after hearing the click, and then quickly release it.
3. After the calibration stops, check the recording. If it looks similar to figure 31A.3, proceed to the data recording section, by clicking **Continue.** If not, check all connections, and then click **Redo Calibration.**

FIGURE 31A.3 Sample calibration recording.

Source: Courtesy of and © *BIOPAC* Systems Inc. www.biopac.com

PROCEDURE C: Recording

1. The subject should remain seated and relaxed, with eyes closed, holding the hand switch upright in the dominant hand.
2. All members of the group should read and become familiar with the procedure, before starting to record data. The recording portion of lab exercise 31A will consist of four segments, in the following order:
   A. Stimuli (clicks) presented at random intervals (1–10 seconds); dominant hand
   B. Stimuli presented at fixed intervals (every 4 seconds); dominant hand

C. Stimuli presented at random intervals (1–10 seconds); nondominant hand

D. Stimuli presented at fixed intervals (every 4 seconds); nondominant hand

3. Click **Record** to begin segment A (random-interval, dominant hand). A series of 10 stimuli (clicks) will be presented to the subject at intervals ranging from 1 to 10 seconds. The subject should press the button on the hand switch as soon as possible after hearing each click, and then release the button quickly.

4. The recording will stop automatically after 10 clicks. The data window should look similar to figure 31A.4, with a reaction pulse displayed after each event marker (a triangle at the top of the data window). If similar, click **Continue** to proceed to step 5. If not, click **Redo.** In some cases, the subject may not have pressed the button hard enough for the system to record a pulse. If no response has been recorded for one or more clicks, you should redo the segment.

FIGURE 31A.4 Sample recording of data for random-interval stimulus presentation.

Source: Courtesy of and © *BIOPAC* Systems Inc. www.biopac.com

5. Click **Record** to begin segment B (fixed-interval, dominant hand). This time, a series of 10 clicks will be presented every 4 seconds. The subject should press the button on the hand switch as in step 3.

6. The recording will stop automatically after 10 clicks. The data window should display a reaction pulse for each of the 10 stimuli, and it should look similar to figure 31A.4. If it does, click **Continue** to proceed to step 7. If not, click **Redo.**

7. The subject should now transfer the hand switch to the nondominant hand.

8. Click **Record** to begin segment C (random-interval, nondominant hand). Press the button on the hand switch as above. Click **Redo** if necessary; otherwise, click **Continue** to proceed to step 9.

9. Click **Record** to begin segment D (fixed-interval, nondominant hand). Again, press the button on the hand switch as above. Click **Redo** if necessary; otherwise click **Done** to end the recording portion of the lab exercise.

10. A pop-up window will now appear with several options. Click either **Analyze current data file,** if you are going to conduct your data analysis now, or **Quit,** if you are going to analyze the file at a later time. Then click **OK.** The data from the lab exercise will be saved by the BIOPAC program.

11. When you are sure that you have obtained complete data for the lab exercise, you can unplug the hand switch and headphones.

PROCEDURE D: Data Analysis

1. You can either **Analyze the current data file** now, or conduct the analysis later, by using the **Review Saved Data** mode. When you begin the data analysis, the data window should look similar to figure 31A.5.

FIGURE 31A.5 Sample reaction time data and journal.

Source: Courtesy of and © *BIOPAC* Systems Inc. www.biopac.com

2. Note the channel number designation: CH 1—hand switch.

3. Across the top of the recording, you will see three boxes for each channel: channel number, measurement type, and result. Note the setting for the measurement box: CH 1 should be set on **Delta T,** which measures the change in time between the beginning and end of the selected area.

4. Set up your data window so that you are only viewing the first event marker and the first reaction pulse of Segment A. You can spread out the data, or "zoom in" on the first pulse, by clicking anywhere in the time line at the bottom of the data window. This will bring up a text box, in which you can choose the "start" and "end" times to display in the window. After choosing these times (perhaps from 3 to 10 seconds), click **OK.**

5. Your data window should now look similar to figure 31A.6. This view will be used to manually calculate the reaction time between the first event marker and the first reaction pulse. Using the I-beam cursor, select the area between the first event marker and the beginning edge of the first reaction pulse. See figure 31A.7 for proper selection of the area. The event marker represents the stimulus (the click), and the beginning of the reaction pulse represents the point at which the button was pressed. Record this reaction time (**Delta T**) measurement in question 1 in Part A of Laboratory Assessment 31A.

FIGURE 31A.7 Selection of the area for manual calculation of reaction time.

Source: Courtesy of and © *BIOPAC* Systems Inc. www.biopac.com

FIGURE 31A.6 Data window displaying the first event marker and reaction pulse.

Source: Courtesy of and © *BIOPAC* Systems Inc. www.biopac.com

6. Compare the manually calculated reaction time (**Delta T**) to the first reaction time shown in the journal. The two numbers should be close. This manual calculation is conducted in order for you to confirm that the journal results are correct.
7. Copy the data from the journal into Table 31A.1 in the laboratory assessment. Note: The reaction times provided in the journal are shown in seconds. When you enter the data into the table, enter them in milliseconds (msec). For example, 0.215 seconds = 215 msec.
8. Then proceed to fill out the rest of the tables, following the directions provided in the laboratory assessment.

NOTES

Name _____

Date _____

Section _____

The Ⓐ corresponds to the indicated Learning Outcome(s) Ⓞ found at the beginning of the Laboratory Exercise.

Reaction Time: BIOPAC© Exercise

Subject Profile

Name _____

Age _____   Gender: Male / Female

PART A: Data and Calculations Assessments

1. Manual Calculation of Reaction Time for the First Stimulus

   Measure the reaction time for the first stimulus of the lab exercise.   **Delta T** = _____ msec

2. Reaction Time Results for One Subject (copy out of journal) A1 A2

Table 31A.1   Reaction Times for One Subject

| Stimulus Number | REACTION TIMES (MSEC) | | | |
| | Dominant Hand | | Nondominant Hand | |
	Random-Interval	Fixed-Interval	Random-Interval	Fixed-Interval
1				
2				
3				
4				
5				
6				
7				
8				
9				
10				
Mean				

3. Comparison of Reaction Time to Stimulus Presentation Number in Five Subjects

Copy the appropriate data for your subject from Table 31A.1 into line 1 of Table 31A.2. This table will be filled out with data from segments A (random-interval, dominant hand) and B (fixed-interval, dominant hand) of the lab exercise. Next, obtain corresponding data from four other subjects to complete the table. Then calculate the means for each column of data.

Table 31A.2 Reaction Times by Stimulus Presentation Number

Name of Subject	RANDOM-INTERVAL REACTION TIMES (MSEC) (DOMINANT HAND)			FIXED-INTERVAL REACTION TIMES (MSEC) (DOMINANT HAND)		
	Stimulus #1	Stimulus #5	Stimulus #10	Stimulus #1	Stimulus #5	Stimulus #10
1.						
2.						
3.						
4.						
5.						
Mean						

4. Group Summary of Mean Reaction Times in Five Subjects

Fill in line 1 of Table 31A.3 with your subject's mean reaction times from the last row of Table 31A.1 (Note: The data in Table 31A.1 are not in the same order as the data in Table 31A.3; be careful to copy the data into the correct column.) Next, obtain data from the same four other subjects to complete the table. Then calculate the means for each column of data.

Table 31A.3 Mean Reaction Times for a Group of Five Subjects

Name of Subject	RANDOM-INTERVAL REACTION TIMES (MSEC)		FIXED-INTERVAL REACTION TIMES (MSEC)	
	Dominant Hand	Nondominant Hand	Dominant Hand	Nondominant Hand
1.				
2.				
3.				
4.				
5.				
Mean				

5. Variance and Standard Deviation Calculations

The formulas for variance and standard deviation follow. You will be able to complete these statistical calculations by following the directions in Tables 31A.4 and 31A.5.

$$Variance = \frac{1}{n-1} \sum_{j=1}^{n} (X_j - \bar{X})^2$$

Definition of variables:

n = number of subjects

X_j = mean reaction time for each subject

$\bar{X}$ = group mean (same for all subjects)

$\sum_{j=1}^{n}$ = sum of all subjects' data

$Standard\ Deviation = \sqrt{Variance}$

Calculate the variance and standard deviation for the group of five subjects, using the data from Table 31A.3. To fill out Table 31A.4, copy the data from Table 31A.3, column 3 (random-interval, nondominant hand), into Table 31A.4, column 2 (Mean Reaction Time for Subject, X_j). Also copy the group mean into column 3. Then complete the calculations as indicated. When you calculate the deviation from the mean in column 4, enter absolute values; negative signs should be eliminated. These statistical calculations will indicate the variability of reaction times obtained from the five subjects in the random-interval, nondominant hand segment of the lab exercise.

Table 31A.4   Calculation of Variance and Standard Deviation for Random-Interval Reaction Times for the Nondominant Hand in a Group of Five Subjects

Name of Subject	ENTER Mean Reaction Time for Subject X_j	ENTER Group Mean $\bar{X}$	CALCULATE Deviation from Mean $X_j - \bar{X}$	CALCULATE Deviation from Mean² $(X_j - \bar{X})^2$
1.				
2.				
3.				
4.				
5.				
Sum of data for all five subjects: (Add the five numbers in column 5) $\sum_{j=1}^{n} (X_j - \bar{X})^2$				
Variance = Sum x 0.25 or $\frac{1}{n-1}$				
Standard Deviation = Square Root of Variance				

To fill in Table 31A.5, copy the data from Table 31A.3, column 5 (fixed-interval, nondominant hand), into Table 31A.5, column 2 (Mean Reaction Time for Subject, X_j). Then complete the calculations as indicated. These statistical calculations will indicate the reaction time variability obtained from the five subjects in the fixed-interval, nondominant hand segment of the lab exercise.

Table 31A.5 Calculation of Variance and Standard Deviation for Fixed-Interval Reaction Times for the Nondominant Hand in a Group of Five Subjects

Name of Subject	ENTER Mean Reaction Time for Subject X_j	ENTER Group Mean $\bar{X}$	CALCULATE Deviation from Mean $X_j - \bar{X}$	CALCULATE Deviation from Mean² $(X_j - \bar{X})^2$
1.				
2.				
3.				
4.				
5.				
Sum of data for all five subjects: (Add the five numbers in column 5) $\sum\limits_{j=1}^{n} (X_j - \bar{X})^2$				
Variance = Sum × 0.25 or $\dfrac{1}{n-1}$				
Standard Deviation = Square Root of Variance				

PART B: Assessments

1. **a.** What are the components of a typical reflex arc? List them in order. _____
_____
_____

    b. How does a neural pathway for a voluntary reaction differ from a spinal reflex arc? _____
_____
_____

2. What is the difference between a reflex response and a voluntary reaction to a stimulus? Which would you expect to have a shorter response time? Why? _____
_____
_____
_____
_____

3. From the results in Table 31A.1, what was the range of your subject's reaction time? _____
_____

4. From the results in Table 31A.1, which condition showed the shorter mean reaction time in your dominant hand, fixed-interval or random-interval? Did you expect these results? Why or why not? **A1** _____

_____

_____

5. From the results in Table 31A.1, which hand showed the shorter mean reaction time, the dominant or nondominant hand? Did you expect these results? Why or why not? **A2** _____

_____

_____

6. From the results in Table 31A.2, did the reaction times get longer or shorter as the stimuli progressed from #1 to #10? Did you expect these results? Why or why not? **A1** _____

_____

_____

_____

7. **a.** From the group means in Table 31A.3, which condition showed the shorter reaction time, random-interval or fixed-interval? **A1** _____

   b. Which hand had the shorter reaction time, dominant or nondominant? **A2** _____

   c. Did the group results agree with the results from your subject? _____

8. From the calculations of standard deviation in Tables 31A.4 and 31A.5, which condition showed the greater variability in reaction times, random-interval or fixed-interval? Did you expect these results? Why or why not? **A1** **A4** _____

_____

_____

NOTES

EXERCISE 31B

Electroencephalography I:
BIOPAC© Exercise

Purpose of the Exercise

To record an EEG from an awake, resting subject with eyes open and closed, and to identify and examine alpha, beta, delta, and theta components of an EEG.

MATERIALS NEEDED

Computer system (MAC OS X 10.5–10.8, or PC running Windows XP, Vista, 7, or 8)

BIOPAC Student Lab software version 4.0 or above

BIOPAC Data Acquisition Unit MP36, MP35, or MP45

BIOPAC electrode lead set (SS2L)

BIOPAC disposable vinyl electrodes (EL503), 3 electrodes per subject

BIOPAC electrode gel (GEL1) and abrasive pad (ELPAD) or alcohol wipes

Suggested: Lycra swim cap or supportive wrap (such as 3M Coban™ self-adhering support wrap) to press electrodes against head for improved contact

SAFETY

▶ The BIOPAC electrode lead set is safe and easy to use and should be used only as described in the procedures section of the laboratory exercise.

▶ The electrode lead clips are color-coded. Make sure they are connected to the properly placed electrode as demonstrated in figure 31B.2.

▶ The vinyl electrodes are disposable and meant to be used only once. Each subject should use a new set.

Learning Outcomes

After completing this exercise, you should be able to:

 O1 Identify alpha, beta, delta, and theta components of an EEG.

O2 Distinguish variations in these components due to differences in the level of stimulation, that is, eyes closed or open.

The **O** corresponds to the assessments **A** indicated in the Laboratory Assessment for this Exercise.

Pre-Lab

Carefully read introductory material and examine the entire lab. Be familiar with structures and functions of the brain, and the physiology of action potentials from lecture or the textbook.

Nerve impulses generated in the neurons of the nervous system are responsible for controlling many vital functions in the body, such as thought processes, responses to stimuli, higher mental functions (like problem solving), and the control of muscle contraction. Nerve impulses are electrical events that transmit information from neurons to other neurons, muscle cells, or glandular cells. This information is then used to accomplish various functional tasks throughout the body.

Neurons respond to certain types of stimuli by generating *action potentials,* which are accomplished by transporting sodium (Na^+) and potassium (K^+) ions across their plasma membranes in a particular sequence. This neural response leads to changes in the charge both inside the neurons and in the extracellular fluid surrounding them. This movement of ions in the extracellular fluid of the cerebral cortex can be detected by placing electrodes on the surface of the scalp and measuring the voltage between them. The recording of electrical activity in the cerebral cortex, via electrodes placed on the scalp, is called an *electroencephalogram (EEG).* An EEG is a safe, noninvasive, and painless method of recording the sum of electrical activity in many thousands of neurons in the cerebral cortex. It is important to realize that an EEG is not a record of action potentials occurring in an individual neuron.

An EEG can be used to assess states of consciousness (awake, concentrating, asleep) and various brain disorders, and to establish brain death. The major brain waves found in EEGs are of four types, called *alpha, beta, delta,* and *theta* waves. Each type of brain wave is characterized by a specific range of frequencies and amplitudes. Figure 31B.1 shows the typical appearance of the four types of brain waves.

The *frequency* of a brain wave is the number of complete wave cycles that occurs each second; it is measured in Hertz (Hz, or cycles per second). Frequency is an indication of the state of alertness; therefore, the higher the wave frequency, the more alert and attentive the person is.

The *amplitude* of a brain wave is the height, or magnitude, of the wave; it is measured in microvolts (μV, or one millionth of a volt). Amplitude indicates the degree of synchrony, or simultaneous firing of many neurons underneath the recording electrodes. Therefore, the higher the amplitude, the more neurons are firing simultaneously (in synchrony). It is important to understand that synchrony is not a measure of alertness; in fact, some of the highest amplitudes are found in delta waves, which occur mainly during the deeper stages of sleep. The lowest amplitudes are usually found in beta waves, which are dominant during high levels of attentiveness. This can be explained by the fact that asynchronous firing of neurons leads to a "canceling out" of positive and negative voltages occurring in different neurons at the same time.

All four types of brain waves can usually be observed in the EEG of a normal adult, although different wave forms become predominant in the various states of consciousness,

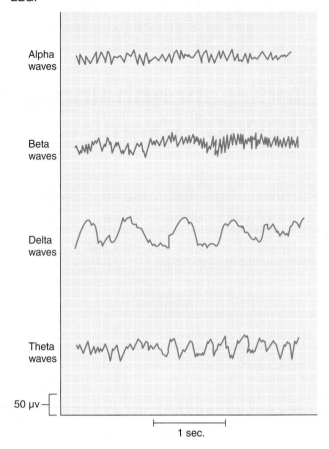

FIGURE 31B.1   The four types of brain waves of the EEG.

including the different stages of sleep. There is also quite a bit of individual variation in the characteristics of brain waves. This variation is due to a variety of factors, including gender, age, personality, the type of activity a person is engaged in, the area of the brain being used for the EEG, and various disease states of the brain. Table 31B.1 shows the average frequencies and amplitudes of the four types of brain waves, along with a short description of the state(s) of consciousness in which they are predominant.

In this laboratory exercise, you will be using the bipolar recording method. In this method, two electrodes are placed over the same region of the cerebral cortex, perhaps 2–3" apart, and the difference in the electrical potential (voltage) between the electrodes is measured. A third electrode is placed over the mastoid process as a "ground," or reference point for the body's baseline voltage.

The BIOPAC program will separate your EEG into the four types of brain waves. You will be able to observe all four types and measure the frequencies of the waves, as well as the standard deviation (variability) of wave amplitudes. You will also demonstrate *alpha block,* a replacement of alpha waves with lower-amplitude beta waves that occurs when closed eyes are suddenly opened.

Table 31B.1 Frequencies, Amplitudes, and State(s) of Consciousness Associated with the Four Types of Brain Waves

Brain Wave	Average Frequency (Hz)	Average Amplitude (µV)	State(s) of Consciousness
Alpha	8–13	~50 (can vary from 20–200)	Awake, resting, with eyes closed; decrease when concentrating
Beta	14–30 (sometimes higher)	5–10 (can be as high as 30)	Awake, alert, with eyes open, paying attention or concentrating
Delta	1–4	~100 (can vary from 20–200)	Deep stages of sleep; some forms of severe brain disease
Theta	4–7	5–10 (can be as high as 60–70)	More commonly observed in children, but often seen in adults, especially while concentrating or under stress, or during sleep; some forms of brain disease

PROCEDURE A: Setup

1. With your computer turned **ON** and the BIOPAC MP36/35 unit turned **OFF,** plug the electrode lead set (SSL2) into Channel 1. If you are using the MP45, the USB cable must be connected, and the ready light must be on.
2. Turn on the MP36/35 Data Acquisition Unit.
3. Selection of the subject is important; choose someone with easy access to his/her scalp so that you can be sure to make good contact between the skin and the electrode. With the subject sitting in a relaxed position, place the electrodes on the same side of the head, as in figure 31B.2. The ground electrode (black) is attached over the mastoid process. To help ensure good contact, make sure the area to be in contact with the electrodes is clean by wiping with the abrasive pad (ELPAD) or alcohol wipe. A small amount of BIOPAC electrode gel (GEL1) may also be used to make better contact if necessary. For optimal adhesion, hold the electrodes against the skin for about a minute after placement. Wrap the subject's head with a swim cap or supportive wrap to secure the electrodes in place.
4. Start the BIOPAC Student Lab Program for L03-Electroencephalography (EEG) I.
5. Designate a filename for the subject's data and click **OK.**

PROCEDURE B: Calibration

1. The subject should be in a sitting position with eyes closed. Click **Calibrate.** The calibration will stop automatically after 8 seconds.
2. The calibration recording should look similar to figure 31B.3. If it does, click **Continue** to proceed to the EEG recording. If not, click **Redo Calibration.**

FIGURE 31B.2 Electrode placement for EEG experiment.

Source: Courtesy of and © *BIOPAC* Systems Inc. www.biopac.com

(a)   (b)

FIGURE 31B.3 Sample calibration recording.

Source: Courtesy of and © *BIOPAC* Systems Inc. www.biopac.com

PROCEDURE C: Recording

1. Before beginning the recording phase, you should designate a **Director** (who will instruct the subject to open or close eyes) and a **Recorder** (who will insert markers by pressing the F4 key when eyes are opened, and the F5 key when eyes are closed.)

2. The data for this exercise will be recorded over a 1-minute period. During the first 20 seconds, the subject will have his/her eyes closed. During the next 20 seconds, the subject should have eyes open, and for the last 20 seconds, eyes closed again. Make sure all group members understand the procedure before proceeding to step 3.

3. Click on **Record.** The Director should then instruct the subject to remain relaxed and change eye condition as follows:

   a. 0–20 seconds, eyes closed

   b. 21–40 seconds, eyes open

   c. 41–60 seconds, eyes closed

   The Recorder should insert markers at these times: 21 seconds when eyes are opened (F4 key), and 41 seconds when eyes are closed again (F5 key).

4. Click on **Suspend.** If the recording is similar to figure 31B.4, click **Done,** and go to step 5. If there are large fluctuations or baseline drift in the recording, or if the EEG is flat, click on **Redo.** This will erase the current data. Check electrode placement, and make sure that the subject does not make any excessive movements during the new recording. After completing the new recording, click **Suspend,** and then **Done.**

5. A pop-up window will appear with several options. Click either **Analyze current data file,** if you are going to conduct your data analysis now, or **Quit,** if you are going to analyze the file at a later time. Then click **OK.** The data will be saved by the BIOPAC program.

PROCEDURE D: Data Analysis

1. You can either analyze the current file now or save it and do the analysis later, using the **Review Saved Data**

FIGURE 31B.5 EEG recording showing four different types of brain waves.

Source: Courtesy of and © *BIOPAC* Systems Inc. www.biopac.com

mode. When you enter the analysis mode, the program will separate the EEG into alpha, beta, delta, and theta waves, and display them in the data window. The window should look similar to figure 31B.5. If it is similar, the subject may now remove the electrodes.

2. Note the channel number designations: CH 1—EEG (hidden), CH 40—alpha, CH 41—beta, CH 42—delta, CH 43—theta.

3. Across the top of the recordings you will see three boxes for each channel: channel number, measurement type, and the result. Note the settings for the measurement boxes: CH 40–CH 43 should all be set on **Stddev** and SC on **Freq.** The program will automatically calculate the standard deviation or frequency for the selected area.

4. Using the I-beam cursor in the upper right corner, select the area from time 0–20 seconds (to the first marker). See figure 31B.6 for an illustration of how to select the area for the first 20 seconds. Record the result for each channel in table 31B.2 of Part A of Laboratory Assessment 31B. The standard deviation of amplitude will not provide actual wave amplitudes, but rather the variability in those amplitudes.

5. Repeat step 4 for the time period from 20–40 seconds (between the first and second markers) and record the results in table 31B.2.

FIGURE 31B.4 Sample EEG recording.

Source: Courtesy of and © *BIOPAC* Systems Inc. www.biopac.com

---

FIGURE 31B.6 Proper selection of the area from 0 to 20 seconds for Standard Deviation measurement.

Source: Courtesy of and © *BIOPAC* Systems Inc. www.biopac.com

6. Repeat step 4 for the time period from 40–60 seconds (from the second marker to the end) and record the results in table 31B.2.
7. Now set up the data window, so you can just see a 4- to 5-second interval in the first 20 seconds of the recording. This can be done by clicking anywhere in the horizontal scale (the numbers just above the word "seconds" at the bottom of the data window). A box will pop up in the middle of the screen, called **Set Screen Horizontal Axis.** Change the start time to 2 and the end time to 6. Then click **OK.** This will spread out the data so you can see individual waves and measure their frequency.
8. Now begin to measure wave frequencies. In order for the program to measure **FREQ** of each type of wave form, you have to select each type and measure it individually. First select alpha waves for **FREQ** measurement by clicking on the word "alpha" along the left side of the data window. Now SC (Selected Channel) will display only alpha wave frequencies.
9. Use the I-beam cursor to select an area that represents one cycle in the alpha wave. One cycle runs from one peak to the next peak (refer to figure 31B.7). Record the result in table 31B.3 of the laboratory assessment.
10. Repeat for two other nearby alpha cycles and record the results. Calculate the mean and record it in table 31B.3.
11. Repeat steps 9 and 10 for each of the other waveforms (beta, delta, theta) and record the results in table 31B.3.

FIGURE 31B.7 Proper selection of one alpha wave for frequency measurement.

Source: Courtesy of and © *BIOPAC* Systems Inc. www.biopac.com

(Remember to click on the word "beta," delta," or "theta" before beginning to measure **FREQ** for each of these wave types.)
12. If you would like to view the entire recording after completing **FREQ** measurements, open the **Display** menu. Click on **Autoscale Horizontal** and then **Autoscale Waveforms.** Now the entire one-minute recording will be visible in the data window.
13. When finished, you can save the data on the hard drive, a disk, or a network drive, and then exit the system by clicking **File, Quit.**
14. Complete Part B of the laboratory assessment.

NOTES

Name _____

Date _____

Section _____

The Ⓐ corresponds to the indicated Learning Outcome(s) Ⓞ found at the beginning of the Laboratory Exercise.

Electroencephalography I: BIOPAC© Exercise

Subject Profile

Name _____

Age _____

Gender: Male / Female

PART A: Data and Calculations Assessments

1. EEG Variability in Amplitude Measurements

Table 31B.2 Standard Deviation [stddev]

Rhythm	CH Measurement	Eyes Closed	Eyes Open	Eyes Re-closed
Alpha	40 Stddev			
Beta	41 Stddev			
Delta	42 Stddev			
Theta	43 Stddev			

2. EEG Frequency Measurements

Table 31B.3 Frequency [Hz] Measurements

Rhythm	CH Measurement	Cycle 1	Cycle 2	Cycle 3	Mean
Alpha	SC Freq				
Beta	SC Freq				
Delta	SC Freq				
Theta	SC Freq				

PART B: Assessments

Complete the following:

1. Examine the alpha and beta waveforms for changes between the "eyes closed" state and the "eyes open" state. (A2)

    a. Did desynchronization of the alpha rhythm occur when the eyes were open? How do you know?

    _____

    _____

    _____

    _____

    b. Did the beta rhythm change in the "eyes open" state? If so, how did it change? (A2)

    _____

    _____

    _____

    _____

2. Examine the delta and theta rhythms. Was there a change in delta and theta activity when the eyes were open? Explain your observation. (A1)

    _____

    _____

    _____

3. Define the following terms: (A1)

    a. Alpha rhythm

    _____

    _____

    b. Beta rhythm

    _____

    _____

    c. Delta rhythm

    _____

    _____

    d. Theta rhythm

    _____

    _____

LABORATORY EXERCISE 32

Dissection of the Sheep Brain

Purpose of the Exercise

To observe the major features of the sheep brain and to compare these features with those of the human brain.

MATERIALS NEEDED

Dissectible model of human brain
Preserved sheep brain
Dissecting tray
Dissection instruments
Long knife
Disposable gloves

For Demonstration Activity:
Frontal sections of sheep brains

SAFETY

▶ Wear disposable gloves when handling the sheep brains.
▶ Save or dispose of the brains as instructed.
▶ Wash your hands before leaving the laboratory.

Learning Outcomes AP|R

After completing this exercise, you should be able to:

O1 Locate the major structures of the sheep brain.

O2 Summarize differences and similarities between the sheep brain and the human brain.

O3 Locate the most developed cranial nerves of the sheep brain.

The **O** corresponds to the assessments **A** indicated in the Laboratory Assessment for this Exercise.

Pre-Lab

Carefully read the introductory material and examine the entire lab. Be familiar with basic structures and functions of the meninges, brain, and cranial nerves from the previous laboratory exercises. Answer the pre-lab questions.

Pre-Lab Questions Select the correct answer for each of the following questions:

1. The spinal cord of the sheep has a _____ orientation.
   a. vertical   b. transverse
   c. horizontal   d. sagittal

2. The gyri and sulci of the sheep brain possess the closely attached
   a. pia mater.   b. arachnoid mater.
   c. subarachnoid space.   d. dura mater.

3. By bending the cerebellum and medulla oblongata downward, the _____ should be visible.
   a. vagus nerve   b. optic nerve
   c. pituitary gland   d. pineal gland

4. The sheep brain has _____ pairs of attached cranial nerves.
   a. 2   b. 10
   c. 12   d. 31

5. The superior and inferior colliculi compose the
   a. pineal gland.   b. corpora quadrigemina.
   c. cerebrum.   d. cerebellum.

6. Sulci and gyri are surface features of the cerebrum of the sheep brain.
   a. True ✓   b. False _____

7. The cerebellum of the sheep brain has deep gray matter and superficial white matter.
   a. True ✓   b. False _____

Mammalian brains have many features in common. Human brains may not be available, so sheep brains often are dissected as an aid to understanding mammalian brain structure. However, the adaptations of the sheep differ from the adaptations of the human, so comparisons of their structural features may not be precise. The sheep is a quadruped; therefore, the spinal cord is horizontal, unlike the vertical orientation in a bipedal human. Preserved sheep brains have a different appearance and are firmer than those that are removed directly from the cranial cavity; this is caused by the preservatives used.

PROCEDURE: Dissection of the Sheep Brain

As you examine a sheep brain, contemplate any differences and similarities between the sheep brain and the human brain. Several questions in the laboratory assessment address comparisons of sheep and human brains.

1. Obtain a preserved sheep brain and rinse it thoroughly in water to remove as much of the preserving fluid as possible.
2. Examine the surface of the brain for the presence of meninges. (The outermost layers of these membranes may have been mostly lost during removal of the brain from the cranial cavity.) Locate the following:

    dura mater—the thick, opaque outer layer that adheres to the inside of the skull

    arachnoid mater—the weblike, semitransparent middle layer spanning the fissures of the brain

    pia mater—the thin, delicate, innermost layer that adheres to the gyri and sulci of the surface of the brain

3. Remove any remaining dura mater by cutting and pulling it gently from the surface of the brain.
4. Position the brain with its ventral surface down in the dissecting tray. Study figure 32.1, and locate the following structures on the specimen:

    longitudinal fissure
    cerebral hemispheres
    gyri
    sulci
    frontal lobe
    parietal lobe
    temporal lobe
    occipital lobe
    cerebellum
    medulla oblongata
    spinal cord

5. Gently separate the cerebral hemispheres along the longitudinal fissure and expose the transverse band of white fibers within the fissure that connects the hemispheres. This band is the *corpus callosum.*
6. Bend the cerebellum and medulla oblongata slightly downward and away from the cerebrum (fig. 32.2). This will expose the *pineal gland* in the upper midline and the *corpora quadrigemina*, which consists of four rounded structures, called *colliculi*, associated with the midbrain.

FIGURE 32.1 Dorsal surface of the sheep brain.

FIGURE 32.2 Gently bend the cerebellum and medulla oblongata away from the cerebrum to expose the pineal gland of the diencephlon and the corpora quadrigemina of the midbrain.

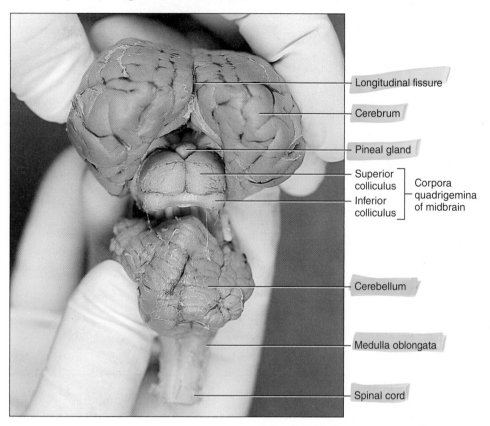

Longitudinal fissure

Cerebrum

Pineal gland

Superior colliculus ⎤
                   ⎥ Corpora
Inferior colliculus ⎦ quadrigemina
                     of midbrain

Cerebellum

Medulla oblongata

Spinal cord

7. Position the brain with its ventral surface upward. Study figures 32.3 and 32.4, and locate the following structures on the specimen:

longitudinal fissure
olfactory bulbs
optic nerves
optic chiasma
optic tract
mammillary body
infundibulum (pituitary stalk)
midbrain
pons

8. Although some of the cranial nerves may be missing or are quite small and difficult to find, locate as many of the following as possible, using figures 32.3 and 32.4 as references:

oculomotor nerves
trochlear nerves
trigeminal nerves
abducens nerves
facial nerves
vestibulocochlear nerves
glossopharyngeal nerves
vagus nerves
accessory nerves
hypoglossal nerves

FIGURE 32.3   Lateral surface of the sheep brain.

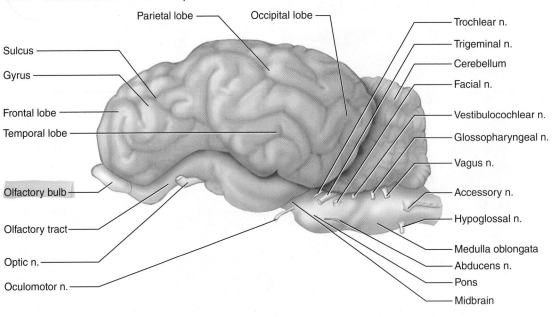

Parietal lobe
Occipital lobe
Sulcus
Gyrus
Frontal lobe
Temporal lobe
Olfactory bulb
Olfactory tract
Optic n.
Oculomotor n.
Trochlear n.
Trigeminal n.
Cerebellum
Facial n.
Vestibulocochlear n.
Glossopharyngeal n.
Vagus n.
Accessory n.
Hypoglossal n.
Medulla oblongata
Abducens n.
Pons
Midbrain

FIGURE 32.4   Ventral surface of the sheep brain.

Longitudinal fissure
Olfactory bulb
Optic chiasma
Optic tract
Optic n.
Infundibulum (pituitary stalk)
Mammillary body
Oculomotor n.
Midbrain
Trochlear n.
Pons
Trigeminal n.
Abducens n.
Facial n.
Medulla oblongata
Vestibulocochlear n.
Glossopharyngeal n.
Vagus n.
Hypoglossal n.
Accessory n.
Spinal cord

9. Using a long, sharp knife, cut the sheep brain along the midline to produce a median section. Study figures 32.2 and 32.5, and locate the following structures on the specimen:

cerebrum
olfactory bulb
corpus callosum
cerebellum
  - white matter
  - gray matter
lateral ventricle—one in each cerebral hemisphere
third ventricle—within diencephalon
fourth ventricle—between brainstem and cerebellum
diencephalon
  - optic chiasma
  - infundibulum
  - pituitary gland—this structure may be missing
  - mammillary body—single in sheep
  - thalamus
  - hypothalamus
  - pineal gland of epithalmus

midbrain
  - corpora quadrigemina
    • superior colliculus
    • inferior colliculus
pons
medulla oblongata

DEMONSTRATION ACTIVITY

Observe a frontal section from a sheep brain (fig. 32.6). Note the longitudinal fissure, gray matter, white matter, corpus callosum, lateral ventricles, third ventricle, and thalamus. Compare this to the image of the frontal section of the human brain (fig. 32.7). Note differences and similarities between the two.

10. Dispose of the sheep brain as directed by the laboratory instructor.
11. Complete Parts A, B, and C of Laboratory Assessment 32.

FIGURE 32.5 Median section of the right half of the sheep brain dissection.

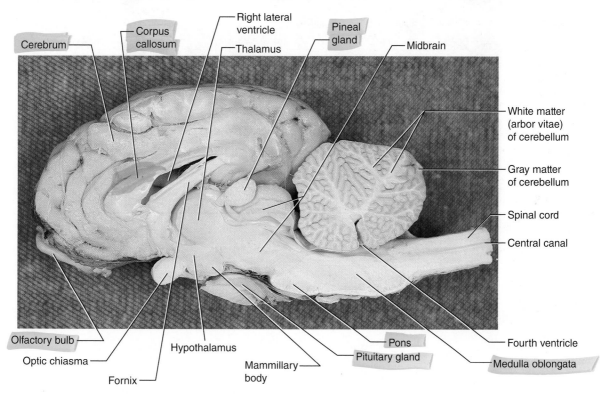

FIGURE 32.6 Frontal section of a sheep brain.

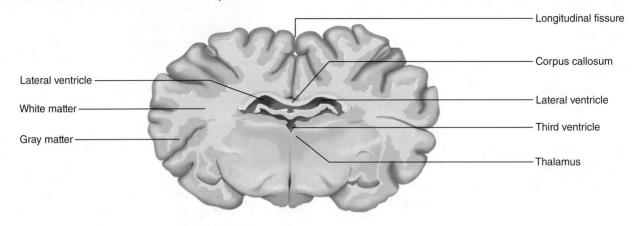

Longitudinal fissure

Corpus callosum

Lateral ventricle

White matter

Gray matter

Lateral ventricle

Third ventricle

Thalamus

FIGURE 32.7 Frontal section of a human brain.

Longitudinal fissure

Corpus callosum

Lateral ventricle

Third ventricle

Thalamus

White matter

Gray matter

1 in

1 cm

Name _____

Date _____

Section _____

The ⒶB corresponds to the indicated Learning Outcome(s) ⒪
found at the beginning of the Laboratory Exercise.

LABORATORY
ASSESSMENT
32

Dissection of the Sheep Brain

PART A: Assessments

Answer the following questions:

1. Describe the location of any meninges observed to be associated with the sheep brain. **A1** _____
   _____
   _____

2. How do the relative sizes of the sheep and human cerebral hemispheres differ? **A2** *Human cerebral*
   hemispheres are larger than sheep cerebral hemispheres.

3. How do the gyri and sulci of the sheep cerebrum compare with the human cerebrum in numbers? **A2** _____
   _____
   _____

4. What is the significance of the differences you noted in your answers for questions 2 and 3? **A2** _____
   _____
   _____

5. What difference did you note in the structures of the sheep cerebellum and the human cerebellum? **A2** _____
   _____
   _____

6. How do the sizes of the olfactory bulbs of the sheep brain compare with those of the human brain? **A2** _____
   _____
   _____

7. Based on their relative sizes, which of the cranial nerves seems to be most highly developed in the sheep brain? **A3**
   _____
   _____
   _____

8. What is the significance of the observations you noted in your answers for questions 6 and 7? **A2** _____
   _____
   _____

PART B: Assessments

CRITICAL THINKING ASSESSMENT

Prepare a list of at least six features to illustrate ways in which the brains of sheep and humans are similar.   A2

1. _____

2. _____

3. _____

4. _____

5. _____

6. _____

Interpret the significance of these similarities.   A2 _____

_____

PART C: Assessments

Identify the features indicated in the median section of the sheep brain in figure 32.8.

FIGURE 32.8   Label the features of this median section of the sheep brain by placing the correct numbers in the spaces provided.   A1

_4_ Cerebellum	_8_ Hypothalamus	_3_ Pineal gland
_1_ Cerebrum	_7_ Medulla oblongata	_9_ Pons
_2_ Corpus callosum	_6_ Midbrain	_5_ Thalamus

General Senses

Purpose of the Exercise

To review the characteristics of sensory receptors and general senses and to demonstrate some of the general senses associated with the skin.

MATERIALS NEEDED

Marking pen (washable)

Millimeter ruler

Bristle or sharp pencil

Forceps with fine points or another two-point discriminator device

Blunt metal probes

Three beakers (250 mL)

Warm tap water or 45°C (113°F) water bath

Cold water (ice water)

Thermometer

For Demonstration Activity:

Prepared microscope slides of tactile (Meissner's) and lamellated (Pacinian) corpuscles

Compound light microscope

Learning Outcomes

After completing this exercise, you should be able to:

O1 Associate types of sensory receptors with general senses throughout the body.

O2 Determine and record the distribution of mechanoreceptors and thermoreceptors in various regions of the skin.

O3 Measure the two-point threshold of various regions of the skin.

The **O** corresponds to the assessments **A** indicated in the Laboratory Assessment for this Exercise.

Pre-Lab

Carefully read the introductory material and examine the entire lab. Be familiar with basic structures and functions of the receptors associated with general senses from the lecture or the textbook. Answer the pre-lab questions.

Pre-Lab Questions Select the correct answer for each of the following questions:

1. General senses include all of the following *except*
   a. touch.
   b. vision.
   c. temperature.
   d. pain.

2. Thermoreceptors are associated with
   a. deep pressure.
   b. light touch.
   c. tissue trauma.
   d. temperature changes.

3. When receptors are continuously stimulated, the sensations may fade away; this phenomenon is known as
   a. tolerance.
   b. perception.
   c. adaptation.
   d. fatigue.

4. Encapsulated nerve endings include
   a. tactile corpuscles.
   b. pain receptors.
   c. cold receptors.
   d. warm receptors.

5. A lamellated corpuscle is stimulated by
   a. light touch.
   b. deep pressure.
   c. warm temperatures.
   d. cold temperatures.

6. Free nerve endings function as pain, warm, and cold receptors.
   a. True ✓
   b. False _____

7. Lamellated corpuscles are located in the epidermis of the skin.
   a. True _____
   b. False ✓
   located in dermis

Sensory receptors are sensitive to changes that occur within the body and its surroundings. Each type of receptor is particularly sensitive to a distinct kind of environmental change and is much less sensitive to other forms of stimulation. When receptors are stimulated, they initiate nerve impulses that travel into the central nervous system. The raw form in which these receptors send information to the brain is called *sensation.* The way our brains interpret this information is called *perception.*

The sensory receptors found widely distributed throughout skin, muscles, joints, and visceral organs are associated with **general senses.** These senses include touch, pressure, temperature, pain, and the senses of muscle movement and body position. Receptors (muscle spindles) associated with muscle movements were included as part of Laboratory Exercise 29.

The general senses associated with the body surface can be classified according to the stimulus type. *Mechanoreceptors* include tactile (Meissner's) corpuscles, which are stimulated by light touch, stretch, or vibration, and lamellated (Pacinian) corpuscles, which are stimulated by deep pressure, stretch, or vibration. *Thermoreceptors* include those associated with temperature changes. Warm receptors are most sensitive to temperatures between 25°C (77°F) and 45°C (113°F). Cold receptors are most sensitive to temperatures between 10°C (50°F) and 20°C (68°F). *Nociceptors* are pain receptors that respond to tissue trauma, which may include a cut or pinch, extreme heat or cold, or excessive pressure. Tests for pain receptors are not included as part of this laboratory exercise.

Receptors possess structural differences at the dendrite ends of sensory neurons. Those with unencapsulated free nerve endings include pain, cold, and warm receptors; those with encapsulated nerve endings include tactile and lamellated corpuscles.

A sensation may fade away when receptors are continuously stimulated. This is known as *sensory adaptation,* like adjusting to a room temperature or an odor of our environment. However, sensory adaptation to pain is not as prevalent because impulses may continue into the CNS for longer periods of time. The importance of perceiving pain not only alerts us to a change in our environment and potential injury, but it also allows us to detect body abnormalities so that proper adjustments can be taken.

Sensory receptors that are more specialized and confined to the head are associated with **special senses.** Laboratory Exercises 34 through 38 describe the special senses.

PROCEDURE A: Receptors and General Senses

1. Reexamine the introduction to this laboratory exercise and study table 33.1.
2. Complete Part A of Laboratory Assessment 33.

DEMONSTRATION ACTIVITY

Observe the tactile (Meissner's) corpuscle with the microscope set up by the laboratory instructor. This type of receptor is abundant in the superficial dermis in outer regions of the body, such as in the fingertips, soles, lips, and external genital organs. It is responsible for the sensation of light touch. (See fig. 33.1.)

Observe the lamellated (Pacinian) corpuscle in the second demonstration microscope. This corpuscle is composed of many layers of connective tissue cells and has a nerve fiber in its central core. Lamellated corpuscles are numerous in the hands, feet, joints, and external genital organs. They are responsible for the sense of deep pressure (fig. 33.2). How are tactile and lamellated corpuscles similar?

_____

_____

How are they different?

_____

_____

Table 33.1 Receptors Associated with General Senses of Skin

Receptor Category	Respond to	Receptor Examples
Mechanoreceptors	Touch, stretch, vibration, pressure	Encapsulated tactile (Meissner's) corpuscle; encapsulated lamellated (Pacinian) corpuscle
Thermoreceptors	Temperature changes	Free nerve endings for warm temperatures; free nerve endings for cold temperatures
Nociceptors	Intense mechanical, chemical, or temperature stimuli; tissue trauma	Free nerve endings for pain

FIGURE 33.1 Tactile (Meissner's) corpuscles, such as this one, are responsible for the sensation of light touch (250×).

- Epidermis
- Tactile corpuscle in dermal papilla
- Dermis

FIGURE 33.2 Lamellated (Pacinian) corpuscles, such as this one, are responsible for the sensation of deep pressure (100×).

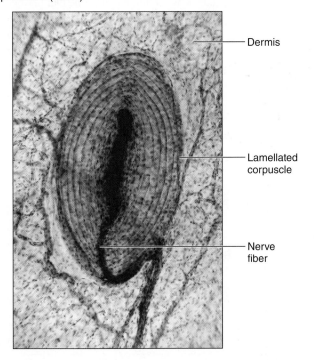

- Dermis
- Lamellated corpuscle
- Nerve fiber

PROCEDURE B: Sense of Touch

1. Investigate the distribution of touch receptors in your laboratory partner's skin. To do this, follow these steps:
   a. Use a marking pen and a millimeter ruler to prepare a square with 2.5 cm on each side on the skin on your partner's inner wrist, near the palm.
   b. Divide the square into smaller squares with 0.5 cm on a side, producing a small grid.
   c. Ask your partner to rest the marked wrist on the tabletop and to keep his or her eyes closed throughout the remainder of the experiment.
   d. Press the end of a bristle on the skin in some part of the grid, using just enough pressure to cause the bristle to bend. A sharp pencil could be used as an alternate device.
   e. Ask your partner to report whenever the touch of the bristle is felt. Record the results in Part B of the laboratory assessment.
   f. Continue this procedure until you have tested twenty-five different locations on the grid. Move randomly through the grid to help prevent anticipation of the next stimulation site.
2. Test two other areas of exposed skin in the same manner, and record the results in Part B of the laboratory assessment.
3. Answer the questions in Part B of the laboratory assessment.

PROCEDURE C: Two-Point Threshold

1. Test your partner's ability to recognize the difference between one and two points of skin being stimulated simultaneously. To do this, follow these steps:
   a. Have your partner place a hand with the palm up on the table and close his or her eyes.
   b. Hold the tips of a forceps tightly together and gently touch the skin on your partner's fingertip (fig. 33.3).
   c. Ask your partner to report if it feels like one or two points are touching the finger.
   d. Allow the tips of the forceps to spread so they are 1 mm apart, press both points against the skin simultaneously, and ask your partner to report as before.
   e. Repeat this procedure, allowing the tips of the forceps to spread more each time until your partner can feel both tips being pressed against the skin. The minimum distance between the tips of the forceps when both can be felt is called the *two-point threshold.* As soon as you are able to distinguish two points, two separate receptors are being stimulated instead of only one receptor (fig. 33.3). Areas of skin with a small threshold are more sensitive than areas with large two-point thresholds.
   f. Record the two-point threshold for the skin of a fingertip in Part C of the laboratory assessment.

FIGURE 33.3 Two-point threshold determination test. Example (a) stimulates a single sensory neuron and (b) stimulates two different sensory neurons.

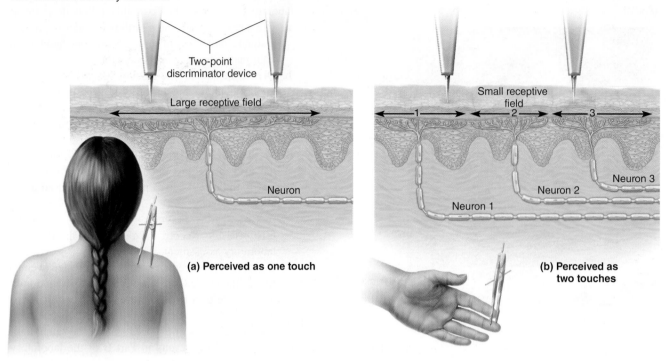

(a) Perceived as one touch

(b) Perceived as two touches

2. Repeat this procedure to determine the two-point threshold of the palm, the back of the hand, the back of the neck, the forearm, and the leg. Record the results in Part C of the laboratory assessment.
3. Answer the questions in Part C of the laboratory assessment.

PROCEDURE D: Sense of Temperature

1. Investigate the distribution of *warm thermoreceptors* in your partner's skin. To do this, follow these steps:
   a. Mark a square with 2.5 cm sides on your partner's palm.
   b. Prepare a grid by dividing the square into smaller squares, 0.5 cm on a side.
   c. Have your partner rest the marked palm on the table and close his or her eyes.
   d. Heat a blunt metal probe by placing it in a beaker of warm (hot) water (about 40–45°C/ 104–113°F) for a minute or so. (*Be sure the probe does not get so hot that it burns the skin.*) Use a thermometer to monitor the appropriate warm water from the tap or the water bath.
   e. Wipe the probe dry and touch it to the skin on some part of the grid.
   f. Ask your partner to report if the probe feels warm. Then record the results in Part D of the laboratory assessment.
   g. Keep the probe warm, and repeat the procedure until you have randomly tested several different locations on the grid.

2. Investigate the distribution of *cold receptors* by repeating the procedure. Use a blunt metal probe that has been cooled by placing it in ice water for a minute or so. Record the results in Part D of the laboratory assessment.
3. Answer the questions in Part D of the laboratory assessment.

LEARNING EXTENSION ACTIVITY

Prepare three beakers of water of different temperatures. One beaker should contain warm water (about 40°C/104°F), one should be room temperature (about 22°C/72°F), and one should contain cold water (about 10°C/50°F). Place the index finger of one hand in the warm water and, at the same time, place the index finger of the other hand in the cold water for about 2 minutes. Then, simultaneously move both index fingers into the water at room temperature. What temperature do you sense with each finger? How do you explain the resulting perceptions?

_____

_____

_____

_____

_____

_____

Name _____

Date _____

Section _____

The (A) corresponds to the indicated Learning Outcome(s) (O) found at the beginning of the Laboratory Exercise.

General Senses

PART A: Receptors and General Senses Assessments

Complete the following statements:

1. Whenever tissues are damaged, _____ receptors are likely to be stimulated. (A1)

2. Receptors that are sensitive to temperature changes are called _____. (A1)

3. A sensation may seem to fade away when receptors are continuously stimulated as a result of _____ adaptation. (A1)

4. Tactile (Meissner's) corpuscles are responsible for the sense of light _____. (A1)

5. Lamellated (Pacinian) corpuscles are responsible for the sense of deep _____. (A1)

6. _____ receptors are most sensitive to temperatures between 25°C (77°F) and 45°C (113°F). (A1)

7. _____ receptors are most sensitive to temperatures between 10°C (50°F) and 20°C (68°F). (A1)

8. Widely distributed sensory receptors throughout the body are associated with _____ senses in contrast to special senses. (A1)

PART B: Sense of Touch Assessments

1. Record a + to indicate where the bristle was felt and a *0* to indicate where it was not felt. (A2)

Skin of wrist

2. Show the distribution of touch receptors in two other regions of skin. (A2)

Region tested _____   Region tested _____

3. Answer the following questions:

   a. How do you describe the pattern of distribution for touch receptors in the regions of the skin you tested? (A2)

   _____

   _____

   b. How does the concentration of touch receptors seem to vary from region to region? (A2) _____

   _____

PART C: Two-Point Threshold Assessments

1. Record the two-point threshold in millimeters for skin in each of the following regions: (A3)

   Fingertip _____

   Palm _____

   Back of hand _____

   Back of neck _____

   Forearm _____

   Leg _____

2. Answer the following questions:

   a. What region of the skin tested has the greatest ability to discriminate two points? (A3)_____

   b. What region of the skin tested has the least sensitivity to this test? (A3)_____

   c. What is the significance of these observations in questions *a* and *b*? (A3) _____

   _____

   _____

PART D: Sense of Temperature Assessments

1. Record a + to indicate where warm was felt and a *0* to indicate where it was not felt. (A2)

   Skin of palm

2. Record a + to indicate where cold was felt and a *0* to indicate where it was not felt. (A2)

   Skin of palm

3. Answer the following questions:

   a. How do temperature receptors appear to be distributed in the skin of the palm? (A2) _____

   _____

   b. Compare the distribution and concentration of warm and cold receptors in the skin of the palm. (A2) _____

   _____

   _____

LABORATORY EXERCISE 34

Smell and Taste

Purpose of the Exercise

To review the structures of the organs of smell and taste and to investigate the abilities of smell and taste receptors to discriminate various chemical substances.

MATERIALS NEEDED

For Procedure A: Sense of Smell (Olfaction)
Compound light microscope
Prepared microscope slide of olfactory
 epithelium
Set of substances in stoppered bottles:
 cinnamon, sage, vanilla, garlic powder, oil
 of clove, oil of wintergreen, and perfume

For Procedure B: Sense of Taste (Gustation)
Compound light microscope
Prepared microscope slide of the tongue
 surface with taste buds
Paper cups (small)
Cotton swabs (sterile; disposable)
5% sucrose solution (sweet)
5% NaCl solution (salt)
1% acetic acid or unsweetened lemon juice (sour)
0.5% quinine sulfate solution or 0.1%
 Epsom salt solution (bitter)
1% monosodium glutamate (MSG) solution
 (umami)

For Learning Extension Activity:
Pieces of apple, potato, carrot, and onion or
 packages of mixed flavors of LifeSavers

SAFETY

▶ Be aware of possible food allergies when
 selecting test solutions and foods to taste.
▶ Prepare fresh solutions for use in
 Procedure B.
▶ Wash your hands before starting the
 taste experiment.
▶ Wear disposable gloves when performing
 taste tests on your laboratory partner.
▶ Use a clean cotton swab for each test. Do
 not dip a used swab into a test solution.
▶ Dispose of used cotton swabs and paper
 towels as directed.
▶ Wash your hands before leaving the
 laboratory.

Learning Outcomes AP|R

After completing this exercise, you should be able to:

○1 Identify and describe the anatomical structures and functions
 associated with the sense of smell.

○2 Indicate the cranial nerves and the regions of the brain involved with
 the perception of smell and taste.

○3 Record recognized odors and the time needed for olfactory sensory
 adaptation to occur.

○4 Identify and describe the anatomical structures and functions
 associated with the sense of taste.

○5 Locate the distribution of taste receptors on the surface of the
 tongue and mouth cavity.

The ○ corresponds to the assessments Ⓐ indicated in the Laboratory
Assessment for this Exercise.

Pre-Lab

Carefully read the introductory material and examine the entire lab.
Be familiar with the basic structures and functions of the receptors
associated with smell and taste from the lecture or the textbook.
Answer the pre-lab questions.

Pre-Lab Questions Select the correct answer for each of the following
questions:

1. Receptor cells for taste are located
   a. only on the tongue.
   b. on the tongue, oral cavity, and pharynx.
   c. in the oral cavity except the tongue.
   d. on the tongue and nasal passages.

2. The primary olfactory interpretation center is located in the
   a. oral cavity.   b. inferior nose.
   c. temporal lobe   d. brainstem.
      of the cerebrum.

3. Which of the following is *not* considered a recognized taste?
   a. mint   b. salty
   c. sweet   d. umami

4. Taste interpretation occurs in the _____ of the cerebrum.
   a. frontal   b. temporal
   c. occipital   d. insula

5. Sour sensations are produced from
   a. acids.   b. sugars.
   c. ionized inorganic salts.   d. alkaloids.

6. The facial, glossopharyngeal, and vagus cranial nerves conduct impulses from taste receptors to the brain.

   a. True _____   **b.** False _____

7. Olfactory nerves are located in the foramina of the inferior nasal concha.

   a. True _____   **b.** False _____

The senses of smell (olfaction) and taste (gustation) are dependent upon modified sensory neurons, called *chemoreceptors,* that are stimulated by various chemicals dissolved in liquids. The receptors of smell are found in the olfactory organs, which are located in the superior parts of the nasal cavity and in a portion of the nasal septum. The receptors of taste occur in the taste buds, which are sensory organs primarily found on the surface of the tongue. Chemicals are considered odorless and tasteless if receptor sites for them are absent.

Olfactory receptor cells are actually neurons surrounded by pseudostratified columnar epithelial cells, called supporting cells. Their apical ends are covered with cilia, also called olfactory hairs, embedded in the mucus of the superior nasal cavity. These neurons form the *olfactory nerve* (cranial nerve I). In order to detect an odor, the molecules must first dissolve in the mucus before they bind to the receptor sites of the cilia (olfactory hairs). When receptor cells temporarily bind to an odorant molecule, it results in an action potential over the olfactory neurons passing through the foramina of the cribriform plate (of the ethmoid bone) and nerve fibers in the olfactory bulb. While several areas of the brain are involved as interpreting centers for smell, the *primary olfactory cortex* is located in the *temporal lobe* of the cerebrum. Although humans have the ability to distinguish thousands of different odors, coded by genes, various odors can result from combinations of the receptor cells stimulated. We become sensory adapted to an odor very quickly. *Sensory adaptation* occurs when there is a decreased response or interpretation of a certain stimulus. Exposure to a different substance can be quickly noticed.

Taste receptor cells are located in *taste buds* on the tongue, but receptor cells are also distributed in other areas of the oral cavity and pharynx. A taste bud contains *taste (gustatory) cells* (receptor cells), *basal cells,* and *supporting cells.* Taste cells are frequently damaged by friction and temperature changes of foods in the oral cavity, and they have a life span of only 7 to 10 days. Fortunately, the basal cells are stem cells that replace the taste receptor cells that have died. Taste cells possess terminal microvilli, called *taste hairs,* that project through a taste pore on the epithelium of the tongue. The taste hairs function as sensory receptor surfaces for substances we are able to taste. Taste sensations are grouped into five recognized categories: sweet, sour, salt, bitter, and umami. The *sweet* sensation is produced from sugars, the *sour* sensation from acids, the *salt* sensation from ionized inorganic salts, the *bitter* sensation from alkaloids and spoiled foods, and the *umami* sensation from aspartic

and glutamic acids or a derivative such as monosodium glutamate. Three cranial nerves conduct impulses from the taste buds. The *facial nerve* (*VII*) conducts sensory impulses from the anterior two-thirds of the tongue; the *glossopharyngeal nerve* (*IX*) from the posterior tongue; and the *vagus nerve* (*X*) from the pharyngeal region. When a molecule binds to a receptor cell, a neurotransmitter is secreted and an action potential occurs over a sensory neuron. The sensory information continues through the medulla oblongata and the thalamus to the *insula* of the cerebrum. The insula serves as the *primary gustatory cortex* for the perception of taste. Sensory adaptation to taste sensations occurs rather quickly, which helps explain the reason the first few bites of a particular food have the most vivid flavor.

The senses of smell and taste function closely together, because substances that are tasted often are smelled at the same moment, and they play important roles in the selection of foods. The taste and aroma of foods are also influenced by appearance, texture, temperature, and the person's mood.

PROCEDURE A: Sense of Smell (Olfaction)

1. Obtain a prepared microscope slide of an olfactory epithelium and observe it under high-power magnification. The olfactory receptor cells are spindle-shaped, bipolar neurons with spherical nuclei. They also have six to eight cilia (olfactory hairs) at their apical ends. The supporting cells are pseudostratified columnar epithelial cells. However, in this region the tissue lacks goblet cells (fig. 34.1).

2. Reexamine the introduction to this laboratory exercise and study figure 34.2.

3. Complete Part A of Laboratory Assessment 34.

4. Test your laboratory partner's ability to recognize the odors of the bottled substances available in the laboratory. To do this, follow these steps:

   a. Have your partner keep his or her eyes closed.

   b. Remove the stopper from one of the bottles, and hold it about 4 cm under your partner's nostrils for about 2 seconds.

   c. Ask your partner to identify the odor, and then replace the stopper.

   d. Record your partner's response in Part B of the laboratory assessment.

   e. Repeat steps *b–d* for each of the bottled substances.

5. Repeat the preceding procedure, using the same set of bottled substances, but present them to your partner in a different sequence. Record the results in Part B of the laboratory assessment.

6. Wait 10 minutes and then determine the time it takes for your partner to experience olfactory sensory adaptation. To do this, follow these steps:

   a. Ask your partner to breathe in through the nostrils and exhale through the mouth.

   b. Remove the stopper from one of the bottles, and hold it about 4 cm under your partner's nostrils.

FIGURE 34.1 **AP|R** Micrograph of olfactory
epithelium (250×). Olfactory receptors have cilia (olfactory hairs) at their apical ends.

Lamina propria connective tissue

Olfactory epithelium (organ of smell)

Nasal cavity

Cilia (olfactory hairs) of receptor cells

c. Keep track of the time that passes until your partner is no longer able to detect the odor of the substance.

d. Record the result in Part B of the laboratory assessment.

e. Wait 5 minutes and repeat this procedure, using a different bottled substance.

f. Test a third substance in the same manner.

g. Record the results as before.

7. Complete Part B of the laboratory assessment.

PROCEDURE B: Sense of Taste (Gustation)

1. Obtain a prepared microscope slide of the tongue surface with taste buds and observe it under low-power and high-power magnifications. Observe the oval-shaped taste bud and the surrounding epithelial cells. The taste pore, an opening into the taste bud, may be filled with taste hairs (microvilli). Within the taste bud there are supporting cells and thinner taste-receptor cells, which often have lightly stained nuclei (figs. 34.3 and 34.4).

2. Reexamine the introduction to this laboratory exercise and study figure 34.5.

3. Complete Part C of the laboratory assessment.

4. Map the distribution of the receptors for the primary taste sensations on your partner's tongue. To do this, follow these steps:

   a. Ask your partner to rinse his or her mouth with water and then partially dry the surface of the tongue with a paper towel.

   b. Moisten a clean cotton swab with 5% sucrose solution, and touch several regions of your partner's tongue with the swab.

   c. Each time you touch the tongue, ask your partner to report what sensation is experienced.

   d. Test the tip, sides, and back of the tongue in this manner.

   e. Test some other representative areas inside the oral cavity such as the cheek, gums, and roof of the mouth (hard and soft palate) for a sweet sensation.

   f. Record your partner's responses in Part D of the laboratory assessment.

FIGURE 34.2 **AP|R** Structures associated with smell.

Olfactory tract

Olfactory bulb

Cribriform plate of ethmoid bone

Olfactory area of nasal cavity

Superior nasal concha

Nasal cavity

Olfactory bulb

Frontal bone

Nerve fibers within olfactory tract

Olfactory nerve fibers

Cribriform plate of ethmoid bone

Olfactory epithelium

Mucus

Supporting cells

Cilia (olfactory hairs)

Olfactory receptor cells

Airflow in nasal cavity containing odor molecules

g. Have your partner rinse his or her mouth and dry the tongue again, and repeat the preceding procedure, using each of the other four test solutions—NaCl, acetic acid, quinine or Epsom salt, and MSG solution. Be sure to use a fresh swab for each test substance and dispose of used swabs and paper towels as directed.

5. Complete Part D of the laboratory assessment.

FIGURE 34.3 Micrograph of tongue surface with taste buds (225x). The arrows mark two of several taste buds viewed in this segment of the tongue.

FIGURE 34.4 Micrograph of a taste bud (400x). Taste receptors are found in taste buds.

Nucleus of supporting cell

Taste hairs

Taste pore

Nucleus of taste receptor cell

Taste bud

FIGURE 34.5 Structures associated with taste: (a) tongue, (b) papillae, and (c) taste bud.

Papillae

(a) AP|R Tongue

Taste buds

(b) AP|R Papillae

Connective tissue

Supporting cell

Taste receptor cell

Sensory nerve fiber

Basal cell

Stratified squamous epithelial tissue

Taste hair (microvillus)

Taste pore

(c) AP|R Taste bud

LEARNING EXTENSION ACTIVITY

Test your laboratory partner's ability to recognize the tastes of apple, potato, carrot, and onion. (A package of mixed flavors of LifeSavers is a good alternative.) To do this, follow these steps:

1. Have your partner close his or her eyes and hold the nostrils shut.
2. Place a small piece of one of the test substances on your partner's tongue.
3. Ask your partner to identify the substance without chewing or swallowing it.
4. Repeat the procedure for each of the other substances.

How do you explain the results of this experiment?

_____

_____

NOTES

LABORATORY
ASSESSMENT
34

Name _____

Date _____

Section _____

The **A** corresponds to the indicated Learning Outcome(s) **O** found at the beginning of the Laboratory Exercise.

Smell and Taste

PART A: Assessments

Complete the following statements:

1. The distal ends of the olfactory neurons are covered with hairlike _____. **A1**

2. Before gaseous substances can stimulate the olfactory receptors, they must be dissolved in _____ that surrounds the cilia. **A1**

3. The axons of olfactory receptors pass through small openings in the _____ of the ethmoid bone. **A2**

4. The primary olfactory cortex for interpreting smell is located in the _____ of the cerebrum. **A2**

5. Olfactory sensations usually fade rapidly as a result of _____. **A1**

6. A chemical would be considered _____ if a person lacks a particular receptor site on the cilia of the olfactory neurons. **A1**

PART: Sense of Smell Assessments

1. Record the results (as +, if recognized; as *0*, if unrecognized) from the tests of odor recognition in the following table:

| Substance Tested | ODOR REPORTED | |
	First Trial	Second Trial

2. Record the results of the olfactory sensory adaptation time in the following table: **A3**

Substance Tested	Adaptation Time in Seconds

3. Complete the following:

    a. How do you describe your partner's ability to recognize the odors of the substances you tested? **A1**

_____

_____

    b. Compare your experimental results with those of others in the class. Did you find any evidence to indicate that individuals may vary in their ability to recognize odors? Explain your answer. **A1**

_____

_____

_____

CRITICAL THINKING ASSESSMENT

Does the time it takes for sensory adaptation to occur seem to vary with the substances tested? Explain your answer. **A3**

_____

_____

_____

PART C: Assessments

Complete the following statements:

  1. Taste is interpreted in the _____ of the cerebrum. **A2**

  2. The opening to a taste bud is called a _____. **A4**

  3. The _____ of a taste cell are its sensitive part. **A4**

  4. The facial, _____, and vagus cranial nerves conduct impulses related to the sense of taste. **A2**

  5. Substances that stimulate taste cells bind with _____ sites on the surfaces of taste hairs. **A4**

  6. Sour receptors are mainly stimulated by _____. **A4**

  7. Salt receptors are mainly stimulated by ionized inorganic _____. **A4**

  8. Alkaloids usually have a _____ taste. **A4**

PART D: Sense of Taste Assessments

1. *Taste receptor distribution.* Record a + to indicate where a taste sensation seemed to originate and a *0* if no sensation occurred when the spot was stimulated. A5

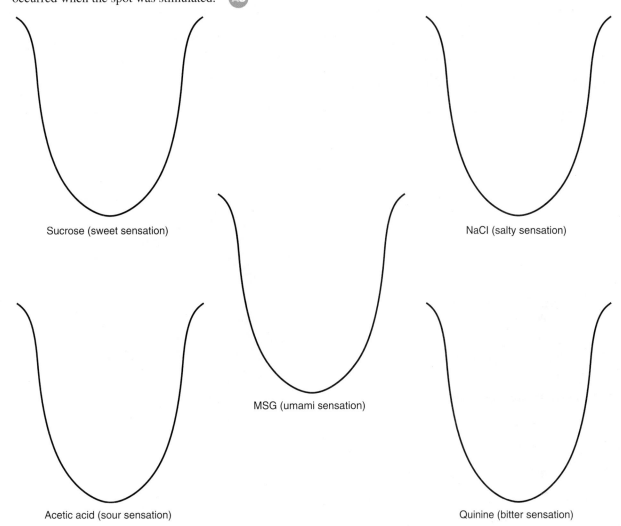

Sucrose (sweet sensation)

NaCl (salty sensation)

MSG (umami sensation)

Acetic acid (sour sensation)

Quinine (bitter sensation)

2. Complete the following:

   a. Describe how each type of taste receptor is distributed on the surface of your partner's tongue. A5

   _____

   _____

   _____

   _____

   b. Describe other locations inside the mouth where any sensations of sweet, salty, sour, bitter, or umami were located. A5

   _____

   _____

   _____

   _____

c. How does your taste distribution map on the tongue compare to those of other students in the class? A5

_____

_____

_____

_____

3. Identify the structures associated with a taste bud in figure 34.6. A4

FIGURE 34.6 Label this diagram of structures associated with a taste bud by placing the correct numbers in the spaces provided.

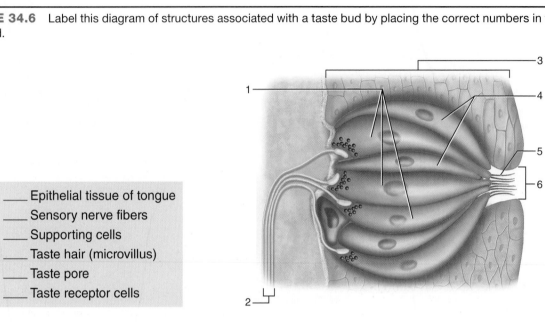

____ Epithelial tissue of tongue
____ Sensory nerve fibers
____ Supporting cells
____ Taste hair (microvillus)
____ Taste pore
____ Taste receptor cells

Eye Structure

Purpose of the Exercise

To review the structure and function of the eye and to dissect a mammalian eye.

MATERIALS NEEDED

Dissectible eye model

Compound light microscope

Prepared microscope slide of a mammalian eye (sagittal section)

Sheep or beef eye (fresh or preserved)

Dissecting tray

Dissecting instruments—forceps, sharp scissors, and dissecting needle

Disposable gloves

For Learning Extension Activity:

Ophthalmoscope

SAFETY

▶ Do not allow light to reach the macula lutea during the eye exam for longer than one-second intervals.

▶ Wear disposable gloves when working on the eye dissection.

▶ Dispose of tissue remnants and gloves as instructed.

▶ Wash the dissecting tray and instruments as instructed.

▶ Wash your laboratory table.

▶ Wash your hands before leaving the laboratory.

Learning Outcomes AP|R

After completing this exercise, you should be able to:

01 Identify the major structures of an eye.

02 Describe the functions of the structures of an eye.

03 Trace the structures through which light passes as it travels from the cornea to the retina.

04 Dissect a mammalian eye and locate its major features.

The Ⓞ corresponds to the assessments Ⓐ indicated in the Laboratory Assessment for this Exercise.

Pre-Lab

Carefully read the introductory material and examine the entire lab. Be familiar with the basic structures and functions of parts of the eye from the lecture or the textbook. Answer the pre-lab questions.

Pre-Lab Questions Select the correct answer for each of the following questions:

1. The cornea and the sclera compose the _____ layer.
   a. outer
   b. middle
   c. inner
   d. gel

2. We are able to see color because the eye contains
   a. melanin.
   b. an optic disc.
   c. rods.
   d. cones

3. The perception of vision occurs in the
   a. optic nerve.
   b. occipital lobe of the cerebrum.
   c. retina.
   d. frontal lobe of the cerebrum.

4. Which of the following is *not* part of the middle eye layer?
   a. choroid
   b. conjunctiva
   c. ciliary body
   d. iris

5. The area of our eye where visual acuity is best is the
   a. macula lutea.
   b. optic disc.
   c. fovea centralis.
   d. pupil.

6. Which of the following extrinsic skeletal muscles rotates the eyeball superiorly and medially?
   a. superior rectus
   b. superior oblique
   c. inferior rectus
   d. inferior oblique

7. The conjunctiva covers the superficial surface of the cornea.
   a. True _____
   b. False _____

8. Tears from the lacrimal gland eventually flow through a nasolacrimal duct into the nasal cavity.
   a. True _____
   b. False _____

The special sense of vision involves the complex structure of the eye, accessory structures associated with the orbit, the optic nerve, and the brain. The eyebrow, eyelids, conjunctiva, and lacrimal apparatus protect the eyes. The vascular *conjunctiva,* a mucous membrane, covers all of the anterior surface of the eyeball except the cornea, and it lines the eyelids. Its mucus prevents the eye from drying out, and the lacrimal fluid from the *lacrimal (tear) gland* cleanses the eye and prevents infections due to the antibacterial agent lysozyme in its secretions. Two voluntary muscles involve control of the eyelids, while six extrinsic muscles allow voluntary eye movements in multiple directions.

Three layers (tunics) compose the main structures of the eye. The *outer (fibrous) layer* includes the protective white *sclera* and a transparent anterior portion, the *cornea,* which allows light to enter the eye cavities. The *middle (vascular) layer* includes the heavily pigmented *choroid,* the *ciliary body* that controls the shape of the *lens,* and the *iris* with its smooth muscle fibers that control the amount of light passing through its central opening, the *pupil.* The inner layer includes pigmented and neural components of the *retina* and the optic nerve, which contains sensory neurons that conduct nerve impulses from the rods and cones. A large posterior cavity is filled with a transparent gel, the *vitreous humor.* A smaller anterior cavity has anterior and posterior chambers separated at the iris, which is filled with a clear *aqueous humor.*

The photoreceptor cells, the *rods* and *cones* of the retina, can absorb light to produce visual images. The more numerous rods can be stimulated in dim light and are most abundant in the peripheral retina. The cones are stimulated only in brighter light, but they allow us to perceive sharper images, and they allow us to see them in color because there are three types of cones, with each type most sensitive to wavelengths peaking in the blue, green, or red portion of the visible light spectrum. Located at the posterior center of the retina is a region containing mostly cones, called the *macula lutea* ("yellow spot"). A tiny pit in the center of the macula lutea, the *fovea centralis,* contains only cones and represents our best image location. Slightly to the medial side of the macula lutea is the *optic disc,* where the nerve fibers of the retina leave the eye to become parts of the optic nerve. It is also referred to as the "blind spot" because rods and cones are absent at this location. Sensory impulses from the optic nerves pass through the thalamus and eventually arrive at the primary visual cortex in the occipital lobe of the cerebrum, where the perception of visual images occurs.

PROCEDURE A: Structure and Function of the Eye

1. Reexamine the introduction to this laboratory exercise.
2. Use figures 35.1, 35.2, 35.3, and table 35.1 as references as you progress through this procedure.
3. Examine the dissectible model of the eye and locate the following features:

  eyelid (palpebra)—anterior protection of eyes

  conjunctiva—a mucous membrane

  tarsal glands—secrete oily lubricant

  orbicularis oculi—closes eyelids

FIGURE 35.1 Accessory structures of the eye (a) sagittal section of orbit (b) lacrimal apparatus.

(a) Lateral view

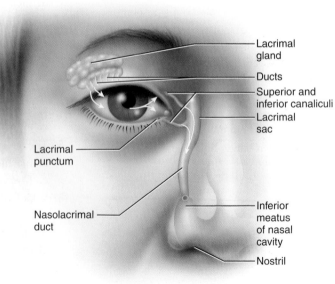

(b) Anterior view

FIGURE 35.2 Extrinsic muscles of the right eyeball (a) lateral view and (b) anterior view. The arrows indicate the direction of the pull by each of the six eyeball muscles.

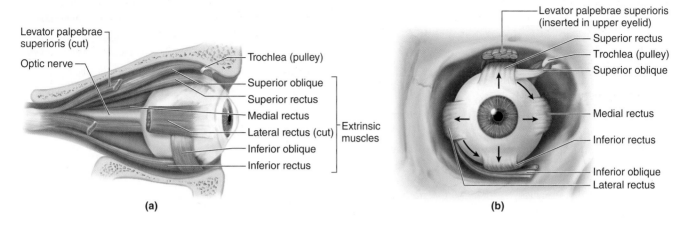

(a)

(b)

FIGURE 35.3 Structures of the eye (sagittal section).

levator palpebrae superioris—elevates upper eyelid

lacrimal apparatus (fig. 35.1b)

- lacrimal gland—secretes tears
- lacrimal puncta—two small pores into canaliculi
- canaliculi—passageways to lacrimal sac
- lacrimal sac—collects tears from cannaliculi
- nasolacrimal duct—drainage to nasal cavity

extrinsic muscles (fig. 35.2 and table 35.1)

- superior rectus
- inferior rectus
- medial rectus
- lateral rectus
- superior oblique
- inferior oblique

Table 35.1 Muscles Associated with the Eyelids and Eyes

Skeletal Muscles	Action	Innervation	Smooth Muscles	Action	Innervation
Muscles of the eyelids			Ciliary muscles	Relax suspensory ligaments	Oculomotor nerve (III) parasympathetic fibers
Orbicularis oculi	Closes eye	Facial nerve (VII)	Iris, circular muscles	Constrict pupil	Oculomotor nerve (III) parasympathetic fibers
Levator palpebrae superioris	Opens eye	Oculomotor nerve (III)	Iris, radial muscles	Dilate pupil	Sympathetic fibers
Extrinsic muscles of the eyeballs					
Superior rectus	Rotates eyeball superiorly and medially	Oculomotor nerve (III)			
Inferior rectus	Rotates eyeball inferiorly and medially	Oculomotor nerve (III)			
Medial rectus	Rotates eyeball medially	Oculomotor nerve (III)			
Lateral rectus	Rotates eyeball laterally	Abducens nerve (VI)			
Superior oblique	Rotates eyeball inferiorly and laterally	Trochlear nerve (IV)			
Inferior oblique	Rotates eyeball superiorly and laterally	Oculomotor nerve (III)			

trochlea (pulley)—fibrocartilage ring for superior oblique muscle

outer (fibrous) layer (tunic)
- sclera—white protective layer of dense fibrous connective tissue
- cornea—transparent anterior portion continuous with sclera

middle (vascular) layer (tunic)
- choroid—rich blood vessel area; contains melanocytes
- ciliary body—thickest part of middle layer
  - ciliary processes—secrete aqueous humor
  - ciliary muscles—change lens shape
  - suspensory ligaments—hold lens in position
- iris—smooth muscles change pupil diameter
- pupil—opening in center of iris allowing light passage

inner layer (tunic)
- retina—sensory layer with rods and cones
  - macula lutea—high cone density area
  - fovea centralis—area of best visual acuity
  - optic disc—location where optic nerve leaves eye
  - orra serrata—serrated anterior margin of sensory retina
- optic nerve—contains sensory neurons

lens—transparent; focuses light onto retina

anterior cavity—contains aqueous humor
- anterior chamber—between cornea and iris
- posterior chamber—between iris and lens
- aqueous humor—clear, watery fluid; fills anterior and posterior chambers and pupil

posterior cavity
- vitreous humor—transparent, gel-like filler

4. Obtain a microscope slide of a mammalian eye section, and locate as many of the features in step 3 as possible.

5. Using low-power magnification, observe the posterior portion of the eye wall, and locate the sclera, choroid, and retina (fig. 35.4).

6. Using high-power magnification, examine the retina portion of the eye wall in more detail and note its layered structure. Locate the following:

> **nerve fibers leading to the optic nerve**— innermost layer of the retina
>
> **layer of ganglion cells**
>
> **layer of bipolar neurons**
>
> **nuclei of rods and cones**
>
> **receptor ends of rods and cones**
>
> **pigmented epithelium**—outermost layer of the retina

7. Complete Part A of Laboratory Assessment 35.

FIGURE 35.4 The cells of the retina are arranged in distinct layers (75×).

Direction of light

Front of eye

Retina

Vitreous humor
Nerve fibers to optic nerve
Nuclei of ganglion cells

Nuclei of bipolar neurons

Nuclei of rods and cones
Receptor ends of rods and cones
Pigmented epithelium of retina

Choroid

Sclera

Back of eye

LEARNING EXTENSION ACTIVITY

Use an ophthalmoscope to examine the interior of your laboratory partner's eye. This instrument consists of a set of lenses held in a rotating disc, a light source, and some mirrors that reflect the light into the test subject's eye (fig. 35.5).

The examination should be conducted in a dimly lit room. Have your partner seated and staring straight ahead at eye level. Move the rotating disc of the ophthalmoscope so that the *O* appears in the lens selection window. Hold the instrument in your right hand with the end of your index finger on the rotating disc. Direct the light at a slight angle from a distance of about 15 cm into the subject's right eye (fig. 35.6). The light beam should pass along the inner edge of the pupil. Look through the instrument, and you should see a reddish, circular area—the interior of the eye. Rotate the disc of lenses to higher values until sharp focus is achieved.

Move the ophthalmoscope to within about 5 cm of the eye being examined, *being very careful that the instrument does not touch the eye,* and again rotate the lenses to sharpen the focus (fig. 35.6). Locate the optic disc and the blood vessels that pass through it. Also locate the macula lutea by having your partner stare directly into the light of the instrument (fig. 35.7). *Do not allow light to reach the macula lutea region for longer than one-second intervals.*

Examine the subject's iris by viewing it from the side and by using a lens with a +15 or +20 value.

FIGURE 35.5 The ophthalmoscope is used to examine the interior of the eye.

Eyepiece

Rotating disc

Lens selection window

Rheostat control for light intensity

Handle of ophthalmoscope

FIGURE 35.6 Examine the retina with the use of an ophthalmoscope. This is the proper position in which to examine the right eye of the subject. Follow the directions in the Learning Extension Activity to examine the left eye.

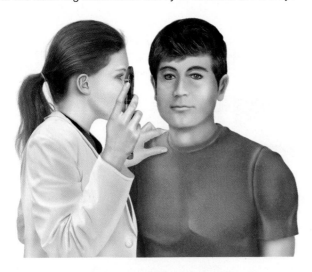

FIGURE 35.7 Right retina as viewed through the pupil using an ophthalmoscope. Only small portions of the retina can be viewed at a time.

— Retinal blood vessels (emerge from optic disc)

— Optic disc (blind spot)

— Fovea centralis (region of best visual acuity)

— Macula lutea region

Lateral   Medial

FIGURE 35.8 Prepare a frontal section of the eye.

— Margin of cornea

PROCEDURE B: Eye Dissection

1. Obtain a mammalian eyeball, place it in a dissecting tray, and dissect it as follows:

   a. Trim away the fat and other connective tissues but leave the stubs of the *extrinsic muscles* and the *optic nerve.* This nerve projects outward from the posterior region of the eyeball.

   b. The *conjunctiva* lines the inside of the eyelid and is reflected over the anterior surface of the eye, except for the cornea. Lift some of this thin membrane away from the eye with forceps and examine it.

   c. Locate and observe the *cornea, sclera,* and *iris.* Also note the *pupil* and its shape. The cornea from a fresh eye is transparent; when preserved, it becomes opaque.

   d. Use sharp scissors to make a frontal section of the eye. To do this, cut through the wall about 1 cm from the margin of the cornea and continue all the way around the eyeball. Try not to damage the internal structures of the eye (fig. 35.8).

   e. Gently separate the eyeball into anterior and posterior portions. Usually, the jellylike vitreous humor will remain in the posterior portion, and the lens may adhere to it. Place the parts in the dissecting tray with their contents facing upward (fig. 35.9).

   f. Examine the anterior portion of the eye, and locate the *ciliary body,* which appears as a dark, circular structure. Also note the *iris* and the *lens* if it remained in the anterior portion. The lens is normally attached to the ciliary body by many suspensory ligaments, which appear as delicate, transparent threads that detach as a lens is removed (fig. 35.9).

   g. Use a dissecting needle to gently remove the lens, and examine it. If the lens is still transparent, hold it up and look through it at something in the distance and note that the lens inverts the image. The lens of a preserved eye is usually too opaque for this experience. If the lens of the human eye becomes opaque, the defect is called a cataract (fig. 35.10).

   h. Examine the posterior portion of the eye. Note the *vitreous humor.* This jellylike mass helps to hold the lens in place anteriorly and helps to hold the *retina* against the choroid.

FIGURE 35.9 Internal structures of the preserved beef eye after dissection.

Choroid (reflected)

Retina (partly detached)

Sclera

Tapetum fibrosum (iridescent)

Cornea (posterior surface)

Pupil (opening in center of iris)

Iris

Ciliary body

Posterior portion

Anterior portion

Vitreous humor (transparent)

Lens

FIGURE 35.10 Human lens showing a cataract and one type of lens implant (intraocular lens replacement) on tips of fingers to show their relative size. A normal lens is transparent.

Lens implant

Small wire used to suspend the implant

Human lens showing a cataract

i. Carefully remove the vitreous humor and examine the retina. This layer will appear as a thin, nearly colorless to cream-colored membrane that detaches easily from the choroid coat. Compare the structures identified to figure 35.9.

j. Locate the *optic disc*—the point where the retina is attached to the posterior wall of the eyeball and where the optic nerve originates. There are no receptor cells in the optic disc, so this region is also called the "blind spot."

k. Note the iridescent area of the choroid beneath the retina. This colored surface in ungulates (mammals having hoofs) is called the *tapetum fibrosum*. It reflects light back through the retina, an action thought to aid the night vision of some animals. The tapetum fibrosum is lacking in the human eye.

2. Discard the tissues of the eye as directed by the laboratory instructor.

3. Complete Parts B and C of the laboratory assessment.

CRITICAL THINKING ACTIVITY

A strong blow to the head might cause the retina to detach. From observations made during the eye dissection, explain why this can happen.

_____

_____

_____

_____

LABORATORY
ASSESSMENT
35

Name _____

Date _____

Section _____

The A corresponds to the indicated Learning Outcome(s) O found at the beginning of the Laboratory Exercise.

Eye Structure

PART A: Assessments

Identify the features of the eye indicated in figures 35.11 and 35.12. A1

FIGURE 35.11   Label the structures in the sagittal section of the eye.

1 _____

2 _____

3 _____

Pupil

4 _____

5 _____

6 _____

7 _____

FIGURE 35.12 **AP|R** Sagittal section of the eye (5×). Identify the numbered features by placing the correct numbers in the spaces provided.

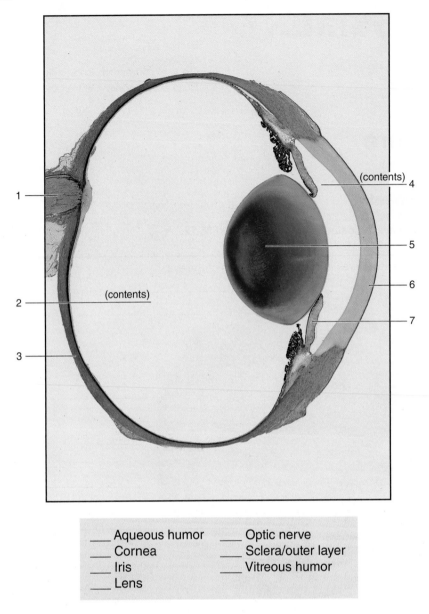

1
2 (contents)
3
4 (contents)
5
6
7

___ Aqueous humor   ___ Optic nerve
___ Cornea   ___ Sclera/outer layer
___ Iris   ___ Vitreous humor
___ Lens

PART B: Assessments

Match the terms in column A with the descriptions in column B. Place the letter of your choice in the space provided.

(A1) (A2)

Column A	Column B
a. Aqueous humor	_____ **1.** Posterior five-sixths of middle (vascular) layer
b. Choroid	_____ **2.** White part of outer (fibrous) layer
c. Ciliary muscles	_____ **3.** Transparent anterior portion of outer layer
d. Conjunctiva	_____ **4.** Inner lining of eyelid
e. Cornea	_____ **5.** Secretes tears
f. Iris	_____ **6.** Fills posterior cavity of eye
g. Lacrimal gland	_____ **7.** Area where optic nerve exits the eye
h. Optic disc	_____ **8.** Smooth muscle that controls the pupil size and light entering the eye
i. Retina	_____ **9.** Fills anterior and posterior chambers of the anterior cavity of the eye
j. Sclera	_____ **10.** Contains photoreceptor cells called rods and cones
k. Suspensory ligament	_____ **11.** Connects lens to ciliary body
l. Vitreous humor	_____ **12.** Cause lens to change shape

Complete the following:

13. List the structures and fluids through which light passes as it travels from the cornea to the retina. (A3) _____

_____

_____

_____

14. List three ways in which rods and cones differ in structure or function. (A2) _____

_____

_____

_____

_____

PART C: Assessments

Complete the following:

1. Which layer/tunic of the eye was the most difficult to cut? (A4) _____

2. What kind of tissue do you think is responsible for this quality of toughness? (A4) _____

3. How do you compare the shape of the pupil in the dissected eye with your own pupil? (A4) _____

_____

_____

4. Where was the aqueous humor in the dissected eye? (A4) _____

_____

5. What is the function of the dark pigment in the choroid? (A2) _____

_____

_____

6. Describe the lens of the dissected eye. (A4) _____

_____

_____

_____

7. Describe the vitreous humor of the dissected eye. (A4) _____

_____

_____

LABORATORY
EXERCISE 36

Visual Tests and Demonstrations

Purpose of the Exercise

To demonstrate the tests for visual acuity, astigmatism, accommodation, color vision, the blind spot, and reflexes of the eye.

MATERIALS NEEDED

Snellen eye chart
3" × 5" card (plain)
3" × 5" card with any word typed in center
Astigmatism chart
Meterstick
Metric ruler
Pen flashlight
Ichikawa's or Ishihara's color plates for color blindness test

Learning Outcomes

After completing this exercise, you should be able to:

O1 Conduct and interpret the tests used to evaluate visual acuity, astigmatism, the ability to accommodate for close vision, and color vision.

O2 Describe four conditions that can lead to defective vision.

O3 Demonstrate and describe the blind spot, photopupillary reflex, accommodation pupillary reflex, and convergence reflex.

O4 Summarize the results of the performed visual tests and demonstrations.

The **O** corresponds to the assessments **A** indicated in the Laboratory Assessment for this Exercise.

Pre-Lab

Carefully read the introductory material and examine the entire lab. Be familiar with common eye tests and defects from lecture or the textbook. Answer the pre-lab questions.

Pre-Lab Questions Select the correct answer for each of the following questions:

1. As a person ages, the elasticity of the lens
   a. decreases.
   b. increases.
   c. remains unchanged.
   d. improves.

2. Visual acuity is measured using a
   a. metric ruler.
   b. pen flashlight.
   c. Snellen eye chart.
   d. astigmatism chart.

3. Color blindness is a characteristic in __C.__ of males.
   a. 0%
   b. 0.4%
   c. 7%
   d. 10%

4. The blind spot is located at the _____ of the eye.
   a. pupil
   b. vitreous humor
   c. aqueous humor
   d. optic disc

5. Astigmatism results from a defect in the
   a. control of the pupil size.
   b. curvature of the cornea or lens.
   c. rods of the retina.
   d. cones of the retina.

6. Only males can inherit the color blindness condition.
   a. True _____
   b. False _✔_

7. Nearsightedness is also called myopia.
   a. True _✔_
   b. False _____

8. A convex lens can be used to correct hyperopia.
   a. True _____
   b. False _____

385

Normal vision (emmetropia) results when light rays from objects in the external environment are refracted by the cornea and lens of the eye and focused onto the photoreceptors of the retina. If a person has *nearsightedness (myopia),* the image focuses in front of the retina because the eyeball is too long; nearby objects are clear, but distant objects are blurred. If a person has *farsightedness (hyperopia),* the image focuses behind the retina because the eyeball is too short; distant objects are clear, but nearby objects are blurred. A concave (diverging) lens is required to correct vision for nearsightedness; a convex (converging) lens is required to correct vision for farsightedness (fig. 36.1).

If a defect occurs from unequal curvatures of the cornea or lens, a condition called *astigmatism* results. Focusing in different planes cannot occur simultaneously. Corrective cylindrical lenses allow proper refraction of light onto the retina to compensate for the unequal curvatures.

As part of a natural aging process, the lens's elasticity decreases, which degrades our ability to focus on nearby objects. The near point of accommodation test is used to determine the ability to accommodate. Often a person will use "reading glasses" to compensate for this condition.

Hereditary defects in color vision result from a lack of certain cones needed to absorb certain wavelengths of light. Special color test plates are used to diagnose color blindness.

If you wear glasses, perform the tests with and without the corrective lenses. If you wear contact lenses, it is not necessary to run the tests under both conditions; however, indicate in the laboratory assessment that all tests were performed with contact lenses in place.

PROCEDURE A: Visual Tests

Perform the following visual tests using your laboratory partner as a test subject. If your partner usually wears glasses, test each eye with and without the glasses.

1. *Visual acuity test.* Visual acuity (sharpness of vision) can be measured by using a Snellen eye chart (fig. 36.2). This chart consists of several sets of letters in different sizes printed on a white card. The letters near the top of the chart are relatively large, and those in each lower set become smaller. At one end of each set of letters is an acuity value in the form of a fraction. One of the sets near the bottom of the chart, for example, is marked 20/20. The normal eye can clearly see these letters from the standard distance of 20 feet and thus is said to have 20/20 vision. The letter at the top of the chart is marked 20/200. The normal eye can read letters of this size from a distance of 200 feet. Thus, an eye able to read only the top letter of the chart from a distance of 20 feet is said

FIGURE 36.1 Normal eye compared to two common refractive defects. (a) The focal point is on the retina in the normal eye (emmetropia); (b) the focal point is behind the retina in hyperopia; (c) the focal point is before the retina in myopia. A convex lens can correct hyperopia; a concave lens can correct myopia.

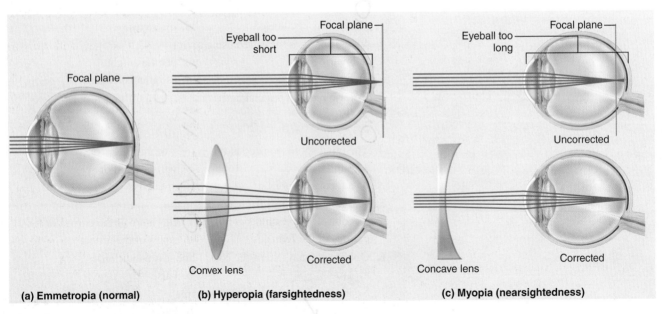

(a) Emmetropia (normal)   (b) Hyperopia (farsightedness)   (c) Myopia (nearsightedness)

FIGURE 36.2 The Snellen eye chart looks similar to this but is somewhat larger.

FIGURE 36.3 Astigmatism is evaluated using a chart such as this one.

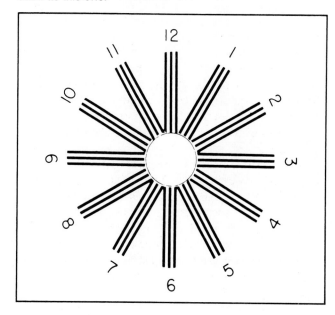

to have 20/200 vision. This person has less than normal vision. A line of letters near the bottom of the chart is marked 20/15. The normal eye can read letters of this size from a distance of 15 feet, but a person might be able to read it from 20 feet. This person has better than normal vision.

To conduct the visual acuity test, follow these steps:

a. Hang the Snellen eye chart on a well-illuminated wall at eye level.

b. Have your partner stand 20 feet in front of the chart, gently cover your partner's left eye with a 3" × 5" card, and ask your partner to read the smallest set of letters possible using the right eye.

c. Record the visual acuity value for that set of letters in Part A of Laboratory Assessment 36.

d. Repeat the procedure, using the left eye for the test.

2. *Astigmatism test.* Astigmatism is a condition that results from a defect in the curvature of the cornea or lens. As a consequence, some portions of the image projected on the retina are sharply focused, but other portions are blurred. Astigmatism can be evaluated by using an astigmatism chart (fig. 36.3). This chart consists of sets of black lines radiating from a central spot like the spokes of a wheel. To a normal eye, these lines appear sharply focused and equally dark; however, if the eye has an astigmatism some sets of lines appear sharply focused and dark, whereas others are blurred and less dark.

To conduct the astigmatism test, follow these steps:

a. Hang the astigmatism chart on a well-illuminated wall at eye level.

b. Have your partner stand 20 feet in front of the chart, and gently cover your partner's left eye with a 3" × 5" card. Ask your partner to focus on the spot in the center of the radiating lines using the right eye and report which lines, if any, appear more sharply focused and darker.

c. Repeat the procedure, using the left eye for the test.

d. Record the results in Part A of the laboratory assessment.

3. *Accommodation test. Accommodation* is the changing of the shape of the lens that occurs when the normal eye is focused for near (close) vision. It involves a reflex in which muscles of the ciliary body are stimulated to contract, releasing tension on the suspensory ligaments that are fastened to the lens capsule. This allows the capsule

FIGURE 36.4 **AP|R** The ciliary muscle and the lens are involved during accommodation. (a) The ciliary muscle is relaxed and the lens more flattened during distant vision; (b) the ciliary muscle contracts and the lens becomes more spherical during near vision.

Ciliary muscle relaxed

Suspensory ligament taut

Lens thins (more flattened)

(a) Lens shape during distant vision (emmetropia)

Ciliary muscle contracted

Suspensory ligament relaxed

Lens thickens (more spherical)

(b) Lens shape during near vision (accommodation)

to rebound elastically, causing the surface of the lens to become more spherical (convex) (fig. 36.4).

The ability to accommodate is likely to decrease with age because the tissues involved tend to lose their elasticity. This common aging condition, called *presbyopia,* can be corrected with reading glasses or bifocal lenses.

To evaluate the ability to accommodate, follow these steps:

  a. Hold the end of a meterstick against your partner's chin so that the stick extends outward at a right angle to the plane of the face (fig. 36.5).
  b. Have your partner close the left eye.
  c. Hold a 3" × 5" card with a word typed in the center at the distal end of the meterstick.

  d. Slide the card along the stick toward your partner's open right eye, and locate the *point closest to the eye* where your partner can still see the letters of the word sharply focused. This distance is called the *near point of accommodation,* and it tends to increase with age (table 36.1).
  e. Repeat the procedure with the right eye closed.
  f. Record the results in Part A of the laboratory assessment.
4. *Color vision test.* Some people exhibit defective color vision because they lack certain cones, usually those sensitive to the reds or greens. This trait is an X-linked (sex-linked) inheritance, so the condition is more prevalent in males (7%) than in females (0.4%). People who

FIGURE 36.5 To determine the near point of accommodation, slide the 3" × 5" card along the meterstick toward your partner's open eye until it reaches the closest location where your partner can still see the word sharply focused.

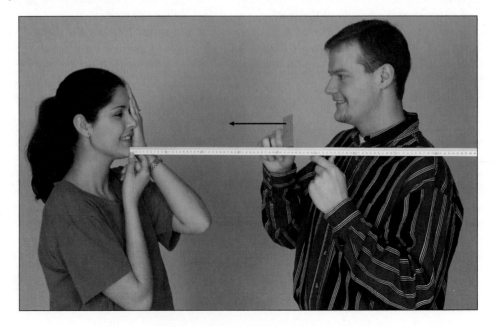

Table 36.1 Near Point of Accommodation

Age (years)	Average Near Point (cm)
10	7
20	10
30	13
40	20
50	45
60	90

lack or possess decreased sensitivity to the red-sensitive cones possess protanopia color blindness; those who lack or possess decreased sensitivity to green-sensitive cones possess deuteranopia color blindness. The color blindness condition is often more of a deficiency or a weakness than one of blindness. Laboratory Exercise 61 describes the genetics for color blindness.

To conduct the color vision test, follow these steps:

a. Examine the color test plates in Ichikawa's or Ishihara's book to test for any red-green color vision deficiency. Also examine figure 36.6.

b. Hold the test plates approximately 30 inches from the subject in bright light. All responses should occur within 3 seconds.

FIGURE 36.6 Sample test for red-green color blindness. A person with normal vision sees a number 74. A person with red-green color blindness does not see a 74 as the red and green colors are too similar. For accurate testing, the original plates should be used.

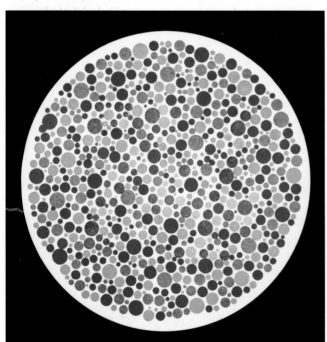

c. Compare your responses with the correct answers in Ichikawa's or Ishihara's book. Determine the percentage of males and females in your class who exhibit any color-deficient vision. If an individual exhibits color-deficient vision, determine if the condition is protanopia or deuteranopia.

d. Record the class results in Part A of the laboratory assessment.

5. Complete Part A of the laboratory assessment.

PROCEDURE B: Visual Demonstrations

Perform the following demonstrations with the help of your laboratory partner.

1. *Blind-spot demonstration.* There are no photoreceptors in the *optic disc,* located where the nerve fibers of the retina leave the eye and enter the optic nerve. Consequently, this region of the retina is commonly called the *blind spot.*

To demonstrate the blind spot, follow these steps:

a. Close your left eye, hold figure 36.7 about 34 cm away from your face, and stare at the "+" sign in the figure with your right eye.

b. Move the figure closer to your face as you continue to stare at the "+" until the dot on the figure suddenly disappears. This happens when the image of the dot is focused on the optic disc. Measure the right eye distance using a metric ruler or a meterstick.

c. Repeat the procedures with your right eye closed. This time stare at the dot with your left eye, and the "+" will disappear when the image falls on the optic disc. Measure the distance.

d. Record the results in Part B of the laboratory assessment.

CRITICAL THINKING ACTIVITY

Explain why small objects are not lost from our vision under normal visual circumstances.

_____

_____

_____

_____

FIGURE 36.7 Blind-spot demonstration.

2. *Photopupillary reflex.* The smooth muscles of the iris control the size of the pupil. For example, when the intensity of light entering the eye increases, a photopupillary reflex is triggered, and the circular muscles of the iris are stimulated to contract. As a result, the size of the pupil decreases, and less light enters the eye. This reflex involves the autonomic nervous system (ANS) and protects the photoreceptor cells of the retina from damage. In this autonomic reflex, the retina is the *receptor;* the *optic nerve* conducts the sensory impulses to the control center in the brain; the *oculomotor nerve* conducts motor impulses to the circular, smooth muscle fibers of the iris (the *effector*). The response, constriction of the pupil, is a result of impulses from the *parasympathetic division* of the ANS. (In contrast, during dim light situations, impulses over the *sympathetic division* of the ANS result in stimulation of radial, smooth muscle fibers and dilation of the pupil.)

To demonstrate this photopupillary reflex, follow these steps:

a. Ask your partner to sit with his or her hands thoroughly covering his or her eyes for 2 minutes.

b. Position a pen flashlight close to one eye with the light shining on the hand that covers the eye.

c. Ask your partner to remove the hand quickly.

d. Observe the pupil and note any change in its size.

e. Have your partner remove the other hand, but keep that uncovered eye shielded from extra light.

f. Observe both pupils and note any difference in their sizes.

3. *Accommodation pupillary reflex.* The pupil constricts as a normal accommodation reflex response to focusing on close objects. This reduces peripheral light rays and allows the refraction of light to occur closer to the center portion of the lens, resulting in a clearer image.

To demonstrate the accommodation pupillary reflex, follow these steps:

a. Have your partner stare for several seconds at a dimly illuminated object in the room that is more than 20 feet away.

b. Observe the size of the pupil of one eye. Then hold a pencil about 25 cm in front of your partner's face, and have your partner stare at it.

c. Note any change in the size of the pupil.

4. *Convergence reflex.* The eyes converge as a normal convergence response to focusing on close objects. The convergence reflex orients the eyeballs toward the object so the image falls on the fovea centralis of both eyes.

To demonstrate the convergence reflex, follow these steps:

a. Repeat the procedure outlined for the accommodation pupillary reflex.

b. Note any change in the position of the eyeballs as your partner changes focus from the distant object to the pencil.

5. Complete Part B of the laboratory assessment.

Name _____

Date _____

Section _____

The Ⓐ corresponds to the indicated Learning Outcome(s) Ⓞ found at the beginning of the Laboratory Exercise.

Visual Tests and Demonstrations

PART A: Assessments

1. Visual acuity test results:

Eye Tested	Acuity Values
Right eye	
Right eye with glasses (if applicable)	
Left eye	
Left eye with glasses (if applicable)	

2. Astigmatism test results:

Eye Tested	Darker Lines
Right eye	
Right eye with glasses (if applicable)	
Left eye	
Left eye with glasses (if applicable)	

3. Accommodation test results: Ⓐ①

Eye Tested	Near Point (cm)
Right eye	
Right eye with glasses (if applicable)	
Left eye	
Left eye with glasses (if applicable)	

4. Color vision test results: (A1)

Condition	Males			Females		
	Class Number	Class Percentage	Expected Percentage	Class Number	Class Percentage	Expected Percentage
Normal color vision			93			99.6
Deficient red-green color vision			7			0.4
Protanopia (lack red-sensitive cones)			less-frequent type			less-frequent type
Deuteranopia (lack green-sensitive cones)			more-frequent type			more-frequent type

5. Complete the following:

a. What is meant by 20/70 vision? (A2) _The normal eye can see the line from 70 feet._

b. What is meant by 20/10 vision? (A2) _The normal eye can see the line from 10 ft._

c. What visual problem is created by astigmatism? (A2) _Some portions of the image projected are sharply focused while others are blurred._

d. Why does the near point of accommodation often increase with age? (A2) _The tissue involved tends to lose their elasticity as age increases._

e. Describe the eye defect that causes color-deficient vision. (A2) _It is an inherited defect. Sex-related._

PART B: Assessments

1. Blind-spot results: (A3)

   a. Right eye distance _____

   b. Left eye distance _____

2. Complete the following:

   a. Explain why an eye has a blind spot. (A3) _There are no photoreceptors in the optic disc._

   b. Describe the photopupillary reflex. (A3) _As intensity of light entering the eye increases, photopupillary reflex is triggered and the circular muscles of the iris are stimulated to contract._

c. What difference did you note in the size of the pupils when one eye was exposed to bright light and the other eye was shielded from the light? (A3) _exposed to bright light pupil contracted._

eye shielded from light - dilated.

d. Describe the accommodation pupillary reflex. (A3) _pupil constricts as a normal accommodation reflex response to focusing on close objects. Reduces peripheal light rays._

e. Describe the convergence reflex. (A3) _Convergence reflex orients the eyeballs toward the object so the image falls on the fovea centralis of both eyes._

3. Summarize the vision of the person tested based upon the visual tests and demonstrations conducted in this laboratory exercise. Include information on suspected structural defects, genetic disorders, and aging conditions. (A4) _____

_____
_____
_____
_____
_____

NOTES

EXERCISE 37

Ear and Hearing

Purpose of the Exercise

To review the structural and functional characteristics of the ear and to conduct some ordinary hearing tests.

Dissectible ear model

Watch that ticks

Tuning fork (128 or 256 cps)

Rubber hammer

Cotton

Meterstick

For Demonstration Activities:

Compound light microscope

Prepared microscope slide of cochlea (section)

Audiometer

Learning Outcomes AP|R

After completing this exercise, you should be able to:

O1 Identify the major structures of the ear.

O2 Trace the pathway of sound vibrations from the tympanic membrane to the hearing receptors.

O3 Describe the functions of the structures of the ear.

O4 Conduct four ordinary hearing tests and summarize the results.

The **O** corresponds to the assessments **A** indicated in the Laboratory Assessment for this Exercise.

Pre-Lab

Carefully read the introductory material and examine the entire lab. Be familiar with the basic structures and functions of the ear associated with hearing from lecture or the textbook. Answer the pre-lab questions.

Pre-Lab Questions Select the correct answer for each of the following questions:

1. Hearing is interpreted in the _____ lobe of the cerebrum.
   - **a.** frontal
   - **b.** occipital
   - **c.** parietal
   - **d.** temporal

2. Sound loudness is measured in
   - **a.** pitch.
   - **b.** frequencies.
   - **c.** decibels.
   - **d.** hertz.

3. The middle ear bones articulate from tympanic membrane to oval window in which of the following sequences?
   - **a.** malleus, incus, stapes
   - **b.** incus, stapes, malleus
   - **c.** stapes, malleus, incus
   - **d.** malleus, stapes, incus

4. The _____ test is done to assess possible conduction deafness by comparing bone and air conduction.
   - **a.** Weber
   - **b.** Rinne
   - **c.** auditory acuity
   - **d.** sound localization

5. Which of the following structures is part of the inner ear?
   - **a.** tympanic membrane
   - **b.** cochlea
   - **c.** stapes
   - **d.** pharyngotympanic tube

6. The pharyngotympanic tube connects the outer ear to the inner ear.
   - **a.** True _____
   - **b.** False _____

7. The cochlear nerve serves as the hearing branch of the vestibulocochlear nerve.
   - **a.** True _____
   - **b.** False _____

8. Endolymph is located within the cochlear duct of the cochlea.
   - **a.** True _____
   - **b.** False _____

The ear is composed of outer (external), middle, and inner (internal) parts. The large external *auricle* gathers sound waves (vibrations) of air and directs them into the *external acoustic meatus* to the *tympanic membrane* (eardrum). The middle ear *auditory ossicles* (small bones) conduct and amplify the waves from the tympanic membrane to the *oval window*. To allow the same pressure in the outer and middle ear, the pharyngotympanic (auditory) tube connects the middle ear to the pharynx. Equalizing pressure is important for proper vibration of the eardrum. The sound waves are further conducted through the fluids (perilymph and endolymph) of the *cochlea* in the inner ear. Perilymph is located within the scala vestibuli and the scala tympani; endolymph occupies the cochlear duct, which contains the *spiral organ*. Finally, some hair cells of the spiral organ are stimulated. As certain hair cells (receptor cells) are stimulated, these receptors initiate nerve impulses that are conducted over the *cochlear branch* of the *vestibulocochlear nerve* into the brain. When nerve impulses arrive at the *primary auditory cortex* of the temporal lobe of the cerebrum, the impulses are interpreted and the sensations of hearing are created (fig. 37.1).

The human hearing range is approximately 20 to 20,000 waves per second, known as *hertz* (*Hz*). We perceive these different frequencies as pitch. Different frequencies stimulate particular hair cells in specific regions of the spiral organ, allowing us to interpret different pitches. An additional quality of sound is loudness. The intensity of the sound is measured in units called *decibels* (*dB*). Permanent damage can happen to the hair cells of the spiral organ if frequent and prolonged exposures occur over 90 decibels.

Although the ear is considered here as a hearing organ, it also has important functions for equilibrium (balance). The inner ear semicircular ducts and vestibule have equilibrium functions. The ear and equilibrium will be studied in Laboratory Exercise 38.

PROCEDURE A: Structure and Function of the Ear AP|R

1. Use figures 37.2, 37.3, and 37.4 as references as you progress through this procedure.
2. Examine the dissectible model of the ear and locate the following features:

   outer (external) ear
   - auricle (pinna)—funnel for sound waves
   - external acoustic meatus (auditory canal)—2.5 cm S-shaped tube
   - tympanic membrane (eardrum)—semitransparent; vibrates upon sound wave arrival

   middle ear
   - tympanic cavity—air-filled chamber in temporal bone
   - auditory ossicles (fig. 37.5)—transmit vibrations to oval window
     - malleus (hammer)—first ossicle to vibrate
     - incus (anvil)—middle ossicle to vibrate

   - stapes (stirrup)—final ossicle to vibrate at oval window
   - oval window—opening at lateral wall of inner ear
   - round window—membrane-covered opening of inner ear wall
   - pharyngotympanic tube (auditory tube; eustachian tube)—connects middle ear to pharynx

   inner (internal) ear
   - bony labyrinth—bony chambers filled with perilymph; includes the cochlea, semicircular canals, and vestibule
   - membranous labyrinth—membranous chambers filled with endolymph; follows pattern of bony labyrinth
   - cochlea—converts sound waves to nerve impulses at spiral organ
   - semicircular ducts—equilibrium organ
   - vestibule—equilibrium organ with two chambers
     - utricle
     - saccule

   vestibulocochlear nerve VIII—cranial nerve
   - vestibular nerve—balance branch
   - cochlear nerve—hearing branch

3. Complete Parts A, B, and C of Laboratory Assessment 37.

DEMONSTRATION ACTIVITY

Observe the section of the cochlea in the demonstration microscope. Locate one of the turns of the cochlea, and using figures 37.4 and 37.6 as a guide, identify the *scala vestibuli, cochlear duct* (*scala media*), *scala tympani, vestibular membrane, basilar membrane,* and the *spiral organ* (*organ of Corti*).

PROCEDURE B: Hearing Tests

Perform the following tests in a quiet room, using your laboratory partner as the test subject.

1. *Auditory acuity test.* To conduct this test, follow these steps:
   a. Have the test subject sit with eyes closed.
   b. Pack one of the subject's ears with cotton.
   c. Hold a ticking watch close to the open ear, and slowly move it straight out and away from the ear.
   d. Have the subject indicate when the sound of the ticking can no longer be heard.
   e. Use a meterstick to measure the distance in centimeters from the ear to the position of the watch.
   f. Repeat this procedure to test the acuity of the other ear.
   g. Record the test results in Part D of the laboratory assessment.

FIGURE 37.1 Auditory pathways. The sensory receptors are hair cells of the spiral organ in the cochlea. The sensory information continues through the medulla oblongata, pons, midbrain, thalamus, and to the primary auditory cortex in the temporal lobe, where sound is perceived.

FIGURE 37.2 AP|R Major structures of the ear. The three divisions of the ear are roughly indicated by the dashed lines.

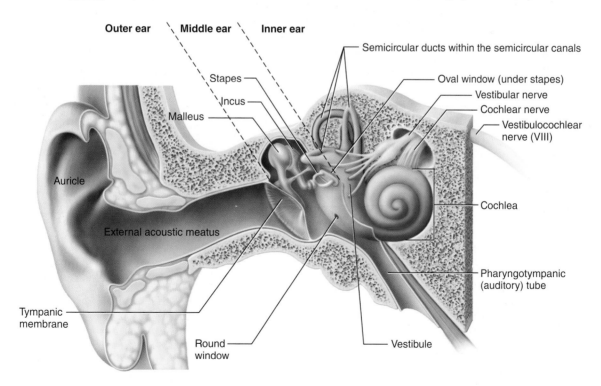

FIGURE 37.3 Structures of the middle and inner ear.

- Semicircular duct
- Endolymphatic sac
- Utricle in vestibule
- Saccule in vestibule
- Semicircular canals:
  - Anterior
  - Posterior
  - Lateral
- Apex of cochlea
- Cochlea:
  - Scala tympani
  - Cochlear duct
  - Scala vestibuli
- Temporal bone
- Auditory ossicles:
  - Malleus
  - Incus
  - Stapes (in oval window)
- Round window
- Tympanic cavity

Color code:
Perilymph — green
Endolymph — blue

FIGURE 37.4 Anatomy of the cochlea: (a) section through cochlea; (b) view of one turn of cochlea; (c) view of spiral organ (organ of Corti).

- Oval window
- Section through one turn of cochlea
- Cochlear nerve

(a)

- Scala vestibuli (with perilymph)
- Scala tympani (with perilymph)
- Spiral organ (organ of Corti)
- Cochlear duct (with endolymph)

(b) AP|R

- Endolymph
- Tectorial membrane
- Hairs (stereocilia)
- Inner hair cell
- Supporting cell
- Cochlear nerve fibers
- Basilar membrane
- Outer hair cells

(c) AP|R

FIGURE 37.5 Middle ear bones (auditory ossicles) superimposed on a penny to show their relative size. The arrangement of the malleus, incus, and stapes in the enlarged photograph resembles the articulations in the middle ear.

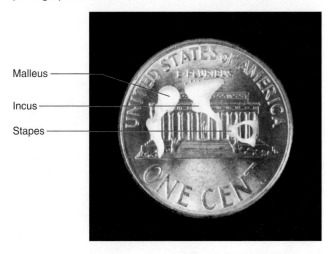

Malleus

Incus

Stapes

FIGURE 37.6 A section through the cochlea (22×).

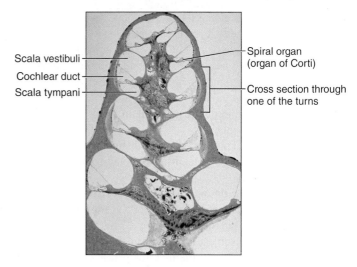

Scala vestibuli

Cochlear duct

Scala tympani

Spiral organ (organ of Corti)

Cross section through one of the turns

2. *Sound localization test.* To conduct this test, follow these steps:
   a. Have the subject sit with eyes closed.
   b. Hold the ticking watch somewhere within the audible range of the subject's ears, and ask the subject to point to the watch.
   c. Move the watch to another position and repeat the request. In this manner, determine how accurately the subject can locate the watch when it is in each of the following positions: in front of the head, behind the head, above the head, on the right side of the head, and on the left side of the head.
   d. Record the test results in Part D of the laboratory assessment.

3. *Rinne test.* This test is done to assess possible conduction deafness by comparing bone and air conduction. To conduct this test, follow these steps:
   a. Obtain a tuning fork and strike it with a rubber hammer, or on the heel of your hand, causing it to vibrate.
   b. Place the end of the fork's handle against the subject's mastoid process behind one ear. Have the prongs of the fork pointed downward and away from the ear, and be sure nothing is touching them (fig. 37.7a). The sound sensation is that of bone conduction. If no sound is experienced, nerve deafness exists.
   c. Ask the subject to indicate when the sound is no longer heard.
   d. Then quickly remove the fork from the mastoid process and position it in the air close to the opening of the nearby external acoustic meatus (fig. 37.7b).

FIGURE 37.7 Rinne test: (a) first placement of vibrating tuning fork until sound is no longer heard; (b) second placement of tuning fork to assess air conduction.

(a)

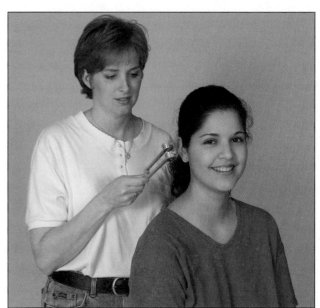

(b)

If hearing is normal, the sound (from air conduction) will be heard again; if there is conductive impairment, the sound will not be heard. Conductive impairment involves outer or middle ear defects. Hearing aids can improve hearing for conductive deafness because bone conduction transmits the sound into the inner ear. Surgery could possibly correct this type of defect.

e. Record the test results in Part D of the laboratory assessment.

4. *Weber test.* This test is used to distinguish possible conduction or sensory deafness. To conduct this test, follow these steps:

a. Strike the tuning fork with the rubber hammer.

b. Place the handle of the fork against the subject's forehead in the midline (fig. 37.8).

c. Ask the subject to indicate if the sound is louder in one ear than in the other or if it is equally loud in both ears.

　　If hearing is normal, the sound will be equally loud in both ears. If there is conductive impairment, the sound will appear louder in the affected ear. If some degree of sensory (nerve) deafness exists, the sound will be louder in the normal ear. The impairment involves the spiral organ or the cochlear nerve. Hearing aids will not improve sensory deafness.

d. Have the subject experience the effects of conductive impairment by packing one ear with cotton and

FIGURE 37.8　Weber test.

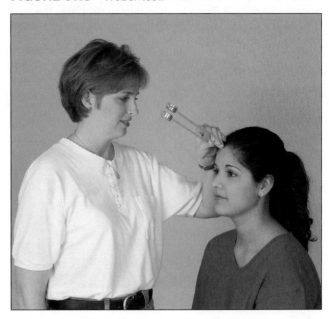

repeating the Weber test. Usually, the sound appears louder in the plugged (impaired) ear because extraneous sounds from the room are blocked out.

e. Record the test results in Part D of the laboratory assessment.

5. Complete Part D of the laboratory assessment.

CRITICAL THINKING ACTIVITY

Ear structures from the outer ear into the inner ear are progressively smaller. Using results obtained from the hearing tests, explain this advantage.

_____

_____

_____

DEMONSTRATION ACTIVITY

Ask the laboratory instructor to demonstrate the use of the audiometer. This instrument produces sound vibrations of known frequencies transmitted to one or both ears of a test subject through earphones. The audiometer can be used to determine the threshold of hearing for different sound frequencies, and, in the case of hearing impairment, it can be used to determine the percentage of hearing loss for each frequency.

ASSESSMENT
37

Name _____

Date _____

Section _____

The Ⓐ corresponds to the indicated Learning Outcome(s) Ⓞ found at the beginning of the Laboratory Exercise.

Ear and Hearing

PART A: Assessments

1. Identify the features of the ear indicated in figures 37.9 and 37.10.

FIGURE 37.9 Label the structures associated with the ear. **A1**

———— 1 ————

———— 2 ————

———— 3 ————

———— 4 ————

———— 5 ————

Stapes

Terms:
Cochlea
Oval window
Semicircular duct
Vestibule
Vestibulocochlear nerve

2. Label the structures indicated in the micrograph of the spiral organ in figure 37.11.

FIGURE 37.11 AP|R Label the structures associated with this spiral organ region of a cochlea, using the terms provided. (300×). A1

Endolymph ———— 1 ————

———— 2 ————

———— 3 ————

———— 4 ————

Perilymph ———— 5 ————

Terms:
Basilar membrane
Cochlear duct
Hair cells
Scala tympani
Tectorial membrane

PART B: Assessments

Number the following structures (1–9) to indicate their respective positions in relation to the pathway of the sound vibrations. Assign number 1 to the most superficial portion of the outer ear. (A2)

_____ Auricle (air vibrations within)

_____ External acoustic meatus (air vibrations within)

_____ Basilar membrane of spiral organ within cochlea

_____ Hair cells of spiral organ

_____ Incus

_____ Malleus

_____ Oval window

_____ Stapes

_____ Tympanic membrane

PART C: Assessments

Match the terms in column A with the descriptions in column B. Place the letter of your choice in the space provided. (A1) (A3)

Column A	Column B
a. Bony labyrinth	_____ 1. Auditory ossicle attached to tympanic membrane
b. Cochlear duct	_____ 2. Air-filled space containing auditory ossicles within middle ear
c. External acoustic meatus	_____ 3. Contacts hairs of hearing receptors
d. Malleus	_____ 4. Leads from oval window to apex of cochlea and contains perilymph
e. Membranous labyrinth	_____ 5. S-shaped tube leading to tympanic membrane
f. Pharyngotympanic (auditory) tube	_____ 6. Tube within cochlea containing spiral organ and endolymph
g. Scala tympani	_____ 7. Cone-shaped, semitransparent membrane attached to malleus
h. Scala vestibuli	_____ 8. Auditory ossicle attached to oval window
i. Stapes	_____ 9. Chambers containing endolymph within bony labyrinth
j. Tectorial membrane	_____ 10. Bony chambers of inner ear in temporal bone
k. Tympanic cavity	_____ 11. Connects middle ear and pharynx
l. Tympanic membrane (eardrum)	_____ 12. Extends from apex of cochlea to round window and contains perilymph

PART D: Assessments

1. Results of auditory acuity test:

Ear Tested	Audible Distance (cm)
Right	
Left	

2. Results of sound localization test:

Actual Location	Reported Location
Front of the head	
Behind the head	
Above the head	
Right side of the head	
Left side of the head	

3. Results of experiments using tuning forks:

Test	Left Ear (Normal or Impaired)	Right Ear (Normal or Impaired)
Rinne		
Weber		

4. Summarize the results of the hearing tests you conducted on your laboratory partner.

_____

_____

_____

_____

_____

_____

_____

LABORATORY
EXERCISE 38

Ear and Equilibrium

Purpose of the Exercise
To review the structure and function of the organs of equilibrium and to conduct some tests of equilibrium.

MATERIALS NEEDED

Swivel chair

Bright light

For Demonstration Activity:
Compound light microscope
Prepared microscope slide of semicircular duct (cross section through ampulla)

SAFETY ⚠️

▶ Do not pick subjects who have frequent motion sickness.

▶ Have four people surround the subject in the swivel chair in case the person falls from vertigo or loss of balance.

▶ Stop your experiment if the subject becomes nauseated.

Learning Outcomes
After completing this exercise, you should be able to:

O1 Locate the organs of static and dynamic equilibrium and describe their functions.

O2 Explain the role of vision in the maintenance of equilibrium.

O3 Conduct and record the results of the Romberg and Bárány tests of equilibrium.

The **O** corresponds to the assessments **A** indicated in the Laboratory Assessment for this Exercise.

Pre-Lab
Carefully read the introductory material and examine the entire lab. Be familiar with the basic structures and functions of the ear associated with equilibrium from lecture or the textbook. Answer the pre-lab questions.

Pre-Lab Questions Select the correct answer for each of the following questions:

1. The sense organs associated with equilibrium are within the
   - **a.** outer ear.
   - **b.** middle ear.
   - **c.** inner ear.
   - **d.** tympanic membrane.

2. The _____ nerve conducts impulses associated with equilibrium.
   - **a.** trochlear
   - **b.** trigeminal
   - **c.** facial
   - **d.** vestibulocochlear

3. Hair cell receptors associated with dynamic equilibrium are located within the _____
   - **a.** utricle.
   - **b.** crista ampullaris.
   - **c.** saccule.
   - **d.** cochlea.

4. Otoliths are composed of
   - **a.** calcium carbonate.
   - **b.** carbon.
   - **c.** bone tissue.
   - **d.** stereocilia.

5. Which of the following is *not* associated with rotational movements?
   - **a.** crista ampullaris
   - **b.** otoliths
   - **c.** ampulla
   - **d.** semicircular duct

6. Eye twitching movements characteristic during rotational movements are called nystagmus.
   - **a.** True _____
   - **b.** False _____

7. Otoliths are located within the semicircular ducts of the inner ear.
   - **a.** True _____
   - **b.** False _____

The sense of equilibrium involves two sets of sensory organs. One set helps to maintain the stability of the head and body when they are motionless or during linear acceleration and produces a sense of static (gravitational) equilibrium. The other set is concerned with balancing the head and body when angular acceleration produces a sense of dynamic (rotational) equilibrium.

The sense organs associated with the sense of *static equilibrium* are located within the *vestibules* of the inner ears. Two chambers, the *utricle* and *saccule,* contain receptors called *maculae.* Each macula is composed of hair cells embedded within a gelatinous otolithic membrane. Embedded within the otolithic membrane are numerous tiny calcium carbonate "ear stones" called *otoliths.* As we change our head position, gravitational forces allow the shift of the otoliths to bend the stereocilia of the *hair cells* of the macula, resulting in stimulation of sensory vestibular neurons. Stimulation of the maculae also occurs during linear acceleration. Horizontal acceleration (as occurs when riding in a car) involves the utricle; vertical acceleration (as occurs when riding in an elevator) involves the saccule. As a result of static equilibrium, recognition of movements such as falling are detected and postural adjustments are accomplished.

The sense organs associated with the sense of *dynamic equilibrium* are located within the *ampullae* of the three *semicircular ducts* of the inner ear. Each membranous semicircular duct is located within a *semicircular canal* of the temporal bone. A small elevation within each ampulla possesses the *crista ampullaris.* Each crista ampullaris contains *hair cells* with stereocilia embedded within a gelatinous cap

DEMONSTRATION ACTIVITY

Observe the cross section of the semicircular duct through the ampulla in the demonstration microscope. Note the crista ampullaris (fig. 38.6) projecting into the lumen of the membranous labyrinth, which in a living person is filled with endolymph. The space between the membranous and bony labyrinths is filled with perilymph.

called the *cupula.* During rotational movements, the cupula is bent, which stimulates hair cells and sensory neurons of the vestibular nerve. Because the three semicircular ducts are in different planes, rotational movements in any direction result in stimulation of the associated hair cells. Impulses from vestibular neurons of the semicircular ducts result in reflex movements of the eye. During rotational movements, characteristic twitching movements of the eyes called *nystagmus* occur, often accompanied by dizziness (vertigo).

Impulses from inner ear receptors travel over the vestibular neurons of the vestibulocochlear nerve and include processing destinations in the brainstem, cerebellum, thalamus, and cerebrum (fig. 38.1). The sense of equilibrium works subconsciously to initiate appropriate corrections to body position and movements. Additional senses work in conjunction with the inner ear and equilibrium. Stretch receptors in muscles and tendons, touch, and vision complement the inner ear for equilibrium and proper body adjustments.

FIGURE 38.1 Vestibular pathways. The sensory receptors are hair cells within the vestibule and semicircular ducts. Sensory information over vestibular neurons continues through the brainstem, cerebellum, thalamus, and cerebrum of the brain. Conscious awareness of the body position and movement is processed in the cerebrum.

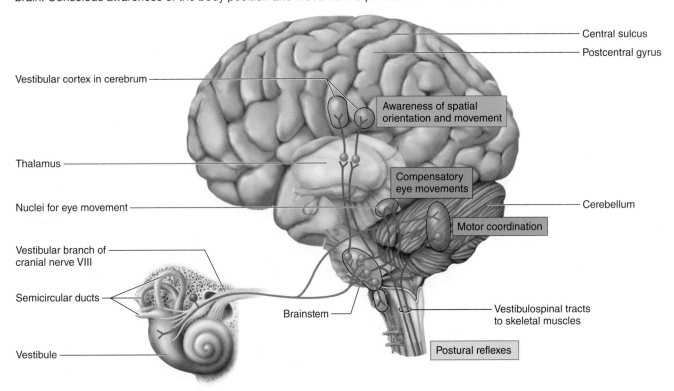

406

FIGURE 38.2 Structures of static equilibrium; (a) utricle and saccule each contain a macula; (b) macula and otolithic membrane orientation when body is upright; (c) macula and otolithic membrane changes when the body is tilted.

Utricle
Maculae
Saccule

Body upright
(a)

Otoliths
Otolithic membrane

Stereocilia
Hair cell
Vestibular neuron
(b)

Body tilted

Stereocilia of hair cells bend

Otoliths composed of calcium carbonate add additional weight to otolithic membrane

Otolithic membrane sags

Gravitational force
(c)

PROCEDURE A: Organs of Equilibrium

1. Re-examine the introduction to this laboratory exercise.
2. Study figures 38.2 and 38.3, which present the structures and functions of static equilibrium.
3. Study figures 38.4 and 38.5, which present the structures and functions of dynamic equilibrium.
4. Complete Part A of Laboratory Assessment 38.

FIGURE 38.3 Particularly large otoliths are found in a freshwater drum (*Aplodinotus grunniens*). They have been used for lucky charms and jewelry. Human otoliths are microscopic size.

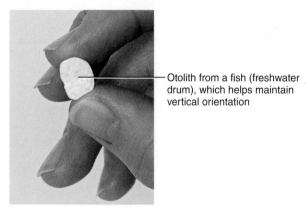

Otolith from a fish (freshwater drum), which helps maintain vertical orientation

FIGURE 38.4 Structures of dynamic equilibrium; (a) each semicircular duct has an ampulla containing a crista ampullaris; (b) crista ampullaris when the body is stationary; (c) changes in crista ampullaris during body rotation.

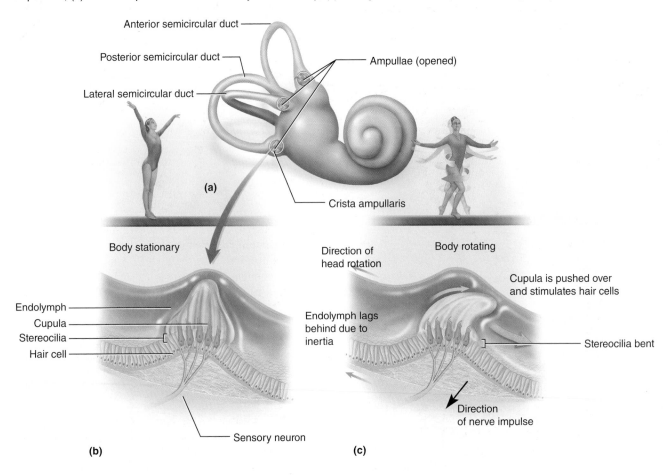

FIGURE 38.5 Semicircular canal superimposed on a penny to show its relative size. A semicircular duct of the same shape would occupy the semicircular canal.

FIGURE 38.6 A micrograph of a crista ampullaris (1,400×). The crista ampullaris is located within the ampulla of each semicircular duct.

PROCEDURE B: Tests of Equilibrium

Use figures 38.2 and 38.4 as references as you progress through the various tests of equilibrium. Perform the following tests, using a person as a test subject who is not easily disturbed by dizziness or rotational movement. Also have some other students standing close by to help prevent the test subject from falling during the tests. *The tests should be stopped immediately if the test subject begins to feel uncomfortable or nauseated.*

1. *Vision and equilibrium test.* To demonstrate the importance of vision in the maintenance of equilibrium, follow these steps:
   a. Have the test subject stand erect on one foot for 1 minute with his or her eyes open.
   b. Observe the subject's degree of unsteadiness.
   c. Repeat the procedure with the subject's eyes closed. *Be prepared to prevent the subject from falling.*
   d. Answer the questions related to the vision and equilibrium test in Part B of the laboratory assessment.
2. *Romberg test.* The purpose of this test is to evaluate how the organs of static equilibrium in the vestibule enable one to maintain balance (fig. 38.2). To conduct this test, follow these steps:
   a. Position the test subject close to a chalkboard with the back toward the board.
   b. Place a bright light in front of the subject so that a shadow of the body is cast on the board.
   c. Have the subject stand erect with feet close together and eyes staring straight ahead for 3 minutes.
   d. During the test, make marks on the chalkboard along the edge of the shadow of the subject's shoulders to indicate the range of side-to-side swaying.
   e. Measure the maximum sway in centimeters and record the results in Part B of the laboratory assessment.
   f. Repeat the procedure with the subject's eyes closed.
   g. Position the subject so one side is toward the chalkboard.
   h. Repeat the procedure with the eyes open.
   i. Repeat the procedure with the eyes closed.
      The Romberg test is used to evaluate a person's ability to integrate sensory information from proprioceptors and receptors within the organs of static equilibrium and to relay appropriate motor impulses to postural muscles. A person who shows little unsteadiness when standing with feet together and eyes open, but who becomes unsteady when the eyes are closed, has a positive Romberg test.
3. *Bárány test.* The purpose of this test is to evaluate the effects of rotational acceleration on the semicircular ducts and dynamic equilibrium (fig. 38.4). To conduct this test, follow these steps:
   a. Have the test subject sit on a swivel chair with his or her eyes open and focused on a distant object, the head tilted forward about 30°, and the hands gripped firmly to the seat. Position four people around the chair for safety. *Be prepared to prevent the subject and the chair from tipping over.*
   b. Rotate the chair ten rotations within 20 seconds.
   c. Abruptly stop the movement of the chair. The subject will still have the sensation of continuous movement and might experience some dizziness (vertigo).
   d. Have the subject look forward and immediately note the nature of the eye movements and their direction. (Such reflex eye twitching movements are called *nystagmus.*) Also note the time it takes for the nystagmus to cease. Nystagmus will continue until the cupula is returned to an original position.
   e. Record your observations in Part B of the laboratory assessment.
   f. Allow the subject several minutes of rest, then repeat the procedure with the subject's head tilted nearly 90° onto one shoulder.
   g. After another rest period, repeat the procedure with the subject's head bent forward so that the chin is resting on the chest.
      In this test, when the head is tilted about 30°, the lateral semicircular ducts receive maximal stimulation, and the nystagmus is normally from side to side. When the head is tilted at 90°, the superior ducts are stimulated, and the nystagmus is up and down. When the head is bent forward with the chin on the chest, the posterior ducts are stimulated, and the nystagmus is rotary.
4. Complete Part B of the laboratory assessment.

NOTES

Name _____

Date _____

Section _____

The (A) corresponds to the indicated Learning Outcome(s) (O) found at the beginning of the Laboratory Exercise.

Ear and Equilibrium

PART A: Assessments

Complete the following statements:

1. The organs of static equilibrium are located within two expanded chambers within the vestibule called the _____ and the saccule. (A1)

2. All of the balance organs are found within the _____ bone of the skull. (A1)

3. Otoliths are small grains composed of _____. (A1)

4. Sensory impulses travel from the organs of equilibrium to the brain on vestibular neurons of the _____ nerve. (A1)

5. The sensory organ of a semicircular duct lies within a small elevation called the _____. (A1)

6. The sensory organ within the ampulla of a semicircular duct is called a _____. (A1)

7. The _____ of this sensory organ consists of a dome-shaped gelatinous cap. (A1)

8. The vestibular cortex in the _____ of the brain processes awareness of body position and movement. (A1)

PART B: Tests of Equilibrium Assessments

1. Vision and equilibrium test results:

   a. When the eyes are open, what sensory organs provide information needed to maintain equilibrium? (A2)

   _____

   _____

   b. When the eyes are closed, what sensory organs provide such information? (A2)

   _____

   _____

2. Romberg test results:

    a. Record the test results in the following table: (A3)

Conditions	Maximal Movement (cm)
Back toward board, eyes open	
Back toward board, eyes closed	
Side toward board, eyes open	
Side toward board, eyes closed	

    b. Did the test subject's unsteadiness increase when the eyes were closed? _____ What is the significance of this observation? (A2) _____

_____

_____

    c. Why would you expect a person with impairment of the organs of equilibrium to become more unsteady when the eyes are closed? (A2) _____

_____

_____

3. Bárány test results:

    a. Record the test results in the following table: (A3)

Position of Head	Description of Eye Movements	Time for Movement to Cease
Tilted 30° forward		
Tilted 90° onto shoulder		
Tilted forward, chin on chest		

    b. Summarize the results of this test. (A3)

_____

_____

_____

CRITICAL THINKING ASSESSMENT

What additional sensory information would you expect persons with impairment of organs of equilibrium to use to supplement their relative lack of some sensory information?

_____

_____

_____

LABORATORY
EXERCISE 39

Endocrine Structure and Function

Purpose of the Exercise

To review the structure, function, and microscopic anatomy of major endocrine glands. To measure and compare the effect of cooling on metabolic rate in animals that differ in their ability to release thyroid hormone.

MATERIALS NEEDED

Human torso model

Compound light microscope

Prepared microscope slides of the following:

   Pituitary gland

   Thyroid gland

   Parathyroid gland

   Adrenal gland

   Pancreas

For Learning Extension Activity:

Water bath equipped with temperature control mechanism set at 37°C (98.6°F)

Laboratory thermometer

Ph.I.L.S. 4.0, Lesson 19 Endocrine Function: Thyroid Gland and Metabolic Rate

Learning Outcomes

After completing this exercise, you should be able to:

O1 Name and locate the major endocrine glands.

O2 Sketch and label tissue sections from the pituitary gland, thyroid gland, parathyroid glands, adrenal glands, and pancreas.

O3 Associate the principal hormones secreted by each of the major endocrine glands.

O4 Explain the role of the hypothalamus in controlling the release of thyroid hormone.

O5 Explain the relationship of thyroid hormone levels, metabolic rate, body temperature, and oxygen consumption.

O6 Interpret changes in oxygen consumption that occur with decreases in environmental temperature.

O7 Apply the concepts of thyroid hormone levels and their effects on metabolic rate, oxygen consumption, and temperature regulation to the human body.

The **O** corresponds to the assessments **A** indicated in the Laboratory Assessment for this Exercise.

Pre-Lab

Carefully read the introductory material and examine the entire lab. Be familiar with basic structures and functions of the pituitary, thyroid, parathyroid, adrenal, and pancreas glands from lecture or the textbook. Answer the pre-lab questions.

Pre-Lab Questions Select the correct answer for each of the following questions:

1. Endocrine glands secrete _____ into the blood.
   a. neurotransmitters   **b. true hormones**
   c. paracrine substances   d. autocrine substances

2. True hormones influence
   a. blood cells.   b. the secreting cells.
   c. neighboring cells.   **d. target cells around the body.**

3. Which of the following endocrine glands is also an exocrine (ductile secretion) gland?
   a. pancreas   b. thyroid
   c. thymus   d. adrenal

4. Which two endocrine glands have the closest proximity to each other?
   a. pancreas; thymus   b. thymus; thyroid
   c. thyroid; parathyroid   d. pituitary; thyroid

5. Which endocrine gland stores hormones synthesized by the hypothalamus?
   a. posterior pituitary
   b. anterior pituitary
   c. adrenal cortex
   d. adrenal medulla
6. The hypothalamus, pineal gland, and pituitary gland are located within the cranial cavity.
   a. True ___✓___
   b. False _____
7. Insulin is secreted by beta cells of the pancreatic islets.
   a. True ___✓___
   b. False _____
8. The adrenal medulla secretes the hormones aldosterone and cortisol.
   a. True ___✓___
   b. False ___✓___ X

The endocrine system consists of *ductless glands* that act together with parts of the nervous system to help control body activities. The endocrine glands secrete regulatory molecules, called *hormones,* into the internal environment of interstitial fluid. Hormones are then transported in the *blood* and influence *target cells* around the body. A target cell is any cell that expresses a receptor that will bind the hormone. At target cells, hormones bind to *receptors,* resulting in the activation of processes leading to functional changes within the target cells. In this way, hormones influence the rate of metabolic reactions, the transport of substances through cell membranes, the regulation of water and electrolyte balances, and many other functions.

Some of the glands included in this laboratory exercise have ducts and are known as exocrine glands. These glands are mixed endocrine and exocrine. The pancreas, ovaries, and testes have ducts and function within systems other than the endocrine system. Additional secreted substances that are regulatory, although not considered true hormones, include paracrine and autocrine secretions. *Paracrine secretions* influence only neighboring cells; *autocrine secretions* influence only the secreting cell itself.

The nervous system and the endocrine system are both regulatory and involve chemical secretions and receptor sites. The neurons of the nervous system secrete neurotransmitters. Receptor sites for neurotransmitters are on the postsynaptic cell, resulting in rapid responses of a brief duration unless additional stimulation occurs. In contrast, the endocrine cells secrete hormones transported by the blood to receptor sites on target cells. The responses from the hormones last longer and might even continue for hours to days even without additional hormonal secretions. By controlling cellular processes, endocrine glands play important roles in the maintenance of homeostasis.

You walk outside on a cold winter day. As your body temperature begins to drop, specific physiological processes will be activated to help maintain your body temperature. Initially, increased amounts of thyroid hormone will be released from the thyroid gland. (Temperature receptors in your skin send neural signals to the hypothalamus to trigger the release of thyrotropin-releasing hormone [TRH]; TRH stimulates the anterior pituitary to release thyroid-stimulating hormone [TSH]; and TSH stimulates the thyroid gland to release the thyroid hormones, T_3 and T_4.) Thyroid hormone is able to help maintain body temperature by increasing the metabolic rate of almost all cells, especially neurons. Neurons are specifically stimulated to increase their number of Na^+/K^+ pumps. Since there are more Na^+/K^+ pumps, more energy is converted (chemical energy of ATP is converted to mechanical energy of pumping Na^+ and K^+), more heat is produced (second law of thermodynamics), and body temperature is maintained. If your body temperature continues to decrease, the hypothalamus will initiate shivering, involuntary contractions of skeletal muscle. Shivering increases body temperature as the muscles contract and generate heat. However, what occurs when you walk outside on a cold winter day if your thyroid gland is impaired and insufficient thyroid hormone is released? Will your body temperature drop or will you begin to shiver sooner to maintain body temperature?

PROCEDURE: Endocrine Gland Histology

1. Study the location of the major endocrine glands indicated in figure 39.1.
2. Examine the human torso model and locate the following:

   hypothalamus

   pituitary stalk (infundibulum)

LEARNING EXTENSION ACTIVITY

The secretions of endocrine glands are usually controlled by negative feedback systems. As a result, the concentrations of hormones in body fluids remain relatively stable, although they will fluctuate slightly within a normal range.

Similarly, the mechanism used to maintain the temperature of a laboratory water bath involves negative feedback. In this case, a temperature-sensitive thermostat in the water allows a water heater to operate whenever the water temperature drops below the thermostat's set point. Then, when the water temperature reaches the set point, the thermostat causes the water heater to turn off (a negative effect), and the water bath begins to cool again.

Use a laboratory thermometer to monitor the temperature of the water bath in the laboratory. Measure the temperature at regular intervals until you have recorded ten readings. What was the lowest temperature you recorded? _____ The highest temperature? _____ What was the average temperature of the water bath? _____ How is the water bath temperature control mechanism similar to a hormonal control mechanism in the body?

_____

_____

_____

_____

FIGURE 39.1 AP|R Major endocrine glands.

(handwritten annotations on figure)
- (last figure) 627
- Pineal gland
- Hypothalamus
- Pituitary gland — 700
- (first. fig.) 5
- Thyroid gland
- Thymus
- Adrenal glands 3 (first fig.)
- Pancreas
- Parathyroid glands
- Posterior view
- Trachea
- Ovary (female)
- Testis (male)

adrenal glands deep to pancreas

pituitary gland (hypophysis)
- anterior lobe (adenohypophysis)
- posterior lobe (neurohypophysis)

thyroid gland

parathyroid glands

adrenal glands (suprarenal glands)
- adrenal medulla
- adrenal cortex

pancreas

pineal gland

thymus

ovaries

testes

3. Examine the microscopic tissue sections of the following glands, and identify the features described:

Pituitary gland. The pituitary gland is located in the sella turcica of the sphenoid bone. To examine the pituitary tissue, follow these steps:

a. Observe the tissues using low-power magnification (fig. 39.2).

b. Locate the *pituitary stalk,* the *anterior lobe* (the largest part of the gland), and the *posterior lobe.* The anterior lobe makes hormones and appears glandular; the posterior lobe stores and releases hormones made by the hypothalamus and is nervous in appearance.

c. Observe an area of the anterior lobe with high-power magnification (fig. 39.3). Locate a cluster of relatively large cells and identify some *acidophil cells,* which contain pink-stained granules, and some *basophil cells,* which contain blue-stained granules. These acidophil and basophil cells are the

415

FIGURE 39.2 **AP|R** Micrograph of the pituitary gland (6×).

— Hypothalamus

— Pituitary stalk

— Posterior lobe

— Anterior lobe

— Sphenoid bone

FIGURE 39.3 **AP|R** Micrograph of the anterior lobe of the pituitary gland (240×).

— Basophil

— Acidophils

FIGURE 39.5 **AP|R** Anterior view of thyroid gland and associated structures.

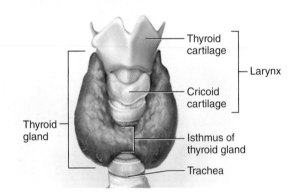

— Thyroid cartilage

— Larynx

— Cricoid cartilage

Thyroid gland —

— Isthmus of thyroid gland

— Trachea

FIGURE 39.4 **AP|R** Micrograph of the posterior lobe of the pituitary gland (240×).

— Unmyelinated nerve fibers

— Pituicytes (glial cells)

hormone-secreting cells. The hormones secreted include thyroid-stimulating hormone (TSH), adrenocorticotropic hormone (ACTH), prolactin (PRL), growth hormone (GH), follicle-stimulating hormone (FSH), and luteinizing hormone (LH).

d. Observe an area of the posterior lobe with high-power magnification (fig. 39.4). Note the numerous unmyelinated nerve fibers present in this lobe. Also locate some *pituicytes,* a type of glial cell, scattered among the nerve fibers. The posterior lobe stores and releases antidiuretic hormone (ADH) and oxytocin (OT), which are synthesized in the hypothalamus.

e. Prepare labeled sketches of representative portions of the anterior and posterior lobes of the pituitary gland in Part A of Laboratory Assessment 39.

Thyroid gland. Locate the thyroid gland and associated structures (fig. 39.5). To examine the thyroid tissue, follow these steps:

a. Use low-power magnification to observe the tissue (fig. 39.6). Note the numerous *follicles,* each of which consists of a layer of cells surrounding a colloid-filled cavity.

b. Observe the tissue using high-power magnification (fig. 39.6). The cells forming the wall of a follicle are simple cuboidal epithelial cells. These cells

FIGURE 39.6 **AP|R**   Micrograph of thyroid gland (300×).

Follicular cells secrete thyroid hormone (T₃ and T₄)

Colloid of thyroglobulin stores thyroid hormone

Parafollicular cells or C cells secrete calcitonin

Follicle

secrete thyroid hormone (T$_3$ and T$_4$). The parafollicular cells secrete calcitonin.

c. Prepare a labeled sketch of a representative portion of the thyroid gland in Part A of the laboratory assessment.

Parathyroid gland. Typically, four parathyroid glands are attached on the posterior surface of the thyroid gland. Locate the parathyroid glands and associated structures (fig. 39.7). To examine the parathyroid tissue, follow these steps:

a. Use low-power magnification to observe the tissue (fig. 39.8). The gland consists of numerous tightly packed secretory cells.
b. Switch to high-power magnification and locate two types of cells—a smaller form (chief cells) arranged in cordlike patterns and a larger form (oxyphil cells) that have distinct cell boundaries and are present in clusters. The *chief cells* secrete parathyroid hormone (PTH). The function of the *oxyphil cells* is not well understood.

c. Prepare a labeled sketch of a representative portion of the parathyroid gland in Part A of the laboratory assessment.

Adrenal gland. Locate an adrenal gland and associated structures (fig. 39.9). To examine the adrenal tissue, follow these steps:

a. Use low-power magnification to observe the tissue (fig. 39.10). Note the thin capsule that covers the gland. Just beneath the capsule, there is a relatively thick *adrenal cortex*. The central portion of the gland is the *adrenal medulla*. The cells of the cortex are in three poorly defined layers. Those of the outer layer (*zona glomerulosa*) are arranged irregularly; those of the middle layer (*zona fasciculata*) are in long cords; and those of the inner layer (*zona reticularis*) are arranged in an interconnected network of cords. The most noted corticosteroids secreted from the cortex include aldosterone from the zona glomerulosa, cortisol from the zona fasciculata, and

FIGURE 39.7   Posterior view of parathyroid glands and associated structures.

Pharynx

Lobes of thyroid gland

Parathyroid glands

Esophagus

Trachea

FIGURE 39.8   Micrograph of the parathyroid gland (65×).

Blood vessel

Oxyphil cells

Chief cells

Adipose cell

417

FIGURE 39.9 AP|R Adrenal gland: (a) frontal section with associated structures; (b) diagram of the layers of the adrenal cortex and adrenal medulla.

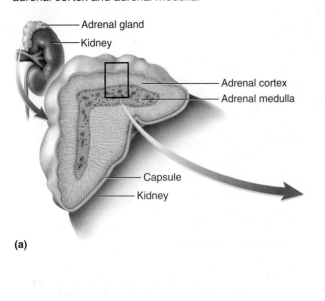

Adrenal gland
Kidney
Adrenal cortex
Adrenal medulla
Capsule
Kidney

(a)

Capsule
Zona glomerulosa
Adrenal cortex
Zona fasciculata
Zona reticularis
Adrenal medulla

(b)

FIGURE 39.10 AP|R Micrograph of the adrenal cortex and the adrenal medulla (75×).

Capsule
Zona glomerulosa
Zona fasciculata
Cortex
Zona reticularis
Medulla
Chromaffin cells

estrogens and androgens from the zona reticularis. The cells of the medulla (chromaffin cells) are relatively large, irregularly shaped, and often occur in clusters. The medulla cells secrete epinephrine and norepinephrine.

b. Using high-power magnification, observe each of the layers of the cortex and the cells of the medulla.

c. Prepare labeled sketches of representative portions of the adrenal cortex and medulla in Part A of the laboratory assessment.

Pancreas. Locate the pancreas and associated structures (fig. 39.11). To examine the tissues of the pancreas, follow these steps:

a. Use low-power magnification to observe the tissue (fig. 39.12). The gland largely consists of deeply stained exocrine cells arranged in clusters around secretory ducts. These exocrine cells (acinar cells) secrete pancreatic juice rich in digestive enzymes. There are circular masses of differently stained cells scattered throughout the gland. These clumps of cells constitute the *pancreatic islets (islets of Langerhans),* and they represent the endocrine portion of the pancreas. The *beta cells* secrete insulin, the *alpha cells* secrete glucagon, and the *delta cells* secrete somatostatin (growth hormone-inhibiting hormone).

b. Examine an islet, using high-power magnification. Unless special stains are used, it is not possible to distinguish alpha, beta, and delta cells.

c. Prepare a labeled sketch of a representative portion of the pancreas in Part A of the laboratory assessment.

4. Complete Parts B and C of the laboratory assessment.

FIGURE 39.11 AP|R (a) Pancreas and associated structures. (b) Diagram of a pancreatic islet.

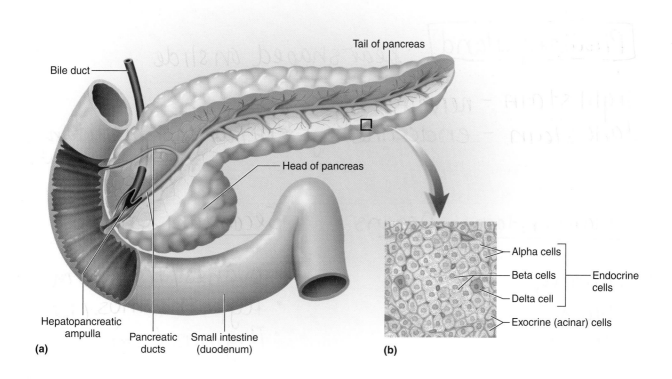

(a)

Bile duct

Tail of pancreas

Head of pancreas

Hepatopancreatic ampulla

Pancreatic ducts

Small intestine (duodenum)

(b)

Alpha cells

Beta cells — Endocrine cells

Delta cell

Exocrine (acinar) cells

FIGURE 39.12 AP|R Micrograph of the pancreas (200×).

Pancreatic islet (islet of Langerhans) composed of endocrine cells that secrete hormones into the blood

Exocrine (acinar) cells that secrete digestive enzymes into pancreatic ducts

2/2/17

Pituitary Gland   pear shaped on slide

lightstain - nervous system
dank stain - endocrine system (anterior pituitary)

Primary Endocrine Organs   Secondary Organs

• Pituitary   • Ovaries/Testosterone (R.)
• Thyroid   • Hypothalmus (N)
• Adrenal   • Thymus (Im)

• Pineal sands (pink/purple blotch) on slide
• Thyroid (pink staining lumen)

Pancreas   (posterior to the stomach)
   • Both endocrine/exocrine
      (insulin, glucagon)

Thymus

   • looks like piece of steak on slide
   • Hassal's body / Hassal's corpusele
         -swirl (very colorful)

(2/2/17)
picture 1 — Pineal Gland

picture 2 - Thymus Gland

Zona Glomerulosa
Fasciculata
Reticularis

Name _____

Date _____

Section _____

The (A) corresponds to the indicated Learning Outcome(s) (O) found at the beginning of the Laboratory Exercise.

Endocrine Structure and Function

PART A: Assessments

Sketch and label representative portions of the following endocrine glands: (A2)

Pituitary gland (_____×) (anterior lobe)	Pituitary gland (_____×) (posterior lobe)
Thyroid gland (_____×)	Parathyroid gland (_____×)
Adrenal gland (_____×) (cortex and medulla)	Pancreas (_____×)

PART B: Assessments

Match the endocrine gland in column A with a characteristic of the gland in column B. Place the letter of your choice in the space provided. (A1) (A3)

Column A	Column B
a. Adrenal cortex	_____ **1.** Located in sella turcica of sphenoid bone
b. Adrenal medulla	__*d.*__ **2.** Contains alpha, beta, and delta cells
c. Hypothalamus	_____ **3.** Contains colloid-filled cavities
d. Pancreatic islets	_____ **4.** Attached to posterior surface of thyroid gland
e. Parathyroid gland	_____ **5.** Secretes corticosteroids
f. Pituitary gland	_____ **6.** Attached to pituitary gland by a stalk
g. Thymus	_____ **7.** Gland inside another gland near kidneys
h. Thyroid gland	_____ **8.** Located in mediastinum

PART C: Assessments

Complete the following:

1. Name six hormones secreted by the anterior lobe of the pituitary gland. (A3) _____

   _____

2. Name two hormones secreted by the posterior lobe of the pituitary gland. (A3) _____

3. Name two thyroid hormones secreted from the follicular cells. (A3) _____

4. Name the hormone secreted from the parafollicular cells of the thyroid gland. (A3) _____

5. Name two hormones secreted by the adrenal medulla. (A3) _____

6. Name the most important corticosteroid secreted by the zona fasciculata cells of the adrenal cortex. (A3) _____

LESSON 19 | Endocrine Function
Thyroid Gland and Metabolic Rate

The endocrine system maintains homeostasis by secreting hormones that affect the actions of cells around the body. Metabolic reactions are a prime target of endocrine hormones, as demonstrated by the actions of thyroxine, the active hormone produced by the thyroid gland.

Thyroxine increases metabolic activity in nearly all the cells of the body. When body temperature decreases, thyroxine secretion increases. The increase in metabolic rate can be inferred from oxygen consumption. In this simulation, you will measure the rate of oxygen consumption, over a range of temperatures, of a normal mouse and a mouse that has been given propylthiouracil (PTU) to reduce thyroid gland function. You will create a graph that will clearly depict the effect of temperature on metabolic rate as controlled by the presence or absence of thyroxine secretion.

Before you begin, familiarize yourself with the following concepts (use your textbook as a reference):

- Thyroid gland
- Thyroxine
- Metabolic rate and oxygen consumption
- Metabolic rate and temperature regulation

State your hypothesis regarding the relationship between thyroxine secretion from the thyroid gland and metabolic rate.

_____

_____

_____

_____

1. Open Lesson 19 in Ph.I.L.S., Endocrine Function: Thyroid Gland and Metabolic Rate (fig. 39.13).

2. Read the objectives and introduction and take the pre-laboratory quiz.

3. After completing the pre-laboratory quiz, read through the wet lab.

4. The laboratory exercise will open when you click CONTINUE after completing the wet lab.

5. Print a copy of the final graph you recorded from the virtual computer screen (fig. 39.14).

6. Follow the instructions provided in the laboratory exercise. Using the graph you printed, answer the following questions as you proceed.

FIGURE 39.13 Opening screen for the laboratory exercise on Endocrine Function: Thyroid Gland and Metabolic Rate.

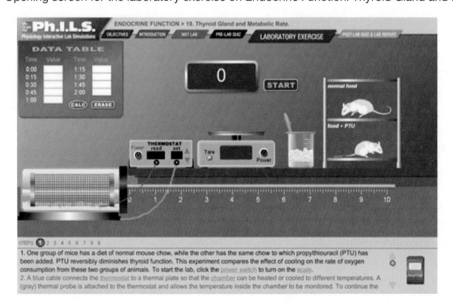

FIGURE 39.14 Example: Graph to show the effect of thyroid function and temperature on oxygen consumption in mice.

O₂/min

Temperature °C

Legend: Normal 24.2 g; +PTU 23.4 g

Interpreting Results

1. Using the graphs you printed, first use the recording for the "Norm" mouse to answer these questions.
   a. How many lines are on the graph? _____
   b. Examine the change in oxygen consumption as the environmental temperature decreases from 24°C to 18°C. Is oxygen consumption increasing, decreasing, or staying the same?
   _____
   c. Given the relationship of oxygen consumption and metabolic rate, metabolic rate is
   _____
   (increasing, decreasing, or staying the same).
   d. This part of the graph can be explained by
   _____
   (the release of thyroid hormone or shivering).
   e. Examine the change in oxygen consumption as the environmental temperature decreases from 18°C to 8°C. Is oxygen consumption increasing, decreasing, or staying the same?
   _____

   f. Given the relationship of oxygen consumption and metabolic rate, metabolic rate is
   _____
   (increasing, decreasing, or staying the same).
   g. This part of the graph can be explained by
   _____
   (the release of thyroid hormone or shivering).

2. Use the graph you generated for the thyroid hormone–deficient mouse (PTU treated) to answer the following questions. A4 A5 A6
   a. How many lines are on the graph? _____
   b. Examine the change in oxygen consumption as the environmental temperature decreases from 24°C to 18°C. Is oxygen consumption increasing, decreasing, or staying the same?
   _____
   c. Given the relationship of oxygen consumption and metabolic rate, metabolic rate is
   _____
   (increasing, decreasing, or staying the same).
   d. This part of the graph can be explained by
   _____
   (the impaired release of thyroid hormone or shivering).
   e. Examine the change in oxygen consumption as the environmental temperature decreases from 18°C to 8°C. Is oxygen consumption increasing, decreasing, or staying the same?
   _____

   f. Given the relationship of oxygen consumption and metabolic rate, metabolic rate is
   _____
   (increasing, decreasing, or staying the same).
   g. This part of the graph can be explained by
   _____
   (the impaired release of thyroid hormone or shivering).

3. Click the BOTH button to display the graphs generated for both animals. A4 A5 A6
   a. For both animals, explain the relationship of temperature and oxygen consumption.
   _____
   _____
   _____
   _____
   _____

When is the greatest amount of oxygen consumed—at cooler or warmer temperatures?

_____

b. Compare the red and black lines between 18°C and 24°C. Which is steeper? _____
Which animal has a greater oxygen consumption?

_____

Which animal has a higher metabolic rate?

_____

c. The point of intersection of the two black lines (results from the mouse with normal functioning thyroid gland) represents the point at which the animal begins to shiver. At what temperature does this occur? _____

d. The point of intersection of the two red lines (results from the mouse with malfunctioning thyroid gland) represents the point at which the animal begins to shiver. At what temperature does this occur? _____

e. Which animals begin to shiver first? _____
Explain. _____

_____

_____

_____

_____

CRITICAL THINKING ASSESSMENT A7

Would you predict your thyroid hormone level to increase, decrease, or stay the same in the winter? _____
Predict the results (increased, decreased, stay the same) for the following for a person with a hyperthyroid condition (overactive thyroid) and one with a hypothyroid condition (underactive thyroid):

Variable	Change Predicted with Hyperthyroidism	Change Predicted with Hypothyroidism
Thyroid hormone level		
Metabolic rate		
Respiratory rate		
Body temperature		
Body weight		

Which individual, one with a hyperthyroid or a hypothyroid condition, would be more likely to shiver when exposed to

decreases in environmental temperature? _____

NOTES

Diabetic Physiology

Purpose of the Exercise

To understand the disease of diabetes mellitus, to observe behavioral changes that occur during insulin shock and recovery from insulin shock, and to examine microscopically the pancreatic islets.

MATERIALS NEEDED

500 mL beakers

Live small fish (1" to 1.5" goldfish, guppy, rosy red feeder, or other)

Small fish net

Insulin (regular U-100) (HumulinR in 10 mL vials has 100 units/mL; store in refrigerator—do not freeze)

Syringes for U-100 insulin

10% glucose solution

Clock with second hand or timer

Compound light microscope

Prepared microscope slides of the following:

  Normal human pancreas (stained for alpha and beta cells)

  Human pancreas of a diabetic

SAFETY

▶ Review all the safety guidelines inside the front cover.

▶ Wear disposable gloves when handling the fish and syringes.

▶ Dispose of the used syringe and needle in the puncture-resistant container.

▶ Wash your hands before leaving the laboratory.

Learning Outcomes AP|R

After completing this exercise, you should be able to:

01 Identify the cells of the pancreatic islets and the hormones they produce and secrete.

02 Compare the causes, symptoms, and treatments for type 1 and type 2 diabetes mellitus.

03 Examine and record the behavioral changes caused by insulin shock.

04 Examine and record the recovery from insulin shock when sugar is provided to cells.

05 Distinguish tissue sections of normal pancreatic islets from those indicating diabetes mellitus, and sketch both types.

The O corresponds to the assessments A indicated in the Laboratory Assessment for this Exercise.

Pre-Lab

Carefully read the introductory material and examine the entire lab. Be familiar with diabetes mellitus from lecture or the textbook. Answer the pre-lab questions.

Pre-Lab Questions Select the correct answer for each of the following questions:

1. The _____ cells of pancreatic islets produce insulin.
   a. alpha
   b. beta
   c. delta
   d. F

2. Type 1 diabetes mellitus usually
   a. occurs in those over 40.
   b. afflicts 90% of diabetics.
   c. has a gradual onset.
   d. has a rapid onset.

3. Most of the insulin-secreting cells of a pancreas are
   a. scattered evenly throughout the pancreatic islet.
   b. near the periphery of the pancreatic islet.
   c. near the central portion of the pancreatic islet.
   d. among the exocrine cells of the pancreas.

4. Beta cells of a pancreatic islet occupy about _____ of the islet.
   a. 1%
   b. 5%
   c. 15%
   d. 80%

5. Which of the following has the *least* influence in the onset of type 2 diabetes mellitus?
   a. sleep patterns
   b. obesity
   c. heredity
   d. lack of exercise

The primary "fuel" for cells is sugar (glucose). Insulin produced by the beta cells of the pancreatic islets has a primary regulation role in the transport of blood sugar across the plasma membranes of most cells of the body. Some individuals do not produce enough insulin; others might not have adequate transport of glucose into the body cells when these target cells lose insulin receptors. This rather common disease is called *diabetes mellitus*. Diabetes mellitus is a functional disease that can have its onset either during childhood or later in life. (A different functional disease, *diabetes insipidus,* results from a deficiency of antidiuretic hormone with symptoms of copious urine production, but not hyperglycemia.)

The protein hormone insulin has several functions: it stimulates the liver and skeletal muscles to synthesize glycogen from glucose; it inhibits the conversion of noncarbohydrates into glucose; it promotes the facilitated diffusion of glucose across plasma membranes of target cells possessing insulin receptors (adipose tissue, cardiac muscle, and skeletal muscle); it decreases blood sugar (glucose); it increases protein synthesis; and it promotes fat storage in adipose cells. Normally, as a person's blood sugar increases during nutrient absorption after a meal, the rising glucose levels directly stimulate beta cells of the pancreas to secrete insulin. Insulin prevents sudden surges in blood glucose (hyperglycemia) by promoting glycogen synthesis in the liver and increasing glucose entry into muscle and adipose cells.

Between meals and during sleep, blood glucose levels decrease and insulin decreases. When insulin concentration decreases, more glucose is available to enter cells that lack insulin receptors. Brain cells, liver cells, kidney cells, and red blood cells either lack insulin receptors or express limited insulin receptors, and they absorb and utilize glucose without the need for insulin. These cells depend upon blood glucose concentrations to enable cellular respiration and ATP production. When insulin is given as a medication for diabetes mellitus, blood glucose levels may drop drastically (hypoglycemia) if the individual does not eat and/or has been vigorously exercising. This can have significant negative effects on the nervous system, resulting in *insulin shock*.

Type 1 diabetes mellitus usually has a rapid onset relatively early in life, but it can occur in adults. Because this type of diabetes mellitus occurs when insulin is no longer produced, it is sometimes referred to as *insulin-dependent diabetes mellitus (IDDM)*. Affecting about 10% to 15% of diabetics, this autoimmune disorder occurs when antibodies from the immune system destroy beta cells, resulting in a decrease in insulin production. The treatment for type 1 diabetes mellitus is to administer insulin, usually by injection.

Type 2 diabetes mellitus has a gradual onset, usually diagnosed in people over age 40. It is sometimes called *non-insulin-dependent diabetes mellitus (NIDDM)*. This disease results as body cells become less sensitive to insulin or lose insulin receptors and thus cannot respond to insulin even if insulin levels remain normal, a condition called *insulin resistance*. Risk factors for the onset of type 2 diabetes mellitus, which afflicts about 85% to 90% of diabetics, include heredity, obesity, and lack of exercise. Obesity is a major indicator for predicting type 2 diabetes; a weight loss of a mere 10 pounds will increase the body's sensitivity to insulin as a means of prevention. The treatments for type 2 diabetes mellitus include a diet that avoids foods that stimulate insulin production, weight control, exercise, and medications. (The causes and treatments for type 2 diabetes closely resemble those for coronary artery disease.)

PROCEDURE A: Insulin Shock

If too much insulin is present relative to the amount of glucose available to the cells, insulin shock can occur. In this procedure, insulin shock will be created in a fish by introducing insulin into the water with a fish. Insulin will be absorbed into the blood of the gills of the fish. When insulin in the blood reaches levels above normal, rapid hypoglycemia occurs from the increased insulin absorption; consequently, the brain cells obtain less of the much-needed glucose, and epinephrine is secreted. The cumulative effect of complications during insulin shock is a rapid heart rate, anxiety, sweating, mental disorientation, impaired vision, dizziness, convulsions, and possible unconsciousness. Although these complications take place in humans, some of them are visually noticeable in a fish that is in insulin shock. The most observable components of insulin shock in the fish are related to behavioral changes, including rapid and irregular movements of the entire fish, along with increased gill cover and mouth movements.

1. Reexamine the introduction to this laboratory exercise.
2. Complete the table in Part A of Laboratory Assessment 40.
3. A fish can be used to observe normal behavior, induced insulin shock, and recovery from insulin shock. The pancreatic islets are located in the pyloric ceca or other scattered regions in fish. The behavioral changes of fish when they experience insulin shock and recovery mimic those of humans. Observe a fish in 200 mL of aquarium water in a 500 mL beaker. Make your observations of swimming, gill cover (operculum), and mouth movements for 5 minutes. Record your observations in Part B of the laboratory assessment.
4. Add 200 units of room-temperature insulin slowly, using the syringe, into the water with the fish to induce insulin shock. Record the time or start the timer when the insulin is added. Watch for changes of the swimming, gill cover, and mouth movements as insulin diffuses into the blood at the gills. Signs of insulin shock might include rapid and irregular movements. Record

your observations of any behavioral changes and the time of the onset in Part B of the laboratory assessment.

5. After definite behavioral changes are observed, use the small fish net to move the fish into another 500 mL beaker containing 200 mL of a 10% room-temperature glucose solution. Record the time when the fish is transferred into this container. Make observations of any recovery of normal behaviors, including the amount of time involved. Record your observations in Part B of the laboratory assessment.

6. When the fish appears to be fully recovered, return it to its normal container. Dispose of the syringe according to directions from your laboratory instructor.

7. Complete Part B of the laboratory assessment.

PROCEDURE B: Pancreatic Islets

1. Examine a normal stained pancreas tissue under low-power and high-power magnification (fig. 40.1). Locate the clumps of cells that constitute the *pancreatic islets (islets of Langerhans)* that represent the endocrine portions of the organ. A pancreatic islet contains four distinct cells that secrete different hormones: *alpha cells,* which secrete glucagon, *beta cells,* which secrete insulin, *delta cells,* which secrete somatostatin, and *F cells,* which secrete pancreatic polypeptide. Some of the specially stained pancreatic islets allow for distinguishing the different types of cells (figs. 40.2 and 40.3). The normal pancreatic islet has about 80% beta cells concentrated in the central region of the islet, about 15% alpha cells near the periphery of an islet, about 5% delta cells, and a small number of F cells. There are complex interactions among the four hormones affecting blood sugar, but in general, glucagon will increase blood sugar, whereas insulin will lower blood sugar; thus, these two hormones are considered antagonistic in their function. Somatostatin acts as a paracrine secretion to inhibit alpha and beta cell secretions, and pancreatic polypeptide inhibits secretions from delta cells and digestive enzyme secretions from the pancreatic acini.

2. Prepare a labeled sketch of a normal pancreatic islet in Part C of the laboratory assessment. Include labels for alpha cells and beta cells.

FIGURE 40.1 AP|R Micrograph of a stained normal pancreas (200x)

— Pancreatic islet (Islet of Langerhans)

— Exocrine (acinar) cells

FIGURE 40.2 AP|R Micrograph of a specially stained pancreatic islet surrounded by exocrine (acinar) cells (500×). The insulin-containing beta cells are stained dark purple.

— Secretory granules of exocrine (acinar) cells

— Capillary

— Stained beta cells within pancreatic islet

FIGURE 40.3 AP|R Micrograph of a pancreatic islet using immunofluorescence stains to visualize alpha and beta cells. Only the areas of interest are visible in the fluorescence microscope.

Stained beta cells

Stained alpha cells

3. Examine the pancreas of a diabetic under high-power magnification (fig. 40.4). Locate a pancreatic islet and note the number of beta cells as compared to a normal number of beta cells in the normal pancreatic islet.

4. Prepare a labeled sketch of a pancreatic islet that shows some structural changes due to diabetes mellitus. Place the sketch in Part C of the laboratory assessment.

5. Complete Part C of the laboratory assessment.

FIGURE 40.4 Micrograph of pancreas showing indications of changes from diabetes mellitus (100×).

Exocrine (acinar) cells

Pancreatic islet

Areas of fibrosis and amyloid replacement of islet cells

Name _____

Date _____

Section _____

The corresponds to the indicated Learning Outcome(s) O found at the beginning of the Laboratory Exercise.

Diabetic Physiology

PART A: Assessments

Complete the missing parts of the table on diabetes mellitus: **A2**

Characteristic	Type 1 Diabetes	Type 2 Diabetes
Onset age	Early age or adult	
Onset of symptoms		Slow
Percentage of diabetics		85–90%
Natural insulin levels	Below normal	
Beta cells of pancreatic islets		Not destroyed
Pancreatic islet cell antibodies	Present	
Risk factors of having the disease	Heredity	
Typical treatments	Insulin administration	
Untreated blood sugar levels		Hyperglycemia

PART B: Assessments

1. Record your observations of normal behavior of the fish.

   a. Swimming movements: _____
   _____

   b. Gill cover movements: _____
   _____

   c. Mouth movements: _____
   _____

2. Record your observations of the fish after insulin was added to the water.

   a. Recorded time that insulin was added: _____
   b. Swimming movements: _____
   _____

   c. Gill cover movements: _____
   _____

   d. Mouth movements: _____
   _____

   e. Elapsed time until the insulin shock symptoms occurred: _____

3. Record your observations after the fish was transferred into a glucose solution. (A4)

    a. Recorded time that fish was transferred into the glucose solution: _____

    b. Describe any changes in the behavior of the fish that indicate a recovery from insulin shock. _____

    _____

    _____

    _____

    c. Elapsed time until indications of a recovery from insulin shock occurred: _____

PART C: Assessments

1. Match the pancreatic islet hormone in column A to its proper description in column B. Place the letter of your choice in the space provided. (Answers will be used more than once.) (A1)

Column A	Column B
a. Insulin	_____ **1.** Inhibits pancreatic digestive enzyme secretion
b. Glucagon	_____ **2.** Released in response to low blood sugar levels
c. Somatostatin	_____ **3.** Released in response to high blood sugar levels
d. Pancreatic polypeptide	_____ **4.** Inhibits or modulates alpha and beta cell secretions
	_____ **5.** Secreted by beta cells
	_____ **6.** Secreted by delta cells
	_____ **7.** Secreted by alpha cells
	_____ **8.** Secreted by F cells
	_____ **9.** Hyperglycemic hormone
	_____ **10.** Hypoglycemic hormone

2. Sketch and label representative portions of the following pancreatic islets: (A5)

         Pancreatic islet (_____×)   Pancreatic islet (_____×)

                (normal)   (showing changes from diabetes mellitus)

3. Describe the differences that you observed between a normal pancreatic islet and a pancreatic islet of a person with diabetes mellitus. **A5** _____

_____

_____

_____

CRITICAL THINKING ASSESSMENT

Justify the importance for a type 1 diabetic to regulate insulin administration, meals, and exercise. **A2**

_____

_____

_____

_____

_____

CRITICAL THINKING ASSESSMENT

A patient came to the hospital complaining of excessive urination and intense thirst and hunger. Clinical tests of the blood and urine were conducted to confirm diagnosis of diabetes mellitus. Explain what these clinical tests would show and why. **A2**

_____

_____

_____

_____

_____

NOTES

Blood Cells

Purpose of the Exercise

To review the characteristics of blood cells, to examine them microscopically, and to perform a differential white blood cell count.

MATERIALS NEEDED

Compound light microscope

Prepared microscope slides of human blood (Wright's stain)

Colored pencils

For Demonstration Activity:

Mammal blood other than human or contaminant-free human blood is suggested as a substitute for collected blood

Microscope slides (precleaned)

Sterile disposable blood lancets

Alcohol swabs (wipes)

Slide staining rack and tray

Wright's stain

Distilled water

Disposable gloves

For Learning Extension Activity:

Prepared slides of pathological blood, such as eosinophilia, leukocytosis, leukopenia, and lymphocytosis

SAFETY ⚠

▶ It is important that students learn and practice correct procedures for handling body fluids. Consider using either mammal blood other than human or contaminant-free blood that has been tested and is available from various laboratory supply houses. Some of the procedures might be accomplished as demonstrations only. If student blood is used, it is important that students handle only their own blood.

▶ Use an appropriate disinfectant to wash the laboratory tables before and after the procedures.

▶ Wear disposable gloves and safety glasses when handling blood samples.

▶ Clean the end of a finger with an alcohol swab before the puncture is performed.

▶ Use the sterile blood lancet only once.

▶ Dispose of used lancets and blood-contaminated items in an appropriate container (never use the wastebasket).

▶ Wash your hands before leaving the laboratory.

Learning Outcomes AP|R

After completing this exercise, you should be able to:

01 Identify and sketch red blood cells, five types of white blood cells, and platelets.

02 Describe the structure and function of red blood cells, white blood cells, and platelets.

03 Perform and interpret the results of a differential white blood cell count.

The ⓞ corresponds to the assessments Ⓐ indicated in the Laboratory Assessment for this Exercise.

Pre-Lab due 2/9/17

Carefully read the introductory material and examine the entire lab. Be familiar with <u>RBCs</u>, <u>WBCs</u>, and <u>platelets</u> from lecture or the textbook. Answer the pre-lab questions.

Pre-Lab Questions Select the correct answer for each of the following questions:

1. Which of the following have significant functions mainly during bleeding?
   - a. red blood cells
   - b. white blood cells
   - ⓒ platelets
   - d. plasma

2. Which of the following is among the agranulocytes?
   - ⓐ monocyte
   - b. neutrophil
   - c. eosinophil
   - d. basophil

3. Which white blood cell has the greatest nuclear variations?
   - a. monocyte
   - ⓑ neutrophil
   - c. eosinophil
   - d. basophil

4. A _____ lacks a nucleus. *erythrocytes*
   - ⓐ red blood cell
   - b. lymphocyte
   - c. monocyte
   - d. basophil

5. Which cell has a large nucleus that fills most of the cell?
   - a. red blood cell
   - b. platelet
   - c. eosinophil
   - ⓓ lymphocyte

6. Which leukocyte is the most abundant in a normal differential count?
   a. basophil
   b. monocyte
   c. neutrophil
   d. lymphocyte

7. Eosinophil numbers typically increase during allergic reactions.
   a. True ✓
   b. False _____

8. Erythrocytes are also called granulocytes because granules are visible in their cytoplasm when using Wright's stain.
   a. True _____
   b. False ✓

⚠ WARNING

Because of the possibility of blood infections being transmitted from one student to another if blood slides are prepared in the classroom, it is suggested that commercially prepared blood slides be used in this exercise. The instructor, however, may wish to demonstrate the procedure for preparing such a slide. Observe all safety procedures for this lab.

DEMONSTRATION ACTIVITY

To prepare a stained blood slide, follow these steps:

1. Obtain two precleaned microscope slides. Avoid touching their flat surfaces.
2. Thoroughly wash hands with soap and water and dry them with paper towels. Don disposable gloves except on the hand of the person with the finger to be lanced.
3. Cleanse the end of the middle finger with an alcohol swab and let the finger dry in the air.
4. Remove a sterile disposable blood lancet from its package without touching the sharp end.
5. Puncture the skin on the side near the tip of the middle finger with the lancet and properly discard the lancet.
6. Wipe away the first drop of blood with the alcohol swab. Place a drop of blood about 2 cm from the end of a clean microscope slide. Cover the lanced finger location with a bandage.
7. Use a second slide to spread the blood across the first slide, as illustrated in figure 41.1. Discard the slide used for spreading the blood in the appropriate container.
8. Place the blood slide on a slide staining rack and let it dry in the air.
9. Put enough Wright's stain on the slide to cover the smear but not overflow the slide. Count the number of drops of stain that are used.
10. After 2–3 minutes, add an equal volume of distilled water to the stain and let the slide stand for 4 minutes. From time to time, gently blow on the liquid to mix the water and stain.
11. Flood the slide with distilled water until the blood smear appears light blue.
12. Tilt the slide to pour off the water, and let the slide dry in the air.
13. Examine the blood smear with low-power magnification, and locate an area where the blood cells are well distributed. Observe these cells, using high-power magnification and then an oil immersion objective if one is available.

FIGURE 41.1 To prepare a blood smear: (a) place a drop of blood about 2 cm from the end of a clean slide; (b) hold a second slide at about a 30–45° angle to the first one, allowing the blood to spread along its edge; (c) push the second slide over the surface of the first so that it pulls the blood with it; (d) observe the completed blood smear. The ideal smear should be 1.5 inches in length, be evenly distributed, and contain a smooth, feathered edge.

Blood is a type of connective tissue whose cells are suspended in a liquid extracellular matrix called *plasma*. Plasma is composed of water, proteins, nutrients, electrolytes, hormones, wastes, and gases. The cells, or formed elements, are formed mainly in red bone marrow, and they include *erythrocytes* (*red blood cells; RBCs*), *leukocytes* (*white blood cells*; *WBCs*), and some cellular fragments called *platelets* (*thrombocytes*). The formed elements compose about 45% of the total blood volume; the plasma composes approximately 55% of the blood volume.

Red blood cells contain hemoglobin and transport gases between the body cells and the lungs, white blood cells defend the body against infections, and platelets play an important role in stoppage of bleeding (hemostasis).

Laboratories in modern hospitals and clinics use updated hematology analyzers for evaluations of the blood characteristics (fig. 41.2). The more traditional procedures that you perform in Laboratory Exercises 41, 42, and 43 will help you better understand the methodology and purpose of measuring these blood characteristics. On occasion, a doctor might question a blood test result from the hematology blood analyzer and request a retest of the blood. Additional verification of a test result might be performed using traditional procedures. A self-diagnosis should never be made as a result of a test conducted in the biology laboratory. Always obtain proper medical exams and treatments from medical personnel.

PROCEDURE A:
Types of Blood Cells

1. Refer to figures 41.3, 41.4, and 41.5 as an aid in identifying the various types of blood cells. Study the functions of the blood cells listed in table 41.1.

FIGURE 41.2 Modern hematology analyzer being used in the laboratory of a clinic.

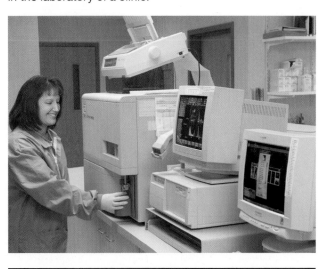

FIGURE 41.3 Red blood cells (erythrocytes) have a biconcave shape and function in the transport of oxygen and carbon dioxide.

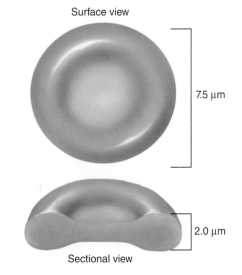

Surface view

7.5 μm

2.0 μm

Sectional view

FIGURE 41.4 AP|R Micrograph of a blood smear using Wright's stain (500×).

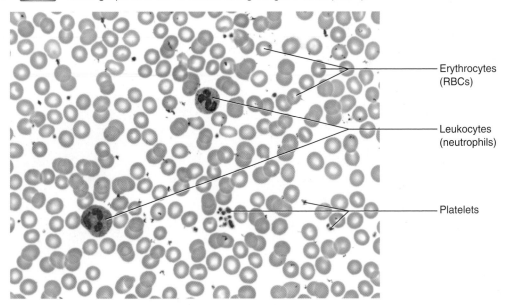

Erythrocytes (RBCs)

Leukocytes (neutrophils)

Platelets

FIGURE 41.5 Micrographs of blood cells illustrating some of the numerous variations of each type. Appearance characteristics for each cell pertain to a thin blood film using Wright's stain (1,000×).

Neutrophils (3 of many variations)
•Fine light-purple granules
•Nucleus single to five lobes (highly variable)
•Immature neutrophils, called bands, have a single C-shaped nucleus
•Mature neutrophils, called segs, have a lobed nucleus
•Often called polymorphonuclear leukocytes when older

Eosinophils (3 of many variations)
•Coarse reddish granules
•Nucleus usually bilobed

Basophils (3 of many variations)
•Coarse deep blue to almost black granules
•Nucleus often almost hidden by granules

Lymphocytes (3 of many variations)
•Slightly larger than RBCs
•Thin rim of nearly clear cytoplasm
•Nearly round nucleus appears to fill most of cell in smaller lymphocytes
•Larger lymphocytes hard to distinguish from monocytes

Monocytes (3 of many variations)
•Largest WBC; 2–3x larger than RBCs
•Cytoplasm nearly clear
•Nucleus round, kidney-shaped, oval, or lobed

Platelets (several variations)
•Cell fragments
•Single to small clusters

Erythrocytes (several variations)
•Lack nucleus (mature cell)
•Biconcave discs
•Thin centers appear almost hollow

Use the prepared slide of blood and locate each of the following:

red blood cell (erythrocyte)

white blood cell (leukocyte)

- granulocytes
  - neutrophil
  - eosinophil
  - basophil
- agranulocytes
  - lymphocyte
  - monocyte

platelet (thrombocyte)

2. In Part A of Laboratory Assessment 41, prepare sketches of single blood cells to illustrate each type. Pay particular attention to the relative size, nuclear shape, and color of granules in the cytoplasm (if present). The sketches should be accomplished using either the high-power

Table 41.1 Cellular Components of Blood

Component	Function
Red blood cell (erythrocyte)	Transports oxygen and carbon dioxide
White blood cell (leukocyte)	Destroys pathogenic microorganisms and parasites, removes worn cells, and provides immunity
Granulocytes—have granular cytoplasm	
1. Neutrophil	Phagocytizes bacteria
2. Eosinophil	Destroys parasites and helps control inflammation and allergic reactions
3. Basophil	Releases heparin (an anticoagulant) and histamine (a blood vessel dilator)
Agranulocytes—lack granular cytoplasm	
1. Monocyte	Phagocytizes dead or dying cells and microorganisms
2. Lymphocyte	Provides immunity; produces antibodies; destroys foreign cells and infected cells with viruses
Platelet (thrombocyte)	Helps control blood loss from injured blood vessels; needed for blood clotting

LEARNING EXTENSION ACTIVITY

Obtain a prepared slide of pathological blood that has been stained with Wright's stain. Perform a differential white blood cell count, using this slide, and compare the results with the values for normal blood listed in table 41.2. What differences do you note?

_____

_____

_____

_____

objective or the oil immersion objective of the compound light microscope.

3. Complete Part B of the laboratory assessment.

PROCEDURE B: Differential White Blood Cell Count

A differential white blood cell count is performed to determine the percentage of each of the various types of white blood cells present in a blood sample. The test is useful because the relative proportions of white blood cells may change in particular diseases, as indicated in table 41.2. Neutrophils, for example, usually increase during bacterial infections, whereas eosinophils may increase during certain parasitic infections and allergic reactions.

1. To make a differential white blood cell count, follow these steps:
   a. Using high-power magnification or an oil immersion objective, focus on the cells at one end of a prepared blood slide where the cells are well distributed.
   b. Slowly move the blood slide back and forth, following a path that avoids passing over the same cells twice (fig. 41.6).
   c. Each time you encounter a white blood cell, identify its type and record it in Part C of the laboratory assessment.
   d. Continue searching for and identifying white blood cells until you have recorded 100 cells in the data table. *Percent* means "parts of 100" for each type of white blood cell, so the total number observed is equal to its percentage in the blood sample.

2. Complete Part C of the laboratory assessment.

FIGURE 41.6 Move the blood slide back and forth to avoid passing the same cells twice.

Table 41.2 Differential White Blood Cell Count

Cell Type	Normal Value (percent)	Elevated Levels May Indicate
Neutrophil	54–62	Bacterial infections, stress
Lymphocyte	25–33	Mononucleosis, whooping cough, viral infections
Monocyte	3–9	Malaria, tuberculosis, fungal infections
Eosinophil	1–3	Allergic reactions, autoimmune diseases, parasitic worms
Basophil	<1	Cancers, chicken pox, hypothyroidism

(handwritten annotations: "large nucleus" under Monocyte; "orange stain" under Eosinophil)

NOTES

Name _____

Date _____

Section _____

The Ⓐ corresponds to the indicated Learning Outcome(s) Ⓞ found at the beginning of the Laboratory Exercise.

LABORATORY
ASSESSMENT
41

Blood Cells

PART A: Assessments

Sketch a single blood cell of each type in the following spaces. Use colored pencils to represent the stained colors of the cells. Label any features that can be identified. Ⓐ1

Red blood cell (____ ×)

Neutrophil (____ ×)

Lymphocyte (____ ×)

Monocyte (____ ×)

Eosinophil (____ ×)

Basophil (____ ×)

Platelet (____ ×)

PART B: Assessments

Complete the following statements:

1. Red blood cells (RBCs) are also called _____. A2

2. The shape of a red blood cell can be described as a _____ disc. A2

3. The functions of red blood cells are _____. A2

4. _____ is the oxygen-carrying substance in a red blood cell. A2

5. A mature red blood cell cannot reproduce because it lacks the _____ that was extruded during late development. A2

6. White blood cells are also called _____. A2

7. White blood cells with granular cytoplasm are called _____. A2

8. White blood cells lacking granular cytoplasm are called _____. A2

9. Polymorphonuclear leukocyte is another name for a _____ with a segmented nucleus. A2

10. Normally, the most numerous white blood cells are _____. A2

11. White blood cells with coarse reddish cytoplasmic granules are called _____. A2

12. _____ are normally the least abundant of the white blood cells. A2

13. _____ are the largest of the white blood cells. A2

14. _____ are small agranulocytes that have relatively large, round nuclei with thin rims of cytoplasm. A2

15. Small cell fragments that function to prevent blood loss from an injury site are called _____. A2

PART C: Assessments

1. *Differential White Blood Cell Count Data Table.* As you identify white blood cells, record them in the table by using a tally system, such as ⠍⠇⠞ ||. Place tally marks in the "Number Observed" column and total each of the five WBCs when the differential count is completed. Obtain a total of all five WBCs counted to determine the percentage of each WBC type. **A3**

Type of WBC	Number Observed	Total	Percent
Neutrophil			
Lymphocyte			
Monocyte			
Eosinophil			
Basophil			
		Total of column	

2. How do the results of your differential white blood cell count compare with the normal values listed in table 41.2? **A3**

_____

_____

_____

_____

_____

CRITICAL THINKING ASSESSMENT

Explain the difference between a differential white blood cell count and a total white blood cell count.

_____

_____

_____

_____

_____

3. Identify the blood cells indicated in figure 41.7.

FIGURE 41.7 Label the specific blood cells on this micrograph of a stained blood smear (400×).

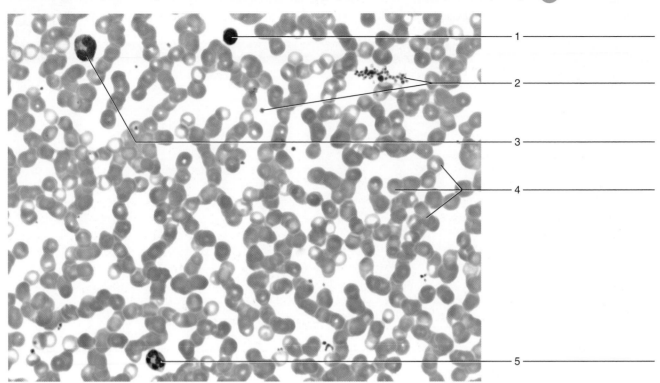

1 _____

2 _____

3 _____

4 _____

5 _____

LABORATORY
EXERCISE 42

Blood Testing

Purpose of the Exercise

To perform and interpret the blood tests used to determine hematocrit, hemoglobin content, coagulation, and cholesterol level. To measure and compare the percent saturation of hemoglobin when the partial pressure of oxygen and the pH are changed.

MATERIALS NEEDED

Mammal blood other than human or contaminant-free human blood is suggested as a substitute for collected blood

For Procedure A:

Heparinized microhematocrit capillary tube

Sealing clay (or Critocaps)

Microhematocrit centrifuge

Microhematocrit reader

For Procedure B:

Tallquist test kit

Hemoglobinometer

Lens paper

Hemolysis applicator

Sterile disposable blood lancets

Alcohol swabs (wipes)

Disposable gloves

For Procedure C:

Small triangular file

Capillary tube (nonheparinized)

Timer

For Procedure D:

Simulated blood kit for cholesterol determination*

* Kit contains all materials needed and is available from several biological supply companies.

Ph.I.L.S. 4.0: Lesson 34 Blood: pH and Hb-Oxygen Binding

SAFETY ⚠

▸ It is important that students learn and practice correct procedures for handling body fluids. Consider using either mammal blood other than human or contaminant-free blood that has been tested and is available from various laboratory supply houses. Some of the procedures might be accomplished as

demonstrations only. If student blood is used, it is important that students handle only their own blood.

▸ Use an appropriate disinfectant to wash the laboratory tables before and after the procedures.

▸ Wear disposable gloves and safety goggles when handling blood samples.

▸ Clean the end of a finger with alcohol swabs before the puncture is performed.

▸ Use the sterile blood lancet only once.

▸ Dispose of used lancets and blood-contaminated items in an appropriate container (never use the wastebasket).

▸ Wash your hands before leaving the laboratory.

Learning Outcomes

After completing this exercise, you should be able to:

01 Determine the hematocrit, hemoglobin, coagulation, and cholesterol level in a blood sample.

02 Evaluate the results and significance of the blood tests compared to normal values.

03 Identify the blood tests performed in this exercise that could indicate anemia.

04 Construct a graph and explain the relationship of the percent transmittance of light to the partial pressure of oxygen.

05 Explain the relationship of percent transmittance of light to percent saturation of hemoglobin.

06 Explain the relationship of partial pressure of oxygen to percent saturation of hemoglobin.

07 Interpret the influence of pH on the oxygen dissociation curve.

08 Apply the concepts of how pH affects the percent saturation of hemoglobin to the human body.

The **O** corresponds to the assessments **A** indicated in the Laboratory Assessment for this Exercise.

Pre-Lab *due 2/9/17*

Carefully read the introductory material and examine the entire lab. Be familiar with hematocrit, hemoglobin, coagulation, and cholesterol from lecture or the textbook. Visit www.mhhe.com/martinseries3 for LabCam videos. Answer the pre-lab questions.

Pre-Lab Questions Select the correct answer for each of the following questions:

1. The _____ is the percentage of red blood cells in whole blood.
   a. hemoglobin **(b.)** hematocrit **c.** buffy coat **d.** coagulation

2. Oxyhemoglobin is formed when oxygen binds to
   a. the hematocrit.
   b. the red blood cell membrane.
   c. globin of hemoglobin.
   d. iron of hemoglobin.

3. Coagulation culminates in the formation of insoluble threads of
   a. fibrin.
   b. fibrinogen.
   c. thrombin.
   d. platelets.

4. Coagulation should normally occur within
   a. 10 seconds.
   b. 1 minute.
   c. 8 minutes.
   d. 1 hour.

5. In order to determine the hematocrit, blood is placed in a
   a. coagulation tube.
   b. hemoglobinometer.
   c. centrifuge.
   d. movable slide.

6. A normal blood total cholesterol level is less than 200 mg/dL.
   a. True _____
   b. False _____

7. The high-density lipoproteins (HDLs) are sometimes considered the "bad" type of cholesterol.
   a. True _____
   b. False _____

As an aid in identifying various disease conditions, tests are often performed on blood to determine how its composition compares with normal values. These tests commonly include measuring the hematocrit (red blood cell percentage), hemoglobin content, coagulation, and cholesterol.

When a blood sample is collected in a heparinized (to prevent coagulation) capillary tube and left standing, the heavier cellular components settle to the bottom of the tube. Spinning the tube in a centrifuge accelerates this process. The red layer at the bottom portion of the tube represents the compacted red blood cells (RBCs). The percentage of compacted RBCs in the blood sample is called the *hematocrit,* normally about 45% of the entire volume. A thin, whitish layer, known as the *buffy coat,* forms on the top of the compacted RBC layer. This contains the white blood cells and platelets, and it represents less than 1% of the total volume. The remaining straw-colored upper portion of the contents contains the plasma, the remaining 55% of the volume.

The contents within the red blood cell (RBC) are about one-third the protein *hemoglobin (Hb)*. When you take a breath of air into your lungs, the inspired oxygen diffuses across the respiratory membrane of the alveoli and capillary beds of your lungs into your blood, where it enters an RBC and binds to a hemoglobin molecule. The hemoglobin molecule is composed of four protein subunits (two alpha and two beta), each containing a heme molecule with an atom of iron (Fe). Each iron binds to one oxygen molecule (O_2). As a result, each hemoglobin molecule can bind to as many as four oxygen molecules. When oxygen binds to hemoglobin, it forms a bright red *oxyhemoglobin* and is transported to systemic capillaries, where it is released from the hemoglobin to diffuse into your body tissues. When oxygen detaches from the hemoglobin in the capillaries of the tissues, it becomes a darker red and is called *deoxyhemoglobin,* which can appear bluish when viewed through the skin's blood vessel walls.

The amount of oxygen bound to hemoglobin molecules is expressed as the *percent (%) saturation of hemoglobin* (i.e., what percentage of the iron-binding sites of hemoglobin is bound with oxygen). The percent saturation of hemoglobin is dependent on a number of variables, the most important being the partial pressure of oxygen (similar to concentration of oxygen in the air). The relationship of percent saturation of hemoglobin to partial pressure of oxygen can be graphed to produce an *oxygen dissociation curve.* Other variables that influence percent saturation of hemoglobin include temperature, carbon dioxide levels, and pH. The change of any of these variables will trigger a "shift" in the oxygen dissociation curve. A *"shift right"* indicates that for any given partial pressure of oxygen, there is a decrease in the percent saturation of hemoglobin. A *"shift left"* indicates that for any given partial pressure of oxygen, there is an increase in the percent saturation of hemoglobin. For example, during physical activity, there is more oxygen being consumed and more carbon dioxide and heat being produced by your body's tissues, which would result in the percent saturation of hemoglobin shifting to the right, unloading more oxygen.

A *complete blood count (CBC)* measures white blood cell count, red blood cell count (including average RBC size), hematocrit, and hemoglobin concentration (including Hb amount per RBC). A CBC is used to detect or monitor various health conditions. An increased hematocrit may be a result of dehydration, living at higher altitudes, smoking, emphysema, or the abuse of erythropoietin (EPO) by athletes for blood doping purposes. Decreased values of hematocrits or hemoglobin levels could be attributed to anemia, hemorrhage, dietary deficiencies, kidney disease where EPO production is impaired, infections, heredity, bone marrow cancer, or exposure to ionizing radiation. Anemia is a condition that has many causes, but the consequences (low oxygen levels, low blood osmolarity, and low blood viscosity) put more stress on the cardiovascular system and may result in cardiac failure.

Stoppage of bleeding, called *hemostasis,* involves three defense mechanisms: blood vessel spasm, platelet plug formation, and *coagulation.* The coagulation mechanism involves a series of chain reactions that result in a blood clot. The final stage of blood clot formation occurs when thrombin converts a soluble plasma protein called fibrinogen into insoluble threads of *fibrin.* The fibrin mesh traps cellular components of the blood, stopping the blood loss. Once blood clotting is initiated, it usually takes 3–8 minutes for fibrin to form. Clotting deficiencies could be attributed to low platelet count, leukemia, hemophilia, liver disease, malnutrition, exposure to ionizing radiation, or anticoagulant drugs.

Blood cholesterol is an important component needed for the structure of cellular membranes and the formation of steroid hormones and bile components. However, blood cholesterol higher than recommended levels may increase the chances of cardiovascular diseases and risks of heart attacks and strokes. Hereditary factors are the primary determinant of total blood cholesterol levels. High-cholesterol diets and lack of exercise are among factors that contribute to high cholesterol levels. Low-cholesterol diets, an increase in exercise, and various medications are some ways used to lower the levels of cholesterol.

The clinical significance or conclusive diagnosis of a disease is often difficult to achieve. Most often, several blood and urine characteristics along with other symptoms are assessed collectively in order to make a determination of the abnormality and its potential cause. Any diagnosis should always be obtained by proper medical exams and treatments from a medical professional.

 WARNING

Because of the possibility of blood infections being transmitted from one student to another during blood-testing procedures, it is suggested that the following procedures be performed as demonstrations by the instructor. Observe all safety procedures listed for this lab.

PROCEDURE A: Hematocrit

To determine the hematocrit (percentage of red blood cells) of a whole blood sample, the cells must be separated from the liquid plasma. This separation can be quickly accomplished by placing a tube of blood in a centrifuge. The force created by the spinning motion of the centrifuge causes the cells to be packed into the lower end of the tube. The quantities of cells and plasma can then be measured, and the hematocrit can be calculated. The normal range for men is 42–52%; the normal range for women is 37–47%.

1. Review all of the safety procedures inside the front cover of your laboratory manual and for Laboratory Exercise 42 related to using human blood. To determine the hematocrit in a blood sample, follow these steps:
   a. Thoroughly wash hands with soap and water and dry them with paper towels. Don disposable gloves except on the hand of the person with the finger to be lanced. (If purchased blood is used for this procedure, skip to step 1e.)
   b. Cleanse the end of the middle finger with an alcohol swab and let the finger dry in the air.
   c. Remove a sterile disposable blood lancet from its package without touching the sharp end.
   d. Puncture the skin on the side near the tip of the middle finger with the lancet and properly discard

FIGURE 42.1 Steps of the hematocrit (red blood cell percentage) procedure: (a) load a heparinized capillary tube with blood; (b) plug the blood end of the tube with sealing clay; (c) place the tube in a microhematocrit centrifuge.

(a)

(b)

(c)

the lancet. Wipe away the first drop of blood with the alcohol swab.
   e. Touch a drop of blood with the colored end of a heparinized capillary tube. Hold the tube tilted slightly upward so that the blood will easily move into it by capillary action (fig. 42.1a). To prevent an air bubble, keep the tip in the blood until filled.
   f. Allow the blood to fill about two-thirds of the length of the tube. Cover the lanced finger location with a bandage.
   g. Hold a finger over the tip of the dry end so that blood will not drain out while you seal the blood end. Plug the blood end of the tube by pushing it with a rotating motion into sealing clay or by adding a plastic Critocap (fig. 42.1b).

h. Place the sealed tube into one of the numbered grooves of a microhematocrit centrifuge. The tube's sealed end should point outward from the center and should touch the rubber lining on the rim of the centrifuge (fig. 42.1c).

i. The centrifuge should be balanced by placing specimen tubes on opposite sides of the moving head, the inside cover should be tightened, and the outside cover should be securely fastened.

j. Run the centrifuge for 3–5 minutes.

k. After the centrifuge has stopped, remove the specimen tube. The red blood cells have been packed into the bottom of the tube. The clear liquid on top of the cells is plasma.

l. Use a microhematocrit reader to determine the percentage of red blood cells in the tube. If a microhematocrit reader is not available, measure the total length of the blood column in millimeters (red cells plus plasma) and the length of the red blood cell column alone in millimeters. Divide the red blood cell length by the total blood column length and multiply the answer by 100 to calculate the percentage of red blood cells.

2. Record the test result in Part A of Laboratory Assessment 42.

PROCEDURE B: Hemoglobin Content

Hemoglobin (Hb) is responsible for binding oxygen in the RBC. It is possible that anemia may be a result of low Hb even if the RBC count is normal (normocytic, hypochromic anemia). The hemoglobin content of a blood sample can be measured in several ways, including the *Tallquist method* and the *hemoglobinometer method,* both included in this laboratory exercise.

The Tallquist method uses a special test paper and is a simple, less expensive test that can be done more quickly than the hemoglobinometer method. The hemoglobinometer instrument is designed to compare the color of light passing through a hemolyzed blood sample with a standard color. The results of these tests are expressed in grams of hemoglobin per 100 mL (g/dL). The normal range for men is 13–18 g/dL; the normal range for women is 12–16 g/dL.

Tallquist Method

1. To measure the hemoglobin content of a blood sample, follow these steps:

a. Gather one piece of paper from the Tallquist booklet/kit, a Tallquist hemoglobin scale, and a blood sample.

b. Obtain a drop of blood from a finger by following the directions in Procedure A.

c. Place a large drop of blood on the center of the Tallquist paper so that the drop of blood is larger than the holes in the Tallquist hemoglobin scale.

d. Let the blood dry only long enough to lose its glossy appearance. (It should not dry to a brown color.)

FIGURE 42.2 A simple method of measuring the amount of hemoglobin blood is to compare a piece of Tallquist paper that has been saturated with a sample of blood with a Tallquist color scale.

e. Match the color, using natural light, of the blood on the Tallquist paper with the most accurate match of the color scale provided with the kit (fig. 42.2).

2. Record the test result in Part A of the laboratory assessment.

Hemoglobinometer Method

1. To measure the hemoglobin content of a blood sample, follow these steps:

a. Obtain a hemoglobinometer and remove the blood chamber from the slot in its side.

b. Separate the pieces of glass from the metal clip and clean them with alcohol swabs and lens paper. One of the pieces of glass has two broad, U-shaped areas surrounded by depressions. The other piece is flat on both sides.

c. Obtain a large drop of blood from a finger, by following the directions in Procedure A.

d. Place the drop of blood on one of the U-shaped areas of the blood chamber glass (fig. 42.3a).

e. Stir the blood with the tip of a hemolysis applicator until the blood appears clear rather than cloudy. This usually takes about 45 seconds (fig. 42.3b).

f. Place the flat piece of glass on top of the blood plate and slide both into the metal clip of the blood chamber.

g. Push the blood chamber into the slot on the side of the hemoglobinometer, making sure that it is in all the way (fig. 42.3c).

h. Hold the hemoglobinometer in the left hand with the thumb on the light switch on the underside (fig. 42.3d).

i. Look into the eyepiece and note the green area split in half.

j. Slowly move the slide on the side of the instrument back and forth with the right hand until the two halves of the green area look the same.

FIGURE 42.3 Steps of the hemoglobin content procedure: (a) load the blood chamber with blood; (b) stir the blood with a hemolysis applicator; (c) place the blood chamber in the slot of the hemoglobinometer; (d) match the colors in the green area by moving the slide on the side of the instrument.

(a)

(b)

(c)

(d)

k. Note the value in the upper scale (grams of hemoglobin per 100 mL of blood), indicated by the mark in the center of the movable slide.

2. Record the test result in Part A of the laboratory assessment.

PROCEDURE C: Coagulation

Coagulation time, often called clotting time, is the time from the onset of bleeding until the clot is formed. Injured tissues and platelets release chemicals that trigger the clotting cascade, which ultimately results in thrombin converting soluble fibrinogen into insoluble fibrin. Fibrin is a mesh that traps RBCs and forms the clot (fig. 42.4). Clotting time normally ranges from 3 to 8 minutes. This process is prolonged if the person has clotting deficiencies or is being treated with anticoagulants such as heparin (some is produced by basophils and mast cells), warfarin (Coumadin), or aspirin. In this laboratory exercise we will try to determine the time to the nearest minute.

FIGURE 42.4 Scanning electron micrograph of a blood clot (2,800x). Note that the red blood cells are trapped within a mesh of fibrin threads.

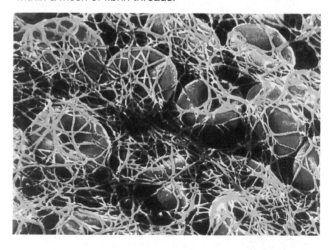

FIGURE 42.5 Steps of the coagulation procedure:
(a) clean break of capillary tube before any fibrin formed;
(b) fibrin spans between the broken capillary tube
segments when coagulation occurs.

(a)

Clot containing fibrin

(b)

1. To determine coagulation time, follow these steps:
   a. Prepare the finger to be lanced by following the directions in Procedure A. Lance the end of a finger to obtain a drop of blood. Wipe away the first drop of blood with the alcohol swab.
   b. Touch a drop of blood with one end of a nonheparinized capillary tube. Hold the tube tilted slightly upward so that the blood will easily move into it by capillary action (fig. 42.1a). Keep the tip in the blood until the tube is nearly filled. If the tube is nearly filled, it will allow enough tube length for breaking it several times.
   c. Place the capillary tube on a paper towel. Cover the lanced finger location with a bandage. Record the time: _____

d. At 1-minute intervals, use the small triangular file and make a scratch on the capillary tube starting near one end of the tube. Hold the tube with fingers on each side of the scratch, the weakened location of the tube, and break the tube away from you, being careful to keep the two pieces close together after the break. Gently pull the two ends of the tube apart while observing carefully to see if it breaks cleanly. If it breaks cleanly, fibrin has not formed yet (fig. 42.5a).
   e. Continue breaking the capillary tube each minute until fibrin is noted spanning the two parts of the capillary tube (fig. 42.5b). Note the time for coagulation.
2. Record the test results in Part A of the laboratory assessment.

PROCEDURE D: Blood Cholesterol Level Determination

The recommended total cholesterol level is less than 200 mg/dL (dL = deciliter), because levels higher than this have been correlated with increased probability of cardiovascular disease. Total blood cholesterol of less than 100 (hypocholesterolemia) is associated with such possible conditions as malnutrition, hyperthyroidism, and depression. Since cholesterol is not water soluble, it is transported in the plasma incorporated with a protein as a low-density lipoprotein (LDL) and a high-density lipoprotein (HDL). The **HDLs** are transported to the liver, where they are degraded and removed from the body, and are therefore sometimes referred to as the "**H**ealthy/good" form. The LDLs are transported to our tissue cells for cellular needs, but if the levels are excessive, there is an increased occurrence of atherosclerosis. Atherosclerosis is the deposition of fatty material within blood vessel walls, resulting in impeded blood flow and increased work by the heart. **LDLs** are sometimes referred to as the "**L**ousy/bad" cholesterol. The recommended levels of HDLs are > 40 mg/dL; recommended levels of LDLs are < 130 mg/dL.

1. Each kit contains cholesterol samples of simulated blood, test strips, color charts, and directions for the particular kit being used. Some available kits contain samples that reflect the use of cholesterol-lowering medications, while others might contain samples that represent diets of various cholesterol levels. Dip a test strip into a sample; then compare the strip to a color chart to determine the cholesterol level of the sample.
2. Record the test results in Part A of the laboratory assessment.
3. Complete Part B of the laboratory assessment that pertains to all procedures performed in this laboratory exercise.

Name _____

Date _____

Section _____

The (A) corresponds to the indicated Learning Outcome(s) (O) found at the beginning of the Laboratory Exercise.

Blood Testing

PART A: Assessments

Blood test data:

Blood Test	Test Results	Normal Values
Hematocrit (mL per 100 mL blood)		Men: 42–52% Women: 37–47%
Hemoglobin content (g per 100 mL blood; g/dL)		Men: 13–18 g/dL Women: 12–16 g/dL
Coagulation		3–8 minutes
Cholesterol	Before reduction program: After reduction program:	Below 200 mg/dL

PART B: Assessments

Complete the following:

1. How does the hematocrit from the blood test compare with the normal value? If the test results were abnormal, what condition would it possibly indicate and why? (A2) _____

_____

_____

2. How does the hemoglobin content from the blood test compare with the normal value? If the test results were abnormal, what condition would it possibly indicate and why? (A2) _____

_____

_____

3. How does the coagulation time from the blood test compare with the normal value? If the test results were abnormal, what condition would it possibly indicate and why? (A2) _____

_____

_____

4. Assess the most beneficial method of lowering cholesterol from samples provided in the simulated blood kit. Explain your reasoning. (A2) _____

_____

_____

CRITICAL THINKING ASSESSMENT

A patient who is taking warfarin (Coumadin) must go off this medicine a few days prior to having surgery. Referencing your laboratory activity on coagulation, explain why this protocol is important. (A2)

_____

_____

_____

_____

CRITICAL THINKING ASSESSMENT

Which blood tests performed in this lab could be used to determine possible anemia? (A3)

_____

_____

LESSON 34 | # Blood
pH and Hb-Oxygen Binding

Oxygen is essential to all tissues for the production of ATP; it is transported throughout the body by red blood cells (RBCs). The RBCs contain the protein hemoglobin (you determined the concentration of Hb in a sample earlier in Procedure B of this lab) that binds oxygen for transport to the tissues. Factors such as partial pressue, temperature, pH, and levels of carbon dioxide can affect the ability of Hb to bind oxygen. In this exercise, you will analyze the effect of various pH conditions on Hb binding.

Hb globin samples were prepared for this lab by inducing hemolysis of the RBCs and centrifuging the samples to remove the membranes. Supernatants were then collected and diluted with phosphate buffers of pH 6.8, 7.4, and 8.0. Using a spectrophotometer, you will measure the transmittance of light (620 nm) through the samples. The more oxygen that is bound to Hb (reduced, oxyhemoglobin), the more light will be absorbed and the less will pass through the sample. The less oxygen bound to Hb (oxidized, deoxyhemoglobin) the lighter in color the blood will be and the less light will be absorbed and the more will pass through the sample.

Before you begin, familiarize yourself with the following concepts (use your textbook as a reference):

- Deoxyhemoglobin and oxyhemoglobin
- Oxygen dissociation curve
- pH effect on Hb binding

State your hypothesis regarding the effect of pH on hemoglobin's ability to bind oxygen.

_____

_____

_____

_____

1. Open Lesson 34 in Ph.I.L.S., Blood: pH and Hb–Oxygen Binding (fig. 42.6).

2. Read the objectives and introduction and take the pre-laboratory quiz.

3. After completing the pre-laboratory quiz, read through the wet lab.

FIGURE 42.6 Opening screen for the laboratory exercise on Blood: pH and Hb-Oxygen Binding.

FIGURE 42.7 Example: Graph to show the effect of pH on the color of sheep hemoglobin set at different partial pressures of oxygen. Lower color values indicate that less oxygen is bound to hemoglobin.

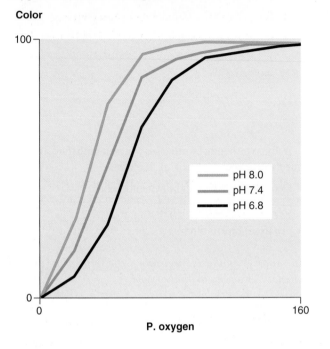

Color

4. The laboratory exercise will open when you click CONTINUE after completing the wet lab.

5. Print a copy of the final graph you recorded from the virtual computer screen (fig. 42.7).

6. Follow the instructions provided in the laboratory exercise. Using the graph you printed, answer the following questions as you proceed.

7. Use the data from the computer simulation collected at a pH of 7.4; graph the partial pressure of oxygen and the percent transmittance of light on the graph below. (Include the title. Label each axis with units. Plot the points and connect). **A4**

8. Examine the graph you produced in question 1. As the partial pressure of oxygen increases, the percent transmittance of light _____. **A4**

9. There is a direct relationship between percent transmittance of light and percent saturation of hemoglobin. So, as the percent transmittance of light increases, the percent saturation of hemoglobin _____ (increases or decreases). Thus, as the partial pressure of oxygen increases, the percent saturation of hemoglobin _____ (increases, decreases, or stays the same). **A5** **A6**

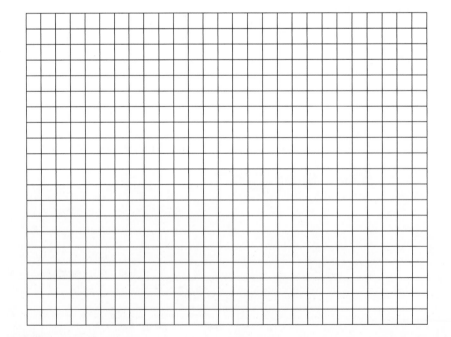

10. Compare the effect of pH on percent transmittance of light by examining the three dissociation curves produced at a pH of 6.8, 7.4, and 8.0 and answer the following questions: **A7**

    a. When comparing the curve produced at 6.8 to the curve produced at 7.4, the curve produced when the pH is 6.8 would be described as _____ (a "shift right" or a "shift left") and at any given partial pressure of oxygen, the percent transmittance of light is _____ (higher or lower) and thus, the percent saturation of hemoglobin would be _____ (higher or lower).

    b. At a given PO_2, hemoglobin has the highest affinity for oxygen (ability to bind oxygen) at this pH _____ and the lowest affinity for oxygen at this pH _____.

CRITICAL THINKING ASSESSMENT **A8**

Hydrogen ion (H^+) levels increase within the body from increased production of lactic acid (during anaerobic cellular respiration) and from increased production of carbon dioxide ($CO_2 + H_2O \longleftrightarrow H^+ + HCO_3^-$). The increased amounts of H^+ bind with the amino acids that compose the protein of hemoglobin, slightly altering hemoglobin's shape. This change in shape interferes with the ability of iron within the hemoglobin molecule to bind the oxygen (decreased affinity or attraction). (This is referred to as the Bohr effect.)

If the tissues of the body are metabolically active, _____ (more or less) H^+ are produced and the pH is _____ (decreased or increased). As a result of the Bohr effect, when hemoglobin reaches the systemic capillaries, the hemoglobin will bind the oxygen with _____ (greater or less) affinity; _____ (more or less) oxygen is released to the cells; the saturation of hemoglobin will be _____ (greater or less), and hemoglobin returns to the lungs with _____ (more or less) oxygen than would occur at a normal pH.

NOTES

Blood Typing

Purpose of the Exercise

To determine and explain the significance of the ABO blood type of a blood sample and an Rh blood-typing test.

MATERIALS NEEDED

For Procedure A:

ABO blood-typing kit

Simulated blood-typing kits are suggested as a substitute for collected blood

For Procedure B:

Microscope slide

Alcohol swabs (wipes)

Sterile blood lancet

Disposable gloves

Toothpicks

Anti-D serum

Slide warming table (Rh blood-typing box)

SAFETY

▶ It is important that students learn and practice correct procedures for handling body fluids. Consider using simulated blood-typing kits or contaminant-free blood that has been tested and is available from various laboratory supply houses. Some of the procedures might be accomplished as demonstrations only. If student blood is used, it is important that students handle only their own blood.

▶ Use an appropriate disinfectant to wash the laboratory tables before and after the procedures.

▶ Wear disposable gloves and safety goggles when handling blood samples.

▶ Clean the end of a finger with an alcohol swab before the puncture is performed.

▶ Use the sterile blood lancet only once.

▶ Dispose of used lancets and blood-contaminated items in an appropriate container (never use the wastebasket).

▶ Wash your hands before leaving the laboratory.

Learning Outcomes

After completing this exercise, you should be able to:

O1 Explain the fundamentals of the ABO blood group and possible adverse reactions from transfusions.

O2 Interpret the ABO type of a blood sample.

O3 Analyze the basis and importance of the Rh blood factor.

O4 Interpret the Rh type of a blood sample.

The **O** corresponds to the assessments **A** indicated in the Laboratory Assessment for this Exercise.

Pre-Lab

Carefully read the introductory material and examine the entire lab. Be familiar with the ABO blood group and the Rh blood group from lecture or the textbook. Answer the pre-lab questions.

Pre-Lab Questions Select the correct answer for each of the following questions:

1. The antigens related to the ABO blood group are located
   a. on the red blood cell membrane.
   b. within the red blood cell nucleus.
   c. within the red blood cell cytosol.
   d. on the red blood cell ribosome.

2. The D antigen related to the Rh factor is present in about _____ of Americans.
   a. 4%   **b.** 38%
   c. 47%   **d.** 85%

3. Blood type _____ is sometimes considered the universal donor within the ABO blood group.
   a. A   **b.** B
   c. AB   **d.** O

4. Blood type _____ is sometimes considered the universal recipient within the ABO blood group.
   a. A   **b.** B
   c. AB   **d.** O

5. Hemolytic disease of the newborn could be of concern when an
   a. Rh-positive fetus and an Rh-positive mother condition occur.
   b. Rh-positive fetus and an Rh-negative mother condition occur.
   c. Rh-negative fetus and an Rh-positive mother condition occur.
   d. Rh-negative fetus and an Rh-negative mother condition occur.

6. Blood type B is the most common blood type found in the United States population.
   a. True _____   **b.** False ✓_____

457

7. An individual with blood type O lacks both RBC antigens A and B.

a. True __✓__   **b.** False _____

Blood typing involves identifying substances called *antigens* present on the outer surface of red blood cell membranes. Although many different antigens are associated with human red blood cells, only a few of them are of clinical importance. These include the antigens of the *ABO group* and those of the *Rh group*.

To determine which antigens are present, a blood sample is mixed with blood-typing sera that contain known types of antibodies. If a particular antibody contacts a corresponding antigen, a reaction occurs, and the red blood cells clump together (*agglutination*). Thus, if blood cells are mixed with serum containing antibodies that react with antigen A and the cells agglutinate, antigen A must be present on the outer surface of those cells.

Although there are many antigens on the red blood cell membranes, the ABO group and the Rh group antigens are of the greatest concern during *transfusions.* Most other antigens cause little if any transfusion reaction. As a final check on blood compatibility before transfusions, an important cross-matching test occurs. In this test, small samples of the blood from the donor and the recipient are mixed to be sure agglutination does not create a transfusion reaction, which would result in red blood cell destruction (hemolysis) and other life-threatening conditions. If surgery is elective, autologous transfusions are promoted, allowing some of the person's own predonated blood from storage to be used during the surgery.

An antigen commonly found on the human RBC membrane was first identified in rhesus monkeys, and it became known as the *Rh factor.* Although many different types of antigens are related to the Rh factor, only the *D antigen* is checked using an anti-D reagent. About 85% of Americans possess the D antigen and are therefore considered to be *Rh-positive* (*Rh⁺*). Those lacking the D antigen are considered *Rh-negative* (*Rh⁻*). If an Rh-negative woman is pregnant carrying an Rh-positive fetus, the mother might obtain some RBCs from the fetus during the birth process or during a miscarriage. As a result, the Rh-negative mother would begin producing

WARNING

anti-D antibodies, creating complications for future pregnancies. This condition is known as *hemolytic disease* of the fetus and newborn (erythroblastosis fetalis). This condition can be prevented with the proper administration of *RhoGAM*, which prevents the mother from producing anti-D antibodies.

PROCEDURE A: ABO Blood Typing

1. Reexamine the introductory material and study figures 43.1, 43.2, and 43.3 and table 43.1.
2. Figure 43.1 illustrates the antigens on the RBC membranes that are compatible with various antibodies in the plasma of the ABO blood group. If an antigen of the RBC is introduced into a person with an incompatible plasma antibody, an adverse reaction occurs between the antigen and the antibody, illustrated as an example in figure 43.2. This is most likely to occur if blood from an incompatible donor was used for the transfusion. Table 43.1 indicates the preferred donors, permissible donors (in limited amounts), and incompatible donors for transfusions. If blood is used from an incompatible donor for a transfusion, clumps of RBCs form rather quickly that can block circulation through smaller blood vessels, resulting in serious consequences (fig. 43.3).

Table 43.1 labels blood type AB a *universal recipient* and blood type O a *universal donor.* This terminology can help you to understand the concept of possible clumping (agglutination) reactions during transfusions. Blood type AB is sometimes referred to as a universal recipient (acceptor) because AB blood lacks anti-A and

Table 43.1 Antigens and Antibodies of the ABO Blood Group and Preferred, Permissible, and Incompatible Donors

Blood Type	RBC Antigens (Agglutinogens)	Plasma Antibodies (Agglutinins)	Preferred Donor Type	Permissible Donor Type in Limited Amounts	Incompatible Donor
A	A	Anti-B	A	O	B, AB
B	B	Anti-A	B	O	A, AB
AB (universal recipient)	A and B	Neither anti-A nor anti-B	AB	A, B, O	None
O (universal donor)	Neither A nor B	Both anti-A and anti-B	O	No alternative types	A, B, AB

FIGURE 43.1 Illustration of four possible combinations of the RBC antigens and plasma antibodies within the ABO blood group.

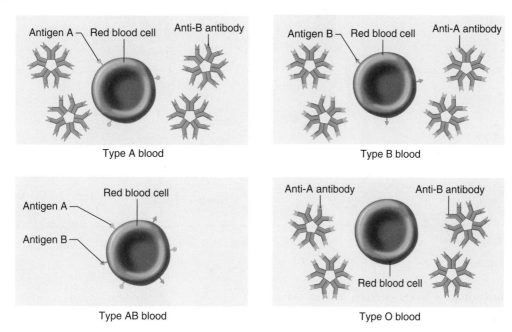

FIGURE 43.2 Illustration of an agglutination (clumping) reaction when antigen A reacts with anti-A antibody from a transfusion of mismatched blood.

FIGURE 43.3 Micrograph of agglutinated RBCs (220x). Clumps of agglutinated RBCs block small blood vessels and circulation to vital tissue.

anti-B antibodies and will not clump with other blood types when used for transfusions. Similarly, blood type O is theoretically referred to as a universal donor as it lacks both A and B antigens (agglutinogens). However, it is not reliable to refer to a blood type as a universal recipient or a universal donor. Under certain conditions, an adverse reaction could occur for reasons that include the number of units used in the transfusions, plasma volumes, infusions of packed RBCs, and other possible agglutinogens that may be present in whole blood.

3. Compete Part A of Laboratory Assessment 43.
4. Perform the ABO blood type test using the blood-typing kit. To do this, follow these steps:
   a. Obtain a clean microscope slide and mark across its center with a wax pencil to divide it into right and left halves. Also write "Anti-A" near the edge of the left half and "Anti-B" near the edge of the right half (fig. 43.4).

FIGURE 43.4 Slide prepared for ABO blood typing.

FIGURE 43.5 Four possible results of the ABO test.

b. Place a small drop of blood on each half of the microscope slide. Work quickly so that the blood will not have time to clot.

c. Add a drop of anti-A serum to the blood on the left half and a drop of anti-B serum to the blood on the right half. Note the color coding of the anti-A and anti-B typing sera. To avoid contaminating the serum, avoid touching the blood with the serum while it is in the dropper; instead, allow the serum to fall from the dropper onto the blood.

d. Use separate toothpicks to stir each sample of serum and blood together, and spread each over an area about as large as a quarter. Dispose of toothpicks in an appropriate container.

e. Examine the samples for clumping of red blood cells (agglutination) after 2 minutes.

f. See table 43.2 and figure 43.5 for aid in interpreting the test results. The frequency of occurrence of ABO blood types varies substantially between ethnic groups. Although there is a very diverse population in the United States, blood type O is the most common and blood type AB the least common.

g. Discard contaminated materials as directed by the laboratory instructor.

5. Record your results and complete Part B of the laboratory assessment.

Table 43.2 ABO Blood-Typing Sera Reactions and U.S. Blood Type Frequency

POSSIBLE REACTIONS		Blood Type	Overall U.S. Frequency
Anti-A Serum	Anti-B Serum		
Clumping (agglutination)	No clumping	Type A	Varies
No clumping	Clumping	Type B	Varies
Clumping	Clumping	Type AB	Least common
No clumping	No clumping	Type O	Most common

Note: The inheritance of the ABO blood groups is described in Laboratory Exercise 61.

PROCEDURE B: Rh Blood Typing

1. Reexamine the introductory material.
2. Complete Part C of the laboratory assessment.
3. To determine the Rh blood type of a blood sample, follow these steps:

   a. Lance the tip of a finger. (See the demonstration procedures in Laboratory Exercise 41 for directions.) Place a small drop of blood in the center of a clean microscope slide. Cover the lanced finger location with a bandage.

   b. Add a drop of anti-D serum to the blood and mix them together with a clean toothpick.

   c. Place the slide on the plate of a slide warming table (Rh blood-typing box) prewarmed to 45°C (113°F) (fig. 43.6).

   d. Slowly rock the table (box) back and forth to keep the mixture moving and watch for clumping (agglutination) of the blood cells. When clumping occurs in anti-D serum, the clumps usually are smaller than those that appear in anti-A or anti-B sera, so they may be less obvious. However, if clumping occurs, the blood is called Rh positive; if no clumping occurs *within 2 minutes,* the blood is called Rh negative.

   e. Discard all contaminated materials in appropriate containers.

4. Record your results and complete Part D of the laboratory assessment.

CRITICAL THINKING ACTIVITY

Judging from your observations of the blood-typing results, suggest which components in the anti-A and anti-B sera caused clumping (agglutination).

_____

_____

_____

FIGURE 43.6 Slide warming table (box) used for Rh blood typing.

Mixture of blood and anti-D serum

Plate heated to 45°C (113°F)

Rh TYPING BOX

ON
OFF

Antigen/Antibody

ABO Blood System

A

B

AB → universal receiver

O → universal donor

*antibody first, then blood

Name _____

Date _____

Section _____

The Ⓐ corresponds to the indicated Learning Outcome(s) Ⓞ found at the beginning of the Laboratory Exercise.

LABORATORY ASSESSMENT
43

Blood Typing

PART A: Assessments

Complete the following statements:

1. The antigens of the ABO blood group are located on the red blood cell _____. Ⓐ1

2. The blood of every person contains one of (how many possible?) _____ combinations of antigens. Ⓐ1

3. Type A blood contains antigen _____. Ⓐ1

4. Type B blood contains antigen _____. Ⓐ1

5. Type A blood contains _____ antibody in the plasma.

6. Type B blood contains _____ antibody in the plasma. Ⓐ1

7. Persons with ABO blood type _____ are sometimes called universal recipients. Ⓐ1

8. Persons with ABO blood type _____ are sometimes called universal donors. Ⓐ1

9. Assume a person with blood type B receives a transfusion of blood type AB. Would the transfusion most likely have successful results? _____ Explain your answer. Ⓐ1

_____

_____

_____

PART B: Assessments

Complete the following:

1. What was the ABO type of the blood tested? Ⓐ2 _____

2. What ABO antigens are present on the red blood cells of this blood type? Ⓐ2 _____

3. What ABO antibodies are present in the plasma of this blood type? Ⓐ2 _____

4. If a person with this blood type needed a blood transfusion, what ABO type(s) of blood could be received safely? Ⓐ1

_____

5. If a person with this blood type donated blood, what ABO blood type(s) could receive the blood safely? Ⓐ1 _____

_____

PART C: Assessments

Complete the following statements:

1. The Rh blood group was named after the _____. A3

2. Of the antigens in the Rh group, the most important is _____. A3

3. If red blood cells lack Rh antigens, the blood is called _____. A3

4. If an Rh-negative person who is sensitive to Rh-positive blood receives a transfusion of Rh-positive blood, the donor's cells are likely to _____. A3

5. An Rh-negative woman who might be carrying a(n) _____ fetus is given an injection of RhoGAM to prevent hemolytic disease of the fetus and newborn. A3

PART D: Assessments

Complete the following:

1. What was the Rh type of the blood tested? A4 _____

2. What Rh antigen is present on the red blood cells of this type of blood? A4 _____

3. If a person with this blood type needed a blood transfusion, what blood type could be received safely? A3

   _____

4. If a person with this blood type donated blood, a person with what blood type could receive the blood safely? A3

   _____

5. Observe the test results shown in figure 43.7 from an individual being tested for the ABO type and the Rh type.
   a. This individual has ABO type _____. A2
   b. This individual has Rh type _____. A4

FIGURE 43.7   Blood test results after adding anti-A serum, anti-B serum, and anti-D serum to separate drops of blood. An unaltered drop of blood is shown as a control; it does not indicate any clumping (agglutination) reaction.

Anti-A serum   Anti-B serum   Anti-D serum   Blood only (control)

Heart Structure

Purpose of the Exercise

To review the structural characteristics of the human heart and to examine the major features of a mammalian heart.

MATERIALS NEEDED

Dissectible human heart model

Dissectible human torso model

Preserved sheep or other mammalian heart

Dissecting tray

Dissecting instruments

Disposable gloves

Long knife

Millimeter ruler

For Learning Extension Assessment:

Colored pencils (red and blue)

SAFETY ⚠

▶ Wear disposable gloves when working on the heart dissection.

▶ Save or dispose of the dissected heart as instructed.

▶ Wash the dissecting tray and instruments as instructed.

▶ Wash your laboratory table.

▶ Wash your hands before leaving the laboratory.

Learning Outcomes AP|R

After completing this exercise, you should be able to:

01 Identify and label the major structural features of the human heart and closely associated blood vessels.

02 Trace blood flow through the heart chambers and the systemic and pulmonary circuits.

03 Describe locations and functions of heart structures.

04 Compare the features of the human heart with those of another mammal.

The **O** corresponds to the assessments **A** indicated in the Laboratory Assessment for this Exercise.

Pre-Lab

Carefully read the introductory material and examine the entire lab. Be familiar with the structure of the heart from lecture or the textbook. Answer the pre-lab questions.

Pre-Lab Questions Select the correct answer for each of the following questions:

1. The _____ is the inferior end of the heart that is bluntly pointed.
   a. base
   b. apex
   c. auricle
   d. sulcus

2. Oxygen-rich blood is located in the
   a. left-side chambers.
   b. right-side chambers.
   c. superior chambers.
   d. inferior chambers.

3. The two superior heart chambers are the
   a. left atrium and left ventricle.
   b. right atrium and right ventricle.
   c. atria. (left + right atrium) atrium
   d. ventricles.

4. Which of the following is an atrioventricular (AV) valve?
   a. aortic
   b. pulmonary
   c. semilunar
   d. mitral

5. The _____ lines the heart chambers.
   a. endocardium
   b. fibrous pericardium
   c. serous pericardium
   d. epicardium

6. Which heart valve has two cusps instead of three cusps?
   a. aortic
   b. mitral
   c. tricuspid
   d. pulmonary

7. Chordae tendineae connect the cusps of the AV valves to papillary muscles of the ventricles.
   a. True ✓
   b. False _____

8. The systemic circuit delivers blood to the lungs and back to the heart.
   a. True _____   b. False _✓_

9. The right and left coronary arteries containing oxygen-rich blood originate from the base of the aorta.
   a. True _✓_   b. False _____

The heart is a muscular pump located within the mediastinum and resting upon the diaphragm. It is enclosed by the lungs, thoracic vertebrae, and sternum, and attached at its superior end (the *base*) are several large blood vessels. Its inferior end extends downward to the left and terminates as a bluntly pointed *apex*.

The heart and the proximal ends of the attached blood vessels are enclosed by a double-layered *pericardium*. The innermost layer of this membrane (visceral layer of serous pericardium) consists of a thin covering closely applied to the surface of the heart, whereas the outer layer (parietal layer of serous pericardium with fibrous pericardium) forms a tough, protective sac surrounding the heart. Between the

parietal and visceral layers of the pericardium is a space, the *pericardial cavity,* that contains a small volume of *serous (pericardial) fluid*.

Functioning as a muscular pump of blood, the heart consists of four chambers: two superior *atria* and two inferior *ventricles*. The *interatrial septum* and the *interventricular septum* prevent oxygen-rich (oxygenated) blood in the left-side chambers from mixing with oxygen-poor (deoxygenated) blood in the right chambers. *Atrioventricular (AV) valves (mitral and tricuspid)* are located between each atrium and ventricle, while *semilunar valves (aortic and pulmonary)* are at the exits of the ventricles into the two large arteries. The heart valves only allow blood flow in one direction.

PROCEDURE A: The Human Heart

As you study the human heart, use a combination of laboratory manual figures and heart models available in the laboratory to locate structures. This laboratory exercise covers basic information that will be very important as you work on the rest of the cardiovascular system exercises.

1. Study figures 44.1, 44.2, 44.3, 44.4, 44.5, and 44.6.

FIGURE 44.1 Anterior view of the human heart.

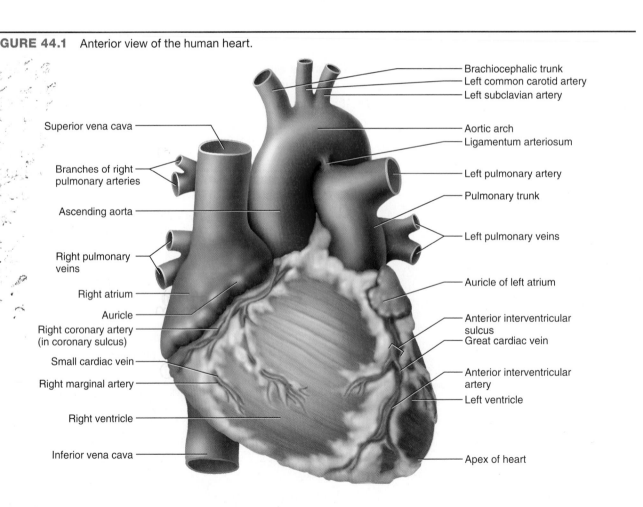

466

FIGURE 44.2 Anterior view of the two coronary arteries and their major branches. The coronary arteries originate from the base of the aorta and carry oxygen-rich blood to the cardiac muscle tissue.

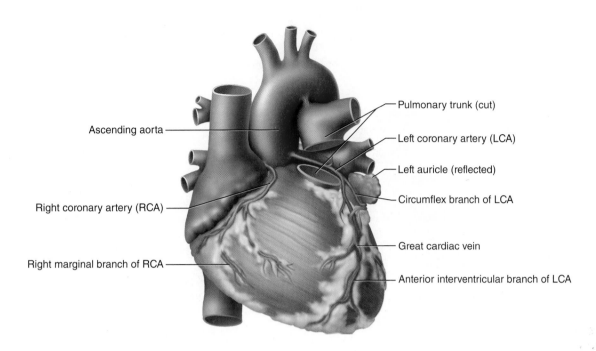

Ascending aorta

Right coronary artery (RCA)

Right marginal branch of RCA

Pulmonary trunk (cut)

Left coronary artery (LCA)

Left auricle (reflected)

Circumflex branch of LCA

Great cardiac vein

Anterior interventricular branch of LCA

FIGURE 44.3 Posterior view of the human heart.

Aorta

Left pulmonary artery

Left pulmonary veins

Left atrium

Great cardiac vein

Coronary sinus

Left ventricle

Apex of heart

Right pulmonary artery

Superior vena cava

Right pulmonary veins

Right atrium

Inferior vena cava

Posterior interventricular sulcus

Middle cardiac vein

Posterior interventricular artery

Right ventricle

Sup. vena cava
right atrium
R AV valve
R ventricle
R semil valve
Pulm. Trunk
Pulm Art
Pulm Vein
Lungs
Left Atrium
L AV valve
L ventricle
L semil. valve
Aorta

FIGURE 44.4 Frontal section of the human heart.

Aortic arch

Ascending aorta

Right pulmonary artery

Left pulmonary artery

Superior vena cava

Pulmonary trunk

Right pulmonary veins

Left pulmonary veins

Interatrial septum

Pulmonary valve

Pectinate muscles

Left atrium

Fossa ovalis

Aortic valve

Right atrium

Mitral valve

Tricuspid valve

Trabeculae carneae

Left ventricle

Papillary muscle

Chordae tendineae

Interventricular septum

Endocardium

Myocardium — Heart wall

Right ventricle

Epicardium

Inferior vena cava

FIGURE 44.5 Superior view of the four heart valves (atria removed).

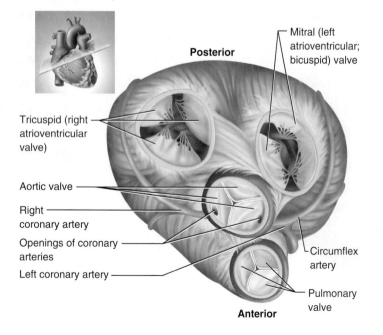

Posterior

Mitral (left atrioventricular; bicuspid) valve

Tricuspid (right atrioventricular valve)

Aortic valve

Right coronary artery

Openings of coronary arteries

Circumflex artery

Left coronary artery

Pulmonary valve

Anterior

FIGURE 44.6 Frontal section of the heart wall and pericardium.

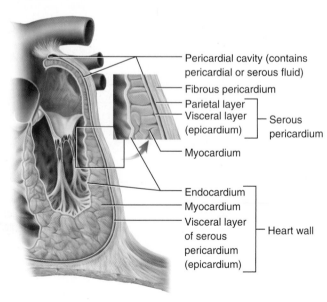

Pericardial cavity (contains pericardial or serous fluid)

Fibrous pericardium

Parietal layer

Visceral layer (epicardium) — Serous pericardium

Myocardium

Endocardium

Myocardium

Visceral layer of serous pericardium (epicardium) — Heart wall

2. Examine the human heart model and locate the following features:

heart
- base of heart—superior region where blood vessels emerge
- apex of heart—inferior, rounded end

3. Examine figure 44.6 of the pericardium and the heart wall. Observe the arrangement of the structures and note the descriptions of the following features:

pericardium (pericardial sac)
- fibrous pericardium—outer layer
- serous pericardium
  - parietal layer—inner fused lining of fibrous pericardium
  - visceral layer (epicardium)

pericardial cavity—between parietal and visceral layers of serous pericardial membranes

myocardium—composed of cardiac muscle tissue (fig. 44.7)

endocardium—lines heart chambers

FIGURE 44.7 Cardiac muscle tissue forms the myocardium of the heart chamber walls. It also composes the muscular ridges (trabeculae carneae of ventricles and pectinate muscles of portions of atria) and papillary muscles of ventricles.

Cardiac muscle tissue (400×)

4. Use the human heart model and examine the structures, locations, and descriptions of the following:

atria—two superior chambers
- right atrium—receives blood from systemic veins
- left atrium—receives blood from pulmonary veins
- auricles—earlike extensions of atria allowing greater expansions for blood volumes

ventricles—two inferior chambers
- right ventricle—pumps blood into pulmonary circuit
- left ventricle—pumps blood into systemic circuit; has thicker myocardium than right ventricle

atrioventricular orifices—openings between atria and ventricles and guarded by AV valves

atrioventricular valves (AV valves)—located between atria and ventricles
- tricuspid valve (right atrioventricular valve)— has three cusps
- mitral valve (left atrioventricular valve; bicuspid valve)—has two cusps

semilunar valves—regulate flow into two large arteries that exit from ventricles
- pulmonary valve—has three pocketlike cusps at entrance of pulmonary trunk
- aortic valve—has three pocketlike cusps at entrance of aorta

chordae tendineae (tendinous cords)—connect AV valve cusps to papillary muscles

papillary muscles—projections from inferior ventricles; anchor AV valve cusps during ventricular contraction

coronary sulcus (atrioventricular sulcus or groove)—encircles heart between atria and ventricles

interventricular sulci—grooves with blood vessels along septum between ventricles
- anterior sulcus
- posterior sulcus

5. Inspect figure 44.8 as a reference to trace the flow of blood through the heart chambers and the systemic and pulmonary circuits. The arrows within the figure indicate the direction of the blood flow within the heart chambers and the two circuits. Use a dissectible human torso model and the heart model and locate the major blood vessels near the heart. Follow the arrows on figure 44.8 to trace the pathway of blood. Observe the figures and the models to take notice of the use of the red and blue colors that pertain to the relative oxygen levels of the blood within those blood vessels and heart chambers. The red color indicates oxygen-rich (oxygenated) blood; the blue color indicates oxygen-poor (deoxygenated) blood. Compare the two circuits in relation to the oxygen content in the arteries and veins. *Arteries carry blood away from the heart, while veins return blood to the heart. Notice that the arteries of the systemic circuit contain oxygen-rich blood; the arteries of the pulmonary circuit contain oxygen-poor blood.* Locate the following blood vessels associated with the heart and those serving the myocardium:

superior vena cava—returns blood to right atrium from superior part of systemic circuit

inferior vena cava—returns blood to right atrium from inferior part of systemic circuit

pulmonary trunk—main artery of pulmonary circuit

pulmonary arteries—main arteries to lungs

pulmonary veins—return blood from pulmonary circuit into left atrium

aorta—main artery of systemic circuit

left coronary artery—originates from base of aorta; has two main branches (figs. 44.2 and 44.5)

- circumflex artery
- anterior interventricular (descending) artery

right coronary artery—originates from base of aorta; has two main branches (figs. 44.2, 44.3, and 44.5)

- posterior interventricular artery
- right marginal artery

FIGURE 44.8 The heart pumps blood into two major circuits. The left heart side pumps blood into the systemic circuit, serving all body tissues except the lungs. The right heart side pumps blood into the pulmonary circuit, serving both lungs. Red indicates oxygen-rich blood; blue indicates oxygen-poor blood. The arrows indicate the direction of blood flow.

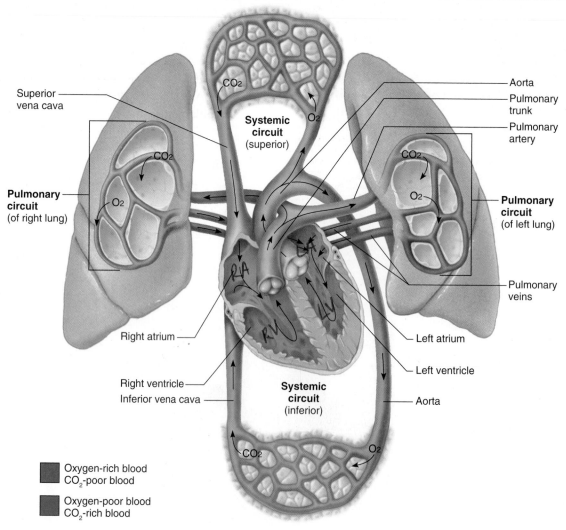

Superior vena cava

Systemic circuit (superior)

Pulmonary circuit (of right lung)

Aorta

Pulmonary trunk

Pulmonary artery

Pulmonary circuit (of left lung)

Pulmonary veins

Right atrium

Left atrium

Right ventricle

Left ventricle

Inferior vena cava

Systemic circuit (inferior)

Aorta

Oxygen-rich blood
CO₂-poor blood

Oxygen-poor blood
CO₂-rich blood

cardiac veins—drain the myocardium into the coronary sinus; include several main veins (figs. 44.1, 44.2, and 44.3)

- great cardiac vein
- middle cardiac vein
- small cardiac vein

coronary sinus—for return of blood from cardiac veins into right atrium (fig. 44.3)

6. Complete Parts A and B of Laboratory Assessment 44.

PROCEDURE B: Dissection of a Sheep Heart

The directions and the figures for this procedure are for a sheep heart. The structures of a sheep heart are similar to those of a human heart and will therefore be used for futher exploration. Although sheep are quadrupeds, human terminology will be used for the sheep heart. The directions and the figures would be comparable if another mammal heart were used for the dissection. Other mammal hearts that might be used include pig, cow, and deer. A cadaver observation would enhance this laboratory experience if it could be arranged.

1. Obtain a preserved sheep heart. Rinse it in water thoroughly to remove as much of the preservative as possible. Also run water into the large blood vessels to force any blood clots out of the heart chambers.

2. Place the heart in a dissecting tray with its anterior surface up (fig. 44.9), and proceed as follows:
   a. Although the relatively thick *pericardial sac* probably is missing, look for traces of this membrane around the origins of the large blood vessels. Review figure 44.6 to help you locate some pericardial sac structures.
   b. Locate the *visceral pericardium,* which appears as a thin, transparent layer on the surface of the heart. Use a scalpel to remove a portion of this layer and expose the *myocardium* beneath. Also note the abundance of fat along the paths of various blood vessels. This adipose tissue occurs in the loose connective tissue that underlies the visceral pericardium.
   c. Identify the following:

      right atrium

      right ventricle

      left atrium

      left ventricle

      coronary (atrioventricular) sulcus

      anterior interventricular sulcus
   d. Carefully remove the fat from the anterior interventricular sulcus and expose the blood vessels that pass along this groove. They include a branch of the *left coronary artery (anterior interventricular artery)* and a *cardiac vein.*

FIGURE 44.9 Anterior surface of sheep heart. The probe passes through the left pulmonary artery and the pulmonary trunk.

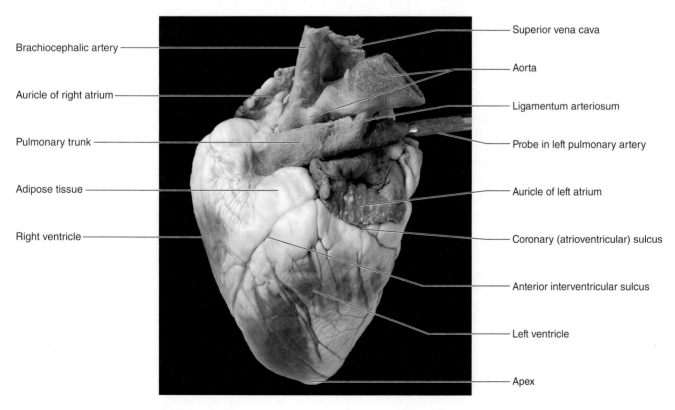

Brachiocephalic artery

Auricle of right atrium

Pulmonary trunk

Adipose tissue

Right ventricle

Superior vena cava

Aorta

Ligamentum arteriosum

Probe in left pulmonary artery

Auricle of left atrium

Coronary (atrioventricular) sulcus

Anterior interventricular sulcus

Left ventricle

Apex

3. Examine the posterior surface of the heart (fig. 44.10), and proceed as follows:
   a. Identify the *coronary sulcus* and the *posterior interventricular sulcus*. The posterior interventricular sulcus contains a *posterior interventricular artery* and a *cardiac vein*.
   b. Locate the stumps of two relatively thin-walled veins that enter the right atrium. Demonstrate this connection by passing a slender probe through them. The upper vessel is the *superior vena cava,* and the lower one is the *inferior vena cava* (fig. 44.10).

4. In order to view the internal heart chambers and heart valves, a frontal (coronal) section of the heart will be prepared using a long knife or a scalpel. Place the base of the heart down in a dissecting tray with the apex pointing upward and the anterior part of the heart facing toward your own body. Make a cut from the apex toward the base while attempting the cut near the center of the heart to separate the heart into equal anterior and posterior portions (fig. 44.11). This cut should expose the four chambers, interventricular septum, and heart valves.

5. Place the anterior portion in the dissecting tray with the anterior surface down, which will allow internal viewing of structures of the anterior portion (section) of the heart (fig. 44.11*a*). Identify as many of the following structures as possible:

   right atrium
   right ventricle
   left atrium
   left ventricle
   interventricular septum
   atrioventricular valves (AV valves)—mitral and tricuspid
   - cusps of valves
   - chordae tendineae
   - papillary muscles
   semilunar valves—aortic and pulmonary

6. Next place the posterior portion in the dissecting tray with the posterior surface down, which will allow internal viewing of structures of the posterior portion (section) of the heart (fig. 44.11*b*). Many of the structures listed in step 5 of Procedure B should also be located. Depending on the major blood vessels that remained on the heart for the frontal section, portions of the stumps of sectioned major blood vessels such as the aorta, pulmonary trunk, brachiocephalic artery, and superior vena cava can be identified (fig. 44.11).

7. Note the muscular ridges and pitted pattern of the *trabeculae carneae* inside the ventricle chambers (fig. 44.11*a*). Examine the muscular ridges of the pectinate muscles inside the right atrium (fig. 44.11*b*).

8. Obtain another sheep heart that has not been dissected for making a transverse cut. Place the heart in the dissecting tray with the posterior of the heart down and the anterior of the heart facing upward. Make a transverse cut starting just inferior to the coronary sulci and

FIGURE 44.10 Posterior surface of sheep heart. The probe passes through the right atrium and the vena cavae.

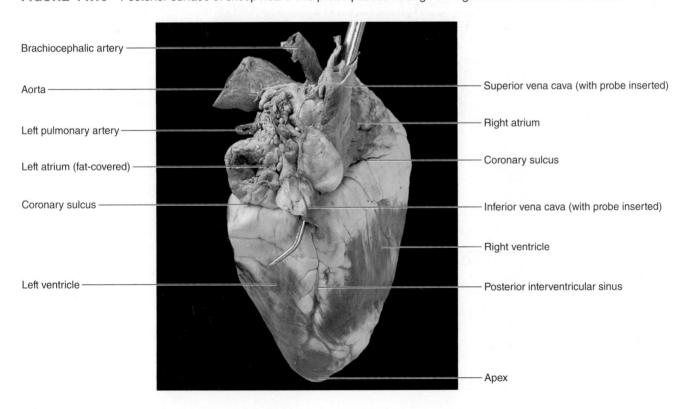

Brachiocephalic artery

Aorta

Left pulmonary artery

Left atrium (fat-covered)

Coronary sulcus

Left ventricle

Superior vena cava (with probe inserted)

Right atrium

Coronary sulcus

Inferior vena cava (with probe inserted)

Right ventricle

Posterior interventricular sinus

Apex

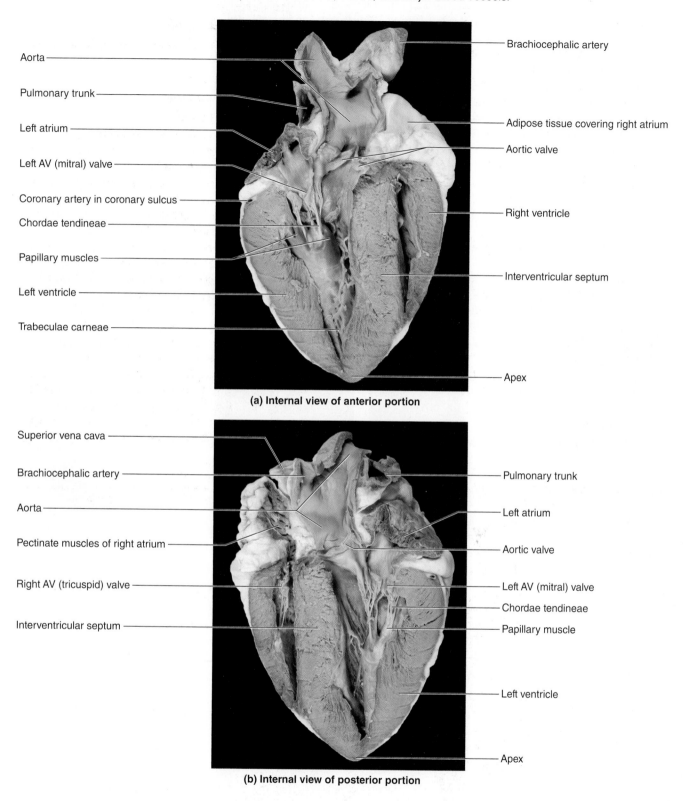

Aorta

Pulmonary trunk

Left atrium

Left AV (mitral) valve

Coronary artery in coronary sulcus

Chordae tendineae

Papillary muscles

Left ventricle

Trabeculae carneae

Brachiocephalic artery

Adipose tissue covering right atrium

Aortic valve

Right ventricle

Interventricular septum

Apex

(a) Internal view of anterior portion

Superior vena cava

Brachiocephalic artery

Aorta

Pectinate muscles of right atrium

Right AV (tricuspid) valve

Interventricular septum

Pulmonary trunk

Left atrium

Aortic valve

Left AV (mitral) valve

Chordae tendineae

Papillary muscle

Left ventricle

Apex

(b) Internal view of posterior portion

continue the cut through both ventricles. The cut will produce a transverse view of the inferior and superior portions of the ventricles near the AV valves (fig. 44.12).

9. Study the transverse view of the inferior piece of the heart (fig. 44.12*a*). Observe the chamber cavities with an internal view toward the apex. Note any strands of the chordae tendineae of the AV valves and the general shape of the two chamber cavities in this transverse view. Use a millimeter ruler and measure the thickness of the outer wall of the left and right ventricles. Record the two measurements in Part C of the laboratory assessment.

10. Study the transverse view of the superior piece of the heart (fig. 44.12*b*). Observe the chamber cavities with an internal view toward the AV valves. Note strands of chordae tendineae, papillary muscles, and cusps of the AV valves.

11. Discard or save the specimens as directed by your laboratory instructor.

12. Complete Part C of the laboratory assessment.

FIGURE 44.12 Transverse section of ventricles of sheep heart near the AV valves.

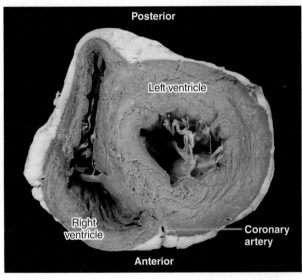

(a) Inferior piece: internal view of chambers toward the apex

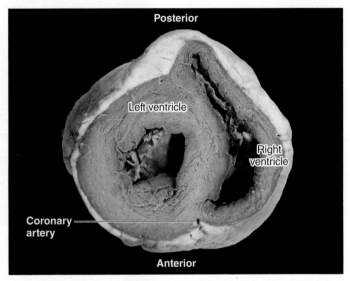

(b) Superior piece: internal view of chambers toward the AV valves

ASSESSMENT
44

Name _____

Date _____

Section _____

The Ⓐ corresponds to the indicated Learning Outcome(s) Ⓞ found at the beginning of the Laboratory Exercise.

Heart Structure

PART A: Assessments

Examine the labeled features and identify the numbered features in figures 44.13, 44.14, and 44.15 of the heart.

FIGURE 44.13 AP|R Identify the features on this anterior view of the heart region of a cadaver, using the terms provided. Ⓐ1

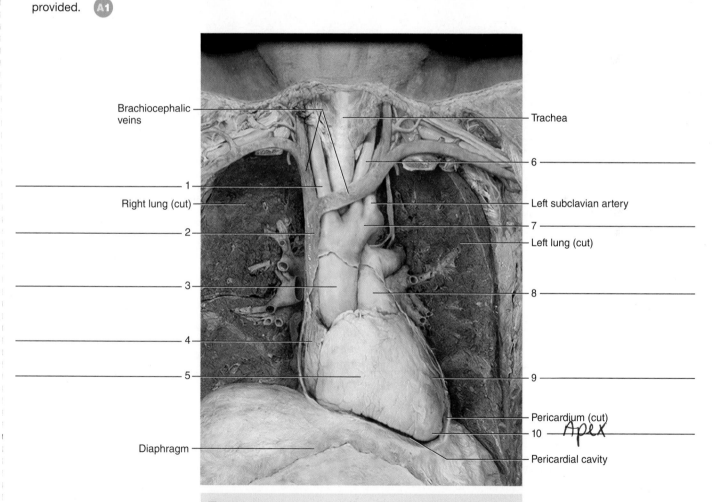

Terms:

Aortic arch	Left ventricle
~~Apex of heart~~	Pulmonary trunk
Ascending aorta	Right atrium
Brachiocephalic trunk	Right ventricle
Left common carotid artery	Superior vena cava

FIGURE 44.14   Identify the features indicated on this anterior view of a frontal section of a human heart model, using the terms provided. (*Note:* The pulmonary valve is not shown on the portion of the model photographed.)   **A1**

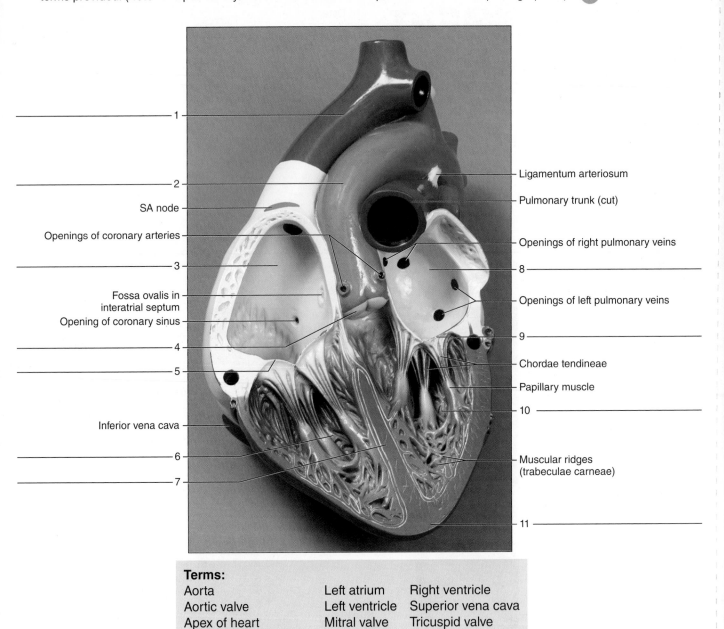

Ligamentum arteriosum

Pulmonary trunk (cut)

Openings of right pulmonary veins

Openings of left pulmonary veins

Chordae tendineae

Papillary muscle

Muscular ridges
(trabeculae carneae)

SA node

Openings of coronary arteries

Fossa ovalis in
interatrial septum

Opening of coronary sinus

Inferior vena cava

1

2

3

4

5

6

7

8

9

10

11

Terms:
Aorta	Left atrium	Right ventricle
Aortic valve	Left ventricle	Superior vena cava
Apex of heart	Mitral valve	Tricuspid valve
Interventricular septum	Right atrium	

LEARNING EXTENSION ASSESSMENT

Use red and blue colored pencils to color the blood vessels in figure 44.15. Use red to illustrate a blood vessel with oxygen-rich blood, and use blue to illustrate a blood vessel with oxygen-poor blood.   **A2**

FIGURE 44.15 Label this frontal section of the human heart. The arrows indicate the direction of blood flow.

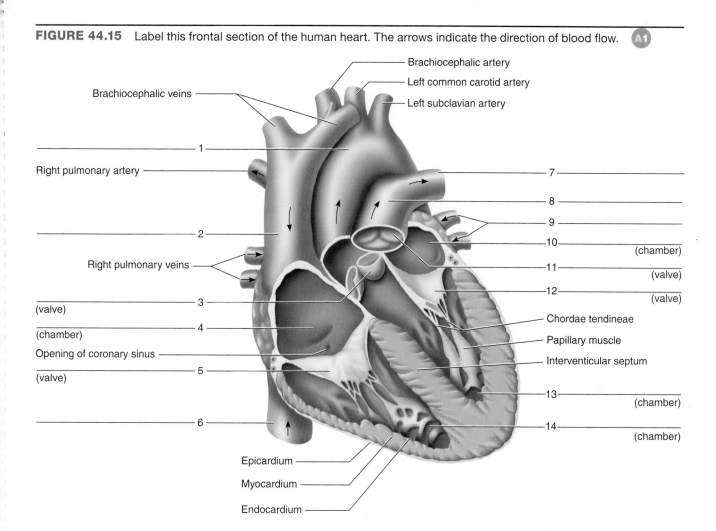

Brachiocephalic artery

Left common carotid artery

Brachiocephalic veins

Left subclavian artery

Right pulmonary artery — 1

7

2

8

9

10
(chamber)

Right pulmonary veins

11
(valve)

12
(valve)

3
(valve)

Chordae tendineae

4
(chamber)

Papillary muscle

Opening of coronary sinus

Interventicular septum

5
(valve)

13
(chamber)

6

14
(chamber)

Epicardium

Myocardium

Endocardium

PART B: Assessments

Match the terms in column A with the descriptions in column B. Place the letter of your choice in the space provided. A2

A3

Column A	Column B

Column A

a. Aorta
b. Cardiac vein
c. Coronary artery
d. Coronary sinus
e. Endocardium
f. Mitral valve
g. Myocardium
h. Papillary muscle
i. Pericardial cavity
j. Pericardial sac
k. Pulmonary trunk
l. Tricuspid valve

Column B

_____ 1. Structure from which chordae tendineae originate

_____ 2. Prevents blood movement from right ventricle to right atrium

_____ 3. Membranes around heart

_____ 4. Prevents blood movement from left ventricle to left atrium

_____ 5. Gives rise to left and right pulmonary arteries

_____ 6. Drains blood from myocardium into right atrium

_____ 7. Inner lining of heart chamber

_____ 8. Layer largely composed of cardiac muscle tissue

_____ 9. Space containing serous fluid to reduce friction during heartbeats

_____ 10. Drains blood from myocardial capillaries

_____ 11. Supplies blood to heart muscle

_____ 12. Distributes blood to body organs (systemic circuit) except lungs

PART C: Assessments

Complete the following:

1. Compare the structure of the left atrioventricular valve with that of the aortic valve. (A4) _____

_____

_____

_____

2. Describe the functions of the chordae tendineae and the papillary muscles. (A4) _____

_____

_____

_____

3. What is the functional significance of the difference in thickness between the wall of the aorta and the wall of the pulmonary trunk? (A4) _____

_____

_____

_____

4. List the correct pathway through which blood must flow in relation to regions of the heart. Assume blood is currently in the vena cava and will eventually enter the aorta. Include the major blood vessels (arteries and veins) attached directly to a heart chamber, the four heart chambers, and the four heart valves in your list. (A2) (A4) _____

_____

_____

_____

_____

_____

5. Describe the general overall shape of the left and right cavities of the ventricles as observed in a transverse section of the heart. (A4) _____

_____

6. What was the measured thickness of the left ventricle wall? _____

What was the measured thickness of the right ventricle wall? (A4) _____

CRITICAL THINKING ASSESSMENT

What is the functional significance of the difference in thickness of the ventricular walls? (A4)

_____

_____

_____

EXERCISE 45

Cardiac Cycle

Purpose of the Exercise

To review the events of a cardiac cycle, to become acquainted with normal heart sounds, and to record an electrocardiogram.

MATERIALS NEEDED

For Procedure A—Heart Sounds:
Stethoscope
Alcohol swabs

For Procedure B—Electrocardiogram:
Electrocardiograph (or other instrument for recording an ECG)
Cot or table
Alcohol swabs
Electrode paste (gel)
Electrodes and cables

Ph.I.L.S. 4.0, Lesson 26 ECG and Heart Function: The Meaning of Heart Sounds

Learning Outcomes

After completing this exercise, you should be able to:

01 Interpret the major events of a cardiac cycle.

02 Associate the sounds produced during a cardiac cycle with the valves closing.

03 Correlate the structures and components of a normal ECG pattern with the phases of a cardiac cycle.

04 Record and interpret an electrocardiogram.

05 Diagram and label the waves of an electrocardiogram (ECG) and correlate with the heart sounds.

06 Integrate the electrical changes and the heart sounds with the cardiac cycle.

The ⓞ corresponds to the assessments Ⓐ indicated in the Laboratory Assessment for this Exercise.

Pre-Lab

Carefully read the introductory material and examine the entire lab. Be familiar with the cardiac cycle, heart sounds, cardiac conduction system, and electrocardiograms from lecture or the textbook. Answer the pre-lab questions.

Pre-Lab Questions Select the correct answer for each of the following questions:

1. The _____ of the conduction system is known as the pacemaker.
   a. SA node   b. AV node
   c. AV bundle   d. left bundle branch

2. The _____ of the conduction system is/are located throughout the ventricular walls.
   a. AV node   b. left bundle branch
   c. right bundle branch   d. Purkinje fibers

3. The first of two heart sounds (lubb) occurs when the
   a. AV valves open.   b. AV valves close.
   c. semilunar valves open.   d. semilunar valves close.

4. One cardiac cycle would consist of
   a. left chamber contractions followed by right chamber contractions.
   b. right chamber contractions followed by left chamber contractions.
   c. atrial chamber contractions followed by ventricular chamber contractions.
   d. ventricular chamber contractions followed by atrial chamber contractions.

A set of atrial contractions while the ventricular walls relax, followed by ventricular contractions while the atrial walls relax, constitutes a *cardiac cycle*. Each cycle generates pressure changes within the heart chambers, resulting in the ejection of blood. Movement only occurs in one direction due to the one-way valves found between the chambers (atrioventricular valves) and between the ventricles and the circulatory circuits. These closing valves produce vibrations in the tissues and thus create the sounds associated with the heartbeat. If backflow of blood occurs when a heart valve is closed, it creates a turbulence noise known as a murmur.

The regulation and coordination of the cardiac cycle involves the *cardiac conduction system*. The pathway of electrical signals originates from the *sinoatrial (SA) node* located in the right atrium near the entrance of the superior vena cava. The stimulation of the heartbeat and the heart rate originate from the SA node, so it is often called the *pacemaker*. As the signals pass through the atrial walls toward the *atrioventricular (AV) node*, contractions of the atria followed by relaxations take place. Once the AV node has been signaled, the rapid continuation of the electrical signals occurs through the *AV bundle* and the *right and left bundle branches* within the interventricular septum and terminates via the *Purkinje fibers* throughout the ventricular walls. Once the myocardium of the ventricles has been stimulated, ventricular contractions followed by relaxations occur, completing one cardiac cycle. During the relaxation of the ventricles they fill with blood for the next cycle.

The sympathetic and parasympathetic subdivisions of the autonomic nervous system influence the activity of the pacemaker under various conditions. An increased rate results from sympathetic responses; a decreased rate results from parasympathetic responses.

A number of electrical changes also occur in the myocardium as it contracts and relaxes. These changes can be detected by using recording electrodes and an instrument called an *electrocardiograph*. The recording produced by the instrument is an *electrocardiogram*, or *ECG (EKG)*. *Depolarization* and *repolarization* electrical events of the cardiac cycle can be observed and interpreted from the ECG graphic recording.

Clinicians can compare a subject's ECG recording to normal ECG patterns to get a noninvasive view of cardiac function. The heart rate can be determined by counting the number of QRS complexes on the ECG during one minute.

PROCEDURE A: Heart Sounds

1. Listen to your heart sounds. To do this, follow these steps:
   a. Obtain a stethoscope and clean its earpieces and the diaphragm using alcohol swabs.
   b. Fit the earpieces into your ear canals so that the angles are positioned in the forward direction.
   c. Firmly place the diaphragm (bell) of the stethoscope on the chest over the fifth intercostal space near the apex of the heart (fig. 45.1) and listen to the sounds. This is a good location to hear the first sound (*lubb*) of a cardiac cycle when the AV valves are closing, which occurs during ventricular *systole* (contraction).
   d. Move the diaphragm to the second intercostal space, just to the left of the sternum, and listen to the sounds from this region. You should be able to hear the second sound (*dupp*) of the cardiac cycle clearly when the semilunar valves are closing, which occurs during ventricular *diastole* (relaxation).
   e. It is possible to hear sounds associated with the aortic and pulmonary valves by listening from the second intercostal space on either side of the sternum. The aortic valve sound comes from the right and the pulmonary valve sound comes from the left. The sound associated with the mitral valve can be heard from the fifth intercostal space at the nipple line on the left. The sound of the tricuspid valve can be heard at the fifth intercostal space closer to the sternum (fig. 45.1). The four points for listening to (auscultating) the four heart valve sounds when they close are not directly superficial to the valve locations. This is because the sounds associated with the closing valves travel in a slanting direction to the surface of the chest wall where the stethoscope placements occur.

2. Inhale slowly and deeply and exhale slowly while you listen to the heart sounds from each of the locations as before. Note any changes that have occurred in the sounds.

3. Exercise moderately outside the laboratory for a few minutes so that other students listening to heart sounds will not be disturbed. After the exercise period, listen to the heart sounds and note any changes that have occurred in them. *This exercise should be avoided by anyone with health risks.*

4. Complete Parts A and B of Laboratory Assessment 45.

FIGURE 45.1 The first sound (lubb) of a cardiac cycle can be heard by placing the diaphragm of a stethoscope over the fifth intercostal space near the apex of the heart. The second sound (dupp) can be heard over the second intercostal space, just left of the sternum. The thoracic regions circled indicate where the sounds of each heart valve are most easily heard.

PROCEDURE B: Electrocardiogram

A single heartbeat is a sequence of precisely timed contractions and relaxations of the four heart chambers initiated from the cardiac conduction system (fig. 45.2). The basic rhythm is controlled from the SA node (pacemaker), but it can be modified by the autonomic nervous system. During atrial contraction (systole), the ventricles are in a state of relaxation (diastole) and are filling with blood forced into them from the atria. Atrial contraction is soon followed by ventricular contraction (systole), during which time atrial relaxation (diastole) occurs. As a result of the synchronized ventricular contractions, blood is forced into the two large arteries that carry blood away from the heart. The aorta receives blood from the left ventricle into the systemic circuit; the pulmonary trunk receives blood from the right ventricle into the pulmonary circuit. This entire sequence of events of a single heartbeat is the cardiac cycle.

Due to the action of the cardiac conduction system, the heart contracts in a coordinated manner. The events of the cardiac conduction system and the cardiac cycle form the basis for the electrocardiogram. The appearance of a normal ECG pattern for one heartbeat is shown in figure 45.3. The ECG components, normal time durations, and corresponding significance of the waves and intervals are placed in table 45.1. An illustration of the relationship of the ECG components with the depolarization and repolarization of the atria followed by the ventricles is presented in figure 45.4.

1. Study figures 45.2, 45.3, and 45.4, and table 45.1.
2. Complete Parts C and D of the laboratory assessment.
3. The laboratory instructor will demonstrate the proper adjustment and use of the instrument available to record an electrocardiogram.

FIGURE 45.2 AP|R The cardiac conduction system pathway, indicated by the arrows, from the SA node to the Purkinje fibers (frontal section of heart). An action potential originates from the SA node and spreads throughout the atria to the AV node. Conduction from the AV node passes through the AV bundle and spreads along bundle branches within the interventricular septum to the Purkinje fibers within the ventricle walls.

FIGURE 45.3 Components of a normal ECG pattern with a time scale.

Table 45.1 ECG Components, Durations, and Significance

ECG Component	Duration in Seconds	Corresponding Significance
P wave	0.06–0.11	Depolarization of atrial fibers
P-Q (P-R) interval	0.12–0.20	Time for cardiac impulse from SA node through AV node
QRS complex	< 0.12	Depolarization of ventricular fibers
S-T segment	0.12	Time for ventricles to contract
Q-T interval	0.32–0.42	Time from ventricular depolarization to end of ventricular repolarization
T wave	0.16	Repolarization of ventricular fibers (ends pattern)

Note: Atrial repolarization is not observed in the ECG because it occurs at the same time that ventricular fibers depolarize. Therefore it is obscured by the QRS complex.

FIGURE 45.4 Depolarization waves (red) followed by repolarization waves (green) of the atria and ventricles during a cardiac cycle, along with the representative segment of the ECG pattern. The yellow arrows indicate the direction the wave is traveling.

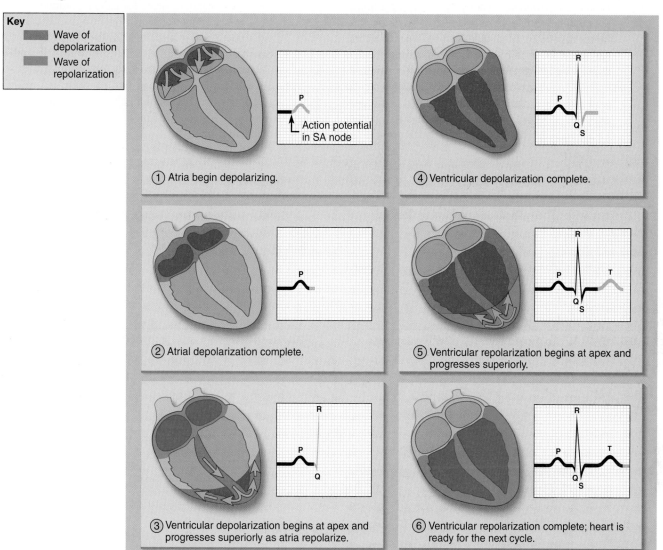

Key
- Wave of depolarization
- Wave of repolarization

① Atria begin depolarizing.
Action potential in SA node

② Atrial depolarization complete.

③ Ventricular depolarization begins at apex and progresses superiorly as atria repolarize.

④ Ventricular depolarization complete.

⑤ Ventricular repolarization begins at apex and progresses superiorly.

⑥ Ventricular repolarization complete; heart is ready for the next cycle.

4. Record your laboratory partner's ECG. To do this, follow these steps:

   a. Have your partner lie on a cot or table close to the electrocardiograph, remaining as relaxed and still as possible.

   b. Scrub the electrode placement locations with alcohol swabs (fig. 45.5). Apply a small quantity of electrode paste (if necessary) to the skin on the insides of the wrists and ankles. (Any jewelry on the wrists or ankles should be removed.)

   c. Spread some electrode paste (if necessary) over the inner surfaces of four plate electrodes and attach one to each of the prepared skin areas (fig. 45.5). Make sure there is good contact between the skin and the metal of the electrodes. The electrode on the right ankle is the grounding system.

   d. Attach the electrodes to the corresponding cables of the ECG recording instrument. When an ECG recording is made, only two electrodes are used at a time, and the instrument allows combinations of electrodes (leads) to be activated. Three standard limb leads placed on the two wrists and the left ankle are used for an ECG. This arrangement has become known as *Einthoven's triangle*,[1] which permits the recording of the potential difference between any two of the electrodes (fig. 45.6). More leads can be done. In a clinical situation, usually a 12–15 lead arrangement is used, which gives the clinician more information.

The standard leads I, II, and III are called bipolar leads because they are the potential difference between two electrodes (a positive and a negative). Lead I measures the potential difference between the right wrist (negative) and the left wrist (positive). Lead II measures the potential difference between the right wrist and the left ankle, and lead III measures the potential difference between the left wrist and the left ankle. The right ankle is always the ground.

---

[1]Willem Einthoven (1860–1927), a Dutch physiologist, received the Nobel prize for Physiology or Medicine for his work with electrocardiograms.

e. Turn on the recording instrument and adjust it as previously demonstrated by the laboratory instructor. The paper speed should be set at 2.5 cm/second. This is the standard speed for ECG recordings.

f. Set the instrument to lead I (right wrist, left wrist electrodes), and record the ECG for 1 minute.

g. Set the instrument to lead II (right wrist, left ankle electrodes), and record the ECG for 1 minute.

h. Set the instrument to lead III (left wrist, left ankle electrodes), and record the ECG for 1 minute.

i. Remove the electrodes and clean the paste from the metal and skin.

j. Use figure 45.3 to label the ECG components of the results from leads I, II, and III. The P-Q interval is often called the P-R interval because the Q wave is often small or absent. The normal P-Q interval is 0.12–0.20 seconds. The normal QRS complex duration is less than 0.10 second.

5. Complete Part E of the laboratory assessment.

FIGURE 45.5   To record an ECG, attach electrodes to the wrists and ankles.

FIGURE 45.6   Standard limb leads for electrocardiograms.

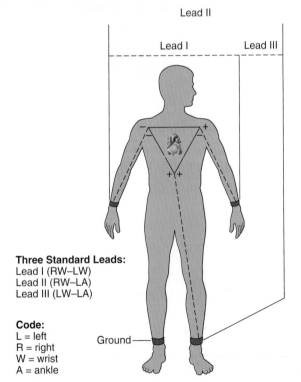

Lead II

Lead I   Lead III

Three Standard Leads:
Lead I (RW–LW)
Lead II (RW–LA)
Lead III (LW–LA)

Code:
L = left
R = right
W = wrist
A = ankle

Ground

Name _____

Date _____

Section _____

The (A) corresponds to the indicated Learning Outcome(s) (O) found at the beginning of the Laboratory Exercise.

Cardiac Cycle

PART A: Assessments

Complete the following statements:

1. The period during which a heart chamber is contracting is called _____. (A1)

2. The period during which a heart chamber is relaxing is called _____. (A1)

3. During ventricular contraction, the AV valves (tricuspid and mitral valves) are _____. (A1)

4. During ventricular relaxation, the AV valves are _____. (A1)

5. The pulmonary and aortic valves open when the pressure in the _____ exceeds the pressure in the pulmonary trunk and aorta. (A1)

6. The first sound of a cardiac cycle occurs when the _____ are closing. (A2)

7. The second sound of a cardiac cycle occurs when the _____ are closing. (A2)

8. The sound created when blood leaks back through an incompletely closed valve is called a _____. (A2)

PART B: Assessments

Complete the following:

1. What changes did you note in the heart sounds when you inhaled deeply? (A2)

_____

_____

2. What changes did you note in the heart sounds following the exercise period? (A2)

_____

_____

PART C: Assessments

Complete the following statements:

1. Normally, the _____ node serves as the pacemaker of the heart. (A3)

2. The _____ node is located in the inferior portion of the interatrial septum. (A3)

3. The fibers that carry cardiac impulses from the interventricular septum into the myocardium are called _____ fibers. (A3)

4. A(n) _____ is a recording of electrical changes occurring in the myocardium during a cardiac cycle. A3

5. The P wave corresponds to depolarization of the muscle fibers of the _____. A3

6. The QRS complex corresponds to depolarization of the muscle fibers of the _____. A3

7. The T wave corresponds to repolarization of the muscle fibers of the _____. A3

8. Why is atrial repolarization not observed in the ECG? A3

_____

_____

Part D Assessments

Identify the heart chambers and conduction system structures in the frontal section of the heart in figure 45.7.

FIGURE 45.7 Label the indicated heart chambers and conduction system structures (anterior view). A3

PART E: Assessments

1. Attach a short segment of the ECG recording from each of the three leads you used, and label the waves of each.

Lead I

Lead II

Lead III

2. What differences do you find in the ECG patterns of these leads? (A4) _____

_____

_____

_____

_____

3. How much time passed from the beginning of the P wave to the beginning of the QRS complex (P-Q interval, or P-R interval) in the ECG from lead I? (A4) _____

4. What is the significance of this P-Q (P-R) interval? (A4) _____

_____

_____

_____

_____

5. How can you determine heart rate from an electrocardiogram? (A4) _____

_____

_____

_____

6. What was your heart rate as determined from the ECG? (A4) _____

CRITICAL THINKING ASSESSMENT

If a person's heart rate were 72 beats per minute, determine the number of QRS complexes that would have appeared on an ECG during the first 30 seconds. (A4)

_____

_____

_____

LESSON 26 | ECG and Heart Function
The Meaning of Heart Sounds

Heart sounds are diagnostic of cardiac function. The sounds are produced when valves close following the initiation of ventricular systole and the beginning of ventricular diastole. In Procedure A of this lab, you heard the first sound (lubb) caused by the closure of the AV or atrioventricular valves and the second sound (dupp) caused by the closure of the semilunar valves. The closure of the valves makes it possible for the blood to flow in only one direction in response to the contraction and relaxation of the ventricles. In Procedure B of this lab you recorded the electrical activity (the ECG) that stimulates the mechanical activity. The ECG directly relates to mechanical events of the cardiac cycle.

In this simulation, you will demonstrate how the sounds made by the heart are matched to the waveforms of the ECG. Table 45.1 provides the physiological significance of each segment and wave of the ECG.

Before you begin, familiarize yourself with the following concepts (use your textbook as a reference):

- Atrioventricular valve function and location(s)

- Semilunar valve function and location(s)

- P, QRS, and T waves of the ECG

- Electrical stimulation of mechanical activity in the heart

State your hypothesis regarding the relationship between heart sounds and the waveforms of a typical electrocardiogram.

_____

_____

_____

_____

1. Open Lesson 26 in Ph.I.L.S., ECG and Heart Function: The Meaning of Heart Sounds.

2. Read the objectives and introduction and take the pre-laboratory quiz.

3. After completing the pre-laboratory quiz, read through the wet lab.

4. The laboratory exercise will open when you click CONTINUE after completing the wet lab (fig. 45.8).

FIGURE 45.8 Opening screen for the laboratory exercise on ECG and Heart Function: The Meaning of Heart Sounds.

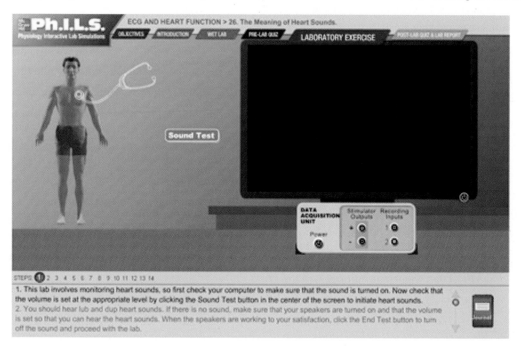

(*Hint:* Remember, there are two sounds, but they will be very close together. When you depress the button to record the heart sounds, make sure to record both sounds).

5. Print a copy of the final graph you recorded from the virtual computer screen.

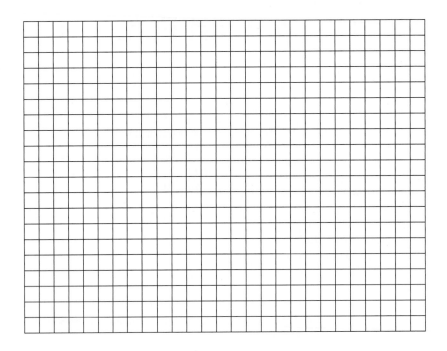

6. Follow the instructions provided in the laboratory exercise. Using the graph you printed, answer the following questions as you proceed.

7. Diagram the graph of the ECG recording from the computer simulation and indicate the heart sounds.

8. Answer the following questions about the graph in question 1.

   a. The QRS wave is correlated with the _____

   _____

   (first heart sound, lubb, or the second heart sound, dupp)?
   b. The T wave is correlated with the _____

   _____

   (first heart sound, lubb, or the second heart sound, dupp)?

9. Complete the following:

   a. Atrial depolarization, causing the _____ wave of the ECG, has triggered contraction of the

atria, and the blood is pumped from the atria into the ventricles. The position of the valves (AV valves open and semilunar valves closed) remains unchanged.

   b. Ventricular depolarization, causing the

   _____ wave of the ECG, has triggered contraction of the ventricles, and the blood is pumped from the ventricles into the arteries (aorta and pulmonary trunk). The _____ valves close, producing the_____ heart sound, and the _____ valves remain open.
   c. Ventricular repolarization, causing the

   _____ wave of the ECG, occurs as the ventricles are relaxing. To prevent the backflow of blood from the arteries into the ventricles, the

   _____ valves close, producing the

   _____ heart sounds.
   d. At what point do the AV valves reopen? _____

   _____

490

If a person has a heart murmur (caused by the improper opening or closing of a heart valve), the _____ (ECG or the heart sounds) will be unusual as the blood is forced through a more narrow opening (stenotic valve), or the blood will backflow through a valve that does not close properly (incompetent valve). **A2**

NOTES

Electrocardiography: BIOPAC© Exercise

Purpose of the Exercise

To record an ECG and calculate heart rate and rhythm changes that occur with changes in body position and breathing.

Computer system (Mac OS X 10.5–10.8, or PC running Windows XP, Vista, 7, or 8)

BIOPAC Student Lab software version 4.0 or above

BIOPAC Acquisition Unit MP36, MP35, or MP45

BIOPAC electrode lead set (SS2L)

BIOPAC disposable vinyl electrodes (EL503), 3 electrodes per subject

BIOPAC electrode gel (GEL1) and abrasive pad (ELPAD) or alcohol wipes

Exercise mat, cot, or lab table with pillow

Chair or stool

SAFETY

▶ The BIOPAC electrode lead set is safe and easy to use and should be used only as described in the procedures section of the laboratory exercise.

▶ The electrode lead clips are color-coded. Make sure they are connected to the properly placed electrode as demonstrated in figure 46.3.

▶ The vinyl electrodes are disposable and are meant to be used only once. Each subject should use a new set.

Learning Outcomes

After completing this exercise, you should be able to:

O1 Record the electrical components of an electrocardiogram.

O2 Correlate electrical events as displayed on the ECG with the mechanical events that occur during the cardiac cycle.

O3 Interpret rate and rhythm changes in the ECG associated with changes in body position, activity level, and breathing.

The **O** corresponds to the assessments **A** indicated in the Laboratory Assessment for this Exercise.

Pre-Lab

Carefully read the introductory material and examine the entire lab content. Be familiar with the cardiac cycle, cardiac conduction system, and electrocardiogram from lecture or the textbook.

The beating of the heart is one of the basic requirements for maintaining life, pumping blood that contains vital oxygen and nutrients to all the active cells of the body. A single heartbeat is a coordinated sequence of events, involving precisely timed contractions and relaxations of the four chambers of the heart. During a heartbeat, the atria, the upper chambers of the heart, contract simultaneously. This action squeezes the blood from the atria into the ventricles, the lower chambers of the heart. During atrial contraction (*systole*), the ventricles are in a state of relaxation (*diastole*) and are filling with blood from the contracting atria. Atrial contraction is soon followed by ventricular contraction, the simultaneous contraction of the two large pumping chambers of the heart. After a short period in which all four heart chambers are relaxed, these events repeat. This entire sequence of events occurs during every heartbeat and is called the *cardiac cycle.*

The heart is a large, cone-shaped pump, consisting mainly of cardiac muscle, one of the three types of muscle tissue in the body. Cardiac muscle differs from skeletal muscle in that it is controlled by specialized cardiac muscle cells right within the heart, rather than by motor neurons. The action of cardiac muscle can be modified by autonomic nerve input, but its basic rhythm is set by specialized cells of the heart, called *pacemaker cells,* found in the small clumps (nodes) of the cardiac conduction system.

The *cardiac conduction system* is a specialized group of cardiac muscle cells that generates and spreads cardiac impulses throughout the heart. The major components of the cardiac conduction system are:

1. **Sinoatrial (SA) node:** This small clump of specialized cardiac muscle cells in the upper wall of the right atrium is called the *pacemaker* of the heart. The SA node sets the rhythm of the heart, since it has the fastest innate firing rate of any conducting cells of the heart. It spreads its impulses across the atria to initiate atrial contraction and to the next component of the cardiac conduction system, the AV node.

2. **Atrioventricular (AV) node:** This tiny clump of conducting cardiac muscle cells in the lower portion of the right atrium receives impulses from the SA node and transports them to the next structure of the conduction system, the AV bundle. The AV node provides a "bottleneck" in the conduction system, which serves to delay the signal from the SA node for about 0.1 seconds before sending it on to the AV bundle. This gives the atria time to complete their contraction and squeeze any remaining blood into the ventricles before the ventricles begin to contract.

3. **Atrioventricular bundle (AV bundle; bundle of His):** This group of conducting cells at the top of the interventricular septum transports impulses from the AV node to the next structures of the conduction system, the left and right bundle branches. This structure distributes impulses toward the left and right ventricles.

4. **Left and right bundle branches:** The bundle branches transport cardiac impulses down the interventricular septum toward the apex of the heart. From these branches, the Purkinje fibers split off.

5. **Purkinje fibers:** Purkinje fibers are large cardiac muscle fibers that split off from the bundle branches and transport cardiac impulses throughout the entire myocardium of the ventricles. These fibers are considered the final component of the cardiac conduction system. From here, the impulses are spread from one contractile cardiac muscle cell to the next via gap junctions in their cell membranes.

Figure 46.1 shows the arrangement of the cardiac conduction system and the locations of its components.

Due to the action of the cardiac conduction system, the heart contracts in a coordinated manner. The atria contract just before the ventricles, providing the ventricles with time to fill completely, before starting their own contraction. This increases the efficiency of the heart as a pump and ensures that the heart contracts as a functional unit almost every time.

The events of the cardiac conduction system and the cardiac cycle form the basis for this lab exercise. An *electrocardiogram* (ECG or EKG) is a recording of the electrical activity derived from the heart during the cardiac cycle. As in the EMG and EEG that you studied in previous lab exercises, the ECG is derived from the sum of the ionic movements that occur during the action potentials of many cardiac muscle cells at the same time. It is not a recording of action potentials in a single cardiac muscle cell. The movement of ions in the extracellular fluids generates current flow, which can be detected by electrodes placed on the skin. ECGs are used not only to observe normal heart action,

FIGURE 46.1 The components of the cardiac conduction system.

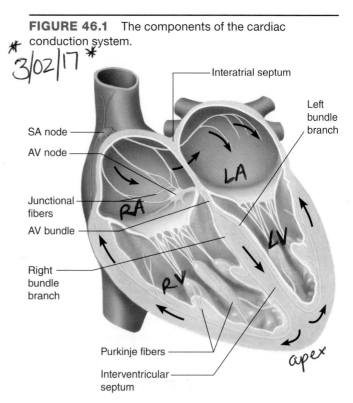

PQRST - alphabet

FIGURE 46.2 Components of the ECG.

Source: Courtesy of and © BIOPAC Systems Inc. www.biopac.com

When atria is contracting

ventricular contraction

ventricular relaxation

FIGURE 46.3 Electrode attachments for lead II.

Source: Courtesy of and © BIOPAC Systems Inc. www.biopac.com

but also to help in the diagnosis of many disorders of the cardiac conduction system and some types of heart disease.

The ECG is recorded in units of mV (millivolts). The typical appearance of an ECG for a single heartbeat is shown in figure 46.2. It consists of three deflections from the isoelectric line (the baseline, or zero-voltage line, which occurs during the short periods in which no voltage is being recorded):

1. **P wave:** The P wave is a positive (+) deflection from the isoelectric line that results from atrial depolarization. This depolarization is caused by the firing of the SA node, and it leads to the immediate contraction of the atria.
2. **QRS complex:** The QRS complex is a large deflection that results from ventricular depolarization. This depolarization of the ventricles is immediately followed by ventricular contraction (systole). The electrical events of atrial repolarization occur simultaneously with those of ventricular depolarization. Therefore, atrial repolarization is "hidden" inside the QRS complex and cannot be detected separately on an ECG.
3. **T wave:** The T wave is a positive deflection that results from ventricular repolarization. Repolarization of the ventricles is immediately followed by ventricular relaxation (diastole).

In addition to the wave components of the ECG, various segments and intervals can also be measured, all with their own significance in the cardiac cycle. A *segment* is a period of time between the end of a wave and the beginning of the next wave; it lies entirely along the isoelectric line. For example, the P-R segment runs from the end of the P wave to the beginning of the QRS complex; it represents the time required for impulses to be conducted from the AV node to the ventricles. An *interval* is a period of time that contains at least one wave and one segment along the isoelectric line. For example, the P-R interval runs from the beginning of the P wave to the beginning of the QRS complex; it represents the time between the beginning of atrial depolarization and the beginning of ventricular depolarization.

ECGs can be conducted by placing electrodes in a variety of positions on the limbs and chest and measuring the voltage between them. There are several possible combinations of electrodes that can measure the ECG, and the results will vary with electrode placement. The particular arrangement of

positive and negative electrodes, along with a ground electrode, that is used to conduct the ECG is called a *lead*. In this lab exercise, you are going to use lead II, in which the positive electrode is on the left ankle, the negative electrode on the right wrist, and the ground electrode on the right ankle. Proper electrode placement for lead II is shown in figure 46.3. Typical lead II ECG values are shown in table 46.1.

PROCEDURE A: Setup

1. With your computer turned **ON** and the BIOPAC MP36/35 unit turned **OFF,** plug the electrode lead set (SSL2) into Channel 1. If you are using the MP45, the USB cable must be connected, and the ready light must be on.
2. Turn on the MP36/35 Data Acquisition Unit.
3. Attach the electrodes to the subject as shown in figure 46.3. To help to ensure good contact, make sure the area to be in contact with the electrodes is clean by wiping with the abrasive pad (ELPAD) or alcohol wipe. A small amount of BIOPAC electrode gel (GEL1) may also be used to make better contact between the sensor in the electrode and the skin. For optimal adhesion, the electrodes should be placed on the skin at least 5 minutes before the calibration procedure.
4. Attach the electrode lead set with the pinch connectors as shown in figure 46.3. Make sure the proper color electrode cable is attached to the appropriate electrode:

Table 46.1 Normal Lead II ECG Values

Phase	Duration (seconds)	Amplitude (millivolts)
P wave	0.06–0.11	<0.25
P-R interval	0.12–0.20	
P-R segment	0.08	
QRS complex (R)	<0.12	0.8–1.2
S-T segment	0.12	
Q-T interval	0.36–0.44	
T wave	0.16	<0.5

white lead on anterior medial surface of right forearm at the wrist; *red lead* on medial surface of left leg at the ankle; and *black lead* (ground) on medial surface of right leg at the ankle.

5. Have the subject lie down in the supine position on the mat, cot, or table and relax. Make sure to position the electrode cables so that they are not pulling on the electrodes. The cable clips may be attached to the subject's clothing for convenience.

6. Start the BIOPAC student lab program for L05-Electrocardiography (ECG) 1, and click **OK**.

7. Designate a unique filename to be used to save the subject's data and click **OK**.

PROCEDURE B: Calibration

1. Make sure the subject is relaxed throughout the calibration procedure. *For all procedures, the subject must be still—no laughing or talking.* Click on **Calibrate**. The calibration will stop automatically after 8 seconds.

2. The calibration recording should look similar to figure 46.4. If it does, click **Continue**. If it does not, check connections and electrode adhesion, and click **Redo Calibration**. If the baseline is not flat (which is called baseline drift), or there are any large spikes or jitters, click **Redo Calibration.**

PROCEDURE C: Recording

This lab exercise consists of recording the ECG under four different conditions (segments): lying down in a supine position, seated (immediately after sitting up), performing deep breathing, and after exercise. In addition to the subject, you will have to designate a **Director** to coordinate the group members and a **Recorder** to insert event markers and time the exercise.

Segment 1—Supine (Lying Down)

1. With the subject lying down and relaxed, click **Record**. A marker labeled "Supine" will come up automatically. Record for 20 seconds.

2. Click **Suspend**. The recording should be similar to figure 46.5. If it is not, click **Redo** and repeat step 1.

3. If correct, click **Continue** to proceed to Segment 2.

FIGURE 46.5 Segment 1 ECG—supine (lying down).

Source: Courtesy of and © BIOPAC Systems Inc. www.biopac.com

Segment 2—Seated (After Sitting Up)

4. Have the subject quickly get up and sit in a chair, with forearms resting comfortably on the chair armrests or thighs. Click **Record**. A marker labeled "Seated" will come up automatically.

5. Record for 20 seconds (seconds 21–40).

6. Click **Suspend**. The recording should be similar to figure 46.6. If it is not, click **Redo** and repeat steps 4 and 5.

7. If correct, click **Continue** to proceed to Segment 3.

Segment 3—Deep Breathing

8. Have the subject remain seated and prepare to take five deep breaths during the Segment 3 recording. Allow the subject to breathe at his/her own resting rate.

9. Click **Record**. A marker labeled "Deep Breathing" will come up automatically. Have the subject inhale and exhale deeply for five cycles (approximately 30 seconds; seconds 41–70). At the beginning of each inhalation, the Recorder should press the F4 key. At the beginning of each exhalation, the Recorder should press the F5 key.

10. After five breaths, click **Suspend**. The recording should be similar to figure 46.7. If it is not, click **Redo** and repeat step 9.

11. If correct, click **Continue** to proceed to Segment 4.

Segment 4—After Exercise

12. Have the subject perform moderate exercise to elevate his/her heart rate, such as jumping jacks or running in place. The electrode cable pinch connectors may

FIGURE 46.4 Sample calibration recording.

Source: Courtesy of and © BIOPAC Systems Inc. www.biopac.com

FIGURE 46.6 Segment 2 ECG—seated (after sitting up).

Source: Courtesy of and © BIOPAC Systems Inc. www.biopac.com

be removed during exercising, but not the electrodes. Immediately after the subject stops exercising, quickly reattach the pinch connectors; then move on to step 13.

13. Click **Record.** A marker labeled "After Exercise" will come up automatically.
14. Record for 60 seconds (seconds 71–130).
15. Click **Suspend.** The recording should be similar to figure 46.8. If it is not, click **Redo** and repeat steps 12–14.
16. If correct, click **Done.**
17. When a pop-up message asks you if you are sure you are done recording, click **Yes.** Remove the electrodes.

PROCEDURE D: Data Analysis

1. You can either analyze the current file or save it and do the analysis later using the **Review Saved Data** mode and selecting the correct file. Note the channel number designations: **CH 1, ECG,** and **CH 40, Heart Rate.** The data window should look similar to figure 46.9.

Heart Rate

2. Note the settings for the measurement boxes across the top of the data window as follows:
   - CH 40—Value
   - CH 1—Delta T (Δ T)
   - CH 1—P-P
   - CH 1—BPM

   These will provide you with the following measurements:

   Value—BPM or beats per minute (heart rate) in the preceding R-R interval

FIGURE 46.8 Segment 4 ECG—after exercise.

Source: Courtesy of and © BIOPAC Systems Inc. www.biopac.com

FIGURE 46.9 Data window for the entire ECG exercise.

Source: Courtesy of and © BIOPAC Systems Inc. www.biopac.com

Δ T—the difference in time between the beginning and the end of the selected area

P-P (Peak to Peak)—measures the difference in amplitude between the maximum and minimum values of the selected area

BPM—the beats per minute; time difference between the beginning and the end of the selected area divided into 60 seconds/minute

3. Set up your display window for optimal viewing of three to five successive beats from Segment 1 (0–20 seconds). Do this by clicking anywhere in the horizontal axis (the numbers just above the word "seconds" at the bottom of the data window). A box will pop up in the middle of the screen, called **Set Screen Horizontal Axis.** Set the start time to 2 and the end time to 8 seconds. Then click **OK.** This will spread out the data so you can see ECG recordings of individual heartbeats.

4. To measure heart rate, click on the I-beam and select any point within the R-R interval, as shown in figure 46.10. The result will be displayed by CH 40, value. Record the heart rate in table 46.2 of Part A of Laboratory Assessment 46.

5. Measure heart rate for two more heartbeats within the 0–20 second range. You can scroll through the segment by clicking on the small square under the word "seconds" on the horizontal axis and dragging the square with your cursor. Record these heart rate measurements in table 46.2. Then take the mean of the three measurements.

FIGURE 46.10 Proper selection of a data point for measuring heart rate.

Source: Courtesy of and © BIOPAC Systems Inc. www.biopac.com

497

6. Now scroll to the other three segments of the lab exercise and repeat steps 4 and 5. Record the data in table 46.2 of the lab assessment.

7. Now hide CH 40. Do this by holding down the Alt key and clicking on the CH 40 square.

Ventricular Systole and Diastole

8. Zoom in on a single cardiac cycle from Segment 1. Using the I-beam tool, measure the duration of ventricular systole and diastole by selecting the appropriate areas, as shown in figure 46.2. To measure ventricular systole, select the area between the peak of an R wave and the point one-third of the way down the descending part of the T wave. To measure ventricular diastole, select the area between the end point of the systole measurement and the peak of the next R wave. The result will be displayed by CH 1, Delta T. Record your results in table 46.3.

9. Now scroll to Segment 4, "After Exercise," Repeat step 8 for Segment 4, and record your results in table 46.3.

Waves, Intervals, and Segments of the ECG

10. Scroll back to Segment 1 (Supine) and observe the ECG of a single cardiac cycle (heartbeat).

11. Using the I-beam tool, measure the durations of the various waves, intervals, and segments of the ECG, as required to fill out table 46.4. See figure 46.11 for an example of how to measure the duration of a P wave. See figure 46.12 for measuring P-R interval. The results for duration measurements will be displayed by CH 1, Delta T. (Refer back to figure 46.2 for proper selection of the areas to measure the intervals and segments.)

FIGURE 46.11 Proper selection of area for measuring the duration of a P wave.

Source: Courtesy of and © BIOPAC Systems Inc. www.biopac.com

FIGURE 46.12 Proper selection of area for measuring the duration of the P-R interval.

Source: Courtesy of and © BIOPAC Systems Inc. www.biopac.com

12. Measure the durations for all waves, intervals, and segments of two more heartbeats in Segment 1 and record the results in table 46.4. Then calculate the means as required in the table.

13. Measure the amplitudes of the P wave, QRS complex, and T wave for three heartbeats in Segment 1. To measure the amplitude of a P wave, for example, use the I-beam tool to select the area between the beginning of the P wave deflection and the highest point of the P wave. The result will be displayed by CH 1, P–P. (The P–P measurement calculates the difference between the maximum and minimum values in the selected area. Therefore, your selection does not have to be perfect; you can simply include the maximum and minimum points in your selection of the P wave, and the P–P measurement will calculate the amplitude correctly.) Record your results in table 46.4, and calculate the means.

14. Scroll to the beginning of Segment 4 (After Exercise).

15. Measure the durations of all waves, intervals, and segments for a single heartbeat, close to the beginning of Segment 4. Record your results in table 46.5.

16. Measure the amplitudes of the P wave, QRS complex, and T wave for a single heartbeat in Segment 4. Record your results in table 46.5.

17. Exit the program.

18. Complete Part B of the laboratory assessment.

ASSESSMENT
46

Name _____

Date _____

Section _____

The Ⓐ corresponds to the indicated Learning Outcome(s) Ⓞ found at the beginning of the Laboratory Exercise.

Electrocardiography:
BIOPAC© Exercise

PART A: Data and Calculations Assessments

Subject Profile

Name _____   Height _____

Age _____

Gender: Male / Female

Complete the following tables with the lesson data indicated and calculate the mean as appropriate. **A1**

1. Heart Rate Measurement

Table 46.2

Segment Condition	Cardiac Cycle 40 Value			Mean (calculate)
	1	2	3	
1: Supine				
2: Seated				
3: Start of inhale				
3: Start of exhale				
4: After exercise				

* If CH 40 was not recorded, use [1] [BPM].

2. Ventricular Systole and Diastole Measurements

Table 46.3

Segment	Duration (sec) 1 Delta T	
	Ventricular Systole	Ventricular Diastole
1: Supine		
4: After exercise		

3a. Waves, Intervals, and Segments of the ECG (Segment 1, Supine)

Table 46.4 Measurements from Three Cardiac Cycles of the Supine Recording

Component of the ECG	Normal Values (based on resting heart rate of 75 BPM)		Duration (sec) 1 Delta T				Amplitude (mV) 1 P-P			
			Cycle 1	Cycle 2	Cycle 3	Mean (calc.)	Cycle 1	Cycle 2	Cycle 3	Mean (calc.)
Waves	Dur. (sec)	Amp. (mV)								
P	.07–.18	< .20								
QRS complex	.06–.12	.10–1.5								
T	.10–.25	< .5								
Intervals	Duration (sec)									
P-R	.12–.20									
Q-T	.32–.36									
R-R	~ .80									
Segments	Duration (sec)									
P-R	.02–.10									
S-T	< .20									
T-P	0–.40									

3b. Waves, Intervals, and Segments of the ECG (Segment 4, After Exercise)

Table 46.5 Measurements from One Cardiac Cycle of the After Exercise Recording

Component of the ECG	Normal Values (based on resting heart rate of 75 BPM)		Duration (sec) 1 Delta T	Amplitude (mV) 1 P-P
Waves	Dur. (sec)	Amp. (mV)		
P	.07–.18	< .20		
QRS complex	.06–.12	.10–1.5		
T	.10–.25	< .5		
Intervals	Duration (sec)			
P-R	.12–.20			
Q-T	.32–.36			
R-R	~ .80			
Segments	Duration (sec)			
P-R	.02–.10			
S-T	< .20			
T-P	0–.40			

PART B: Assessments

Complete the following:

1. Was there a change in heart rate when the subject went from the supine position (Segment 1) to sitting (Segment 2)?

_____

Describe the physiological mechanisms causing this change. A3

_____

_____

_____

_____

2. Was there a difference between the heart rates when the subject was sitting at rest (Segment 2) and just after exercise (beginning of Segment 4)? _____ Explain any differences. A3

_____

_____

_____

_____

3. What changes occurred in the duration of ventricular systole between the supine position (Segment 1) and post-exercise (beginning of Segment 4)? Explain any differences. A3

_____

_____

_____

_____

_____

4. What changes occurred in the duration of ventricular diastole between the supine position (Segment 1) and post-exercise (beginning of Segment 4)? Explain any differences. A3

_____

_____

_____

_____

_____

5. What changes occurred in the cardiac cycle during the respiratory cycle? In other words, did you observe a difference in heart rate between inhaling and exhaling (Segment 3)? Explain.

_____

_____

_____

_____

_____

P - atria contraction
QRS - ventricular contraction
T - ventricular relaxation.

relaxation - ♡ getting filled with blood (D.)
contraction - ♡ contracting out blood (S.)

atrial relaxation, atrial contraction,
ventricle relaxation, ventricle contraction

~~█████~~ 70% of blood goes down to ventricle
from atria

contractile
noncontractile cells w/ property of muscle + nervous
tissue

- SA node sets rhythm of heart

electrical current (page 494)
_____

SA node
↓
AV node
↓
apex & backup sides of ♡

LABORATORY
EXERCISE 47

Blood Vessel Structure, Arteries, and Veins

Purpose of the Exercise

To review the structure and functions of blood vessels and the circulatory pathways, and to locate major arteries and veins.

MATERIALS NEEDED

Compound light microscope

Prepared microscope slides:
  Artery cross section
  Vein cross section

Live frog or goldfish

Frog Ringer's solution

Frog board or heavy cardboard (with a 1-inch hole cut in one corner)

Paper towel

Rubber bands

Masking tape

Dissecting pins

Thread

Dissectible human heart model

Human torso model

Anatomical charts of the cardiovascular system

For Learning Extension Activity:

Ice

Hot plate

Thermometer

SAFETY

▶ Wear disposable gloves when handling the live frogs.

▶ Return the frogs to the location indicated after the experiment.

▶ Wash your hands before leaving the laboratory.

Learning Outcomes AP|R

After completing this exercise, you should be able to:

01 Distinguish and sketch the three major layers in the wall of an artery and a vein.

02 Interpret the types of blood vessels in the web of a frog's foot.

03 Distinguish and trace the pulmonary and systemic circuits.

04 Locate the major arteries in these circuits on a diagram, chart, or model.

05 Locate the major veins in these circuits on a diagram, chart, or model.

The O corresponds to the assessments A indicated in the Laboratory Assessment for this Exercise.

Pre-Lab

Carefully read the introductory material and examine the entire lab. Be familiar with the structure of blood vessels and the major arteries and veins of the systemic and pulmonary circuits from lecture or the textbook. Visit www.mhhe.com/martinseries3 for LabCam videos. Answer the pre-lab questions.

Pre-Lab Questions Select the correct answer for each of the following questions:

1. The middle layer of an artery and vein contains mostly
   a. connective tissue.
   b. simple squamous epithelium.
   c. smooth muscle.
   d. nervous tissue.

2. Arteries carry blood
   a. away from the heart.
   b. toward the heart.
   c. both away from and toward the heart.
   d. only in the systemic circuit.

3. The _____ has the thickest wall.
   a. right atrium   b. left atrium
   c. right ventricle   d. left ventricle

4. Which of the following arteries is *not* a direct branch of the aorta?
   a. coronary artery   b. brachiocephalic trunk
   c. left subclavian artery   d. brachial artery

5. Which of the following veins is *not* located in the neck?
   a. vertebral vein   b. brachiocephalic vein
   c. external jugular vein   d. internal jugular vein

503

6. Which of the following arteries is part of the cerebral arterial circle (circle of Willis)?
  a. anterior communicating artery
  b. external carotid artery
  c. facial artery
  d. popliteal artery

7. A capillary wall is composed of three tunics similar to an artery or a vein.
  a. True _____   **b.** False ✓_____

8. The left ventricle pumps blood into the aorta of the systemic circuit.
  a. True ✓_____   **b.** False _____

The blood vessels form a closed system of tubes that carry blood to and from the heart, lungs, and body cells. These tubes include *arteries* and *arterioles* that conduct blood away from the heart; *capillaries* in which exchanges of substances occur between the blood and surrounding tissues; and *venules* and *veins* that return blood to the heart.

Arteries and veins are composed of three layers or tunics. The inner layer, *tunica interna*, is composed of an endothelium of simple sqamous epithelial cells and the basement membrane. This establishes a smooth lining next to the lumen of the blood vessel. The middle layer, *tunica media*, is composed of smooth muscle. The contraction and relaxation of the smooth muscle modify the size of the lumen, which influences the amount of blood flow through a particular area of the body. The outer layer, *tunica externa*, is composed of connective tissues rich in collagen, providing some support and protection. In general, the arteries have thicker middle and outer layers and elastic laminae, which provide additional support and elasticity to withstand the higher blood pressures. In contrast, veins have much lower blood pressure, contain thinner middle and outer layers, and have occasional valves to prevent backflow. Relatively small arteries are called arterioles; small veins are termed venules. Capillaries connecting arterioles to venules retain only the endothelium and basement membrane.

The blood vessels of the cardiovascular system can be divided into two major pathways—the *pulmonary circuit* and the *systemic circuit*. Within each circuit, arteries transport blood away from the heart. After exchanges of gases, nutrients, and wastes have occurred between the blood and the surrounding tissues, veins return the blood to the heart.

Variations exist in anatomical structures among humans, especially in arteries and veins. The illustrations in this laboratory manual represent normal (meaning the most common variation) anatomy.

PROCEDURE A: Blood Vessel Structure

1. Reexamine the introductory material. Compare the structure of an artery and a vein, illustrated in figure 47.1. Note that arteries and veins possess the same three layers or tunics, but an artery has thicker middle and outer tunics and also contains two elastic laminae. The vein has thinner middle and outer tunics, lacks the two elastic laminae, and has periodic valves.

FIGURE 47.1 Structure of the wall of an artery and a vein.

Direction of blood flow

Direction of blood flow

Valve

Endothelium
Basement membrane

Tunica interna

Tunica media

Tunica externa

Endothelium
Basement membrane
Internal elastic lamina
External elastic lamina

Vein

Artery

FIGURE 47.2 `AP|R` Cross section of a small artery (arteriole) and a small vein (venule) (200×).

Endothelium of inner layer (tunica interna)

Middle layer (tunica media)

Outer layer (tunica externa)

Lumen of venule

Lumen of arteriole

2. Obtain a microscope slide of an artery cross section, and examine it using low-power and high-power magnification (fig. 47.2). Identify the three distinct layers (tunics) of the arterial wall. The inner layer (tunica interna), is composed of an endothelium (simple squamous epithelium) and appears as a wavy line due to an abundance of elastic fibers that have recoiled just beneath it. The middle layer (tunica media) consists of numerous concentrically arranged smooth muscle cells with elastic fibers scattered among them. The outer layer (tunica externa) contains connective tissue rich in collagen fibers.

3. Prepare a labeled sketch of the arterial wall in Part A of Laboratory Assessment 47.

4. Obtain a slide of a vein cross section and examine it as you did the artery cross section (fig. 47.2). Note the thinner wall and larger lumen relative to an artery of comparable locations. Identify the three layers of the wall and prepare a labeled sketch in Part A of the laboratory assessment.

5. Complete Part A of the laboratory assessment.

6. Observe the blood vessels in the webbing of a frog's foot. (As a substitute for a frog, a live goldfish could be used to observe circulation in the tail.) To do this, follow these steps:

   a. Obtain a live frog. Wrap its body in a moist paper towel, leaving one foot extending outward. Secure the towel with rubber bands, but be careful not to wrap the animal so tightly that it could be injured. Try to keep the nostrils exposed.

   b. Place the frog on a frog board or on a piece of heavy cardboard with the foot near the hole in one corner.

   c. Fasten the wrapped body to the board with masking tape.

   d. Carefully spread the web of the foot over the hole and secure it to the board with dissecting pins and thread (fig. 47.3). Keep the web moist with frog Ringer's solution.

   e. Secure the board on the stage of a microscope with heavy rubber bands and position it so that the web is beneath the objective lens.

   f. Focus on the web, using low-power magnification, and locate some blood vessels. Note the movement of the blood cells and the direction of the blood flow. You might notice that red blood cells of frogs are nucleated. Identify an arteriole, a capillary, and a venule.

   g. Examine each of these vessels with high-power magnification.

   h. When finished, return the frog to the location indicated by your instructor. The microscope lenses and stage will likely need cleaning after the experiment.

7. Complete Part B of the laboratory assessment.

FIGURE 47.3 Spread the web of the foot over the hole and secure it to the board with pins and thread.

- Nostrils exposed
- Moist paper towel
- Tape
- Board
- Hole in board

LEARNING EXTENSION ACTIVITY

Investigate the effect of temperature change on the blood vessels of the frog's foot by flooding the web with a small quantity of ice water. Observe the blood vessels with low-power magnification and note any changes in their diameters or the rate of blood flow. Remove the ice water and replace it with water heated to about 35°C (95°F). Repeat your observations. What do you conclude from this experiment?

_____

_____

_____

PROCEDURE B: Paths of Circulation

1. Study figures 47.4 and 47.5, which illustrate the pathway of circulation in the pulmonary and systemic circuits. Examine the major blood vessels of the two circuits that are associated near and connected to the heart chambers.
2. Locate the following blood vessels on the available anatomical charts, the dissectible human heart model, and the human torso model:

    pulmonary circuit (figs. 47.4 and 47.5)
- pulmonary trunk   1
- pulmonary arteries   2
- pulmonary veins   4

    systemic circuit (fig. 47.4)
- aorta   1
- superior vena cava   1
- inferior vena cava   1

PROCEDURE C: The Arterial System

The *pulmonary trunk,* connected to the right ventricle, is the main artery of the pulmonary circuit. The *aorta,* connected to the left ventricle, is the main artery of the systemic circuit. The aorta has a thicker wall than the pulmonary trunk. This allows the aorta to withstand the greater pressure created from the contraction of the very thick myocardium of the left ventricle compared to the right ventricle. In this procedure, trace the pathway of blood vessels as outlined in the various sections. The first branches of the aorta near the aortic valve are the two coronary arteries, which deliver oxygen-rich blood to the myocardium. The brachiocephalic trunk, the left common carotid artery, and the left subclavian artery branch from the aortic arch. One pathway of particular importance to note is the *cerebral arterial circle* (*circle of Willis*) of the arteries at the base of the brain. This *arterial anastomosis* of blood vessels encircles the pituitary gland and the optic chiasma, which allows for alternative routes of blood supply to various parts of the brain.

1. Use figures 47.6, 47.7, 47.8, 47.9, and 47.10 as references to identify the major arteries of the systemic circuit.

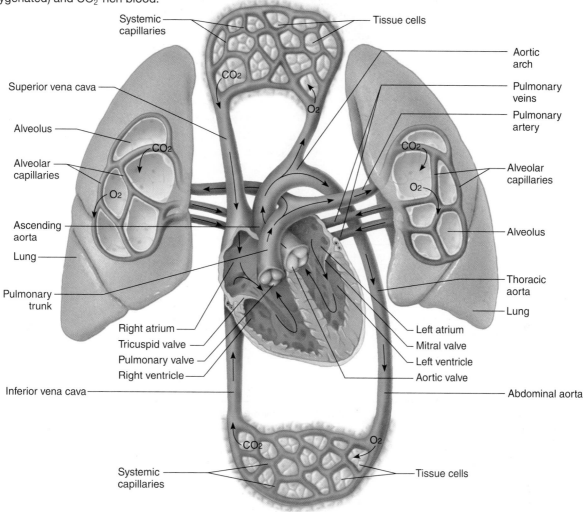

FIGURE 47.4 The major blood vessels associated with the pulmonary and systemic circuits. The red colors indicate locations of oxygen-rich (oxygenated) and CO_2-poor blood. The blue colors indicate locations of oxygen-poor (deoxygenated) and CO_2-rich blood.

FIGURE 47.5 Blood vessels of the pulmonary circuit. The pulmonary arteries carry oxygen-poor blood from the right ventricle to the lung capillaries; the pulmonary veins carry oxygen-rich blood from the lung capillaries to the left atrium. Exchange of carbon dioxide and oxygen occurs at the pulmonary alveoli.

FIGURE 47.6   Arteries supplying the right side of the neck and head. (The clavicle has been removed.)

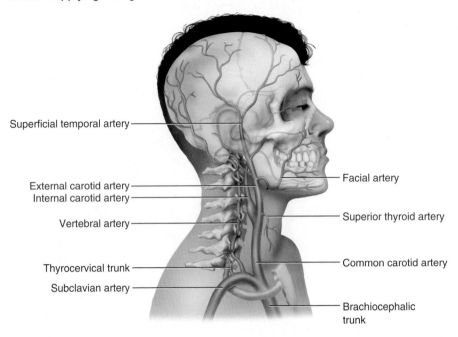

Superficial temporal artery

External carotid artery
Internal carotid artery

Vertebral artery

Thyrocervical trunk

Subclavian artery

Facial artery

Superior thyroid artery

Common carotid artery

Brachiocephalic
trunk

FIGURE 47.7   AP|R   Inferior view of the brain showing the major arteries of the base of the brain, including the cerebral arterial circle.

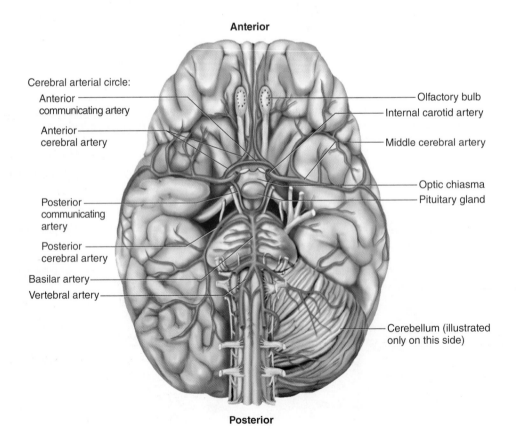

Anterior

Cerebral arterial circle:
Anterior
communicating artery

Anterior
cerebral artery

Posterior
communicating
artery

Posterior
cerebral artery

Basilar artery

Vertebral artery

Olfactory bulb
Internal carotid artery

Middle cerebral artery

Optic chiasma
Pituitary gland

Cerebellum (illustrated
only on this side)

Posterior

FIGURE 47.8 `AP|R` Major arteries of the right shoulder and upper limb.

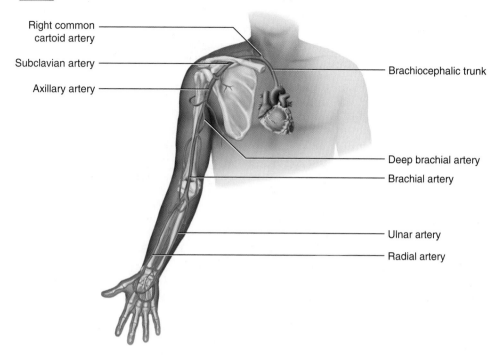

Right common cartoid artery

Subclavian artery

Axillary artery

Brachiocephalic trunk

Deep brachial artery

Brachial artery

Ulnar artery

Radial artery

2. Locate the following arteries on the charts and human torso model:

aorta
- ascending aorta
- aortic arch (arch of aorta)
- thoracic aorta
- abdominal aorta

branches of the aorta
- coronary artery
- brachiocephalic trunk (artery)
- left common carotid artery
- left subclavian artery
- celiac trunk (artery)
  - left gastric artery
  - splenic artery
  - common hepatic artery
- superior mesenteric artery
- renal artery
- inferior mesenteric artery
- median sacral artery
- common iliac artery

arteries to neck, head, and brain
- vertebral artery
- thyrocervical trunk
- common carotid artery

- external carotid artery
  - superior thyroid artery
  - superficial temporal artery
  - facial artery
- internal carotid artery

arteries of base of brain
- vertebral artery
- basilar artery
- internal carotid artery
- cerebral arterial circle (circle of Willis)
  - anterior cerebral artery
  - anterior communicating artery
  - posterior communicating artery
  - posterior cerebral artery
- middle cerebral artery

arteries to shoulder and upper limb
- subclavian artery
- axillary artery
- brachial artery
- deep brachial artery
- ulnar artery
- radial artery

arteries to pelvis and lower limb
- common iliac artery
- internal iliac artery

FIGURE 47.9 `AP|R` (a) Major abdominal, mesenteric, and pelvic arteries (anterior view). (b) Major branches of the celiac trunk (anterior view).

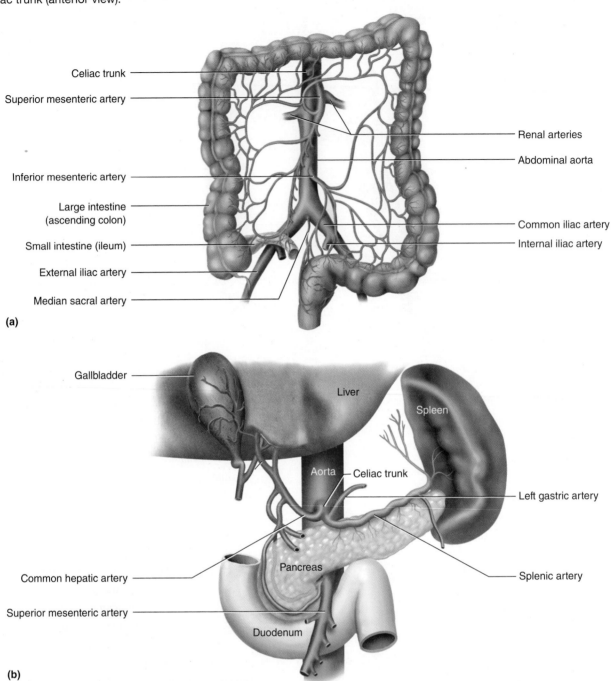

Celiac trunk

Superior mesenteric artery

Renal arteries

Abdominal aorta

Inferior mesenteric artery

Large intestine (ascending colon)

Common iliac artery

Internal iliac artery

Small intestine (ileum)

External iliac artery

Median sacral artery

(a)

Gallbladder

Liver

Spleen

Aorta — Celiac trunk

Left gastric artery

Common hepatic artery

Pancreas

Splenic artery

Superior mesenteric artery

Duodenum

(b)

- external iliac artery
- femoral artery
- deep artery of thigh (deep femoral artery)
- popliteal artery
- anterior tibial artery
- dorsalis pedis artery (dorsal artery of foot)
- posterior tibial artery

3. Complete Part C of the laboratory assessment.

PROCEDURE D: The Venous System

The two venae cavae represent the very large veins returning blood from the superior and inferior portions of the systemic circuit into the right atrium. In this procedure, trace the veins as outlined in the various sections, keeping in mind that the flow of blood from smaller veins converges into larger veins. It should be noted that many of the veins have names corresponding to the arteries that they parallel.

FIGURE 47.10 `AP|R` Major arteries supplying the right lower limb (anterior view).

Lateral ← | → Medial

- Common iliac artery
- External iliac artery
- Inguinal ligament
- Internal iliac artery
- Femoral artery
- Deep artery of thigh (deep femoral artery)
- Popliteal artery
- Posterior tibial artery
- Anterior tibial artery
- Dorsalis pedis artery (dorsal artery of foot)

FIGURE 47.11 Major veins associated with the right side of the head and neck. (The clavicle has been removed.)

- Dural venous sinus
- External jugular vein
- Vertebral vein
- Internal jugular vein
- Subclavian vein
- Brachiocephalic vein

FIGURE 47.12 `AP|R` Anterior view of the veins of the right upper limb. The superficial veins are dark blue and the deep veins are light blue.

- Subclavian vein
- Axillary vein
- Cephalic vein
- Median cubital vein
- Cephalic vein
- Brachial vein
- Basilic vein
- Radial vein
- Ulnar vein
- Basilic vein

1. Use figures 47.11, 47.12, 47.13, and 47.14 as references to identify the major veins of the systemic circuit.
2. Locate the following veins on the charts and the human torso model:

   veins from brain, head, and neck
   - dural venous sinus
   - external jugular vein
   - internal jugular vein
   - vertebral vein
   - subclavian vein
   - brachiocephalic vein
   - superior vena cava

FIGURE 47.13 AP|R Major abdominal, hepatic portal, and pelvic veins (anterior view).

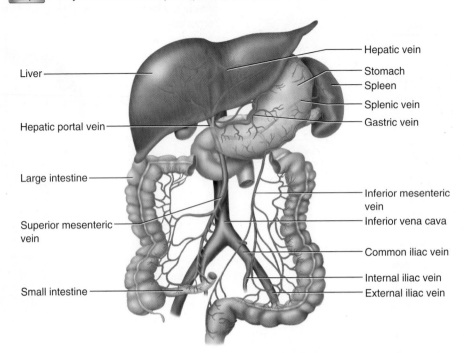

- Hepatic vein
- Liver
- Stomach
- Spleen
- Splenic vein
- Gastric vein
- Hepatic portal vein
- Large intestine
- Inferior mesenteric vein
- Inferior vena cava
- Superior mesenteric vein
- Common iliac vein
- Internal iliac vein
- Small intestine
- External iliac vein

veins from upper limb and shoulder

- radial vein
- ulnar vein
- brachial vein
- basilic vein
- cephalic vein
- median cubital vein (antecubital vein)
- axillary vein
- subclavian vein

veins of the abdominal viscera

- hepatic portal vein
- gastric vein
- superior mesenteric vein
- splenic vein
- inferior mesenteric vein
- hepatic vein
- renal vein

veins from lower limb and pelvis

- anterior tibial vein
- posterior tibial vein
- popliteal vein
- femoral vein
- great (long) saphenous vein
- small (short) saphenous vein
- external iliac vein
- internal iliac vein
- common iliac vein
- inferior vena cava

3. Complete Parts D and E of the laboratory report.

FIGURE 47.14 AP|R Major veins of the right pelvis and lower limb (anterior view).

- Inguinal ligament
- Common iliac vein
- External iliac vein
- Internal iliac vein
- Deep femoral vein
- Femoral vein
- Great saphenous vein
- Popliteal vein
- Small saphenous vein
- Posterior tibial vein
- Anterior tibial vein

Name _____

Date _____

Section _____

The (A) corresponds to the indicated Learning Outcome(s) (O)
found at the beginning of the Laboratory Exercise.

Blood Vessel Structure, Arteries, and Veins

PART A: Assessments

1. Sketch and label a section of an arterial wall next to a section of a venous wall. (A1)

Artery wall (_____×)	Vein wall (_____×)

2. Describe the differences you noted in the structures of the arterial and venous walls. Mention each of the three layers of the wall. (A1) _____

_____

_____

CRITICAL THINKING ASSESSMENT

Explain the functional significance of the differences you noted in the structures of the arterial and venous walls. (A1)

_____

_____

PART B: Assessments

Complete the following:

1. How did you distinguish between arterioles and venules when you observed the vessels in the web of the frog's foot? **A2**

   _____

   _____

2. How did you recognize capillaries in the web? **A2** _____

   _____

3. What differences did you note in the rate of blood flow through the arterioles, capillaries, and venules? **A2**

   _____

   _____

PART C: Assessments

Provide the name of the missing artery in each of the following sequences: **A3**

1. Brachiocephalic trunk, _____, right axillary artery

2. Ascending aorta, _____, descending thoracic aorta

3. Abdominal aorta, _____, ascending colon (right side of large intestine)

4. Brachiocephalic trunk, _____, right external carotid artery

5. Axillary artery, _____, radial artery

6. Common iliac artery, _____, femoral artery

7. Pulmonary trunk, _____, lungs

PART D: Assessments

Provide the name of the missing vein or veins in each of the following sequences: **A3**

1. Right subclavian vein, _____, superior vena cava

2. Anterior tibial vein, _____, femoral vein

3. Internal iliac vein, _____, inferior vena cava

4. Vertebral vein, _____, brachiocephalic vein

5. Popliteal vein, _____, external iliac vein

6. Lungs, _____, left atrium

7. Kidney, _____, inferior vena cava

PART E: Assessments

Label the major arteries and veins indicated in figures 47.15 and 47.16.

FIGURE 47.15 Label the major systemic arteries. **A4**

1

2

3

4

5

6

7

8

9

10

11

12

13

FIGURE 47.16 Label the major systemic veins. A5

1 _____

2 _____

3 _____

4 _____

5 _____

6 _____

7 _____

8 _____

9 _____

10 _____

11 _____

12 _____

LABORATORY
EXERCISE 48

Pulse Rate and Blood Pressure

Purpose of the Exercise

To examine the pulse, to determine the pulse rate, to measure blood pressure, and to investigate the effects of body position and exercise on pulse rate and blood pressure. To investigate the effects of breathing on heart rate and stroke volume.

MATERIALS NEEDED

Clock with second hand

Sphygmomanometer

Stethoscope

Alcohol swabs (wipes)

For Demonstration Activity:

Pulse pickup transducer or plethysmograph

Physiological recording apparatus

Ph.I.L.S. 4.0: Lesson 40 Respiration: Deep Breathing and Cardiac Function

Learning Outcomes

After completing this exercise, you should be able to:

O1 Determine pulse rate and pulse characteristics.

O2 Test the effects of various factors on pulse rate.

O3 Measure and record blood pressure using a sphygmomanometer and determine pulse pressure.

O4 Test the effects of various factors on blood pressure.

O5 Distinguish components of pulse and blood pressure measurements.

O6 Explain the relationship of the heart rate (and stroke volume) to the breathing cycle.

The **O** corresponds to the assessments **A** indicated in the Laboratory Assessment for this Exercise.

Pre-Lab

Carefully read the introductory material and examine the entire lab. Be familiar with characteristics of pulse and blood pressure from lecture or the textbook. Answer the pre-lab questions.

Pre-Lab Questions Select the correct answer for each of the following questions:

1. The _____ is responsible for the pulse wave in the systemic circuit.
   - **a.** right atrium
   - **b.** right ventricle
   - **c.** left atrium
   - **d.** left ventricle

2. Which of the following arteries would *not* be palpated on the surface of the body?
   - **a.** facial artery
   - **b.** dorsalis pedis artery
   - **c.** common iliac artery
   - **d.** common carotid artery

3. The _____ is the most common artery palpated to determine heart rate.
   - **a.** temporal artery
   - **b.** radial artery
   - **c.** popliteal artery
   - **d.** posterior tibial artery

4. The _____ is the standard artery used to determine blood pressure.
   - **a.** brachial artery
   - **b.** common carotid artery
   - **c.** femoral artery
   - **d.** posterior tibial artery

5. Which of the following resting blood pressures would be considered hypertension?
   - **a.** 100/60 mm Hg
   - **b.** 110/75 mm Hg
   - **c.** 120/80 mm Hg
   - **d.** 140/90 mm Hg

517

6. Pulse pressure is the difference between the systolic and diastolic pressures.
   a. True ___✓___   b. False _____
7. The blood pressure in a capillary is higher than in a small artery.
   a. True ___✓___   b. False ___✓___
8. The stroke volume is the amount of blood pumped by a ventricle in one minute.
   a. True ___✓✓___   b. False _____

The sudden surge of blood that enters the arteries each time the ventricles of the heart contract (systole) causes the elastic walls of these vessels to expand. Then, as the ventricles relax (diastole), the arterial walls recoil. This alternate expanding and recoiling of an arterial wall can be palpated (felt) as a *pulse* in any vessel that is near the surface of the body. Because most of the major arteries are deep in the body, there are a limited number of possible locations to palpate the pulse. If any artery is superficial and firm tissue is just beneath the artery, a pulse can be palpated at that location. The number of pulse expansions per minute correlates with the heart rate or cardiac cycles. The left ventricle is the systemic pumping chamber, and it is therefore responsible for the pulse wave and for the blood pressure in the arteries selected for these assessments.

The force exerted by the blood pressing against the inner walls of arteries creates *blood pressure*. This systemic arterial pressure reaches its maximum, called the *systolic pressure*, during contraction of the left ventricle. Because the left ventricle relaxes when it fills with blood again, the blood pressure then drops to its lowest level, called the *diastolic pressure*. Blood pressure is measured in millimeters of mercury (mm Hg) and is expressed as a systolic pressure over diastolic pressure. A normal resting pressure is considered to be 120/80 mm Hg or slightly lower. If pressures during resting conditions are too high, the condition is called hypertension; if pressures are too low, the condition is called hypotension.

The pulse and the systolic and diastolic blood pressures have some close relationships. The difference between the systolic pressure and the diastolic pressure is called the *pulse pressure*, which is related to the natural elasticity and recoil of the arteries. Obtain the pulse pressure by subtracting the diastolic pressure from the systolic pressure. For example, a person with a blood pressure of 120/80 mm Hg would have a normal pulse pressure of 40 mm Hg. The pulse pressure indicates the force exerted upon the arteries from ventricular contraction and indicates general condition of the cardiovascular system. An expanded pulse pressure could be an indication of atherosclerosis and hypertension. There are many factors that influence blood pressure, so it is important to realize that a single measurement does not provide enough information to make conclusions about health.

After blood has passed through the capillaries, the blood pressure established by the heart is no longer sufficient to return blood to the heart via the veins. The skeletal muscle pump of the limbs and the respiratory pump of the torso assist in venous return. When the skeletal muscles of the limbs contract, they squeeze in on the blood in the veins of the limbs, forcing the blood into the torso (valves in the veins prevent the backflow of blood). In the torso there are two cavities, the thoracic cavity and abdominopelvic cavity, divided by the diaphragm. When the diaphragm contracts (along with the external intercostal muscles), the thoracic cavity volume increases and the pressure decreases. Simultaneously, the volume of the abdominopelvic cavity decreases and the pressure increases. This difference in pressure between the abdominopelvic cavity and the thoracic cavity assists the movement of blood into the veins of the thoracic cavity and back into the heart. As a result, more blood enters the heart, and the heart rate increases to pump the additional blood. Less blood enters the heart during expiration, since the thoracic cavity volume decreases and its pressure increases.

PROCEDURE A: Pulse Rate

1. Identify some of the pulse locations indicated in figure 48.1.

FIGURE 48.1 Locations where an arterial pulse can be palpated.

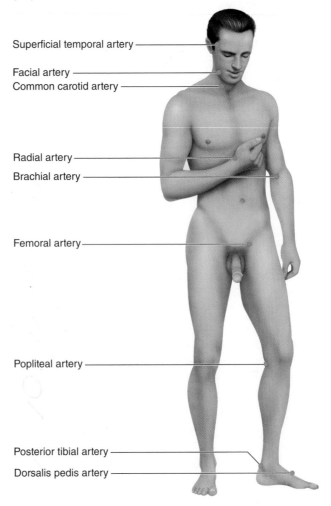

Superficial temporal artery

Facial artery
Common carotid artery

Radial artery
Brachial artery

Femoral artery

Popliteal artery

Posterior tibial artery
Dorsalis pedis artery

2. Examine your laboratory partner's radial pulse. To do this, follow these steps:
   a. Have your partner sit quietly, remaining as relaxed as possible.
   b. Locate the pulse by placing your index and middle fingers over the radial artery on the anterior surface near the lateral side of the wrist (fig. 48.2). Do not use your thumb for sensing the pulse because you may feel a pulse coming from an artery in the thumb. The radial artery is most often used because it is easy and convenient to locate.
   c. To determine the pulse rate, count the number of pulses that occur in 1 minute. This can be accomplished by counting pulses in 30 seconds and multiplying that number by 2. (A pulse count for a full minute, although taking a little longer, would give a better opportunity to detect pulse characteristics and irregularities.)
   d. Note the characteristics of the pulse. That is, can it be described as regular or irregular, strong or weak, hard or soft? The pulse should be regular and the amplitude (magnitude) will decrease as the distance

FIGURE 48.2 Taking a pulse rate from the radial artery.

LEARNING EXTENSION ACTIVITY

Determine the pulse rate and pulse characteristics in two additional locations on your laboratory partner. Locate and record the pulse rate from the common carotid artery and the dorsalis pedis artery (fig. 48.1).

Common carotid artery pulse rate _____

Dorsalis pedis artery pulse rate _____

Compare the amplitude of the pulse characteristic between these two pulse locations, and interpret any of the variations noted. _____

_____

DEMONSTRATION ACTIVITY

If the equipment is available, the laboratory instructor will demonstrate how a photoelectric pulse pickup transducer or plethysmograph can be used together with a physiological recording apparatus to record the pulse. Such a recording allows an investigator to analyze certain characteristics of the pulse more precisely than is possible using a finger to examine the pulse. For example, the pulse rate can be determined very accurately from a recording, and the heights of the pulse waves provide information about the blood pressure.

from the left ventricle increases. The strength of the pulse reveals some indications of blood pressure. Under high blood pressure, the pulse feels very hard and strong; under low blood pressure, the pulse feels weak and can be easily compressed.
   e. Record the pulse rate and pulse characteristics while sitting in Part A of Laboratory Assessment 48.
3. Repeat the procedure and determine the pulse rate and pulse characteristics in each of the following conditions:
   a. immediately after lying down;
   b. 3–5 minutes after lying down;
   c. immediately after standing;
   d. 3–5 minutes after standing quietly;
   e. immediately after 3 minutes of moderate exercise (*omit if the person has health problems*);
   f. 3–5 minutes after exercise has ended.
   g. Record the pulse rates and pulse characteristics in Part A of the laboratory assessment.
4. Complete Part A of the laboratory assessment.

PROCEDURE B: Blood Pressure

Reexamine the introductory material related to blood pressure. Several basic structural and functional factors influence arterial pressure, including *cardiac output, blood volume,* and *peripheral resistance.* Cardiac output is the volume of blood discharged from a ventricle in a minute. The blood ejected from a single ventricle contraction is about 70 mL and is called the *stroke volume.* A 75 beats/minute resting heart rate and a stroke volume of 70 mL would calculate as a cardiac output of 5,250 mL/minute, which is about the average adult blood volume. If either heart rate or stroke volume increases, cardiac output would increase as would blood pressure. Blood volume influences blood pressure. Because blood pressure is directly proportional to blood volume, any volume change can alter blood pressure. Peripheral resistance refers to the amount of friction between the blood and the walls of the blood vessels. Although blood viscosity and blood vessel length influence peripheral resistance, the blood vessel diameter changes are the most significant from vasoconstriction or vasodilation as a result of contraction or relaxation of smooth muscle of the middle tunic. Certain conditions, such as obesity, excess salt intake, inactive lifestyle, stress, smoking, medications, and others can aggravate blood pressures. Be aware of these factors and conditions influencing blood pressure as you take a blood pressure measurement using a sphygmomanometer and a stethoscope.

1. Although blood pressure is present in all blood vessels, the standard location to record blood pressure is the brachial artery. Examine figure 48.3, which indicates various pressure relationships throughout the systemic circuit.

2. Measure your laboratory partner's arterial blood pressure. To do this, follow these steps:

    a. Obtain a sphygmomanometer and a stethoscope.

    b. Clean the earpieces and the diaphragm of the stethoscope with alcohol swabs.

    c. Have your partner sit quietly with upper limb resting on a table at heart level. Have the person remain as relaxed as possible.

    d. Locate the brachial artery at the antecubital space. Wrap the cuff of the sphygmomanometer around the arm so that its lower border is about 2.5 cm above the bend of the elbow. Center the bladder of the cuff in line with the *brachial pulse* (fig. 48.4).

    e. Palpate the *radial pulse.* Close the valve on the neck of the rubber bulb connected to the cuff and pump air from the bulb into the cuff. Inflate the cuff while watching the sphygmomanometer and note the pressure when the pulse disappears. (This is a rough estimate of the systolic pressure.) Immediately deflate the cuff. *Do not leave the cuff inflated for more than 1 minute.*

    f. Position the stethoscope over the brachial artery. Reinflate the cuff to a level 30 mm Hg higher than the

FIGURE 48.4 Blood pressure is commonly measured by using a sphygmomanometer (blood pressure cuff). The use of a column of mercury is the most accurate measurement, but due to environmental concerns, it has been replaced by alternative gauges and digital readouts.

- Aneroid pressure gauge
- Rubber inflatable cuff
- Antecubital space (location of brachial pulse and stethoscope placement)
- Air control valve
- Bulb
- Location of radial pulse

point where the pulse disappeared during palpation (fig. 48.5). When the cuff is inflated above the systolic pressure, no blood flows distal to the cuff, and no sound is heard.

    g. Slowly open the valve of the bulb until the pressure in the cuff drops at a rate of about 2 or 3 mm Hg per second.

    h. Listen for sounds (*Korotkoff sounds*) from the brachial artery. When the systolic pressure becomes greater than the cuff's pressure, the brachial artery is forced open. When the first loud tapping sound is heard, record the reading as the systolic pressure. This indicates the pressure exerted against the arterial wall during systole.

    i. Continue to listen to the sounds as the pressure drops and note the level when the last sound is heard. Record this reading as the diastolic pressure, which measures the constant arterial resistance. When sounds are no longer heard, the cuff's pressure no longer compresses the brachial artery enough to restrict blood flow to the upper limb.

    j. Release all of the pressure from the cuff.

FIGURE 48.3 Pressure changes in blood vessels of the systemic circuit.

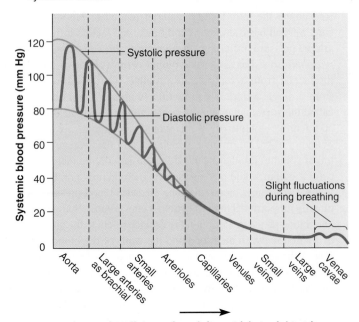

Systolic pressure

Diastolic pressure

Slight fluctuations during breathing

Systemic blood pressure (mm Hg)

Aorta — Large arteries as brachial — Small arteries — Arterioles — Capillaries — Venules — Small veins — Large veins — Venae cavae

Increasing distance from left ventricle to right atrium

FIGURE 48.5 Sphygmomanometer and stethoscope placement on arm while taking blood pressure.

Air control valve

Pressure gauge

Bulb

Inflatable cuff

k. Repeat the procedure until you have two blood pressure measurements from each arm, allowing 2–3 minutes of rest between readings.

l. Average your readings and enter them in the table in Part B of the laboratory assessment.

3. Measure your partner's blood pressure in each of the following conditions:

   a. 3–5 minutes after lying down;

   b. 3–5 minutes after standing quietly;

   c. immediately after 3 minutes of moderate exercise *(omit if the person has health problems);*

   d. 3–5 minutes after exercise has ended.

   e. Record the blood pressures in Part B of the laboratory assessment.

4. Complete Parts B and C of the laboratory assessment.

Sphygmomanometer

Sphyg momanometer

Sphygmomanometer

Sphygmomanometer

Relax: diastole
Contraction: systole

$$\frac{120}{80} \rightarrow \text{normal systolic} \quad mmHg$$
$$\rightarrow \text{normal diastolic}$$

pulse: blood entering of an arterial wall
(expanding & recoiling of arterial wall)

LABORATORY ASSESSMENT

48

Name _Juliana Klehn_

Date _____

Section _____

The (A) corresponds to the indicated Learning Outcome(s) (O) found at the beginning of the Laboratory Exercise.

Pulse Rate and Blood Pressure

$37\frac{2}{74}$

PART A: Assessments

1. Enter your observations of pulse rates and pulse characteristics in the table. *mine* (A1) (A2)

Test Subject	Pulse Rate (beats/min)	Pulse Characteristics
Sitting	74	regular
Lying down	38 (76)	regular
3–5 minutes later	28 (56)	regular
Standing	34 (68)	regular
3–5 minutes later	36 (72)	regular
After exercise	47 (94)	regular
3–5 minutes later	44 (88)	regular

2. Summarize the effects of body position and exercise on the pulse rates and pulse characteristics. (A2) _____

_____

_____

_____

PART B: Assessments

choose one & stay with it

1. Enter the initial measurements of blood pressure from a relaxed sitting position in the table.

Reading	Blood Pressure in Right Arm	Blood Pressure in Left Arm
First	120/80	
Second	120/80	
Average	120/80	

2. What was the average pulse pressure calculated from the blood pressure recorded from the arms? _____

_____

3. Enter your test results in the table. **A4**

Test Subject	Blood Pressure
3–5 minutes after lying down	115/70
3–5 minutes after standing	115/80
After 3 minutes of moderate exercise	130/90
3–5 minutes later	120/85

4. Summarize the effects of body position and exercise on blood pressure. **A4** _____

_____

_____

_____

5. Summarize any correlations between pulse rate and blood pressure from any of the experimental conditions.

_____

_____

_____

PART C: Assessments

Complete the following statements:

1. The maximum pressure achieved during ventricular contraction is called _____ pressure. (A5)

2. The lowest pressure that remains in the arterial system during ventricular relaxation is called _____ pressure. (A5)

3. The pulse rate is equal to the _____ rate. (A5)

4. A pulse that feels full and is not easily compressed is produced by an elevated _____. (A5)

5. The instrument commonly used to measure systemic arterial blood pressure is called a _____. (A5)

6. Blood pressure is expressed in units of _____. (A5)

7. The upper number of the fraction used to record blood pressure indicates the _____ pressure. (A5)

8. The _____ artery in the arm is the standard systemic artery in which blood pressure is measured. (A5)

9. A person with a blood pressure of 134/82 mm Hg would have a pulse pressure of _____ mm Hg. (A5)

CRITICAL THINKING ASSESSMENT

When a pulse is palpated and counted, which pressure (systolic or diastolic) is characteristic at that moment? Explain your answer. (A1)

_____

_____

_____

_____

Ph.I.L.S.
Physiology Interactive Lab Simulations **4.0**

LESSON 40 | # Respiration
Deep Breathing and Cardiac Function

The purpose of this exercise is to observe how heart rate varies during the breathing cycle. You will compare how the volume of air changes in the thoracic cavity during slow, deep breathing with changes in stroke volume and heart rate. In a normal breathing cycle, inhalation increases the volume of the thoracic cavity and decreases pressure. Exhalation decreases the volume of the thoracic cavity and increases pressure.

The heart provides sufficient pressure to push blood through the arteries and capillaries to the veins of the body. As the blood travels through the circulatory system there is a gradual decline in pressure that results in low blood pressure in the veins, especially the veins in the lower extremities. This problem of low blood pressure is corrected by the body using a couple of mechanisms. One mechanism is the contraction of leg muscles that compress the veins and push the blood up toward the hips. A second mechanism takes effect when the inspiratory muscles (sternocleidomastoid, external intercostals, pectoralis minor, internal intercostals, diaphragm) are contracted, producing a negative pressure inside the thoracic cavity. This negative pressure helps draw venous blood into the heart from the lower extremities. This also causes an increase in heart rate. When the expiratory muscles (internal intercostals, rectus abdominis, external oblique) are contracted, positive pressure is created inside the thoracic cavity. This decreases the venous blood flow to the heart and decreases the heart rate.

In this laboratory experiment, you will measure the air volume breathed in and out and the pulse and heart rate of a virtual volunteer using a breathing apparatus and a finger pulse unit. You will observe two complete breathing cycles to acquire your data with which to measure thoracic cavity volume, heart rate, and stroke volume changes.

Before you begin, familiarize yourself with the following concepts (use your textbook as a reference):

- Blood pressure
- Stroke volume
- Pulse rate
- The breathing cycle

State your hypothesis regarding the relationship between the breathing cycle and heart rate, including changes in thoracic cavity volume, stroke volume, and heart rate.

_____

_____

_____

_____

1. Open Lesson 40, Respiration: Deep Breathing and Cardiac Function.

2. Read the objectives and introduction and take the pre-laboratory quiz.

3. After completing the pre-laboratory quiz, read through the wet lab.

4. The laboratory exercise will open when you click Continue after completing the wet lab (fig. 48.6).

5. Print a copy of the final graph you recorded from the virtual computer screen (fig. 48.7).

6. Follow the instructions provided in the laboratory exercise. Using the graph you printed, answer the following questions as you proceed.

7. During inspiration, the breathing curve is _____ (an upward, a downward) deflection, and during expiration the breathing curve is _____ (an upward, a downward) deflection.

8. The heart rate is determined by the _____ (frequency or amplitude) of the pulses.

9. The stroke volume is determined by the _____ (frequency or amplitude) of the pulses.

526

FIGURE 48.6 Opening screen for the laboratory exercise on Respiration: Deep Breathing and Cardiac Function.

FIGURE 48.7 Example: A recording to show the relationship between slow breathing (upper trace) and heart function; finger pulse is on the middle tracing, and heart rate is on the lower.

10. During inspiration:

    a. The heart rate _____
    (increases, decreases, or stays the same).

    b. The stroke volume _____
    (increases, decreases, or stays the same).

11. During expiration:

    a. The heart rate _____
    (increases, decreases, or stays the same).

    b. The stroke volume _____
    (increases, decreases, or stays the same).

12. Cardiac output (CO) is determined by heart rate (HR) and stroke volume (SV): HR x SV = CO. To maintain a consistent cardiac output at rest, as heart rate increases, stroke volume _____, and as heart rate decreases, stroke volume _____. Did you observe this relationship on the graph? _____ Is there a direct or inverse relationship between heart rate and stroke volume when maintaining cardiac output at a set value? _____

CRITICAL THINKING ASSESSMENT

Explain the relationship between inspiration, pressure changes in the thoracic cavity, venous return (movement of blood in veins), heart rate, stroke volume, and cardiac output at rest. A2 A6

_____

_____

_____

_____

_____

NOTES

EXERCISE 49

Lymphatic System

Purpose of the Exercise

To review the gross anatomical structures of the lymphatic system and to observe the microscopic structure of a lymph node, the thymus, the spleen, and a tonsil.

MATERIALS NEEDED

Human torso model

Anatomical chart of the lymphatic system

Compound light microscope

Prepared microscope slides:

  Lymph node section

  Human thymus section

  Human spleen section

  Human tonsil section

Learning Outcomes

After completing this exercise, you should be able to:

O1 Identify the major lymphatic pathways and components of the lymphatic system.

O2 Locate and sketch the major microscopic structures of a lymph node, the thymus, the spleen, and a tonsil.

O3 Describe the structure and function of a lymph node, the thymus, the spleen, and a tonsil.

The **O** corresponds to the assessments **A** indicated in the Laboratory Assessment for this Exercise.

Pre-Lab

Carefully read the introductory material and examine the entire lab. Be familiar with the basic structures and functions of lymphatic pathways, lymph nodes, thymus, spleen, and tonsils from lecture or the textbook. Answer the pre-lab questions.

Pre-Lab Questions Select the correct answer for each of the following questions:

1. Lymph finally flows into lymphatic _____ just before connecting to a blood vessel.
   - **a.** capillaries
   - **b.** collecting ducts
   - **c.** vessels
   - **d.** trunks

2. The right lymphatic duct drains only the _____quadrant of the entire body.
   - **a.** right upper
   - **b.** left upper
   - **c.** right lower
   - **d.** left lower

3. Which of the following areas would contain the fewest lymph nodes?
   - **a.** cervical
   - **b.** axillary
   - **c.** inguinal
   - **d.** plantar

4. The _____ is the largest lymphatic organ.
   - **a.** thymus
   - **b.** pharyngeal tonsil
   - **c.** spleen
   - **d.** popliteal lymph node

5. The thoracic duct drains into the
   - **a.** right internal jugular vein.
   - **b.** right subclavian vein.
   - **c.** left internal jugular vein.
   - **d.** left subclavian vein.

6. Lymph flows faster during periods of exercise.
   - **a.** True _____
   - **b.** False _____

7. The thymus continues to enlarge as a person ages.
   - **a.** True _____
   - **b.** False _____

8. Lymphatic capillaries and blood capillaries are open at both ends.
   - **a.** True _____
   - **b.** False _____

The lymphatic system is closely associated with the cardiovascular system and includes a network of vessels that assist in the circulation of body fluids. Fluid in the lymphatic vessels is *lymph*. As blood flows through capillary beds, fluids continually filter from the plasma of the blood along with some dissolved nutrients, with most of the plasma proteins left behind in the blood. Most of the fluids are reabsorbed back into the blood at the venous ends of the capillary beds; however the amount not reabsorbed becomes part of the interstitial fluids among the tissue cells. Lymphatic vessels provide a necessary pathway to drain the tissue of excess fluid and return it to the blood of the cardiovascular system (fig. 49.1). Without this system, this fluid would accumulate in tissue spaces, producing *edema.*

The *lymphatic capillaries* are microscopic, blind-ended vessels, in which excess interstitial fluid first enters the lymphatic system (fig. 49.2). These capillaries drain into *lymphatic collecting vessels,* then into larger *lymphatic*

FIGURE 49.2 (a) Blind-ended lymphatic capillaries among tissue cells and blood vessel capillaries; (b) uptake of tissue fluid between overlapping endothelial cells of lymphatic capillary.

(a)

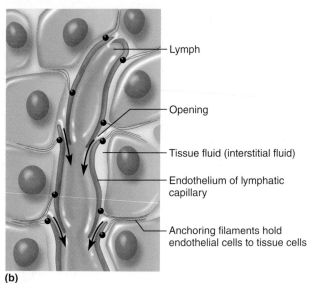

(b)

FIGURE 49.1 AP|R Schematic relationship of fluid movements between the cardiovascular and lymphatic systems.

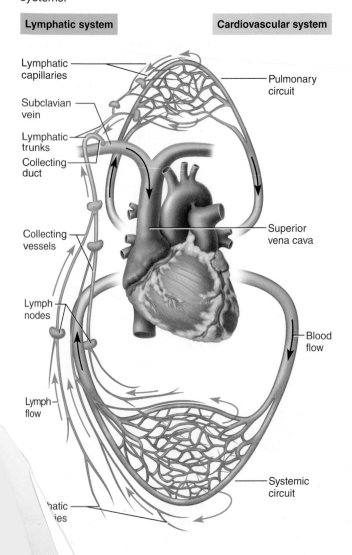

trunks, and finally into one of two *collecting ducts* just before the connections with the subclavian veins (fig. 49.1). As a result of the orientation of valves, only one-way flow of lymph occurs (fig. 49.3). This happens during skeletal muscle contraction and breathing, causing compression upon the vessels that forces fluid through the lymphatic vessels. The larger *thoracic duct* provides drainage for the entire body except for the right upper quadrant. The thoracic duct connects to the left subclavian vein, while a *right lymphatic duct* drains only the right upper quadrant into the right subclavian vein (fig. 49.4).

Along the pathway of the lymphatic vessels, *lymph nodes* are positioned; they are especially concentrated in areas of major joints like axillary and inguinal regions.

FIGURE 49.3 Micrograph of a valve within a lymphatic vessel (60x). The arrows indicate the flaplike valve that helps prevent backflow of lymph.

Valve

FIGURE 49.4 Regions of the body drained by the right lymphatic duct are shaded purple. The unshaded regions of the body are drained by the larger thoracic duct into the left subclavian vein.

Drained by right lymphatic duct

Drained by thoracic duct

The lymph nodes contain concentrations of lymphocytes and macrophages that cleanse the lymph and activate an immune response. Swollen and sore lymph nodes represent areas of infection that occur as our bodies fight off the microorganisms.

Other lymphatic organs are part of the lymphatic system. The red bone marrow is a major production site of white blood cells. The *thymus,* located within the mediastinum, is a location of T lymphocyte maturation. The thymus is proportionately very large in a fetus and in a child, but it begins to atrophy around puberty, and this continues as we age. The *spleen,* the largest lymphatic organ, is located in the left upper quadrant of the abdominopelvic cavity. The spleen is highly vascular and contains large numbers of red blood cells, lymphocytes, and macrophages. Palatine, pharyngeal, and lingual *tonsils* are located in regions of the entrance into the pharynx (throat). Numerous lymphatic nodules within the tonsils contain concentrations of lymphocytes. The organs of the lymphatic system help to defend the tissues against infections by filtering particles from the lymph and by supporting the activities of lymphocytes that furnish immunity against specific disease-causing agents or pathogens.

PROCEDURE A: Lymphatic Pathways

1. Study figures 49.5 and 49.6.
2. Observe the human torso model and the anatomical chart of the lymphatic system, and locate the following features:

   lymphatic collecting vessels—collect lymph from microscopic lymphatic capillaries

   lymph nodes—bean-shaped organs that cleanse lymph and activate immune response

   lymphatic trunks—collect lymph from lymphatic collecting vessels
   - lumbar trunk
   - intestinal trunk
   - bronchomediastinal trunk
   - subclavian trunk
   - jugular trunk

   cisterna chyli—inferior expanded sac of thoracic duct where fatty intestinal lymph collects

   collecting ducts—largest lymphatic vessels; collect lymph from lymphatic trunks
   - thoracic duct—connects to left subclavian vein near left internal jugular vein
   - right lymphatic duct—connects to right subclavian vein near right internal jugular vein

   internal jugular veins

   subclavian veins—receive lymph from collecting ducts

3. Complete Part A of Laboratory Assessment 49.

FIGURE 49.5 AP|R Lymphatic vessels, lymph nodes, and lymphatic organs.

Tonsil

Cervical lymph nodes

Thymus

Axillary lymph nodes

Cisterna chyli

Lymphatic trunks

Spleen

Abdominal and pelvic lymph nodes

Red bone marrow

Inguinal lymph nodes

Popliteal lymph nodes

Lymphatic vessels

FIGURE 49.6 AP|R Lymphatic pathways into the subclavian veins. The arrows indicate the direction of lymph drainage.

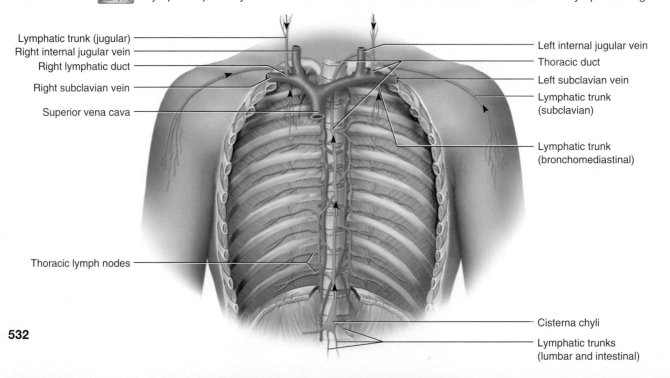

Lymphatic trunk (jugular)

Right internal jugular vein

Right lymphatic duct

Right subclavian vein

Superior vena cava

Left internal jugular vein

Thoracic duct

Left subclavian vein

Lymphatic trunk (subclavian)

Lymphatic trunk (bronchomediastinal)

Thoracic lymph nodes

Cisterna chyli

Lymphatic trunks (lumbar and intestinal)

PROCEDURE B: Lymph Nodes

1. Study the lymph node locations in figures 49.5 and 49.6.
2. Observe the anatomical chart of the lymphatic system and the human torso model, and locate the clusters of lymph nodes in the following regions:

   cervical region
   axillary region
   popliteal region
   inguinal region
   pelvic cavity
   abdominal cavity
   thoracic cavity

3. Palpate the lymph nodes in your cervical region. They are located along the lower border of the mandible and between the ramus of the mandible and the sternocleidomastoid muscle. They feel like small, firm lumps.
4. Study figures 49.7 and 49.8.
5. Obtain a prepared microscope slide of a lymph node and observe it, using low-power magnification (fig. 49.8c). Identify the *capsule* that surrounds the node and is mainly composed of collagen fibers, the *lymphatic nodules* that appear as dense masses near the surface of the node, and the *lymphatic sinuses* that appear as narrow spaces where lymph circulates between the nodules and the capsule.
6. Using high-power magnification, examine a nodule within the lymph node. The nodule contains densely packed *lymphocytes*.

FIGURE 49.7 AP|R Diagram of a lymph node showing internal structures, attached lymphatic vessels, and arrows indicating lymph flow.

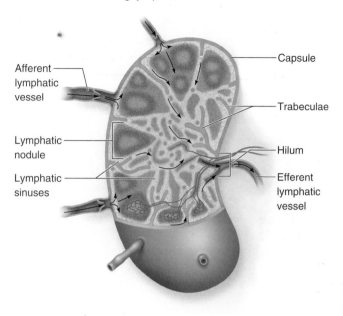

7. Prepare a labeled sketch of a representative section of a lymph node in Part B of the laboratory assessment.
8. Complete Part C of the laboratory assessment.

FIGURE 49.8 APR Lymph nodes: (a) inguinal lymph nodes in a cadaver; (b) photograph of a human lymph node on tip of finger; (c) micrograph of a lymph node (20×).

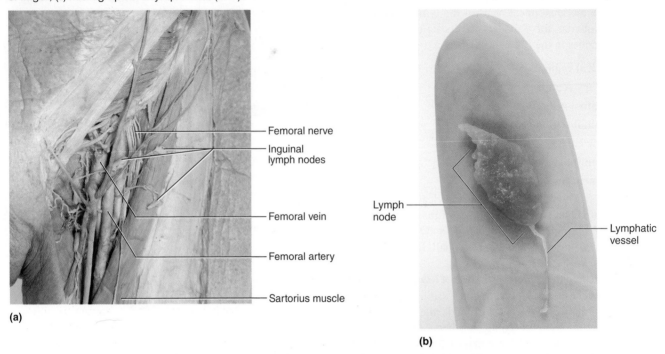

(a)

- Femoral nerve
- Inguinal lymph nodes
- Femoral vein
- Femoral artery
- Sartorius muscle

(b)

Lymph node

Lymphatic vessel

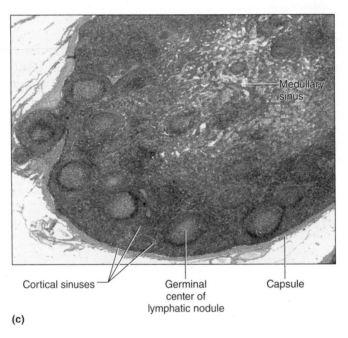

(c)

- Medullary sinus
- Cortical sinuses
- Germinal center of lymphatic nodule
- Capsule

PROCEDURE C: Thymus, Spleen, and Tonsil

1. Locate the thymus, spleen, and tonsils in the anatomical chart of the lymphatic system and on the human torso model.
2. Obtain a prepared microscope slide of human thymus and observe it, using low-power magnification (fig. 49.9). Note how the thymus is subdivided into *lobules* by *septa*

of connective tissue that contain blood vessels. Identify the capsule of loose connective tissue that surrounds the thymus, the outer *cortex* of a lobule composed of densely packed cells and deeply stained, and the inner *medulla* of a lobule composed of loosely packed lymphocytes and epithelial cells and lightly stained.

3. Examine the cortex tissue of a lobule using high-power magnification. The cells of the cortex are composed of densely packed *lymphocytes* among some epithelial cells

FIGURE 49.9   AP|R   Micrograph of a section of the thymus (15×).

Septum of connective tissue

Cortex of lobule

Medulla of lobule

Capsule

Lobule

and *macrophages.* Some of these cortical cells may be undergoing mitosis, so their chromosomes may be visible.

4. Prepare a labeled sketch of a representative section of the thymus in Part B of the laboratory assessment.

5. Obtain a prepared slide of the human spleen and observe it, using low-power magnification (fig. 49.10). Identify the *capsule* of dense connective tissue that surrounds the spleen. The tissues of the spleen include circular *nodules* of *white pulp* enclosed in a matrix of *red pulp.*

6. Using high-power magnification, examine a nodule of white pulp and red pulp. The cells of the white pulp are mainly lymphocytes. Also, there may be an arteriole centrally located in the nodule. The cells of the red pulp are mostly red blood cells with many lymphocytes and

macrophages. The macrophages engulf and destroy cellular debris and bacteria.

7. Prepare a labeled sketch of a representative section of the spleen in Part B of the laboratory assessment.

8. Obtain a prepared microscope slide of a tonsil and observe it, using low-power magnification (fig. 49.11). Identify the surface epithelium, deep invaginated pits called *tonsillar crypts,* and numerous lymphatic nodules. Although the tonsillar crypts often harbor bacteria and food debris, infections are usually prevented within these lymphoid organs.

9. Prepare a labeled sketch of a representative section of a tonsil in Part B of the laboratory assessment.

10. Complete Part D of the laboratory assessment.

FIGURE 49.10   AP|R   Micrograph of a section of the spleen (40×).

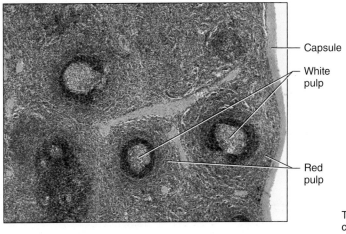

Capsule

White pulp

Red pulp

FIGURE 49.11   AP|R   Micrograph of a section of the palatine tonsil (5×).

Tonsillar crypts

Lymphatic nodules

Pharyngeal epithelium

Pharynx

NOTES

Name _____

Date _____

Section _____

The **A** corresponds to the indicated Learning Outcome(s) **O** found at the beginning of the Laboratory Exercise.

Lymphatic System

PART A: Assessments

Complete the following statements:

1. Lymphatic pathways begin as lymphatic _____ that merge to form lymphatic collecting vessels. **A1**

2. Lymph drainage from collecting ducts enters the _____ veins. **A1**

3. Once tissue (interstitial) fluid is inside a lymphatic capillary, the fluid is called _____. **A1**

4. Lymphatic vessels contain _____ that help prevent the backflow of lymph. **A1**

5. Lymphatic vessels usually lead to lymph _____ that filter the fluid being transported. **A1**

6. The _____ duct is the larger and longer of the two lymphatic collecting ducts. **A1**

PART B: Assessments

Sketch and label the microscopic structures of the following organs: **A2**

Lymph node (_____×)	Thymus (_____×)
Spleen (_____×)	Tonsil (_____×)

PART C: Assessments

Complete the following statements:

1. Lymph nodes contain large numbers of white blood cells called _____ and macrophages that fight invading microorganisms. (A3)

2. The indented region where blood vessels and nerves join a lymph node is called the _____. (A3)

3. Lymphatic _____ that contain germinal centers are the structural units of a lymph node. (A3)

4. The spaces within a lymph node are called lymphatic _____ through which lymph circulates. (A3)

5. Lymph enters a node through a(an) _____ lymphatic vessel. (A3)

6. The lymph nodes associated with the lymphatic vessels that drain the lower limbs are located in the _____ region. (A1)

PART D: Assessments

Complete the following statements:

1. The thymus is located in the _____, anterior to the aortic arch. (A1)

2. The _____ is very large in a child and will atrophy in advanced age. (A3)

3. The _____ is the largest of the lymphatic organs. (A1)

4. A _____ of connective tissue surrounds the spleen. (A3)

5. The tiny islands (nodules) of tissue within the spleen that contain many lymphocytes constitute the _____ pulp. (A3)

6. The _____ pulp of the spleen contains large numbers of red blood cells, lymphocytes, and macrophages. (A3)

7. _____ within the spleen engulf and destroy foreign particles and cellular debris. (A3)

8. The lymphoid organs in the pharynx (throat) that are possible infection sites are collectively called _____. (A3)

LABORATORY EXERCISE 50

Respiratory Organs

Purpose of the Exercise

To review the structure and function of the respiratory organs and to examine the tissues of some of these organs.

MATERIALS NEEDED

Human skull (sagittal section)

Human torso model

Larynx model

Thoracic organs model

Compound light microscope

Prepared microscope slides of the following:

    Trachea (cross section)

    Lung, human (normal)

For Demonstration Activities:

Animal lung with trachea (fresh or preserved)

Bicycle pump

Prepared microscope slides of the following:

    Lung tissue (smoker)

    Lung tissue (emphysema)

Learning Outcomes AP|R

After completing this exercise, you should be able to:

O1 Locate the major organs and structural features of the respiratory system.

O2 Describe the functions of these organs.

O3 Sketch and label features of tissue sections of the trachea and lung.

The **O** corresponds to the assessments **A** indicated in the Laboratory Assessment for this Exercise.

Pre-Lab

Carefully read the introductory material and examine the entire lab. Be familiar with the basic structures and functions of the respiratory organs from lecture or the textbook. Visit www.mhhe.com/martinseries3 for LabCam videos. Answer the pre-lab questions.

Pre-Lab Questions Select the correct answer for each of the following questions:

1. The right lung has ____ lobes; the left lung has ____ lobes.
   - **a.** 3; 3
   - **b.** 2; 2
   - **c.** 3; 2
   - **d.** 2; 3

2. Paranasal sinuses are within the following bones *except* the
   - **a.** maxillary bones.
   - **b.** mandible.
   - **c.** frontal bone.
   - **d.** sphenoid bone.

3. The alveoli are composed of
   - **a.** simple squamous epithelium.
   - **b.** hyaline cartilage.
   - **c.** smooth muscle.
   - **d.** epithelium with cilia.

4. The _____ adheres to the surface of the lung.
   - **a.** pericardium
   - **b.** pleural cavity
   - **c.** parietal pleura
   - **d.** visceral pleura

5. The _____ is the most inferior cartilage of the larynx.
   - **a.** epiglottis
   - **b.** cricoid cartilage
   - **c.** thyroid cartilage
   - **d.** corniculate cartilage

6. Which of the following structures increases the surface area and air turbulence the most during breathing?
   - **a.** nasal meatuses
   - **b.** nasal septum
   - **c.** nasal conchae
   - **d.** nares (nostrils)

7. Which of the following airway tubes would have the smallest lumens?
   - **a.** alveolar ducts
   - **b.** segmental bronchi
   - **c.** lobar bronchi
   - **d.** main bronchi

The respiratory system involves the movement of oxygen from the external environment into our lungs upon inspiration and eventually to our cells, and the movement of carbon dioxide produced by our cells until it exits our body upon expiration. Breathing (pulmonary ventilation) involves the nose, nasal cavity, paranasal sinuses, pharynx, larynx, trachea, and bronchial tree, all serving as passageways for gases into and out of the lungs. The exchange of oxygen and carbon dioxide in the lungs is called *external respiration;* the exchange of oxygen and carbon dioxide in the tissues is called *internal respiration.* The blood transports gases to and from the alveolar sacs and the body cells. The cells utilize the oxygen and produce carbon dioxide in *cellular respiration.*

Larger bronchial tubes possess supportive cartilaginous rings and plates so that they do not collapse during breathing. Smooth muscle is part of the walls of these bronchial tubes. If the smooth muscle of these tubes relaxes, the air passages dilate, which allows a greater volume of air movement. If the smooth muscle of the respiratory tubes contracts, a reduced volume of oxygen enters the lungs. The epithelial lining of the respiratory tubes includes occasional *goblet cells,* which secrete protective mucus that catches dust and potential pathogens. Numerous epithelial cells possess *cilia* that extend into the mucus. The action of the cilia creates a current of mucus that moves any entrapped debris toward the pharynx, which helps prevent respiratory infections. The mucus with any entrapped particles is normally swallowed.

The right lung has three lobes; the left lung has two lobes. A right main bronchus branches into three lobar bronchi, each extending into one of the three right lobes. A left main bronchus branches into two lobar bronchi, each extending into a left lobe. The *hilum* of the lung represents an area where a bronchus and major blood vessels are located. A *visceral pleura* adheres to the surface of each lung, including the fissures between the lobes. A *parietal pleura* covers the internal surface of the thoracic wall and the superior surface of the diaphragm. A narrow potential space, called the *pleural cavity,* is located between the two pleurae and contains pleural fluid secreted by these membranes. This pleural fluid serves as lubrication during breathing movements, and due to the surface tension, it resists separations of the pleurae.

PROCEDURE A: Respiratory Organs

1. Study figures 50.1 and 50.2.
2. Examine the sagittal section of the human skull and the human torso model. Locate the following features:

   nasal cavity—hollow space behind nose
   - naris (nostril)—external opening into nasal cavity
   - nasal septum—divides nasal cavity into right and left portions
   - nasal conchae—increase surface area and air turbulence during breathing
   - nasal meatuses—air passageways along nasal conchae

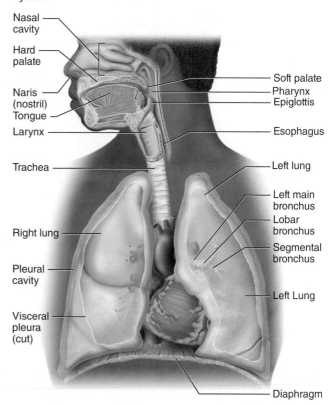

FIGURE 50.1   The major features of the respiratory system.

Labels: Nasal cavity, Hard palate, Naris (nostril), Tongue, Larynx, Trachea, Right lung, Pleural cavity, Visceral pleura (cut), Soft palate, Pharynx, Epiglottis, Esophagus, Left lung, Left main bronchus, Lobar bronchus, Segmental bronchus, Left Lung, Diaphragm

   paranasal sinuses—air-filled spaces that open into nasal cavity; lined with mucous membrane
   - maxillary sinus
   - frontal sinus
   - ethmoidal sinus
   - sphenoidal sinus
3. Examine figures 50.1, 50.2, 50.3, 50.4, and 50.5.
4. Observe the larynx model, the thoracic organs model, and the human torso model. Palpate the larynx as you swallow. Locate the following features:

   pharynx—often called throat
   - nasopharynx—air passageway only
   - oropharynx—air and food passageway
   - laryngopharynx—air and food passageway

   larynx—airway enlargement superior to trachea
   - vocal folds—consist of two sets of folds
     - vestibular folds (false vocal cords)—upper folds but do not produce sounds
     - vocal folds (true vocal cords)—vibrate when exhaling to produce sounds
   - thyroid cartilage ("Adam's apple")—largest hyaline cartilage of larynx; covered by part of thyroid gland; more pronounced in males
   - cricoid cartilage—complete ring of hyaline cartilage of inferior larynx

FIGURE 50.2 **AP|R**   The features of the upper respiratory system (sagittal section): (a) illustration; (b) cadaver.

Frontal sinus

Nasal conchae

Naris (nostril)

Epiglottis

Hyoid bone (cut)

Vestibular fold

Vocal fold (true vocal cord)

Nasal meatuses

Sphenoidal sinus

Nasopharynx

Oropharynx

Laryngopharynx

Esophagus

(a)

Cerebrum

Frontal sinus

Cribriform plate

Nasal conchae:
    Superior
    Middle
    Inferior

Nasal meatuses

Hard palate

Tongue

Larynx:
    Epiglottis
    Vestibular fold
    Vocal fold

Trachea

Nasopharynx

Oropharynx

Laryngopharynx

Vertebral column

Esophagus

(b)

nose → pharynx → larynx → trachea

FIGURE 50.3 `AP|R` Major features of the larynx: (a) anterior view; (b) `AP|R` posterior view; (c) `AP|R` sagittal view.

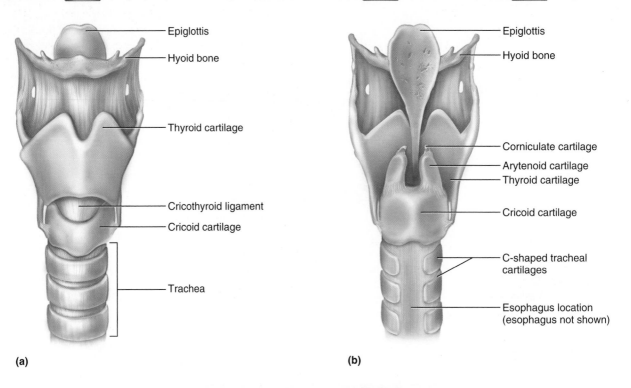

(a)

Epiglottis
Hyoid bone
Thyroid cartilage
Cricothyroid ligament
Cricoid cartilage
Trachea

(b)

Epiglottis
Hyoid bone
Corniculate cartilage
Arytenoid cartilage
Thyroid cartilage
Cricoid cartilage
C-shaped tracheal cartilages
Esophagus location (esophagus not shown)

(c)

Epiglottis
Hyoid bone
Epiglottic cartilage
Fat pad
Thyroid cartilage
Cuneiform cartilage
Corniculate cartilage
Vestibular fold
Vocal cord
Arytenoid cartilage
Cricoid cartilage
Tracheal cartilages

FIGURE 50.4 Superior view of larynx with the glottis open as seen using a laryngoscope.

Anterior

Epiglottis

Glottis

Vestibular fold (false vocal cord)

Vocal fold (true vocal cord)

Tracheal lining

Posterior

FIGURE 50.5 **AP|R** Features of the lower respiratory system (anterior view).

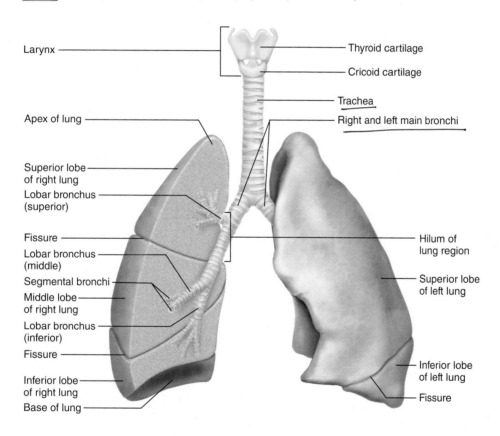

Larynx

Thyroid cartilage

Cricoid cartilage

Trachea

Right and left main bronchi

Apex of lung

Superior lobe of right lung

Lobar bronchus (superior)

Fissure

Lobar bronchus (middle)

Segmental bronchi

Middle lobe of right lung

Lobar bronchus (inferior)

Fissure

Inferior lobe of right lung

Base of lung

Hilum of lung region

Superior lobe of left lung

Inferior lobe of left lung

Fissure

- cricothyroid ligament—connects thyroid and cricoid cartilages
- epiglottis—covers part of larynx opening during swallowing
- epiglottic cartilage—elastic cartilage supporting epiglottis
- cuneiform cartilages—support soft tissues inferior to the epiglottis
- arytenoid cartilages—vocal cords attached to these small cartilages
- corniculate cartilages—small cartilages superior to arytenoid cartilages
- glottis—vocal folds and opening between them

trachea—windpipe anterior to esophagus

bronchial tree—branched airways extending throughout lungs
- right and left main (primary) bronchi—arise from trachea
- lobar (secondary) bronchi—branches of main bronchi into lobes of lungs
- segmental (tertiary) bronchi—branches of lobar bronchi into segments of lungs

bronchioles—smaller branches of airways

lung—location of external respiration between alveoli and blood capillaries
- hilum of lung—medial indentation where main bronchus, blood vessels, and nerves enter lung
- lobes—large regions of lungs
  - superior (upper) lobe—located in both lungs
  - middle lobe—only in right lung
  - inferior (lower) lobe—located in both lungs
- lobules—smallest visible subdivisions of a lung

visceral pleura—serous membrane on surface of lung

parietal pleura—serous membrane on inside of thoracic wall and superior surface of diaphragm

pleural cavity—small space between pleurae containing lubricating serous fluid.

5. Complete Part A of Laboratory Assessment 50.

PROCEDURE B: Respiratory Tissues

1. Obtain a prepared microscope slide of a trachea, and use low-power magnification to examine it. Notice the inner lining of ciliated pseudostratified columnar epithelium and the deep layer of hyaline cartilage, which represents a portion of an incomplete (C-shaped) tracheal ring (fig. 50.6).
2. Use high-power magnification to observe the cilia on the free surface of the epithelial lining. Locate the wineglass-shaped goblet cells, which secrete a protective mucus, in the epithelium (fig. 50.7).
3. Prepare a labeled sketch of a representative portion of the tracheal wall in Part B of the laboratory assessment.
4. Study the internal structures of a lung (fig. 50.8).
5. Obtain a prepared microscope slide of human lung tissue. Examine it, using low-power magnification, and note the many open spaces of the air sacs (alveoli). Look for a bronchiole—a tube with a relatively thick wall and a wavy inner lining. Locate the smooth muscle tissue in the wall of this tube (fig. 50.9). You also may see a section of cartilage as part of the bronchiole wall.
6. Use high-power magnification to examine the alveoli (fig. 50.10). Their walls are composed of simple squamous epithelium. The thin walls of the alveoli and the associated capillaries allow efficient gas exchange of oxygen and carbon dioxide. You also may see sections of blood vessels containing blood cells.
7. Prepare a labeled sketch of a representative portion of the lung in Part B of the laboratory assessment.
8. Complete Part C of the laboratory assessment.

3 parts of respiration

1) Breathing (exchange between atmosphere → Lungs)
   - Inhalation
   - Exhalation

2) External Respiration (Lungs → C.V.S.)

3) Internal Respiration (C.V.S. → cells)

- External Intercostalis
  &
- Diaphragm
responsible muscles for respiration

Boyle's Law
relationship between pressure & volume
↓
"SPACE"

FIGURE 50.6 `AP|R` Micrograph of a section of the tracheal wall (60×).

Hyaline cartilage of (C-shaped) tracheal ring

Seromucous glands in submucosa

Connective tissue

Lumen of trachea

Ciliated pseudostratified columnar epithelium

FIGURE 50.7 `AP|R` Micrograph of ciliated pseudostratified columnar epithelium in the respiratory tract (275×).

Connective tissue

Epithelial cell

Goblet cell

Cilia

Lumen

Nose→ Trachea (Decrease) Cartilage

Alveolis → decrease & down

3 Trends

① What type of epithelia?

② Amount of cartilage? (Increasing or Decreasing?)

③ Smooth muscle (Increase or Decrease?)

trachea → pseudostratified

FIGURE 50.8 Bronchioles, alveoli, and blood vessel network in a lung.

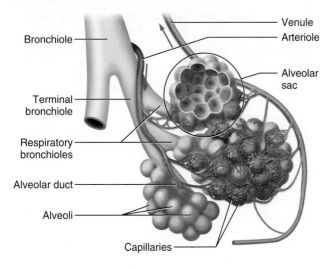

- Bronchiole
- Terminal bronchiole
- Respiratory bronchioles
- Alveolar duct
- Alveoli
- Capillaries
- Venule
- Arteriole
- Alveolar sac

FIGURE 50.10 AP|R Micrograph of human lung tissue (250×).

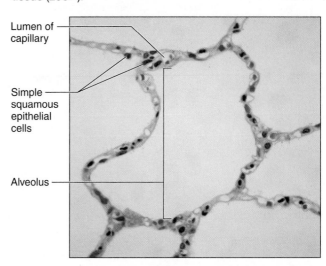

- Lumen of capillary
- Simple squamous epithelial cells
- Alveolus

FIGURE 50.9 AP|R Micrograph of human lung tissue (35×).

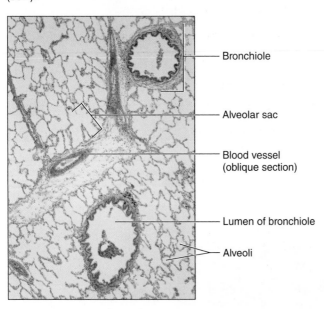

- Bronchiole
- Alveolar sac
- Blood vessel (oblique section)
- Lumen of bronchiole
- Alveoli

Trachea
1st Bronchus
2nd Bronchus
3rd Bronchus
Terminal Bronchiole
Respiratory Duct
Alveolar Sacs
Alveolus

Alveol

Alveolus- structural & functional unit of respiratory system.

ASSESSMENT
50

Name _____

Date _____

Section _____

The Ⓐ corresponds to the indicated Learning Outcome(s) Ⓞ found at the beginning of the Laboratory Exercise.

Respiratory Organs

PART A: Assessments

1. Match the terms in column A with the descriptions in column B. Place the letter of your choice in the space provided. Ⓐ1
Ⓐ2

Column A	Column B
a. Alveolus	_____ 1. Potential space between visceral and parietal pleurae
b. Cricoid cartilage	_____ 2. Most inferior portion of larynx
c. Epiglottis	_____ 3. Air-filled space in skull bone that opens into nasal cavity
d. Glottis	_____ 4. Microscopic air sac for gas exchange
e. Lung	_____ 5. Consists of large lobes
f. Nasal concha	_____ 6. Vocal folds, including the opening between them
g. Pharynx	_____ 7. Fold of mucous membrane containing elastic fibers responsible for sounds
h. Pleural cavity	_____ 8. Increases surface area of nasal mucous membrane
i. Sinus (paranasal sinus)	
j. Vocal fold (true vocal cord)	_____ 9. Passageway for air and food
	_____ 10. Partially covers opening of larynx during swallowing

2. Label the structures indicated in figures 50.11, 50.12, and 50.13.

FIGURE 50.11 Label the features of the respiratory system. Ⓐ1

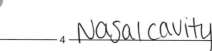

4. _Nasal cavity_

5. _trachea_

1. _Left primary bronchi_

2. _____

3. _Diaphragm_

FIGURE 50.12 Label the features of the upper respiratory system (sagittal section). A1

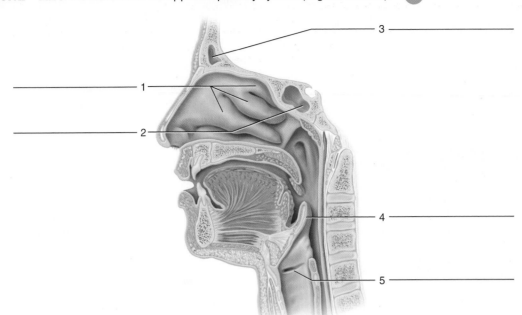

3

1

2

4

5

FIGURE 50.13 AP|R Label the features of the larynx region of a cadaver (lateral view). A1

1

2

3

Laryngopharynx

4

5

Esophagus

6

Anterior ← → Posterior

PART B: Assessments

Sketch and label a portion of the tracheal wall and a portion of lung tissue. (A3)

Tracheal wall (_____×)	Lung tissue (_____×)

PART C: Assessments

Complete the following:

1. What is the function of the mucus secreted by the goblet cells? (A2) _____

    _____

    _____

2. Describe the function of the cilia in the respiratory tubes. (A2) _____

    _____

    _____

3. How is breathing affected if the smooth muscle of the bronchial tree relaxes? (A2) _____

    _____

    _____

4. How is breathing affected if the smooth muscle of the bronchial tree contracts? (A2) _____

    _____

    _____

CRITICAL THINKING ASSESSMENT

What is the functional advantage of the alveolar walls being so thin? **A2**

_____

_____

_____

What affect would pulmonary edema have on this function? **A2**

_____

_____

_____

CRITICAL THINKING ASSESSMENT

Describe the airway of a patient who is having an asthma attack. What would be the desired effect of treating this patient with medications such as beta2-agonists and corticosteroids? **A2**

_____

_____

_____

LABORATORY EXERCISE 51

Breathing and Respiratory Volumes

Purpose of the Exercise

To review the mechanisms of breathing and to measure or calculate certain respiratory air volumes and respiratory capacities.

MATERIALS NEEDED

Lung function model

Spirometer, handheld (dry portable)

Alcohol swabs (wipes)

Disposable mouthpieces

Nose clips

Meterstick

For Learning Extension Activity:

Clock or watch with seconds timer

Ph.I.L.S. 4.0: Lesson 38 Respiration: Altering Airway Volume

SAFETY ⚠

▶ Clean the spirometer with an alcohol swab (wipe) before each use.

▶ Place a new disposable mouthpiece on the stem of the spirometer before each use.

▶ Dispose of the alcohol swabs and mouthpieces according to directions from your laboratory instructor.

Learning Outcomes AP|R

After completing this exercise, you should be able to:

O1 Differentiate between the mechanisms and muscles responsible for inspiration and expiration.

O2 Measure respiratory volumes using a spirometer.

O3 Calculate respiratory capacities using the data obtained from respiratory volumes.

O4 Match the respiratory volumes and respiratory capacities with their definitions.

O5 Explain the changes in tidal volume in response to changes in volume of the anatomic dead space.

O6 Apply concepts covered in this laboratory exercise to the changes that occur when the bronchioles are dilated or constricted by the autonomic nervous system.

The **O** corresponds to the assessments **A** indicated in the Laboratory Assessment for this Exercise.

Pre-Lab

Carefully read the introductory material and examine the entire lab. Be familiar with inspiration, expiration, and respiratory volumes from lecture or the textbook. Visit www.mhhe.com/martinseries3 for LabCam videos. Answer the pre-lab questions.

Pre-Lab Questions Select the correct answer for each of the following questions:

1. The size of the thoracic cavity is increased by contractions of all of the following muscles *except* the
   a. diaphragm.   b. external intercostals.
   c. pectoralis minor.   (d.) external oblique.

2. A _____ is an instrument to measure air volumes during breathing.
   a. flow meter   (b.) spirometer
   c. lung function model   d. capacity meter

3. The _____ is the maximum volume of air that can be exhaled after taking the deepest breath possible.
   a. tidal volume   b. expiratory reserve volume
   (c.) vital capacity   d. total lung capacity

4. Tidal volume is estimated to be about
   (a.) 500 mL.   b. 1,200 mL.
   c. 3,600 mL.   d. 4,800 mL.

5. A normal resting breathing rate is about _____ breaths per minute.
  a. 5–10
  b. 12–15
  c. 16–20
  d. 21–30

6. The contraction of the diaphragm increases the size of the thoracic cavity.
  a. True _____
  b. False _____

7. Vital capacity is the total of tidal volume, expiratory reserve volume, and residual volume.
  a. True _____
  b. False _____

8. Vital capacities gradually decrease as a person continues to age.
  a. True _____
  b. False _____

Breathing, or pulmonary ventilation, involves the movement of air from outside the body through the bronchial tree and into the alveoli and the reversal of this air movement to allow gas (oxygen and carbon dioxide) exchange between air and blood. These movements are caused by changes in the size of the thoracic cavity that result from skeletal muscle contractions. The size of the thoracic cavity during *inspiration* is increased by contractions of the diaphragm, external intercostals, internal intercostals (intercartilaginous part), pectoralis minor, sternocleidomastoid, and scalenes. *Expiration* is aided by contractions of the internal intercostals (interosseous part), rectus abdominis, and external oblique, and from the elastic recoil of stretched tissues (fig. 51.1 and table 51.1).

The volumes of air that move into and out of the lungs during various phases of breathing are called *respiratory (pulmonary) volumes* and *capacities*. A *volume* is a single measurement, whereas *capacity* represents two or more combined volumes. Respiratory volumes can be measured by using an instrument called a *spirometer.* Respiratory capacities can be determined by using various combinations of respiratory volumes. The values obtained vary with a person's age, sex, height, weight, stress, and physical fitness.

With each breath you take, what volume of air reaches the alveoli of your lungs? You take a normal breath of air through your nose or mouth. The volume of air you breathe

---

FIGURE 51.1 Respiratory muscles involved in inspiration and forced expiration (anterior view). Boldface indicates the primary muscles increasing or decreasing the size of the thoracic cavity. Blue arrows indicate the direction of contraction during inspiration that increases the capacity of the thoracic cavity; the green arrows indicate the direction of contraction during forced expiration that decreases the capacity of the thoracic cavity. (The green arrows associated with the diaphragm indicate its ascending direction primarily from the elastic recoil of the stretched tissues during quiet breathing.)

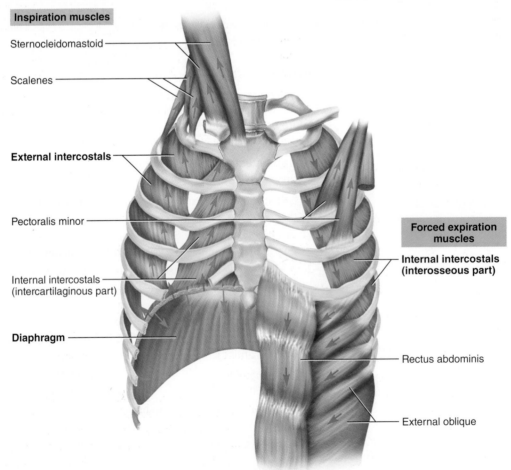

Inspiration muscles
- Sternocleidomastoid
- Scalenes
- **External intercostals**
- Pectoralis minor
- Internal intercostals (intercartilaginous part)
- **Diaphragm**

Forced expiration muscles
- **Internal intercostals (interosseous part)**
- Rectus abdominis
- External oblique

Table 51.1 Muscles of Respiration

Muscles of Inspiration	Origin	Insertion	Inspiration Action
Diaphragm	Costal cartilages and ribs 7–12, xiphoid process, and lumbar vertebrae	Central tendon of diaphragm	Flattens (descends) during contraction; this expands thoracic cavity and compresses abdominal viscera during inspiration
External intercostals	Inferior border of superior rib	Superior border of inferior rib	Elevates ribs; this expands thoracic cavity during inspiration
Sternocleidomastoid	Manubrium of sternum and medial clavicle	Mastoid process of temporal bone	Elevates sternum
Scalenes	Transverse processes of cervical vertebrae C2–C6	Ribs 1 and 2	Elevates ribs 1 and 2
Pectoralis minor	Sternal ends of ribs 3–5	Coracoid process of scapula	Elevates ribs 3–5
Internal intercostals (intercartilaginous part)	Superior border of inferior costal cartilages	Inferior border of superior costal cartilages	Assists in elevating ribs
Muscles of Expiration	**Origin**	**Insertion**	**Expiration Action**
Internal intercostals (interosseous part)	Superior border of inferior rib	Inferior border of superior rib	Depresses ribs; this compresses thoracic cavity during forced expiration
Rectus abdominis	Pubic crest and pubic symphysis	Xiphoid process of sternum and costal cartilages of ribs 5–7	Compresses inferior ribs and abdominal contents, helping to raise diaphragm
External oblique	Anterior surfaces of ribs 5–12	Anterior iliac crest and linea alba by aponeurosis	Compresses abdominal contents, helping to raise diaphragm

in (or out) is called the *tidal volume.* The average tidal volume is 500 mL (imagine the equivalent of a 1/2 liter bottle). This volume of air is moved in through your nose and mouth and through your respiratory passageway, including the nasal cavity, pharynx, larynx, trachea, and bronchial tree, to the alveoli. A certain amount of that air never reaches your alveoli—it is within the respiratory passageway. This air is said to be in the *anatomic dead space.* The average volume of anatomic dead space is 150 mL. If the tidal volume is 500 mL and the volume of the anatomic dead space is 150 mL, what volume of air reaches your alveoli? Understanding the difference between tidal volume and the volume of air that reaches the lungs is important because *only* the air that reaches the alveoli provides oxygen for gas exchange with the blood.

PROCEDURE A: Breathing Mechanisms

The size of the thoracic cavity increases during inspiration by contractions of inspiration muscles, primarily the diaphragm, resulting in increased depth of the thoracic cavity, and external intercostals, widening the thoracic cavity. A combination of synergistic contractions of the sternocleidomastoid, scalenes, pectoralis minor, and the intercartilaginous part of the internal intercostals elevate the sternum and rib cage. During forced expiration, the interosseous part of the internal intercostals depresses the rib cage, narrowing the thoracic cavity. The contractions of the rectus abdominis and the external oblique compress the abdominal viscera, forcing the diaphragm to ascend and thus reducing the size of the thoracic cavity (fig. 51.1).

FIGURE 51.2 Formula for Boyle's Law. P_1 and V_1 represent the original pressure and volume, respectively, and P_2 and V_2 represent the second pressure and volume. The pressure of a gas is inversely proportional to its volume, assuming a constant temperature. The graph shows this relationship.

$$P_1V_1 = P_2V_2$$

Volume V

Pressure P

Air enters the lungs during inspiration and exits the lungs during expiration based upon characteristics of *Boyle's law.* According to Boyle's law, the pressure of a gas is inversely proportional to its volume, assuming a constant temperature (fig. 51.2). In other words, when the volume increases in a container such as the thoracic cavity during inspiration, pressure decreases and air flows into the lungs. In contrast, when volume decreases in a container such as the thoracic cavity during expiration, pressure increases and air flows out of the lungs. Air moves into and out of the lungs in much the same way as can be demonstrated using a syringe (fig. 51.3). When you pull back on the plunger of the syringe, the volume increases in the barrel of the syringe, which decreases the pressure, resulting in air rushing into the barrel of the syringe. In contrast, pushing the plunger into the barrel of the syringe causes the pressure to increase in the barrel of the syringe, and air flows out of the container.

1. Observe the mechanical lung function model (fig. 51.4). It consists of a heavy plastic bell jar with rubber sheeting clamped over its wide open end. Its narrow upper opening is plugged with a rubber stopper through which a Y tube is passed. Small rubber balloons are fastened to the arms of the Y. What happens to the balloons when the rubber sheeting is pulled downward?

_____

_____

What happens when the sheeting is pushed upward?

_____

_____

2. Compare the lung function model with figures 51.1 and 51.3. Note the muscles of inspiration that increase the size of the thoracic cavity, and the muscles of expiration

FIGURE 51.4 A lung function model.

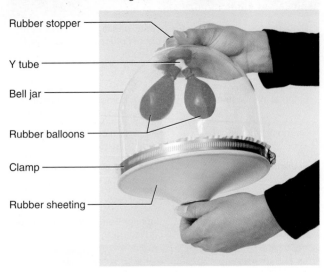

Rubber stopper
Y tube
Bell jar
Rubber balloons
Clamp
Rubber sheeting

that decrease the size of the thoracic cavity. Relate these volume changes to pressure changes and airflow into and out of the lungs.

3. Complete Part A of Laboratory Assessment 51.

PROCEDURE B: Respiratory Volumes and Capacities (This procedure can be in conjunction with Laboratory Exercise 52.)

① Get a handheld spirometer (fig. 51.5). Point the needle to zero by rotating the adjustable dial. Before using the instrument, clean it with an alcohol swab and place a new disposable mouthpiece over its stem. The instrument should be held with the dial upward, and *air should be blown only into the disposable mouthpiece* (fig. 51.6). If air tends to exit the nostrils during exhalation, you can prevent it from doing so by using a nose clip or by pinching your nose when exhaling into the spirometer. Movement of the needle indicates the air volume that leaves the lungs. The exhalation should be slowly and forcefully performed. Too rapid an exhalation can result in erroneous data or damage to the spirometer.

② *Tidal volume (TV)* (about 500 mL) is the volume of air that enters (or leaves) the lungs during a *respiratory cycle* (one inspiration plus the following expiration). *Resting tidal volume* is the volume of air that enters (or leaves)

FIGURE 51.3 Moving the plunger of a syringe as in (a) results in air movements into the barrel of the syringe; moving the plunger of the syringe as in (b) results in air movements out of the barrel of the syringe. Air movements in and out of the lung function model and the lungs during breathing occur for similar reasons.

(a)   (b)

⚠ **WARNING**

If you begin to feel dizzy or light-headed while performing Procedure B, stop the exercise and breathe normally.

FIGURE 51.5 A handheld spirometer can be used to measure respiratory volumes. *(PROPER)*

the lungs during normal, quiet breathing (fig. 51.7). To measure this volume, follow these steps:

a. Sit quietly for a few moments.

b. Position the spirometer dial so that the needle points to zero.

c. Place the mouthpiece between your lips and exhale three ordinary expirations into it after inhaling through the nose each time. *Do not force air out of your lungs; exhale normally.*

d. Divide the total value indicated by the needle by 3 and record this amount as your resting tidal volume on the table in Part B of the laboratory assessment.

3. *Expiratory reserve volume (ERV)* (about 1,200 mL) is the volume of air in addition to the tidal volume that leaves the lungs during forced expiration (fig. 51.7). Chronic obstructive pulmonary disease (COPD) reduces pulmonary ventilation; COPD occurs in emphysema and chronic bronchitis, which are often caused by smoking and inhaling other pollutants. With COPD, the ERV is greatly reduced during physical exertion, and the person becomes easily exhausted from expending more energy just to breathe. To measure this volume, follow these steps:

a. Breathe normally for a few moments. Set the needle to zero.

b. At the end of an ordinary expiration, place the mouthpiece between your lips and exhale all of the air you can force from your lungs through the spirometer. Use a nose clip or pinch your nose to prevent any air from exiting your nostrils.

FIGURE 51.6 Demonstration of use of a handheld spirometer. Air should be blown slowly and forcefully only into a disposable mouthpiece. *Use a nose clip or pinch your nose when measuring expiratory reserve volume and vital capacity volume if air exits the nostrils.*

FIGURE 51.7 Graphic representation of respiratory volumes and capacities.

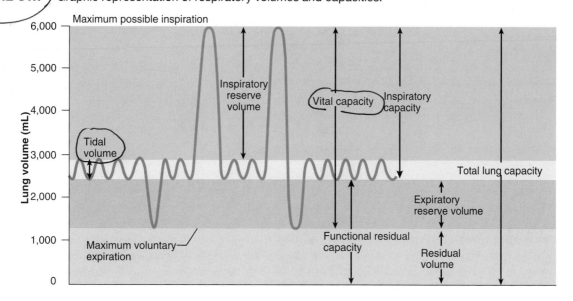

c. Record the results as your expiratory reserve volume in Part B of the laboratory assessment.

4. *Vital capacity (VC)* (about 4,800 mL) is the maximum volume of air that can be exhaled after taking the deepest breath possible (fig. 51.7). To measure this volume, follow these steps:

a. Breathe normally for a few moments. Set the needle at zero.

b. Breathe in and out deeply a couple of times; then take the deepest breath possible.

c. Place the mouthpiece between your lips and exhale all the air out of your lungs, slowly and forcefully. Use a nose clip or pinch your nose to prevent any air from exiting your nostrils.

d. Record the value as your vital capacity in Part B of the laboratory assessment. Compare your result with that expected for a person of your sex, age, and height listed in tables 51.2 and 51.3. Use the meterstick to determine your height in centimeters if necessary or multiply your height in inches times 2.54 to calculate your height in centimeters. Considerable individual variations from the expected will be noted due to parameters other than sex, age, and height, which can include fitness level, health, medications, and others.

5. *Inspiratory reserve volume (IRV)* (about 3,100 mL) is the volume of air in addition to the tidal volume that enters the lungs during forced inspiration (fig. 51.7). Calculate your inspiratory reserve volume by subtracting your tidal volume (TV) and your expiratory reserve volume (ERV) from your vital capacity (VC):

$$IRV = VC - (TV + ERV)$$

6. *Inspiratory capacity (IC)* (about 3,600 mL) is the maximum volume of air a person can inhale following exhalation of the tidal volume (fig. 51.7). Calculate your inspiratory capacity by adding your tidal volume (TV) and your inspiratory reserve volume (IRV):

$$IC = TV + IRV$$

7. *Functional residual capacity (FRC)* (about 2,400 mL) is the volume of air that remains in the lungs following exhalation of the tidal volume (fig. 51.7). Calculate your functional residual capacity (FRC) by adding your expiratory reserve volume (ERV) and your residual volume (RV), which you can assume is 1,200 mL:

$$FRC = ERV + 1,200$$

8. *Residual volume (RV)* (about 1,200 mL) is the volume of air that always remains in the lungs after the most forceful expiration (fig. 51.7). Although it is part of *total lung capacity* (about 6,000 mL), it cannot be measured with a spirometer. The residual air allows gas exchange and the alveoli to remain open during the respiratory cycle.

9. Complete Parts B and C of the laboratory assessment.

LEARNING EXTENSION ACTIVITY

Determine your *minute respiratory volume*. To do this, follow these steps:

1. Sit quietly for a while, and then to establish your breathing rate, count the number of times you breathe in 1 minute. This might be inaccurate because conscious awareness of breathing rate can alter the results. You might ask a laboratory partner to record your breathing rate sometime when you are not expecting it to be recorded. A normal resting breathing rate is about 12–15 breaths per minute.

2. Calculate your minute respiratory volume by multiplying your breathing rate by your tidal volume:

$$\underline{\hspace{2cm}} \times \underline{\hspace{2cm}} = \underline{\hspace{2cm}}$$

(breathing rate)   (tidal volume)   (minute respiratory volume)

3. This value indicates the total volume of air that moves into your respiratory passages during each minute of ordinary breathing.

CRITICAL THINKING ACTIVITY

It can be noted from the data in tables 51.2 and 51.3 that vital capacities gradually decrease with age. Propose an explanation for this normal correlation.

_____

_____

_____

_____

Table 31.2 Predicted Vital Capacities (in milliliters) for Females

HEIGHT IN CENTIMETERS

Age	146	148	150	152	154	156	158	160	162	164	166	168	170	172	174	176	178	180	182	184	186	188	190	192	194
16	2950	2990	3030	3070	3110	3150	3190	3230	3270	3310	3350	3390	3430	3470	3510	3550	3590	3630	3670	3715	3755	3800	3840	3880	3920
18	2920	2960	3000	3040	3080	3120	3160	3200	3240	3280	3320	3360	3400	3440	3480	3520	3560	3600	3640	3680	3720	3760	3800	3840	3880
20	2890	2930	2970	3010	3050	3090	3130	3170	3210	3250	3290	3330	3370	3410	3450	3490	3525	3565	3605	3645	3695	3720	3760	3800	3840
22	2860	2900	2940	2980	3020	3060	3095	3135	3175	3215	3255	3290	3330	3370	3410	3450	3490	3530	3570	3610	3650	3685	3725	3765	3800
24	2830	2870	2910	2950	2985	3025	3065	3100	3140	3180	3220	3260	3300	3335	3375	3415	3455	3490	3530	3570	3610	3650	3685	3725	3765
26	2800	2840	2880	2920	2960	3000	3035	3070	3110	3150	3190	3230	3265	3300	3340	3380	3420	3455	3495	3530	3570	3610	3650	3685	3725
28	2775	2810	2850	2890	2930	2965	3000	3040	3070	3115	3155	3190	3230	3270	3305	3345	3380	3420	3460	3495	3535	3570	3610	3650	3685
30	2745	2780	2820	2860	2895	2935	2970	3010	3045	3085	3120	3160	3195	3235	3270	3310	3345	3385	3420	3460	3495	3535	3570	3610	3645
32	2715	2750	2790	2825	2865	2900	2940	2975	3015	3050	3090	3125	3160	3200	3235	3275	3310	3350	3385	3425	3460	3495	3535	3570	3610
34	2685	2725	2760	2795	2835	2870	2910	2945	2980	3020	3055	3090	3130	3165	3200	3240	3275	3310	3350	3385	3425	3460	3495	3535	3570
36	2655	2695	2730	2765	2805	2840	2875	2910	2950	2985	3020	3060	3095	3130	3165	3205	3240	3275	3310	3350	3385	3420	3460	3495	3530
38	2630	2665	2700	2735	2770	2810	2845	2880	2915	2950	2990	3025	3060	3095	3130	3170	3205	3240	3275	3310	3350	3385	3420	3455	3490
40	2600	2635	2670	2705	2740	2775	2810	2850	2885	2920	2955	2990	3025	3060	3095	3135	3170	3205	3240	3275	3310	3345	3380	3420	3455
42	2570	2605	2640	2675	2710	2745	2780	2815	2850	2885	2920	2955	2990	3025	3060	3100	3135	3170	3205	3240	3275	3310	3345	3380	3415
44	2540	2575	2610	2645	2680	2715	2750	2785	2820	2855	2890	2925	2960	2995	3030	3060	3095	3130	3165	3200	3235	3270	3305	3340	3375
46	2510	2545	2580	2615	2650	2685	2715	2750	2785	2820	2855	2890	2925	2960	2995	3030	3060	3095	3130	3165	3200	3235	3270	3305	3340
48	2480	2515	2550	2585	2620	2650	2685	2715	2750	2785	2820	2855	2890	2925	2960	2995	3030	3060	3095	3130	3160	3195	3230	3265	3300
50	2455	2485	2520	2555	2590	2625	2655	2690	2720	2755	2785	2820	2855	2890	2925	2955	2990	3025	3060	3090	3125	3155	3190	3225	3260
52	2425	2455	2490	2525	2555	2590	2625	2655	2690	2720	2755	2790	2820	2855	2890	2925	2955	2990	3020	3055	3090	3125	3155	3190	3220
54	2395	2425	2460	2495	2530	2560	2590	2625	2655	2690	2720	2755	2790	2820	2855	2885	2920	2950	2985	3020	3050	3085	3115	3150	3180
56	2365	2400	2430	2460	2495	2525	2560	2590	2625	2655	2690	2720	2755	2790	2820	2855	2885	2920	2950	2980	3015	3045	3080	3110	3145
58	2335	2370	2400	2430	2460	2495	2525	2560	2590	2625	2655	2690	2720	2750	2785	2815	2850	2880	2920	2945	2975	3010	3040	3075	3105
60	2305	2340	2370	2400	2430	2460	2495	2525	2560	2590	2625	2655	2685	2720	2750	2780	2810	2845	2875	2915	2940	2970	3000	3035	3065
62	2280	2310	2340	2370	2405	2435	2465	2495	2525	2560	2590	2620	2655	2685	2715	2745	2775	2810	2840	2870	2900	2935	2965	2995	3025
64	2250	2280	2310	2340	2370	2400	2430	2465	2495	2525	2555	2585	2620	2650	2680	2710	2740	2770	2805	2835	2865	2895	2920	2955	2990
66	2220	2250	2280	2310	2340	2370	2400	2430	2460	2495	2525	2555	2585	2615	2645	2675	2705	2735	2765	2800	2825	2860	2890	2920	2950
68	2190	2220	2250	2280	2310	2340	2370	2400	2430	2460	2490	2520	2550	2580	2610	2640	2670	2700	2730	2760	2795	2820	2850	2880	2910
70	2160	2190	2220	2250	2280	2310	2340	2370	2400	2425	2455	2485	2515	2545	2575	2605	2635	2665	2695	2725	2755	2780	2810	2840	2870
72	2130	2160	2190	2220	2250	2280	2310	2335	2365	2395	2425	2455	2480	2510	2540	2570	2600	2630	2660	2685	2715	2745	2775	2805	2830
74	2100	2130	2160	2190	2220	2245	2275	2305	2335	2360	2390	2420	2450	2475	2505	2535	2565	2590	2620	2650	2680	2710	2740	2765	2795

From E. DeF. Baldwin and E. W. Richards, Jr., "Pulmonary Insufficiency 1, Physiologic Classification, Clinical Methods of Analysis, Standard Values in Normal Subjects" in *Medicine* 27:243, © by William & Wilkins. Used by permission.

Table 51.3   Predicted Vital Capacities (in milliliters) for Males

HEIGHT IN CENTIMETERS

Age	146	148	150	152	154	156	158	160	162	164	166	168	170	172	174	176	178	180	182	184	186	188	190	192	194
16	3765	3820	3870	3920	3975	4025	4075	4130	4180	4230	4285	4335	4385	4440	4490	4540	4590	4645	4695	4745	4800	4850	4900	4955	5005
18	3740	3790	3840	3890	3940	3995	4045	4095	4145	4200	4250	4300	4350	4405	4455	4505	4555	4610	4660	4710	4760	4815	4865	4915	4965
20	3710	3760	3810	3860	3910	3960	4015	4065	4115	4165	4215	4265	4320	4370	4420	4470	4520	4570	4625	4675	4725	4775	4825	4875	4930
22	3680	3730	3780	3830	3880	3930	3980	4030	4080	4135	4185	4235	4285	4335	4385	4435	4485	4535	4585	4635	4685	4735	4790	4840	4890
24	3635	3685	3735	3785	3835	3885	3935	3985	4035	4085	4135	4185	4235	4285	4330	4380	4430	4480	4530	4580	4630	4680	4730	4780	4830
26	3605	3655	3705	3755	3805	3855	3905	3955	4000	4050	4100	4150	4200	4250	4300	4350	4395	4445	4495	4545	4595	4645	4695	4740	4790
28	3575	3625	3675	3725	3775	3820	3870	3920	3970	4020	4070	4115	4165	4215	4265	4310	4360	4410	4460	4510	4555	4605	4655	4705	4755
30	3550	3595	3645	3695	3740	3790	3840	3890	3935	3985	4035	4080	4130	4180	4230	4275	4325	4375	4425	4470	4520	4570	4615	4665	4715
32	3520	3565	3615	3665	3710	3760	3810	3855	3905	3950	4000	4050	4095	4145	4195	4240	4290	4340	4385	4435	4485	4530	4580	4625	4675
34	3475	3525	3570	3620	3665	3715	3760	3810	3855	3905	3950	4000	4045	4095	4140	4190	4240	4285	4330	4380	4425	4475	4520	4570	4615
36	3445	3495	3540	3585	3635	3680	3730	3775	3825	3870	3920	3965	4010	4060	4105	4155	4200	4250	4295	4340	4390	4435	4485	4530	4580
38	3415	3465	3510	3555	3605	3650	3695	3745	3790	3840	3885	3930	3980	4025	4070	4120	4165	4210	4260	4305	4350	4400	4445	4495	4540
40	3385	3435	3480	3525	3575	3620	3665	3710	3760	3805	3850	3900	3945	3990	4035	4085	4130	4175	4220	4270	4315	4360	4410	4455	4500
42	3360	3405	3450	3495	3540	3590	3635	3680	3725	3770	3820	3865	3910	3955	4000	4050	4095	4140	4185	4230	4280	4325	4370	4415	4460
44	3315	3360	3405	3450	3495	3540	3585	3630	3675	3725	3770	3815	3860	3905	3950	3995	4040	4085	4130	4175	4220	4270	4315	4360	4405
46	3285	3330	3375	3420	3465	3510	3555	3600	3645	3690	3735	3780	3825	3870	3915	3960	4005	4050	4095	4140	4185	4230	4275	4320	4365
48	3255	3300	3345	3390	3435	3480	3525	3570	3615	3655	3700	3745	3790	3835	3880	3925	3970	4015	4060	4105	4150	4190	4235	4280	4325
50	3210	3255	3300	3345	3390	3430	3475	3520	3565	3610	3650	3695	3740	3785	3830	3870	3915	3960	4005	4050	4090	4135	4180	4225	4270
52	3185	3225	3270	3315	3355	3400	3445	3490	3530	3575	3620	3660	3705	3750	3795	3835	3880	3925	3970	4010	4055	4100	4140	4185	4230
54	3155	3195	3240	3285	3325	3370	3415	3455	3500	3540	3585	3630	3670	3715	3760	3800	3845	3890	3930	3975	4020	4060	4105	4145	4190
56	3125	3165	3210	3255	3295	3340	3380	3425	3465	3510	3550	3595	3640	3680	3725	3765	3810	3850	3895	3940	3980	4025	4065	4110	4150
58	3080	3125	3165	3210	3250	3290	3335	3375	3420	3460	3500	3545	3585	3630	3670	3715	3755	3800	3840	3880	3925	3965	4010	4050	4095
60	3050	3095	3135	3175	3220	3260	3300	3345	3385	3430	3470	3500	3555	3595	3635	3680	3720	3760	3805	3845	3885	3930	3970	4015	4055
62	3020	3060	3110	3150	3190	3230	3270	3310	3350	3390	3440	3480	3520	3560	3600	3640	3680	3730	3770	3810	3850	3890	3930	3970	4020
64	2990	3030	3080	3120	3160	3200	3240	3280	3320	3360	3400	3440	3490	3530	3570	3610	3650	3690	3730	3770	3810	3850	3900	3940	3980
66	2950	2990	3030	3070	3110	3150	3190	3230	3270	3310	3350	3390	3430	3470	3510	3550	3600	3640	3680	3720	3760	3800	3840	3880	3920
68	2920	2960	3000	3040	3080	3120	3160	3200	3240	3280	3320	3360	3400	3440	3480	3520	3560	3600	3640	3680	3720	3760	3800	3840	3880
70	2890	2930	2970	3010	3050	3090	3130	3170	3210	3250	3290	3330	3370	3410	3450	3480	3520	3560	3600	3640	3680	3720	3760	3800	3840
72	2860	2900	2940	2980	3020	3060	3100	3140	3180	3210	3250	3290	3330	3370	3410	3450	3490	3530	3570	3610	3650	3680	3720	3760	3800
74	2820	2860	2900	2930	2970	3010	3050	3090	3130	3170	3200	3240	3280	3320	3360	3400	3440	3470	3510	3550	3590	3630	3670	3710	3740

From E. DeF. Baldwin and E. W. Richards, Jr. "Pulmonary Insufficiency 1. Physiologic Classification, Clinical Methods of Analysis, Standard Values in Normal Subjects" in Medicine 27:243. © by Williams & Wilkins

LABORATORY ASSESSMENT 51

Name _____

Date _____

Section _____

The (A) corresponds to the indicated Learning Outcome(s) (O) found at the beginning of the Laboratory Exercise.

Breathing and Respiratory Volumes

PART A: Assessments

Complete the following statements:

1. When using the lung function model, what part of the respiratory system is represented by: (A1)

   a. The rubber sheeting? _____   c. The Y tube? _____

   b. The bell jar? _____   d. The balloons? ___LUNGS___

2. When the diaphragm contracts, the size of the thoracic cavity _____. (A1)

3. The ribs are raised primarily by contraction of the _____ muscles, which increases the size of the thoracic cavity. (A1)

4. The primary muscles that help to force out more than the normal volume of air, by pulling the ribs downward and inward, are the _____. (A1)

5. We inhale when the diaphragm _____. (A1)

PART B: Assessments

1. Test results for respiratory air volumes and capacities: (A2) (A3)

Respiratory Volume or Capacity	Expected Value* (approximate)	Test Result	Percent of Expected Value (test result/expected value × 100)
Tidal volume (resting) (TV)	500 mL	~~(scribbled)~~ (666)	
Expiratory reserve volume (ERV)	1,200 mL		
Vital capacity (VC)	(enter yours from table 51.2 or 51.3)		
Inspiratory reserve volume (IRV)	3,100 mL		
Inspiratory capacity (IC)	3,600 mL		
Functional residual capacity (FRC)	2,400 mL		

*The values listed are most characteristic for a healthy, tall, young adult male. In general, adult females have smaller bodies and therefore smaller lung volumes and capacities. If your expected value for vital capacity is considerably different than 4,800 mL, your other values will vary accordingly.

2. Complete the following:

   a. How do your test results compare with the expected values? _____

   _____

   _____

   _____

   b. How does your vital capacity compare with the average value for a person of your sex, age, and height?

   _____

   _____

   c. What measurement in addition to vital capacity is needed before you can calculate your total lung capacity?

   _____

3. If your experimental results are considerably different than the predicted vital capacities, propose reasons for the differences. As you write this paragraph, consider factors such as smoking, physical fitness, respiratory disorders, stress, and medications. (Your instructor might have you make some class correlations from class data.)

_____

_____

_____

_____

PART C: Assessments

Match the air volumes in column A with the definitions in column B. Place the letter of your choice in the space provided.

A4

Column A	Column B
a. Expiratory reserve volume	_____ 1. Volume of air in addition to tidal volume that leaves the lungs during forced expiration
b. Functional residual capacity	
c. Inspiratory capacity	_____ 2. Vital capacity plus residual volume
d. Inspiratory reserve volume	_____ 3. Volume of air that remains in lungs after the most forceful expiration
e. Residual volume	
f. Tidal volume	_____ 4. Volume of air that enters or leaves lungs during a respiratory cycle
g. Total lung capacity	
h. Vital capacity	_____ 5. Volume of air in addition to tidal volume that enters lungs during forced inspiration
	_____ 6. Maximum volume of air a person can exhale after taking the deepest possible breath
	_____ 7. Maximum volume of air a person can inhale following exhalation of the tidal volume
	_____ 8. Volume of air remaining in the lungs following exhalation of the tidal volume

LESSON 38 | Respiration
Altering Airway Volume

The purpose of this exercise is to demonstrate how changing the tidal volume, the amount of air that goes in or out of the lungs during a breathing cycle, will affect the alveolar ventilation, the amount of air that enters the alveoli for gas exchange.

During respiration, air moves in and out of the lungs via the airways. At the end of each breathing cycle there is a certain volume of air left in the airways that never makes it to the alveoli called the anatomical dead space. This volume of air cannot be used for gas exchange. The anatomical dead space is usually kept constant. Any change in tidal volume will cause a change in the amount of air that enters the alveoli.

In this laboratory experiment, you will measure the tidal volume of a virtual volunteer using a spirometer. You will measure the volunteer's tidal volume during normal breathing, and then a measurement will be taken while the volunteer breathes through a plastic tube. Breathing through the plastic tube will increase his anatomical dead space.

Before you begin, familiarize yourself with the following concepts (use your textbook as a reference):

- Lung volumes/capacities
- Function of a spirometer
- Breathing mechanism

State your hypothesis concerning what would happen to the anatomical dead space if it were increased. Include effects on tidal volume and the amount of air available for gas exchange.

_____

_____

_____

_____

1. Open Lesson 38, Respiration: Altering Airway Volume.
2. Read the objectives and introduction and take the pre-laboratory quiz.
3. After completing the pre-laboratory quiz, read through the wet lab.
4. The laboratory exercise will open when you click **Continue** after completing the wet lab (fig. 51.8).
5. Follow the instructions provided in the laboratory exercise. Using the data from the Journal and the graph you constructed, complete the following tables and questions as you proceed.

FIGURE 51.8 Opening screen for the laboratory exercise on Respiration: Altering Airway Volume.

6. From the experimental results, complete the following table for the measured tidal volumes comparing breathing apparatus alone and breathing apparatus with a plastic tube. **A2**

Trial	Without Tube	With Tube
Trial # 1		
Trial # 2		
Trial # 3		
Average (mean)		

7. Which setup had the larger average tidal volume? _____ **A5**

8. To maintain normal alveolar ventilation, if the anatomic dead space is increased, the tidal volume will _____ (increase, decrease, or remain the same). **A5**

9. In the setup with the tube attached to the breathing apparatus, did additional air reach the alveoli? _____ **A5**

10. Complete the following table for the effect on each variable if a person breathes with the tube. **A5**

Volume of Air	Change (increase, decrease, or remain the same)
Anatomic dead space	
Tidal volume	
Alveolar ventilation	

CRITICAL THINKING ASSESSMENT A5 A6

If the bronchioles are dilated (when the sympathetic nervous system is activated, for example), the anatomic dead space
_____ (increases, decreases, or remains the same), and tidal volume will
_____ (increase, decrease, or remain the same). In patients with emphysema, the
anatomic dead space increases (since there is a loss of elasticity of lung tissue and the thoracic cage cannot effectively
"be pulled back in"). Predict the changes in breathing of a patient with emphysema as he/she tries to compensate for the
increase in anatomic dead space. _____

_____

_____

_____

_____

3/30/17

A&P II today: Respiratory Systems

1) Ex 51: Breathing & Respiratory Volume
   → Dry Spirometer (by pair)
   → Wet Spirometer (by table)

2) Ex 50: Respiratory Organs
   → Slides
   Normal: Trachea /Lung/Bronchitis
   Pathologic: Emphysema
              Carcinoma
              Anthracosis

3) Sheep Pluera

Spirometry: BIOPAC© Exercise

Purpose of the Exercise

To measure and calculate pulmonary volumes and capacities using the BIOPAC airflow transducer system.

Computer system (Mac OS X 10.5–10.8, or PC running Windows XP, Vista, 7, or 8)

BIOPAC Student Lab software, version 4.0 or above

BIOPAC Data Acquisition Unit (MP36, MP35, or MP45)

BIOPAC Airflow Transducer (SS11LA, with removable/autoclavable head)

BIOPAC Disposable Mouthpiece (AFT2) or optional Autoclavable Mouthpiece (AFT8)

BIOPAC Disposable Nose Clip (AFT3)

BIOPAC Bacteriological Filter (AFT1)

BIOPAC Calibration Syringe, 0.6 liter (AFT6 or AFT6A + AFT11A) or 2 liter (AFT26)

SAFETY

▶ If your lab autoclaves the removable head of the airflow transducer (SS11LA), and you use the autoclavable mouthpieces (AFT8), you will need to use a bacteriological filter (AFT1) only for the calibration step.

▶ If your lab does not autoclave the removable head of the airflow transducer (SS11LA), and you use the disposable mouthpieces (AFT2), you will need a bacteriological filter (AFT1) for the calibration step, and a clean one for each subject.

▶ Each subject will need a disposable mouthpiece (AFT2) or a clean, autoclaved mouthpiece (AFT8).

▶ Each subject will need a disposable nose clip (AFT3).

Learning Outcomes

After completing this exercise, you should be able to:

O1 Examine experimentally and record and/or calculate selected respiratory volumes and capacities.

O2 Compare the measured volumes and capacities with average values.

O3 Compare the volumes and capacities of subjects differing in sex, age, height, and weight.

The **O** corresponds to the assessments **A** indicated in the Laboratory Assessment for this Exercise.

Pre-Lab

Carefully read the introductory material and examine the entire lab content. Be familiar with inspiration, expiration, and respiratory volumes from lecture or the textbook.

Breathing (pulmonary ventilation), one of the most basic processes of life, consists of two phases: inspiration and expiration. *Inspiration* consists of the inhalation of air into the respiratory passages and lungs, which provides vital oxygen (O_2) to the cells of the body. Oxygen is required for the aerobic cellular respiration of food, to break down nutrients into their building blocks, and to generate ATP for energy. Expiration involves the exhalation of air from the lungs into the atmosphere, which rids the body of carbon dioxide (CO_2). Carbon dioxide is produced as a waste product of cellular respiration.

One way in which the breathing process can be studied is through the use of a spirometer. There are various types of spirometers, from small handheld dry spirometers, to wet bell varieties, to computer-linked spirometers. But all spirometers work on the same principle—they all measure certain respiratory volumes, based on different degrees of effort in the breathing process. The respiratory volumes used in spirometry are as follows:

1. *Tidal volume (TV):* The volume of air inspired or expired in one normal, quiet breath (about 500 mL).
2. *Inspiratory reserve volume (IRV):* The volume of air that can be inspired forcefully above a tidal inspiration (about 3,100 mL).
3. *Expiratory reserve volume (ERV):* The volume of air that can be expired forcefully beyond a tidal expiration (about 1,200 mL).
4. *Residual volume (RV):* The volume of air remaining in the lungs after forceful expiration (about 1,200 mL).

Tidal respiration does not require any extra effort, but volumes such as IRV and ERV are defined by the extra effort used to measure them. For example, if you are preparing to blow up a balloon, and you inhale the maximum possible amount of air, the amount you inhale above your tidal inspiration will be your IRV. When you actually blow up the balloon, and you exhale the maximum possible amount of air, the amount beyond your tidal expiration will be your ERV. The residual volume cannot be measured with a spirometer, since it always remains in the lungs; for the purpose of this lab exercise, the default setting for RV by the BIOPAC software is 1.0 L.

In addition to respiratory volumes, several respiratory capacities can also be defined, which can be measured in the lab and/or calculated. Respiratory capacities can be calculated by adding two or more volumes together, as follows:

1. *Inspiratory capacity (IC):* The maximal volume of air that can be inspired after a tidal expiration, IC = TV + IRV (about 3,600 mL).
2. *Expiratory capacity (EC):* The maximal volume of air that can be expired after a tidal inspiration, EC = TV + ERV (about 1,700 mL).
3. *Functional residual capacity (FRC):* The volume of air remaining in the lungs after a tidal expiration, FRC = ERV + RV (about 2,400 mL).

4. *Vital capacity (VC):* The maximal volume of air that can be forcefully expired after a maximal inspiration, VC = TV + IRV + ERV (about 4,800 mL).
5. *Total lung capacity (TLC):* The total volume of air that the lungs and respiratory passages can hold, TLC = TV + IRV + ERV + RV or TLC = VC + RV (about 6,000 mL).

It is important to realize that there is quite a bit of individual variation in respiratory volumes and capacities. This variation is due to differences in sex, height, weight, age, and exercise history. Normal averages are given above, but any reading within 20% of the average is still considered to be in the normal range. Using vital capacity as an example, even though there are several factors determining the VC of an individual, VC can be estimated fairly accurately based on sex, height (H) in cm, and age (A) in years, with the following equations:

Male: $VC = 0.052 H - 0.022 A - 3.60$

Female: $VC = 0.041 H - 0.018 A - 2.69$

This BIOPAC exercise uses a computer-linked spirometry system to measure respiratory volumes. Each subject will breathe into an airflow transducer; the BIOPAC software will then convert airflow into volume. Several volumes and capacities will be measured and/or calculated in this lab exercise. VC will be both measured and estimated from one of the equations given above; the two values will then be compared. Again, remember that RV cannot be measured through spirometry. Figure 52.1 shows a record of volume changes over time for various respiratory volumes and capacities; this is called a *spirogram.*

Spirometry is used in medicine to help in the diagnosis of specific respiratory disorders, determine the extent of a respiratory illness or disease, and assess the respiratory status of a person recovering from an illness, such as pneumonia or an asthma attack. Certain illnesses are associated with characteristic changes in specific respiratory volumes and/or capacities. For example, in people with obstructive lung disease, such as chronic bronchitis, the FRC and RV may be elevated due to an inability to expel the normal amount of air during expiration and excess inflation of the lungs. In cases of restrictive lung disease, such as pulmonary fibrosis, decreased lung compliance makes it difficult to expand the lungs, which results in a decrease in IC and VC.

PROCEDURE A: Setup

1. With your computer **ON** and the BIOPAC MP36/35 unit turned **OFF,** plug the airflow transducer (SS11LA) into Channel 1. If you are using the MP45, the USB cable must be connected, and the ready light must be on.
2. Turn on the MP36/35 Data Acquisition Unit.
3. Start the BIOPAC student lab program for **L12— Pulmonary Function I,** and click **OK.**
4. Designate a filename to be used to save the subject's data, and click **OK.**

FIGURE 52.1 Example of a spirogram, showing respiratory volumes and capacities.

Source: Courtesy of and © BIOPAC Systems Inc. www.biopac.com

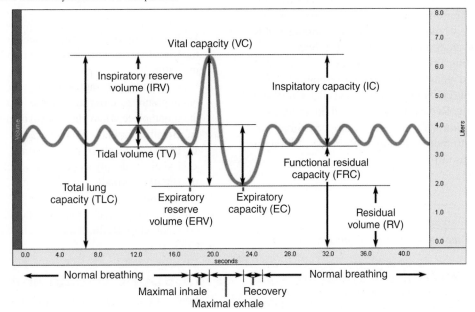

FIGURE 52.2 Sample recording for first calibration stage.

Source: Courtesy of and © BIOPAC Systems Inc. www.biopac.com

5. After typing the filename, a window will pop up, asking for subject details (gender, age, and height). Fill in the information, and click **OK.**

PROCEDURE B: Calibration

1. The calibration will be conducted in two stages. For the first stage, simply hold the airflow transducer still and upright, and click **Calibrate.** This calibration will run for 4–8 seconds and then stop automatically. The recording should look similar to figure 52.2. If it does, click **Continue.** If it does not, click **Redo Calibration.**

2. To prepare for the second calibration stage, connect a bacteriological filter to the calibration syringe, and then connect the calibrations syringe/filter assembly to the airflow transducer, as shown in figure 52.3. Be sure that the assembly is inserted into the side labeled "Inlet." Once you have put the calibration assembly together, always support the calibration syringe by holding it up with your

FIGURE 52.3 Setup for second calibration stage.

Source: Courtesy of and © BIOPAC Systems Inc. www.biopac.com

nondominant hand. Do not hold the assembly by the handle of the airflow transducer; this could damage the calibration syringe.

3. Pull the calibration syringe plunger all the way out, and hold the calibration syringe/filter assembly still and parallel to the floor.

4. In order to conduct the second calibration stage, read these directions before beginning. You will have to push

the syringe plunger in and out five times (a series of ten strokes). This is to mimic normal, quiet breathing, and should be done at this rate: push for 1 second, wait for 2 seconds, pull out for 1 second, wait for 2 seconds, and then repeat the whole cycle. When you understand this procedure, and are ready to begin, click **Calibrate.**

5. After completing five push-and-pull cycles, click **End Calibration.** If your calibration data resembles figure 52.4, remove the calibration syringe, and click **Continue** to proceed to the **Recording** section. If it does not, click **Redo Calibration.** (If you have to redo the calibration, you will have to redo both stages; all of the calibration data will be erased.)

FIGURE 52.4 Sample recording for second calibration stage.

Source: Courtesy of and © BIOPAC Systems Inc. www.biopac.com

FIGURE 52.5 Recording setup, showing airflow transducer, bacteriological filter, and disposable mouthpiece.

Source: Courtesy of and © BIOPAC Systems Inc. www.biopac.com

PROCEDURE C: Recording

1. Insert a clean mouthpiece (and filter, if applicable) into the airflow transducer. Remember, if you are using autoclavable mouthpieces, remove the bacteriological filter. With nose clip on, have the subject begin breathing into the transducer assembly. See figure 52.5 for the recording setup.

2. Have the subject breathe normally for about 20 seconds, making sure to close his/her mouth completely around the mouthpiece. Then click **Record** and have the subject:
   a. Breathe normally for five breaths.
   b. Inhale as deeply as possible once.
   c. Exhale as deeply as possible once.
   d. Breathe normally for five breaths.

3. Click **Stop.**

4. The recording should look similar to figure 52.6. If it does not, click **Redo.** If similar, click **Done.**

5. Click **Yes.**

PROCEDURE D: Data Analysis

1. You can either analyze the current data file, or save it and do the analysis later, using the **Review Saved Data** mode. Note the channel number designations: CH 1: Airflow (hidden) and CH 2: Volume.

2. Note the settings for the measurement boxes, across the top of the data window: CH 2: P–P, CH 2: Max, CH 2: Min, and CH 2: Delta. Following are explanations of these settings:

   P–P: The difference between the maximum and minimum points in the selected area

   Max: Shows the maximum point within the selected area

   Min: Shows the minimum point within the selected area

   Delta: Calculates the amplitude difference between the last and first points in the selected area

3. Measure the subject's *vital capacity* (VC). Click on the I-beam tool. Using the I-beam cursor, select the area from the highest point of the deep inspiration to the lowest point of the deep expiration. Selection of the P–P

FIGURE 52.6 Sample recording for spirometry exercise.

Source: Courtesy of and © BIOPAC Systems Inc. www.biopac.com

area does not have to be perfect; it just has to include the highest and lowest points. See figure 52.7 for an example of proper selection of the area for VC measurement. The result will be displayed by CH 2: P–P. Record your result in Part A of Laboratory Assessment 52.

4. Measure the *tidal volume* (TV), during both normal inspiration and expiration. Use the I-beam tool to make these measurements. To measure the TV during inspiration, select the area between the lowest point before the third normal breath and the highest point of the third breath (valley to peak). For TV during expiration, select the area from the highest point of the third breath to the lowest point at the end of the third breath (peak to valley). See figure 52.8 for an example of proper selection of the area for TV measurement during normal inspiration. Figure 52.9 shows proper selection of the area for TV during expiration. Then repeat the inspiration and expiration measurements for the fourth breath. The results will be displayed by CH 2: P–P. Record your results in table 52.1 of the laboratory assessment.

5. Measure the *inspiratory reserve volume* (IRV). Using the I-beam cursor, select the area from the peak of the fifth normal breath to the peak of the deep inspiration. See figure 52.10 for an example of proper selection of the area for IRV measurement. The result will be displayed by CH 2: Delta. Record your result in table 52.1.

6. Measure the *expiratory reserve volume* (ERV). Using the I-beam cursor, select the area from the lowest point of the deep expiration to the lowest point after the normal breath following it (valley to valley). The result will be displayed by CH 2: Delta. Record your result in table 52.1.

7. Measure the *residual volume* (RV). Using the I-beam cursor, select a small area that includes the lowest point of the deep expiration. The result will be displayed by CH 2: Min. Remember that spirometers cannot actually measure RV. Since the BIOPAC software uses a default setting of 1.0 L for RV, your RV result may be preset for 1.0 L. If this is the case, just record 1.0 L as your measurement result for RV in table 52.1. (Note: Because the RV cannot be measured, it is not possible to measure functional residual capacity (FRC) either. It will be calculated in table 52.1 by adding RV + ERV.)

8. Now measure the *inspiratory capacity* (IC). Using the I-beam cursor, select the area from the lowest point at the end of the fifth normal breath to the highest point of the deep inspiration (valley to peak). The result will be displayed by CH 2: Delta. Record your result in table 52.1.

9. Now measure the *expiratory capacity* (EC). Using the I-beam cursor, select the area from the lowest point of the deep expiration to the highest point of the normal breath following it (valley to peak). The result will

FIGURE 52.7 Proper selection of area for measurement of vital capacity (VC).

Source: Courtesy of and © BIOPAC Systems Inc. www.biopac.com

FIGURE 52.8 Proper selection of area for measuring tidal volume (TV) during inspiration.

Source: Courtesy of and © BIOPAC Systems Inc. www.biopac.com

FIGURE 52.9 Proper selection of area for measuring tidal volume (TV) during expiration.

Source: Courtesy of and © BIOPAC Systems Inc. www.biopac.com

FIGURE 52.10 Proper selection of area for measuring inspiratory reserve volume (IRV).

Source: Courtesy of and © BIOPAC Systems Inc. www.biopac.com

be displayed by CH 2: Delta. Record your result in table 52.1.

10. Finally, measure the *total lung capacity* (TLC). Using the I-beam cursor, select a small area that includes the highest point of the deep inspiration. The result will be displayed by CH 2: Max. Record your result in table 52.1.

11. Now turn to Part A of the laboratory assessment. Complete the calculations for estimated vital capacity and follow the directions to compare measured VC with estimated VC.

12. Complete the calculations for table 52.1 using the measured results you just obtained. It will be interesting to see how close the measured and calculated results are.

13. Complete table 52.2, to see how the subject's measured TV, IRV, and ERV compare with the normal averages.

14. When finished, you can save your data on the hard drive or a network drive and then **Exit** the system.

15. Answer the questions in Part B of the laboratory assessment.

Name _____

Date _____

Section _____

The (A) corresponds to the indicated Learning Outcome(s) (O) found at the beginning of the Laboratory Exercise.

LABORATORY ASSESSMENT
52

Spirometry: BIOPAC© Exercise

PART A: Assessments

1. Vital Capacity

   a. Record the subject's measured vital capacity (VC) as instructed in step 3 of the Data Analysis:

   Subject's measured VC = _____ L.

   b. Estimate the subject's vital capacity (VC) by using the appropriate equation below:

   Male: VC = 0.052 H − 0.022A − 3.60

   Female: VC = 0.041 H − 0.018 A − 2.69

   (H = height in cm, and A = age in years)

   Subject's estimated VC = _____ L.

   c. Perform the following calculation to determine how close the subject's measured VC came to the estimated VC:

   Measured VC / Estimated VC = _____ x 100 = _____ %

2. Measurements and calculations of respiratory volumes and capacities:

   Table 52.1 Measurements and Calculations of Subject's Respiratory Volumes and Capacities

Title	Measurement Result			Calculation
Tidal Volume (TV)	a = 2	P-P	Cycle 3 inhale:	(a + b + c + d) / 4 =
	b = 2	P-P	Cycle 3 exhale:	
	c = 2	P-P	Cycle 4 inhale:	
	d = 2	P-P	Cycle 4 exhale:	
Inspiratory Reserve Volume (IRV)	2	Delta		
Expiratory Reserve Volume (ERV)	2	Delta		
Residual Volume (RV)	2	Min		Default = 1.0 L (*Preference setting*)
Inspiratory Capacity (IC)	2	Delta		TV + IRV =
Expiratory Capacity (EC)	2	Delta		TV + ERV =
Functional Residual Capacity (FRC)				ERV + RV =
Total Lung Capacity (TLC)	2	Max		IRV + TV + ERV + RV =

3. Comparison of subject's measured volumes with average volumes:

Table 52.2 Comparison of Subject's Measured Respiratory Volumes with Average Volumes

Volume Title		Average Volume	Measured Volume
Tidal Volume	TV	Resting subject, normal breathing: TV is approximately 500 mL During exercise: TV can be more than 3 liters	greater than equal to less than
Inspiratory Reserve Volume	IRV	Resting IRV for young adults is males = approximately 3,300 mL females = approximately 1,900 mL	greater than equal to less than
Expiratory Reserve Volume	ERV	Resting ERV for young adults is males = approximately 1,000 mL females = approximately 700 mL	greater than equal to less than

PART B: Assessments

Complete the following:

1. How do the experimental results compare to the expected or normal values presented in the introduction? **A2**

_____

_____

_____

2. Why is the predicted vital capacity dependent upon the height of the subject? **A3**

_____

_____

_____

3. Why might vital capacity and expiratory reserve decrease with age? **A3**

_____

_____

_____

LABORATORY
EXERCISE 53

Control of Breathing

Purpose of the Exercise

To review the muscles that control breathing and to describe the mechanisms that regulate the rate and depth of breathing.

Learning Outcomes

After completing this exercise, you should be able to:

O1 Locate the respiratory areas in the brainstem.

O2 Describe the mechanisms that control and influence breathing.

O3 Select the respiratory muscles involved in inspiration and forced expiration.

O4 Test and record the effect of various factors on the rate and depth of breathing.

The **O** corresponds to the assessments **A** indicated in the Laboratory Assessment for this Exercise.

Pre-Lab

Carefully read the introductory material and examine the entire lab. Be familiar with control of breathing from lecture or the textbook. Answer the pre-lab questions.

Pre-Lab Questions Select the correct answer for each of the following questions:

1. Respiratory areas of the brain include all of the following *except* the
   a. brainstem.
   b. pons.
   c. medulla oblongata.
   d. pineal gland.

2. Breathing rate increases as blood concentrations of
   a. carbon dioxide increase.
   b. carbon dioxide decrease.
   c. hydrogen ions decrease.
   d. oxygen increase.

3. Which of the following muscles *is not* involved in forced expiration?
   a. internal intercostal muscles
   b. rectus abdominis muscles
   c. sternocleidomastoid muscles
   d. external oblique muscles

4. Peripheral chemoreceptors sensitive to low blood oxygen levels are located in the
   a. heart.
   b. aortic arch and carotid arteries.
   c. aorta only.
   d. carotid arteries only.

5. Normal blood pH is
   a. 6.8.
   b. 8.0.
   c. 6.8–8.0.
   d. 7.35–7.45.

6. An increase in the duration of inspirations is the normal response from the inflation reflex.
   a. True _____
   b. False _____

7. The ventral respiratory group of the brainstem respiratory centers is involved with stimulation of the contractions of the diaphragm.
   a. True _____
   b. False _____

Breathing is controlled from regions of the *brainstem respiratory centers,* which control both inspiration and expiration. These areas initiate and regulate nerve impulses that travel to various muscles of breathing, causing rhythmic contractions that can be adjusted to the rate and depth of breathing needed to meet various cellular needs.

PROCEDURE A: Control of Breathing

The brainstem respiratory centers include the *ventral respiratory group (VRG)* and the *dorsal respiratory group (DRG)* in the medulla oblongata, and the *pontine respiratory group (PRG)* in the pons. The VRG is the primary generator of impulses for the respiratory rhythm during quiet breathing. A quiet, more relaxed breathing is called *eupnea.* Impulses from the VRG pass through spinal cord integrating centers and then to the diaphragm via the phrenic nerves and to the intercostal muscles via the intercostal nerves to expand the thoracic cavity volume. An inspiration from contractions of the diaphragm and external intercostal muscles takes about 2 seconds, while expiration takes about 3 seconds. This produces a quiet (resting) breathing rate of about 12–15 breaths per minute. During inspiration, the expiratory neurons of the VRG are inhibited. When inspiratory neurons stop firing, expiratory neurons of the VRG conduct impulses to the muscles for expiration, and the inspiratory muscles relax. There is a delicate balance between inspiratory and expiratory muscles being stimulated and inhibited from the VRG during each respiratory breathing cycle (fig. 53.1).

The DRG receives input from various sources, including the pons, central chemoreceptors, peripheral chemoreceptors, stretch receptors, and irritant receptors. As a result of input into the DRG, messages are relayed to the VRG to adjust the rate and depth of the respiratory rhythm during various changing conditions (fig. 53.1). In essence, the DRG modulates the activity of the VRG.

Neurons in the pons compose the PRG (formerly the pneumotaxic center). Input from higher brain centers to the PRG includes the cerebral cortex, the limbic system, and the hypothalamus. As a result of various messages into the PRG, messages are conducted to the DRG and the VRG, and modifications can occur to the respiratory rhythm. Such modifications to a respiratory rhythm occur during varying situations such as exercise, sleep, and emotional responses (fig. 53.1).

Various factors can influence the respiratory areas and thus affect the rate and depth of breathing. These factors include stretch of the lung tissues, emotional state, and the presence in the blood of certain chemicals, such as carbon dioxide, hydrogen ions, and oxygen. The breathing rate increases primarily as the blood concentration of carbon dioxide or hydrogen ions increases, but much less often as the concentration of oxygen decreases.

The medulla oblongata possesses *central chemoreceptors* that are sensitive to changes in the levels of carbon dioxide and hydrogen ions in the blood and cerebrospinal fluid (CSF). If the levels of carbon dioxide or hydrogen ions increase, the chemoreceptors relay this information to the brainstem respiratory centers. As a result, the rate and depth of breathing increase and more carbon dioxide is exhaled. When the levels decline to a more normal range, breathing rate decreases to normal. Exercise, breath holding, and hyperventilation are examples that considerably alter the blood levels of carbon dioxide and hydrogen ions (fig. 53.1).

Low blood oxygen levels are detected by *peripheral chemoreceptors* located in *carotid bodies* in the carotid arteries and *aortic bodies* in the aortic arch. This information is relayed to the respiratory centers in the brainstem, and the breathing rate increases. In order to trigger this response, the blood oxygen levels must be very low. Therefore, the increased blood levels of carbon dioxide and hydrogen ions have a much greater influence upon the peripheral chemoreceptors for the rate and depth of breathing (fig. 53.2).

The depth of breathing is also influenced by the *inflation reflex.* If the lungs are greatly expanded during forceful breathing, *stretch receptors* within the bronchial tree and the visceral pleura are stimulated. Sensory impulses are conducted to the DRG via the vagus nerves and, as a result, a decrease occurs in the duration of inspirations.

Irritant receptors, located in the respiratory bronchial tree, are stimulated by inhaled dust and various pollutants. When receptors are stimulated, sensory impulses are conducted to the DRG via the vagus nerves. Coughing and respiratory bronchial tree constrictions occur as protective responses to the irritants.

The muscles involved with inspiration and expiration are all voluntary skeletal muscles. Therefore a person has some cerebral conscious influence over the rate and depth of breathing as in talking, singing, and breath holding. However, respiratory centers of the brainstem, chemoreceptors, and certain reflexes represent the primary mechanism of subconscious, automatic control of breathing. The complex coordination of breathing involves considerable integration between respiratory areas of the brain and receptors; however some details of the control of breathing are still obscure.

1. Reexamine figures 53.1 and 53.2. Notice that breathing involves input and responses of the nervous, respiratory, muscular, and skeletal systems. Various environmental factors can also alter the rate and depth of respiratory cycles.
2. Study figure 53.3 and table 53.1. Several skeletal muscles contract during breathing. The principal muscles involved during inspiration are the diaphragm and external intercostals. The sternocleidomastoid, scalenes, and pectoralis minor are synergistic during more forceful inhalation, resulting in greater volumes of air inhaled. During quiet respiration, there is minimal involvement of the expiratory muscles. During forceful expiration, the principal muscles are the internal intercostals, but the rectus abdominis and the external oblique muscles can provide extra force. Contractions of accessory muscles during inspiration or expiration result in extra forces and volumes of air into or out of the lungs.
3. Complete Part A of Laboratory Assessment 53.

FIGURE 53.1 Components of respiratory control. The interactions of the brainstem respiratory centers (VRG, DRG, and PRG) control the respiratory rhythm of inspiration and expiration. Additional inputs from central chemoreceptors, higher brain centers, and peripheral receptors can result in alterations to the rate and depth of breathing.

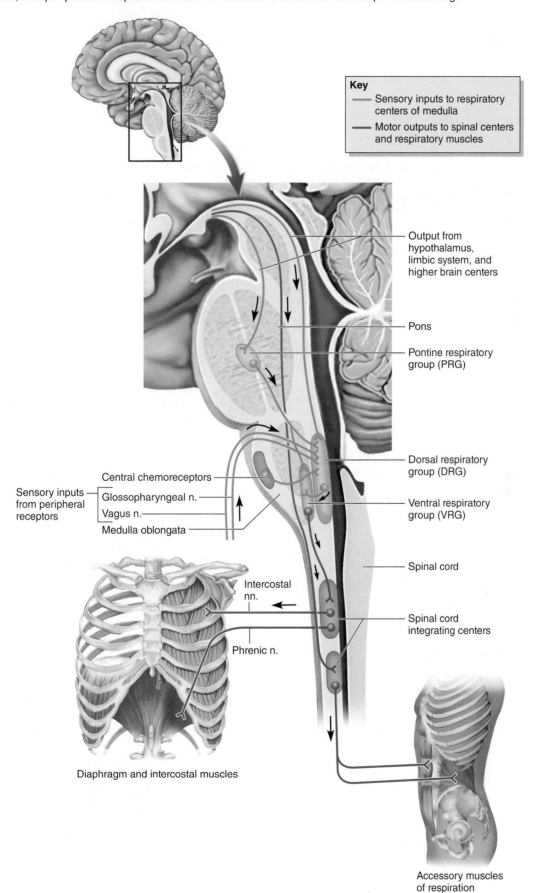

Key
— Sensory inputs to respiratory centers of medulla
— Motor outputs to spinal centers and respiratory muscles

Output from hypothalamus, limbic system, and higher brain centers

Pons

Pontine respiratory group (PRG)

Dorsal respiratory group (DRG)

Ventral respiratory group (VRG)

Spinal cord

Central chemoreceptors

Sensory inputs from peripheral receptors

Glossopharyngeal n.

Vagus n.

Medulla oblongata

Intercostal nn.

Phrenic n.

Spinal cord integrating centers

Diaphragm and intercostal muscles

Accessory muscles of respiration

Table 53.1 Muscles of Respiration

Muscles of Inspiration	Origin	Insertion	Inspiration Action
Diaphragm	Costal cartilages and ribs 7–12, xiphoid process, and lumbar vertebrae	Central tendon of diaphragm	Flattens (descends) during contraction; this expands thoracic cavity and compresses abdominal viscera during inspiration
External intercostals	Inferior border of superior rib	Superior border of inferior rib	Elevates ribs; this expands thoracic cavity during inspiration
Sternocleidomastoid	Manubrium of sternum and medial clavicle	Mastoid process of temporal bone	Elevates sternum
Scalenes	Transverse processes of cervical vertebrae C2–C6	Ribs 1 and 2	Elevates ribs 1 and 2
Pectoralis minor	Sternal ends of ribs 3–5	Coracoid process of scapula	Elevates ribs 3–5
Internal intercostals (intercartilaginous part)	Superior border of inferior costal cartilages	Inferior border of superior costal cartilages	Assists in elevating ribs
Muscles of Expiration	**Origin**	**Insertion**	**Expiration Action**
Internal intercostals (interosseous part)	Superior border of inferior rib	Inferior border of superior rib	Depresses ribs; this compresses thoracic cavity during forced expiration
Rectus abdominis	Pubic crest and pubic symphysis	Xiphoid process of sternum and costal cartilages of ribs 5–7	Compresses inferior ribs and abdominal contents, helping to raise diaphragm
External oblique	Anterior surfaces of ribs 5–12	Anterior iliac crest and linea alba by aponeurosis	Compresses abdominal contents, helping to raise diaphragm

Note: Boldface muscles indicate the principle muscles increasing and decreasing the thoracic cavity.

LEARNING EXTENSION ACTIVITY

A *pneumograph* is a device that can be used together with some type of recording apparatus to record breathing movements. The laboratory instructor will demonstrate the use of this equipment to record various movements, such as those that accompany coughing, laughing, yawning, and speaking.

Devise an experiment to test the effect of some factor, such as hyperventilation, rebreathing air, or exercise, on the length of time a person can hold the breath. *After the laboratory instructor has approved your plan,* carry out the experiment, using the pneumograph and recording equipment. What conclusion can you draw from the results of your experiment?

_____

_____

_____

_____

FIGURE 53.2 The peripheral chemoreceptors involved in control of breathing. Aortic bodies in the aortic arch and carotid bodies in the carotid arteries are sensitive to increased levels of blood carbon dioxide and hyrogen ions as well as decreased blood oxygen levels.

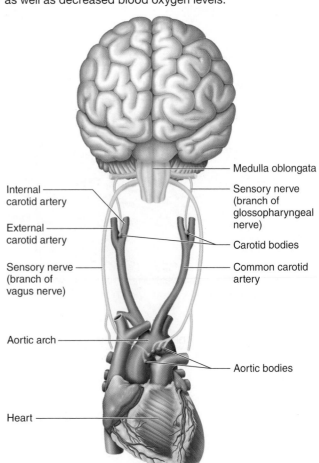

- Internal carotid artery
- External carotid artery
- Sensory nerve (branch of vagus nerve)
- Aortic arch
- Heart
- Medulla oblongata
- Sensory nerve (branch of glossopharyngeal nerve)
- Carotid bodies
- Common carotid artery
- Aortic bodies

FIGURE 53.3 AP|R Respiratory muscles involved in inspiration and forced expiration. The blue arrows indicate the direction of muscle contraction during inspiration; the green arrows indicate the direction of muscle contraction during forced expiration. Boldface indicates the primary muscles involved in breathing.

Inspiration muscles

Sternocleidomastoid

Scalenes

Pectoralis minor

External intercostals

Diaphragm

Forced expiration muscles

Internal intercostals

Rectus abdominis

External oblique

DEMONSTRATION ACTIVITY

When a solution of calcium hydroxide is exposed to carbon dioxide, a chemical reaction occurs, and a white precipitate of calcium carbonate is formed as indicated by the following reaction:

$$Ca(OH)_2 + CO_2 \longrightarrow CaCO_3 + H_2O$$

Thus, a clear water solution of calcium hydroxide (limewater) can be used to detect the presence of carbon dioxide because the solution becomes cloudy if this gas is bubbled through it.

The laboratory instructor will demonstrate this test for carbon dioxide by drawing some air through limewater in an apparatus such as that shown in figure 53.4. Then the instructor will blow an equal volume of expired air through a similar apparatus. (*Note:* A new sterile mouthpiece should be used each time the apparatus is demonstrated.) Watch for the appearance of a precipitate that causes the limewater to become cloudy. Was there any carbon dioxide in the atmospheric air drawn through the limewater?

_____

If so, how did the amount of carbon dioxide in the atmospheric air compare with the amount in the expired air?

_____

_____

FIGURE 53.4 Apparatus used to demonstrate the presence of carbon dioxide in air: (a) atmospheric air is drawn through limewater; (b) expired air is blown through limewater.

Air drawn through short tube

Atmospheric air

Limewater

(a)

Expired air blown through long tube

Limewater

(b)

PROCEDURE B: Factors Affecting Breathing

Perform each of the following tests, using your laboratory partner as a test subject.

1. *Normal breathing.* To determine the subject's normal breathing rate and depth, follow these steps:
   a. Have the subject sit quietly for a few minutes.
   b. After the rest period, ask the subject to count backwards mentally, beginning with five hundred.
   c. While the subject is distracted by counting, watch the subject's chest movements, and count the breaths taken in a minute. If you cannot see the subject's movements while breathing, place your hand on his or her upper back so you can feel the upper body move with each breath. Use this value as the normal breathing rate (breaths per minute).
   d. Note the relative depth of the breathing movements.
   e. Record your observations in the table in Part B of the laboratory assessment.

2. *Effect of hyperventilation.* To test the effect of hyperventilation on breathing, follow these steps:
   a. Seat the subject and *guard to prevent the possibility of the subject falling over.*
   b. Have the subject breathe rapidly and deeply for a maximum of 20 seconds. *If the subject begins to feel dizzy, the hyperventilation should be halted immediately to prevent the subject from fainting from complications of alkalosis. The increased blood pH causes vasoconstriction of cerebral arterioles, which decreases circulation and oxygen to the brain.*
   c. After the period of hyperventilation, determine the subject's breathing rate and judge the breathing depth as before.
   d. Record the results in Part B of the laboratory assessment.

3. *Effect of rebreathing air.* To test the effect of rebreathing air on breathing, follow these steps:
   a. Have the subject sit quietly (approximately 5 minutes) until the breathing rate returns to normal.
   b. Have the subject breathe deeply into a small paper bag that is held tightly over the nose and mouth. *If the subject begins to feel light-headed or like fainting, the rebreathing air should be halted immediately to prevent further acidosis and fainting.*
   c. After 2 minutes of rebreathing air, determine the subject's breathing rate and judge the depth of breathing.
   d. Record the results in Part B of the laboratory assessment.

FIGURE 53.5 The relationship of respiratory rate and blood pH. The normal breathing rate of 12–15 breaths per minute relates to a blood pH within the normal range.

4. *Effect of breath holding.* To test the effect of breath holding on breathing, follow these steps:
   a. Have the subject sit quietly (approximately 5 minutes) until the breathing rate returns to normal.
   b. Have the subject hold his or her breath as long as possible. *If the subject begins to feel light-headed or feel like fainting, breath holding should be halted immediately to prevent further acidosis and fainting.*
   c. As the subject begins to breathe again, determine the rate of breathing and judge the depth of breathing.
   d. Record the results in Part B of the laboratory assessment.

5. *Effect of exercise.* To test the effect of exercise on breathing, follow these steps:
   a. Have the subject sit quietly (approximately 5 minutes) until breathing rate returns to normal.
   b. Have the subject exercise by moderately running in place for 3–5 minutes. *This exercise should be avoided by anyone with health risks.*
   c. After the exercise, determine the breathing rate and judge the depth of breathing.
   d. Record the results in Part B of the laboratory assessment.

6. The importance of the respiratory rate and blood pH relationship is illustrated in figure 53.5. A condition of respiratory acidosis or alkalosis, leading to possible death, can occur from changes in the depth or rate of breathing.

7. Complete Part B of the laboratory assessment.

LABORATORY ASSESSMENT 53

Name _____

Date _____

Section _____

The (A) corresponds to the indicated Learning Outcome(s) (O) found at the beginning of the Laboratory Exercise.

Control of Breathing

PART A: Assessments

Complete the following statements:

1. The respiratory centers are widely scattered throughout the _____ and medulla oblongata of the brainstem. **A1**

2. The _____ respiratory group relays messages to the VRG and the DRG to modify the respiratory rhythm. **A1**

3. The _____ respiratory group within the medulla oblongata generates the basic rhythm of quiet breathing. **A1**

4. Central chemoreceptors are sensitive to changes in the blood concentrations of hydrogen ions and _____. **A2**

5. As the blood concentration of carbon dioxide increases, the breathing rate _____. **A2**

6. As a result of increased breathing, the blood concentration of carbon dioxide is _____. **A2**

7. Peripheral chemoreceptors include aortic bodies and the _____. **A2**

8. The principal muscles of forced expiration are the _____. **A3**

9. The principal muscles of inspiration are the _____ and the external intercostal muscles. **A3**

PART B: Assessments

1. Record the results of your breathing tests in the table. **A4**

Factor Tested	Breathing Rate (breaths/minute)	Breathing Depth (+, + +, + + +)
Normal		
Hyperventilation (shortly after)		
Rebreathing air (shortly after)		
Breath holding (shortly after)		
Exercise (shortly after)		

2. Briefly explain the reason for the changes in breathing that occurred in each of the following cases: A2

    a. Hyperventilation

    _____

    _____

    _____

    _____

    b. Rebreathing air

    _____

    _____

    _____

    _____

    c. Breath holding

    _____

    _____

    _____

    _____

    d. Exercise

    _____

    _____

    _____

    _____

3. Complete the following:

    a. Why is it important to distract a person when you are determining the normal rate of breathing?

    _____

    _____

    b. How can the depth of breathing be measured accurately?

    _____

    _____

CRITICAL THINKING ASSESSMENT

Why is it dangerous for a swimmer to hyperventilate in order to hold the breath for a longer period of time? A2

_____

_____

_____

_____

LABORATORY EXERCISE 54

Digestive Organs

Purpose of the Exercise   AP|R

To review the structure and function of the digestive organs and to examine the tissues of these organs.

MATERIALS NEEDED

Human torso model

Head model, sagittal section

Skull with teeth

Teeth, sectioned

Tooth model, sectioned

Paper cup

Compound light microscope

Prepared microscope slides of the following:

    Sublingual gland (salivary gland)

    Esophagus

    Stomach (fundus)

    Pancreas (exocrine portion)

    Liver

    Small intestine (jejunum)

    Large intestine

Learning Outcomes   AP|R

After completing this exercise, you should be able to:

O1 Sketch and label digestive organ structures in tissue (histological) sections.

O2 Locate the major organs and structural features of the digestive system.

O3 Match digestive organs with their descriptions.

O4 Describe the function(s) of each digestive organ.

The **O** corresponds to the assessments **A** indicated in the Laboratory Assessment for this Exercise.

Pre-Lab

Carefully read the introductory material and examine the entire lab. Be familiar with the basic structures and functions of the digestive organs from lecture or the textbook. Answer the pre-lab questions.

Pre-Lab Questions Select the correct answer for each of the following questions:

1. Which of the following digestive organs possesses villi?
   - **a.** esophagus
   - **b.** stomach
   - **c.** small intestine
   - **d.** large intestine

2. Which is a major function of the oral cavity?
   - **a.** absorb water
   - **b.** mechanical digestion
   - **c.** secrete hydrochloric acid
   - **d.** emulsify fat

3. Bile is produced in the
   - **a.** small intestine.
   - **b.** pancreas.
   - **c.** gallbladder.
   - **d.** liver.

4. The _____ is the exposed part of a tooth.
   - **a.** crown
   - **b.** neck
   - **c.** root
   - **d.** periodontal ligament

5. The _____ sphincter is the muscular valve at the exit of the stomach.
   - **a.** lower esophageal
   - **b.** pyloric
   - **c.** hepatopancreatic
   - **d.** ileocecal

6. The stomach can stretch enough to hold the contents of a large meal because it has
   - **a.** the pyloric antrum.
   - **b.** a fundus region.
   - **c.** gastric folds.
   - **d.** a greater curvature.

7. The _____ lobe is the largest lobe of the liver.
   - **a.** left
   - **b.** quadrate
   - **c.** caudate
   - **d.** right

581

The digestive system includes the organs associated with the *alimentary canal* and several accessory structures. The alimentary canal, a muscular tube, passes through the body from the opening of the mouth to the anus. It includes the mouth, pharynx, esophagus, stomach, small intestine, and large intestine. The canal is adapted to move substances throughout its length. It is specialized in various regions to store, digest, and absorb food materials and to eliminate the residues. The accessory organs, which include the salivary glands, liver, gallbladder, and pancreas, secrete products into the alimentary canal that aid digestive functions.

The *oral cavity* is a primary area for mechanical digestion where food is physically broken down, involving the mastication (chewing) of food using the teeth and jaw muscles. Three pairs of salivary glands secrete mucus and salivary amylase into the oral cavity, where chemical digestion of starch begins. Chemical digestion uses enzymes to hydrolyze larger molecules into smaller absorbable molecules. The *pharynx* and *esophagus* secrete mucus and serve as passageways for the food and liquids to reach the stomach. The contents are forced along by peristaltic waves.

Within the *stomach,* swallowed contents mix with gastric juice, which contains hydrochloric acid and the enzyme pepsin. The hydrochloric acid creates a pH near 2, serving to destroy most ingested bacteria, and it activates pepsin to begin the chemical digestion of proteins. The gastric folds (rugae) allow the stomach to hold large quantities of food.

Only minimal absorption occurs in the stomach. The partially digested content, called chyme, enters the small intestine periodically through the muscular pyloric sphincter.

Within the *small intestine,* chyme is mixed with bile and pancreatic juice. Bile is produced in the liver and then stored in the gallbladder. Bile emulsifies fat, breaking it into smaller particles to allow the enzyme lipase to act upon it more easily. Pancreatic juice contains bicarbonate ions necessary to neutralize acidic chyme, and multiple enzymes for the chemical digestion of carbohydrates, fats, and proteins. The small intestine is nearly 6 meters (21 feet) long and possesses microvilli, villi, and circular folds that help to increase its surface area. Chemical digestion is completed and most nutrient absorption occurs before the contents reach the large intestine.

Contents not absorbed enter the *large intestine* through the ileocecal valve. Within the large intestine, mucus is secreted and water and electrolytes are absorbed. Bacteria that inhabit the large intestine break down some remaining residues, producing some vitamins that are absorbed. Feces composed of water, undigested substances, mucus, and bacteria are formed and stored until elimination.

PROCEDURE A: Oral Cavity and Salivary Glands

1. Study figure 54.1 and the list of structures and descriptions. Examine the head model (sagittal section) and a skull. Locate the following structures:

   oral cavity (mouth)—location of mechanical digestion, some chemical digestion

   vestibule—narrow space between teeth and lips and cheeks

   tongue—muscular organ for food manipulation

     ■ lingual frenulum (frenulum of tongue)—membranous fold connecting tongue and floor of oral cavity

---

FIGURE 54.1 The features of the oral cavity.

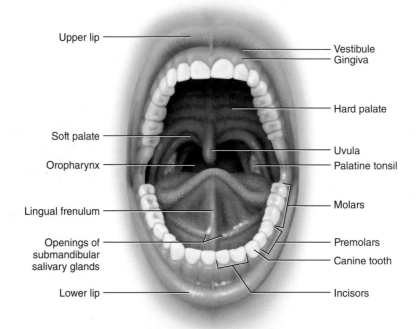

- papillae—contain taste buds

palate—roof of oral cavity

  - hard palate—formed by portions of maxillary and palatine bones
  - soft palate—muscular arch without bone
  - uvula—cone-shaped projection; along with soft palate, helps prevent ingested contents from entering nasal regions during swallowing

palatine tonsils—lymphatic tissues on lateral walls of pharynx

gingivae (gums)—soft tissue that surrounds neck of tooth and alveolar processes

teeth

  - incisors
  - canines (cuspids)
  - premolars (bicuspids)
  - molars

2. Study figure 54.2 and the list of structures and descriptions. Examine a sectioned tooth and a tooth model. Locate the following features:

   crown—tooth portion projection beyond gingivae

   - enamel—hardest substance of body on surface of crown
   - dentin—living connective tissue forming most of tooth

   neck—junction between crown and root

   root—deep portion of tooth beneath surface

   - pulp cavity—central portion of tooth
   - cementum (cement)—thin surface area of bonelike material
   - root canal—contains blood vessels and nerves

FIGURE 54.2 Longitudinal section of a molar.

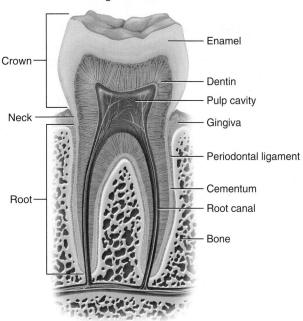

3. Study figures 54.1 and 54.3. Observe the head of the human torso model and locate the following:

   parotid salivary glands—largest glands anterior to ear

   - parotid duct (Stensen's duct)

   submandibular salivary glands—along mandible

   - submandibular duct (Wharton's duct)

   sublingual salivary glands—smallest glands inferior to tongue

   - sublingual ducts (10–12 ducts)

FIGURE 54.3 AP|R The features associated with the salivary glands.

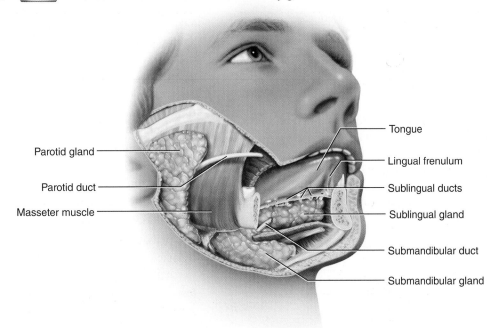

FIGURE 54.4 Micrograph of the sublingual salivary gland (300×).

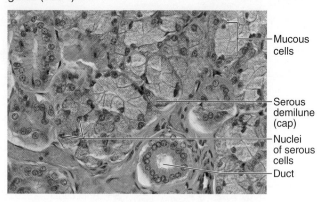

- Mucous cells
- Serous demilune (cap)
- Nuclei of serous cells
- Duct

4. Examine a microscopic section of a sublingual gland using low- and high-power magnification. Note that the mucous cells produce mucus and the serous cells produce enzymes. Serous cells sometimes form caps (demilunes) around mucous cells. Also note any larger secretory duct surrounded by cuboidal epithelial cells (fig. 54.4).
5. Prepare a labeled sketch of a representative section of a salivary gland in Part A of Laboratory Assessment 54.

PROCEDURE B: Pharynx and Esophagus

1. Reexamine figure 50.2 in Laboratory Exercise 50.
2. Observe the human torso model and locate the following features:

   pharynx—connects nasal and oral cavities with larynx
   - nasopharynx—superior to soft palate; without digestive functions
   - oropharynx—posterior to oral cavity
   - laryngopharynx—extends from inferior oropharynx to esophagus

   epiglottis—deflects food and fluids into esophagus when swallowing

   esophagus—straight collapsible tube from pharynx to stomach

   lower esophageal sphincter (cardiac sphincter; gastroesophageal sphincter)—thickened smooth muscle at junction with stomach

3. Have your laboratory partner take a swallow from a cup of water. Carefully watch the movements in the anterior region of the neck. What steps in the swallowing process did you observe?

   _____

   _____

   _____

   _____

FIGURE 54.5 [AP|R] Micrograph of a cross section of the esophagus (10×).

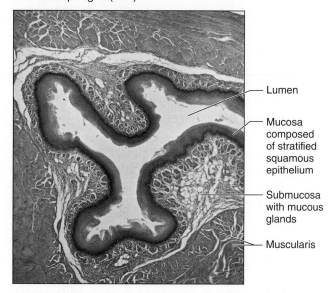

- Lumen
- Mucosa composed of stratified squamous epithelium
- Submucosa with mucous glands
- Muscularis

4. Examine a microscopic section of esophagus wall using low-power magnification (fig. 54.5). The inner mucosa layer is composed of stratified squamous epithelium, and there are two layers of muscle tissue in the muscularis. Peristaltic waves occur from contractions of the muscular layers that move contents through the alimentary canal. Locate some mucous glands in the submucosa layer. They appear as clusters of lightly stained cells. The outer adventitia layer is not shown in this micrograph.
5. Prepare a labeled sketch of the esophagus wall in Part A of the laboratory assessment.

PROCEDURE C: Stomach

1. Study figure 54.6 and the list of structures and descriptions. Observe the human torso model and locate the following features of the stomach:

   gastric folds (rugae)—allow stomach to expand to hold food and liquid contents

   cardia (cardiac region)—region near lower esophageal sphincter

   fundus of stomach (fundic region)—dome-shaped region superior to opening from esophagus

   body of stomach (body region)—large main region

   pyloric part (pyloric region)—region near passage into small intestine
   - pyloric antrum—funnel-like portion
   - pyloric canal—narrow region
   - pylorus—terminal passage into duodenum

   pyloric sphincter—thick, smooth muscle valve that regulates chyme passage into duodenum

   lesser curvature—on superior surface

   greater curvature—on lateral and inferior surface

FIGURE 54.6 AP|R   The major features of the stomach and associated structures (frontal section of internal structures).

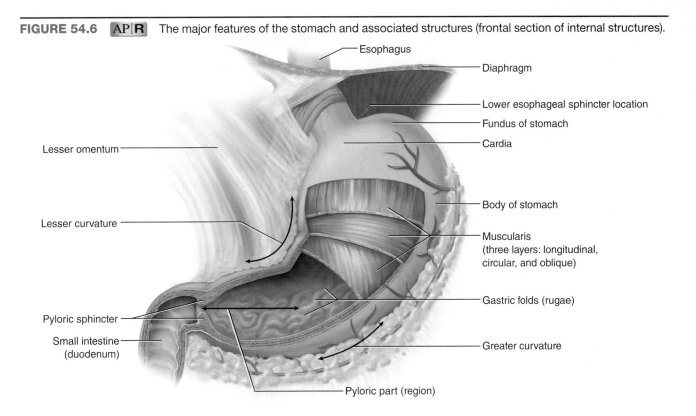

- Esophagus
- Diaphragm
- Lower esophageal sphincter location
- Fundus of stomach
- Cardia
- Lesser omentum
- Body of stomach
- Lesser curvature
- Muscularis (three layers: longitudinal, circular, and oblique)
- Gastric folds (rugae)
- Pyloric sphincter
- Small intestine (duodenum)
- Greater curvature
- Pyloric part (region)

2. Examine a microscopic section of stomach wall using low-power magnification. Note how the inner mucosa layer of simple columnar epithelium dips inward to form gastric pits. The gastric glands are tubular structures that open into the gastric pits. Near the deep ends of these glands, locate the intensely stained (bluish) chief cells and the lightly stained (pinkish) parietal cells (fig. 54.7). The parietal cells secrete hydrochloric acid; the chief cells produce pepsinogen, the inactive form of pepsin.

3. Prepare a labeled sketch of a representative section of the stomach wall in Part A of the laboratory assessment.

FIGURE 54.7 AP|R   Micrograph of the mucosa of the stomach wall (60×).

- Simple columnar epithelium
- Gastric pit
- Mucous cell
- Portion of a gastric gland
- Parietal cell
- Portion of a gastric gland
- Chief cell

PROCEDURE D: Pancreas and Liver

1. Study figures 54.8 and 54.9 and the list of structures and descriptions of the pancreas and the liver. Observe the human torso model and locate the following structures:

   pancreas—a retroperitoneal organ; secretes an alkaline mixture of digestive enzymes and bicarbonate ions (fig. 54.8)

   - tail of pancreas
   - head of pancreas
   - pancreatic duct
   - accessory pancreatic duct

   liver—has four lobes; produces bile (fig. 54.9)

   - right lobe—largest lobe; bare area portion attached to diaphragm
   - left lobe—smaller than right lobe
   - quadrate lobe—minor lobe near gallbladder
   - caudate lobe—minor lobe near vena cava

   gallbladder—stores bile

   hepatic ducts—bile ducts from inferior liver to common hepatic duct

   common hepatic duct—bile duct from liver to cystic duct

   cystic duct—attached to gallbladder

   bile duct—duct from cystic duct to duodenum

   hepatopancreatic sphincter (sphincter of Oddi)—regulates passage of bile and pancreatic juice

2. Examine the pancreas slide using low-power magnification. Observe the exocrine (acinar) cells that secrete pancreatic juice into pancreatic ducts (fig. 54.10). The pancreatic duct secretes pancreatic juice that is regulated, along with bile secretion, by the hepatopancreatic sphincter. The accessory pancreatic duct does not contain a sphincter, so pancreatic secretion is not regulated and is allowed to be secreted without bile.

3. Study figure 54.11. The four liver lobes have structural and functional units, the *hepatic lobules,* which are composed of cuboidal epithelial cells called *hepatocytes.* Blood-filled radiating *hepatic sinusoids* connect to a *central vein* within a hepatic lobule. A *hepatic triad,* composed of small branches of a hepatic portal vein, hepatic artery, and bile ductile, are located at peripheral areas of hepatic lobules. Examine a liver slide using low-power magnification. Locate the structures indicated in figure 54.11*b.*

4. Prepare a labeled sketch of a representative section of the pancreas and the liver in Part A of the laboratory assessment.

FIGURE 54.8 AP|R The major features of the pancreas and the gallbladder associated with the liver and the duodenum of the small intestine.

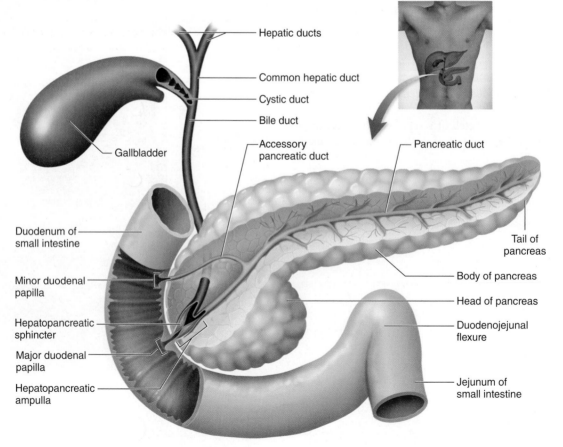

FIGURE 54.9 AP|R Features of the liver: (a) liver location; (b) anterior view; (c) inferior view.

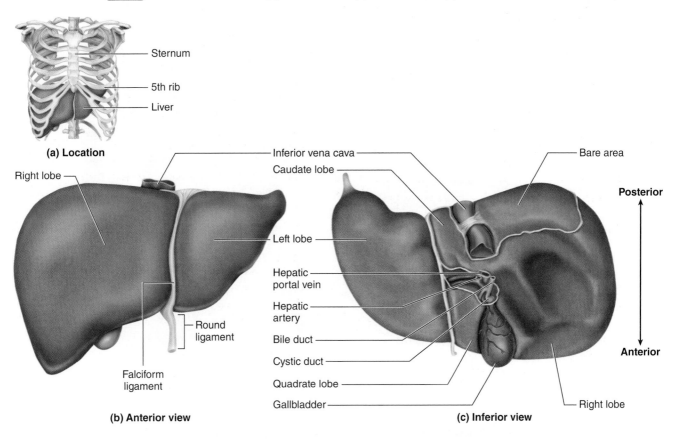

(a) Location

Sternum

5th rib

Liver

Right lobe

Inferior vena cava

Caudate lobe

Bare area

Left lobe

Hepatic portal vein

Hepatic artery

Bile duct

Cystic duct

Quadrate lobe

Gallbladder

Round ligament

Falciform ligament

(b) Anterior view

Posterior

Anterior

Right lobe

(c) Inferior view

FIGURE 54.10 AP|R Micrograph of exocrine portion of pancreas showing acinar cells and small pancreatic ducts (200x). The acinar cells secrete an alkaline pancreatic juice into small ducts that converge into the larger pancreatic duct.

Connective tissue

Ducts

Exocrine acinar cells

Vein

FIGURE 54.11 Microscopic structure of the liver: (a) illustration; (b) **AP|R** micrograph of hepatic lobules (40x).

Hepatocytes

Bile canaliculi

Central vein

Branch of hepatic portal vein

Hepatic triad

Branch of hepatic artery

Bile ductule

Hepatic sinusoid

Connective tissue

(a)

Hepatic triad

Hepatic lobule

Hepatocytes

Hepatic sinusoids

Connective tissue

Central vein

(b)

PROCEDURE E: Small and Large Intestines

1. Study figures 54.12, 54.13, and 54.14 and the list of structures and descriptions. Observe the human torso model and locate each of the following features:

small intestine—tube with small diameter and many loops and coils where chemical digestion and absorption of nutrients occur (fig. 54.12)

- duodenum—first 25 cm (10 inches); shortest portion; located retroperitoneal

- jejunum—next nearly 2.5 m (8 feet); most digestion and absorption occurs here

- ileum—last 3.5 m (12 feet); joins with large intestine

mesentery—fold of peritoneal membrane that suspends and supports abdominal viscera

ileocecal valve (sphincter)—regulates contents entering large intestine

large intestine—tube with large diameter nearly 1.5 m (5 feet) long (fig. 54.13)

FIGURE 54.12 AP|R The features of the small intestine and associated structures (anterior view). (*Note:* The small intestine is pulled aside to expose the ileocecal junction.)

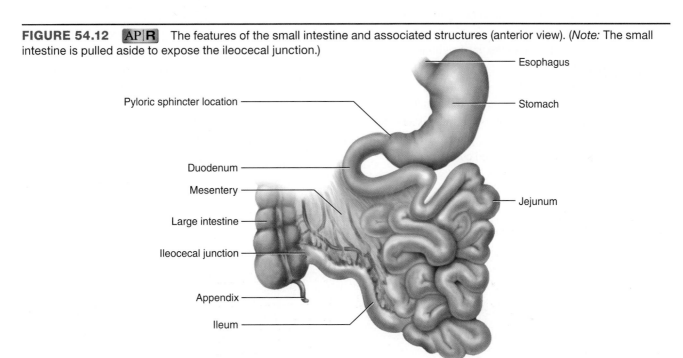

FIGURE 54.13 AP|R The features of the large intestine (anterior view).

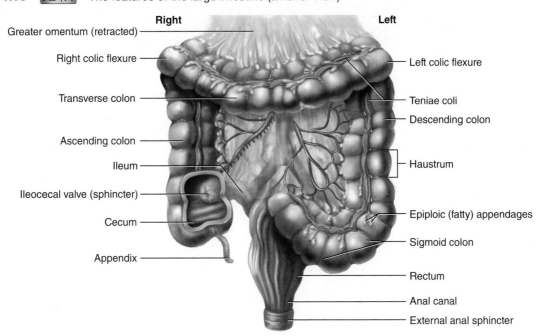

- large intestinal wall—has three unique features
  - teniae coli—three longitudinal bands of smooth muscle
  - haustra—pouches created by muscle tone of teniae coli
  - epiploic appendages—fatty pouches on outer surfaces
- cecum—blind pouch inferior to ileocecal valve
- appendix (vermiform appendix)—2–7 cm wormlike blind pouch attached to cecum (fig. 54.14)
- ascending colon—course on right side from cecum to transverse colon
- right colic (hepatic) flexure—right-angle turn
- transverse colon—course from ascending to descending colon
- left colic (splenic) flexure—right-angle turn
- descending colon—course from transverse to sigmoid colon
- sigmoid colon—S-shaped course to rectum

- rectum—course anterior to sacrum; retains feces until defecation
- anal canal—last few centimeters of large intestine

anal sphincter muscles—encircle anus
- external anal sphincter—the voluntary sphincter of skeletal muscle
- internal anal sphincter—the involuntary sphincter of smooth muscle

anus—external opening

2. Using low-power magnification, examine a microscopic section of small intestine wall. Identify the mucosa, submucosa, muscularis, and serosa. Note the villi that extend into the lumen of the tube. Study a single villus using high-power magnification. Note the core of connective tissue and the covering of simple columnar epithelium that contains some lightly stained goblet cells and microvilli (brush border) projecting from the apical ends of the cells (fig. 54.15). The microvilli and villi projections greatly increase the surface area for absorption of digestive products.

FIGURE 54.14 Normal appendix.

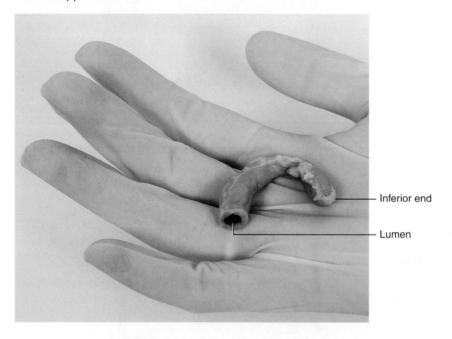

Inferior end

Lumen

FIGURE 54.15   AP|R   Micrograph of the small intestine wall (40×).

Lumen

Villus

Mucus-secreting goblet cells

Simple columnar
epithelium with microvilli

Connective tissue containing
blood vessels and lymphatic
capillary (lacteal) in villus

Mucosa

Submucosa

Muscularis — Circular layer

Longitudinal layer

Serosa

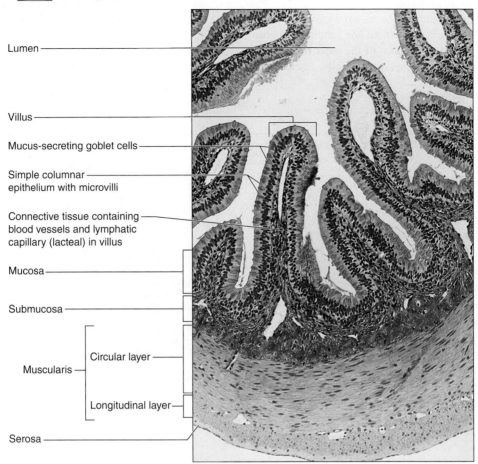

CRITICAL THINKING ACTIVITY

How is the structure of the small intestine better adapted for absorption than that of the large intestine?

_____

_____

3. Prepare a labeled sketch of the wall of the small intestine in Part A of the laboratory assessment.

4. Examine a microscopic section of large intestine wall. Note the lack of villi. Also note the tubular mucous glands that open on the surface of the inner lining and the numerous lightly stained goblet cells. Locate the four layers of the wall (fig. 54.16). The mucus functions as a lubricant and holds the particles of fecal matter together.

5. Prepare a labeled sketch of the wall of the large intestine in Part A of the laboratory assessment.

6. Complete Parts B, C, and D of the laboratory assessment.

FIGURE 54.16 AP|R Micrograph of the large intestine wall (64×).

Name _____

Date _____

Section _____

The (A) corresponds to the indicated Learning Outcome(s) (O) found at the beginning of the Laboratory Exercise.

Digestive Organs

PART A: Assessments

In the space that follows, sketch a representative area of the organ indicated. Label any of the structures observed and indicate the magnification used for each sketch. Add any personal notes in any of the extra space within the box with your sketch that would assist your future ability to recognize the organ and interpret its function. (A1)

Salivary gland (_____×)	Esophagus (_____×)
Stomach wall (_____×)	Pancreas (_____×)

Small intestine wall (_____×)

Large intestine wall (_____×)

Liver (_____×)

PART B: Assessments

Identify the numbered features in figures 54.17, 54.18, 54.19, and 54.20.

FIGURE 54.17 AP|R Label the features of the stomach and nearby regions in this frontal section of a cadaver (anterior view). A2

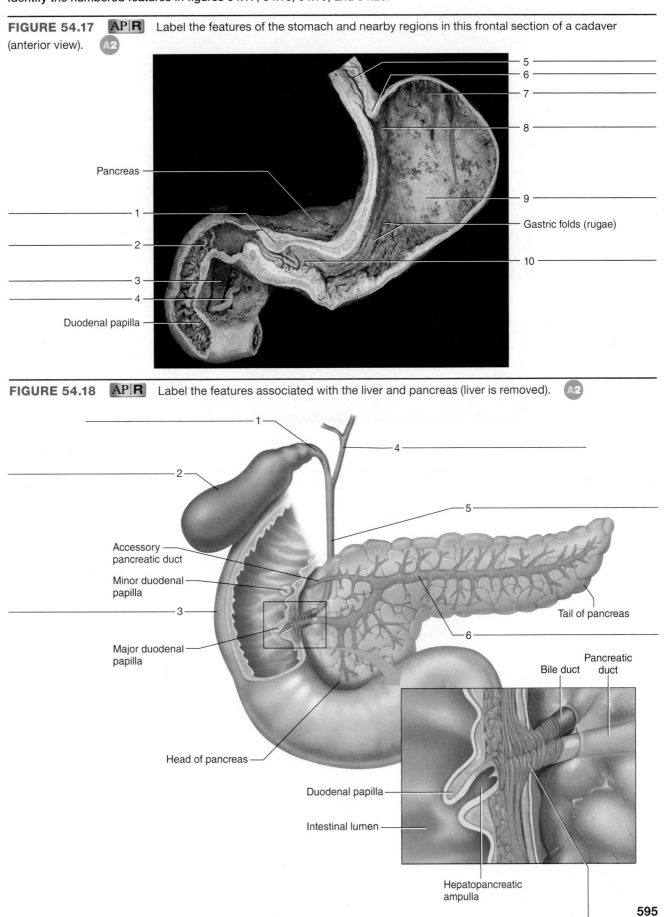

FIGURE 54.18 AP|R Label the features associated with the liver and pancreas (liver is removed). A2

FIGURE 54.19 `AP|R`   Label the digestive structures of this abdominopelvic cavity of a cadaver (anterior view).

Diaphragm (cut)

6

Descending colon

7

8

1

2

3

Ileocecal junction

4

5

Sigmoid colon

FIGURE 54.20   Label the features of the large intestine. **A2**

Right   Left

5

1

6

2

3

4

7

PART C: Assessments

Match the terms in column A with the descriptions in column B. Place the letter of your choice in the space provided.

Column A

a. Cardia
b. Crown
c. Cystic duct
d. Gastric folds
e. Ileum
f. Major duodenal papilla
g. Mucosa
h. Muscular layer
i. Parietal cells
j. Root canal
k. Sinusoids
l. Villi

Column B

_____ 1. Blood-filled channels connected to central vein

_____ 2. Secrete hydrochloric acid into stomach

_____ 3. Last section of small intestine

_____ 4. Region of stomach near lower esophageal sphincter

_____ 5. Contains blood vessels and nerves in a tooth

_____ 6. Responsible for peristaltic waves

_____ 7. Allow stomach to expand

_____ 8. Increase surface area for digested food absorption

_____ 9. Attached to gallbladder

_____ 10. Portion of tooth projecting beyond gingivae

_____ 11. Layer nearest lumen of alimentary canal

_____ 12. Common opening region for bile and pancreatic secretions

PART D: Assessments

Complete the following:

1. Summarize the function(s) of each digestive organ:

Digestive Organ	Function(s)
Oral cavity	
Esophagus	
Stomach	
Small intestine	
Large intestine	

2. Name the valve (sphincter) located between the stomach and the duodenum. **A2** _____

What is the function of this valve? **A4** _____

_____

3. Name the valve located between the small and the large intestines. **A2** _____

What is the function of this valve? **A4** _____

_____

A&P II Today: <u>Digestive System</u> (04/06/17)

1) Handout
   - Digestion of carbs
   - " " proteins
   - " " lipids

2) Concepts / Key terms
   - Polymer / monomers
   - Enzyme / substrate
   - Nutrients
   - Reagents

<u>4 major food groups</u>

- carbs
- protein
- lipids
- nucleic acid

<u>Food Travels</u>

- mouth
- Pharynx
- Esophagus
- Stomach
   - small Intenstine (D, J, I)
   - Large (T, D, A)
   - Rectum
   - Anus

Pancreatin

1) A + B → A - B
   (reactant)   (products)

2) A - B → A + B
   (substrates)   ↑
              enzyme

Enzymes
→ catalyst
→ protein
→ *specific

Factors
Temperature (37°c)
pH (7.4)

Action of a Digestive Enzyme 04/06/17

Purpose of the Exercise

To investigate the action of amylase and the effect of heat on its enzymatic activity.

MATERIALS NEEDED

0.5% amylase solution*	Iodine-potassium-iodide solution
Beakers (50 and 500 mL)	Medicine dropper
Distilled water	Ice
Funnel	Water bath, 37°C (98.6°F)
Pipettes (1 and 10 mL)	Porcelain test plate
Pipette rubber bulbs	Benedict's solution
0.5% starch solution	Hot plates
Graduated cylinder (10 mL)	Test-tube rack
	Thermometer
Test tubes	*For Alternative Procedure Activity:*
Test-tube clamps	Small, disposable cups
Wax marker	Distilled water

*The amylase should be free of sugar for best results; a low-maltose solution of amylase yields good results. See Appendix 1.

SAFETY ⚠

▶ Use only a mechanical pipetting device (never your mouth). Use pipettes with rubber bulbs or dropping pipettes.

▶ Wear safety glasses when working with acids and when heating test tubes.

▶ Use test-tube clamps when handling hot test tubes.

▶ If an open flame is used for heating the test solutions, keep clothes and hair away from the flame.

▶ Review all the safety guidelines inside the front cover.

▶ If student saliva is used as a source of amylase, it is important that students wear disposable gloves and handle only their own materials.

▶ Use an appropriate disinfectant to wash the laboratory tables before and after the procedures.

▶ Dispose of chemicals according to appropriate directions.

▶ Wash your hands before leaving the laboratory.

Learning Outcomes

After completing this exercise, you should be able to:

O1 Test a solution for the presence of starch or the presence of sugar.

O2 Explain the action of amylase.

O3 Test the effects of varying temperatures on the activity of amylase.

The **O** corresponds to the assessments **A** indicated in the Laboratory Assessment for this Exercise.

Pre-Lab

Carefully read the introductory material and examine the entire lab. Be familiar with enzyme structure and activity from lecture or the textbook. Answer the pre-lab questions.

Pre-Lab Questions Select the correct answer for each of the following questions:

1. The enzyme amylase accelerates the reaction of changing
   a. polysaccharides into monosaccharides.
   b. disaccharides into monosaccharides.
   c. starch into disaccharides.
   d. disaccharides into glucose.

2. Amylase originates from
   a. salivary glands and stomach.
   b. salivary glands and pancreas.
   c. esophagus and stomach.
   d. pancreas and intestinal mucosa.

3. The optimum pH for amylase activity is
   a. 0–14.   b. 2.
   c. 6.8–7.0.   d. 7–8.

4. After an enzyme reaction is completed, the enzyme
   a. is still available.   b. was structurally changed.
   c. was denatured.   d. is part of the enzyme-substrate complex.

5. As temperatures decrease, enzyme activity
   a. increases.   b. remains unchanged.
   c. ceases.   d. decreases.

This laboratory exercise is designed to examine the relationship between enzymes and chemical digestion. *Enzymes* are proteins that serve as *biological catalysts*. Acting as a catalyst, an enzyme will modify and rapidly increase the rate of a particular chemical reaction without being consumed in the reaction, which enables the enzyme to function repeatedly. A particular enzyme will catalyze a specific reaction. Various metabolic pathways and chemical digestive processes are accelerated by numerous enzymatic reactions. The **lock-and-key model** of enzyme activity involves an *enzyme* and *substrate* (substance acted upon by the enzyme) temporarily combining as an *enzyme-substrate complex* and ending with a reaction *product* and the original *enzyme* (fig. 55.1).

Enzymes are globular proteins with a three-dimensional structure determined by various chemical bonds including the hydrogen bond. For the enzyme to function as the catalyst, the shape is critical at the active site of the enzyme for the enzyme-substrate complex to take place properly. If the environmental temperature, including the body temperature, rises to a certain level, the three-dimensional shape of the enzyme is modified as hydrogen bonds start breaking and the enzyme becomes *denatured*. Although the primary structure of the amino acid sequence remains unaltered, the secondary and tertiary structures have been altered. A denatured protein could regain its shape unless the three-dimensional shape is destroyed when conditions become too extreme, and the enzyme becomes irreversibly denatured.

The digestive enzyme in salivary secretions is called *salivary amylase.* Pancreatic amylase is secreted among several other pancreatic enzymes. A bacterial extraction of amylase is available for laboratory experiments. This enzyme catalyzes the reaction of changing starch molecules into sugar (disaccharide) molecules, which is the first step in the digestion of complex carbohydrates.

As in the case of other enzymes, amylase is a protein catalyst. Its activity is affected by exposure to certain environmental factors, including various temperatures, pH,

radiation, and electricity. As temperatures increase, faster chemical reactions occur as the collisions of molecules happen at a greater frequency. Eventually, temperatures increase to a point that the enzyme is denatured, and the rate of the enzyme activity rapidly declines. As temperatures decrease, enzyme activity also decreases due to fewer collisions of the molecules; however, the colder temperatures do not denature the enzyme. Normal body temperature provides an environment for enzyme activity near the optimum for enzymatic reactions.

The pH of the environment where enzymes are secreted also has a major influence on the enzyme reactions. The optimum pH for amylase activity is between 6.8 and 7.0, typical of salivary secretions. When salivary amylase arrives in the stomach, hydrochloric acid deactivates the enzyme, diminishing any further chemical digestion of remaining starch in the stomach. However, pancreatic amylase is secreted into the small intestine, where an optimum pH for amylase is once more provided. Other enzymes in the digestive system have different optimum pH ranges of activity compared to amylase. For example, pepsin from stomach secretions has an optimum activity around pH 2, whereas trypsin from pancreatic secretions operates best around pH 7–8.

PROCEDURE A: Amylase Activity

1. Study the lock-and-key model example specific for amylase action on starch digestion (fig. 55.2).
2. Examine the locations where starch is digested into glucose (fig. 55.3).
3. Mark three clean test tubes as *tubes 1, 2,* and *3,* and prepare the tubes as follows (fig. 55.4):

   Tube 1: **Add 6 mL of amylase solution.**

   Tube 2: **Add 6 mL of starch solution.**

   Tube 3: **Add 5 mL of starch solution and 1 mL of amylase solution.**

FIGURE 55.1 Lock-and-key model of an enzyme-catalyzed reaction (**E + S → E–S → P + E**). (Many enzyme-catalyzed reactions, as depicted here, are reversible.) In the forward reaction (dark-shaded arrows), (a) the shapes of the substrate molecules fit the shape of the enzyme's active site. (b) When the substrate molecules temporarily combine with the enzyme, a chemical reaction occurs. (c) The result is a product molecule and an unaltered enzyme. The active site changes shape somewhat as the substrate binds, such that formation of the enzyme-substrate complex is more like a hand fitting into a glove, which has some flexibility, than a key fitting into a lock.

FIGURE 55.2 Lock-and-key model of enzyme action of amylase on starch digestion.

ALTERNATE PROCEDURE ACTIVITY

Human saliva can be used as a source of amylase solutions instead of bacterial amylase preparations. Collect about 5 mL of saliva into a small, disposable cup. Add an equal amount of distilled water and mix them together for the amylase solutions during the laboratory procedures. Be sure to follow all of the safety guidelines.

FIGURE 55.3 Flowchart of starch digestion.

FIGURE 55.4 Test tubes prepared for testing amylase activity.

Add 6 mL amylase solution

Add 6 mL starch solution

Add 5 mL starch solution

Add 1 mL amylase solution

Shake the tubes to mix the contents. Transfer the 3 test tubes from the test-tube rack into the warm water bath (37°C) for 10 minutes.

Tube 1   Tube 2   Tube 3

4. Shake the tubes well to mix the contents and place them in a warm water bath, 37°C (98.6°F), for 10 minutes.

5. At the end of the 10 minutes, test the contents of each tube for the presence of starch. To do this, follow these steps:

   a. Place 1 mL of the solution to be tested in a depression of a porcelain test plate.

   b. Next add 1 drop of iodine-potassium-iodide solution and note the color of the mixture. If the solution becomes blue-black, starch is present.

   c. Record the results in Part A of Laboratory Assessment 55.

6. Test the contents of each tube for the presence of sugar (disaccharides, in this instance). To do this, follow these steps:

   a. Place 1 mL of the solution to be tested in a clean test tube.

   b. Add 1 mL of Benedict's solution.

   c. Place the test tube with a test-tube clamp in a beaker of boiling water for 2 minutes.

   d. Note the color of the liquid. If the solution becomes green, yellow, orange, or red, sugar is present. Blue indicates a negative test, whereas green indicates a positive test with the least amount of sugar, and red indicates the greatest amount of sugar present.

   e. Record the results in Part A of the laboratory assessment.

7. Complete Part A of the laboratory assessment.

PROCEDURE B: Effect of Heat

1. Mark three clean test tubes as *tubes 4, 5,* and *6.*

2. Add 1 mL of amylase solution to each of the tubes and expose each solution to a different test temperature for 3 minutes as follows:

   Tube 4: **Place in beaker of ice water (about 0°C [32°F]).**

   Tube 5: **Place in warm water bath (about 37°C [98.6°F]).**

   Tube 6: **Place in beaker of boiling water (about 100°C [212°F]). Use a test-tube clamp.**

3. *It is important that the 5 mL of starch solution added to tube 4 be at ice-water temperature before it is added to the 1 mL of amylase solution.* Add 5 mL of starch solution to each tube, shake to mix the contents, and return the tubes to their respective test temperatures for 10 minutes.

4. At the end of the 10 minutes, test the contents of each tube for the presence of starch and the presence of sugar by following the same directions as in steps 5 and 6 in Procedure A.

5. Complete Part B of the laboratory assessment.

LEARNING EXTENSION ACTIVITY

Devise an experiment to test the effect of some other environmental factor on amylase activity. For example, you might test the effect of a strong acid by adding a few drops of concentrated hydrochloric acid to a mixture of starch and amylase solutions. Be sure to include a control in your experimental plan. That is, include a tube containing everything except the factor you are testing. Then you will have something with which to compare your results. *Carry out your experiment only if it has been approved by the laboratory instructor.*

LABORATORY ASSESSMENT
55

Name _____

Date _____

Section _____

The (A) corresponds to the indicated Learning Outcome(s) (O) found at the beginning of the Laboratory Exercise.

Action of a Digestive Enzyme

PART A: Assessments

1. Test results:

Tube	Starch	Sugar
1 Amylase solution		
2 Starch solution		
3 Starch-amylase solution		

2. Complete the following:

   a. Explain the reason for including tube 1 in this experiment. _____

   _____

   _____

   _____

   b. What is the importance of tube 2? _____

   _____

   _____

   _____

   c. What do you conclude from the results of this experiment? (A2) _____

   _____

   _____

   _____

PART B: Assessments

1. Test results: (A3)

Tube	Starch	Sugar
4 0°C (32°F)		
5 37°C (98.6°F)		
6 100°C (212°F)		

2. Complete the following:

   a. What do you conclude from the results of this experiment? (A3) _____

     _____

     _____

     _____

     _____

     _____

   b. If digestion failed to occur in one of the tubes in this experiment, how can you tell whether the amylase was destroyed by the factor being tested or the amylase activity was simply inhibited by the test treatment? (A3)

     _____

     _____

     _____

     _____

     _____

CRITICAL THINKING ASSESSMENT

What test result would occur if the amylase you used contained sugar? _____ Would your results be as valid? Explain your answer. (A2)

_____

_____

_____

_____

LABORATORY
EXERCISE 56

Urinary Organs

Purpose of the Exercise

To review the structure of urinary organs and to observe the gross anatomy of a kidney as well as the microscopic nephron.

Learning Outcomes

After completing this exercise, you should be able to:

O1 Identify the structures of a kidney and associated urinary organs.

O2 Identify and sketch the structures of a nephron.

O3 Trace the path of filtrate through a renal nephron.

O4 Trace the path of blood through the renal blood vessels.

O5 Identify and sketch the structures of a ureter and a urinary bladder wall.

O6 Trace the path of urine flow through the urinary system.

The **O** corresponds to the assessments **A** indicated in the Laboratory Assessment for this Exercise.

Pre-Lab

Carefully read the introductory material and examine the entire lab. Be familiar with the structures and functions of the urinary organs from lecture or the textbook. Answer the pre-lab questions.

Pre-Lab Questions Select the correct answer for each of the following questions:

1. When comparing the position of the two kidneys,
   a. they are at the same level.
   b. the right kidney is slightly superior to the left kidney.
   c. the right kidney is slightly inferior to the left kidney.
   d. the right kidney is anterior to the left kidney.

2. Cortical nephrons represent about _____ of the nephrons.
   a. 0 %
   b. 100 %
   c. 20 %
   d. 80 %

3. Which of the following does *not* represent one of the processes in urine formation?
   a. secretion of renin
   b. glomerular filtration
   c. tubular reabsorption
   d. tubular secretion

4. The _____ arteries and veins are located in the corticomedullary junction.
   a. arcuate
   b. renal
   c. cortical radiate
   d. peritubular

5. The _____ is the tube from the kidney to the urinary bladder.
   a. urethra
   b. ureter
   c. renal pelvis
   d. renal column

6. The trigone is a triangular, funnel-like region of the
   a. renal cortex.
   b. renal medulla.
   c. urethra.
   d. urinary bladder.

7. The external urethral sphincter is composed of involuntary smooth muscle.
   a. True _____
   b. False ✓_____
8. Contractions of the detrusor muscle provide the force during micturition (urination).
   a. True ✓_____
   b. False _____

The kidneys are the primary organs of the urinary system. They are located in the abdominal cavity, against the posterior wall and behind the parietal peritoneum (*retroperitoneal*). Masses of adipose tissue associated with the kidneys hold them in place at a vertebral level between T12 and L3. The right kidney is slightly inferior due to the large mass of the liver near its superior border. A variety of functions occur in the kidneys. They remove metabolic wastes from the blood; help regulate blood volume, blood pressure, and pH of blood; control water and electrolyte concentrations; and secrete renin and erythropoietin.

The movements of muscular walls of the ureters force urine by means of *peristaltic waves* into the *urinary bladder,* which temporarily stores urine. The *urethra* conveys urine to the outside of the body.

Each *kidney* contains about 1 million *nephrons,* which serve as the basic structural and functional units of the kidney. A *glomerular capsule, proximal convoluted tubule, nephron loop,* and *distal convoluted tubule* compose the microscopic, multicellular structure of a relatively long nephron tubule. Several nephrons merge and drain into a common *collecting duct.* Approximately 80% of the nephrons are *cortical nephrons* with short nephron loops, whereas the remaining represent *juxtamedullary nephrons,* with long nephron loops extending deeper into the renal medulla. An elaborate network of blood vessels surrounds the entire nephron. *Glomerular filtration, tubular reabsorption,* and *tubular secretion* represent three processes resulting in urine as the final product.

PROCEDURE A: Kidney

1. Study figures 56.1 and 56.2 and the list of structures and descriptions.
2. Observe the human torso model and the kidney model. Locate the following:

   kidneys—paired retroperitoneal organs

   ureters—paired tubular organs about 25 cm long that transport urine from kidney to urinary bladder

   urinary bladder—single muscular storage organ for urine

   urethra—conveys urine from urinary bladder to external urethral orifice

   renal sinus—hollow chamber of kidney

   renal pelvis—funnel-shaped sac at superior end of ureter; receives urine from major calyces

   - major calyces—2–3 major converging branches into renal pelvis; receive urine from minor calyces

FIGURE 56.1 [AP|R] The urinary system and associated structures.

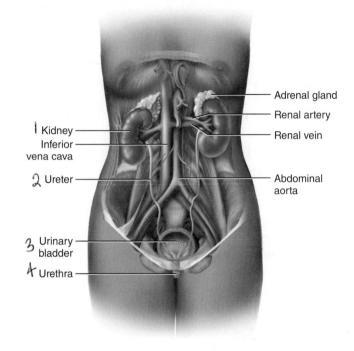

- minor calyces—small subdivisions of major calyces; receive urine from papillary ducts within renal papillae

renal medulla—deep region of kidney

- renal pyramids—6–10 conical regions composing most of renal medulla
- renal papillae—projections at ends of renal pyramids with openings into minor calyces

renal cortex—superficial region of kidney

renal columns—extensions of renal cortical tissue between renal pyramids

nephrons—functional units of kidneys

- cortical nephrons—80% of nephrons close to surface
- juxtamedullary nephrons—20% of nephrons close to medulla

3. Obtain a pig or sheep kidney along with a dissecting tray, dissecting instruments, and disposable gloves, and follow these steps:
   a. Rinse the kidney with water to remove as much of the preserving fluid as possible and place it in a dissecting tray.
   b. Carefully remove any adipose tissue from the surface of the specimen.
   c. Locate the following external anatomical features:

   fibrous (renal) capsule—protective membrane that encloses kidney

   hilum of kidney—indented region containing renal artery and vein and ureter

   renal artery—large artery arising from abdominal aorta

606

FIGURE 56.2 AP|R Frontal section of a kidney: (a) illustration; (b) pig kidney that has a triple injection of latex (*red* in the renal artery, *blue* in the renal vein, and *yellow* in the ureter and renal pelvis).

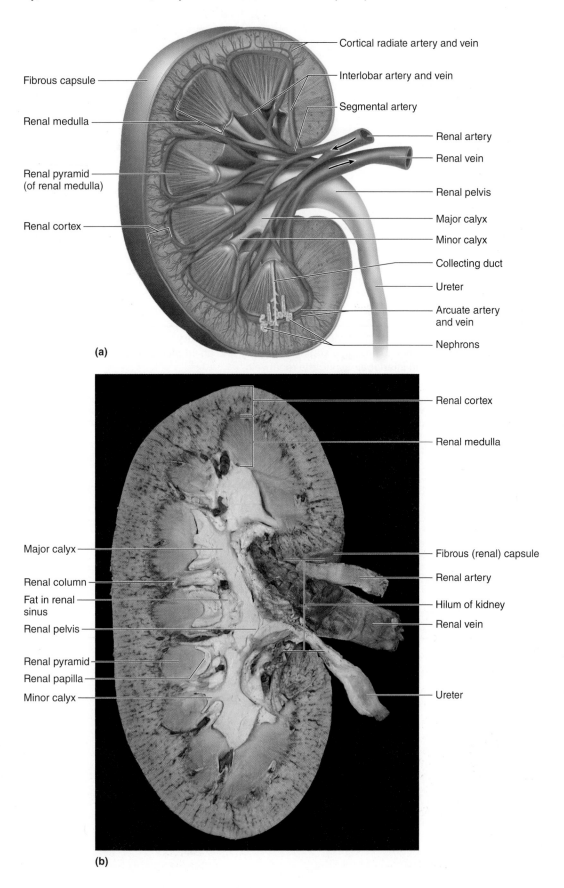

(a)

- Cortical radiate artery and vein
- Interlobar artery and vein
- Segmental artery
- Renal artery
- Renal vein
- Renal pelvis
- Major calyx
- Minor calyx
- Collecting duct
- Ureter
- Arcuate artery and vein
- Nephrons

- Fibrous capsule
- Renal medulla
- Renal pyramid (of renal medulla)
- Renal cortex

(b)

- Major calyx
- Renal column
- Fat in renal sinus
- Renal pelvis
- Renal pyramid
- Renal papilla
- Minor calyx

- Renal cortex
- Renal medulla
- Fibrous (renal) capsule
- Renal artery
- Hilum of kidney
- Renal vein
- Ureter

renal vein—large vein that drains into inferior vena cava

ureter—transports urine from renal pelvis to urinary bladder

    d. Use a long knife to cut the kidney in half longitudinally along the frontal plane, beginning on the convex border.

    e. Rinse the interior of the kidney with water, and using figure 56.2*b* as a reference, locate the following:

    renal pelvis

- major calyces
- minor calyces

    renal cortex

    renal columns

    renal medulla

- renal pyramids
- renal papillae

4. Complete Part A of Laboratory Assessment 56.

PROCEDURE B: Renal Blood Vessels and Nephrons

1. Study figures 56.2*a,* 56.3 and 56.4 illustrating renal blood vessels and structures of a nephron. The arrows within the figures indicate blood flow and tubular fluid flow.

2. Blood flows into the kidney from the renal artery arising off the abdominal aorta; blood leaves the kidney through the renal vein that drains into the inferior vena cava. Vascular subdivisions from the renal artery include segmental arteries → interlobar arteries → arcuate arteries → cortical radiate arteries → afferent arterioles → glomerular capillaries → efferent arterioles → peritubular capillaries (or vasa recta) → cortical radiate veins → arcuate veins → interlobar veins → renal vein. The renal arteries and veins with comparable names are adjacent to each other within the kidney (figs. 56.2*a* and 56.3). The venous pathway of the kidney does not possess corresponding segmental veins.

FIGURE 56.3 Renal blood vessels associated with cortical and juxtamedullary nephrons. The arrows indicate the flow of blood. Blood from the renal artery flows through a segmental artery and an interlobar artery to the arcuate artery. Blood from an arcuate vein flows through an interlobar vein to the renal vein.

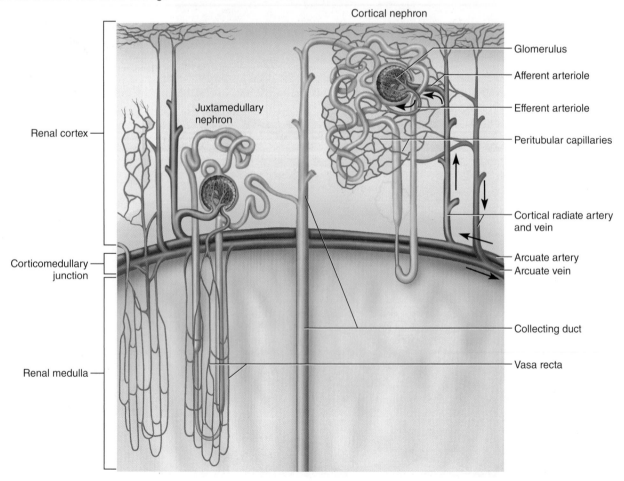

FIGURE 56.4 Structure of a nephron with arrows indicating the flow of tubular fluid. The corticomedullary junction is typical for a cortical nephron.

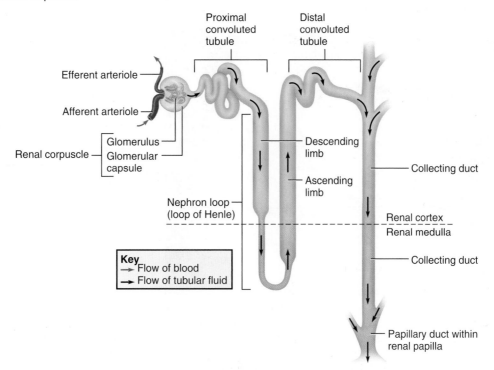

3. Obtain a microscope slide of a kidney section and examine it using low-power magnification. Locate the *renal capsule,* the *renal cortex* (which appears somewhat granular and may be more darkly stained than the other renal tissues), and the *renal medulla* (fig. 56.5).

4. Examine the renal cortex using high-power magnification. Locate a *renal corpuscle.* These structures appear as isolated circular areas. Identify the *glomerulus,* the capillary cluster inside the corpuscle, and the *glomerular (Bowman's) capsule,* which appears as a clear area surrounding the glomerulus. A glomerulus and a glomerular capsule compose a *renal corpuscle.* Also note the numerous sections of renal tubules that occupy

the spaces between renal corpuscles (fig. 56.5*a*). The renal cortex contains the proximal and distal convoluted tubules.

5. Prepare a labeled sketch of a representative section of the renal cortex in Part B of the laboratory assessment.

6. Examine the renal medulla using high-power magnification. Identify longitudinal views of various collecting ducts and nephron loops. These ducts are lined with simple epithelial cells, which vary in shape from squamous to cuboidal (fig. 56.5*b*).

7. Prepare a labeled sketch of a representative section of the renal medulla in Part B of the laboratory assessment.

8. Complete Part C of the laboratory assessment.

FIGURE 56.5 **AP|R** (a) Micrograph of a section of the renal cortex (220×). (b) Micrograph of a section of the renal medulla (80×). (*Note:* With about a million nephrons in a kidney, there are many different orientations visible in micrograph sections of renal cortex and renal medulla.)

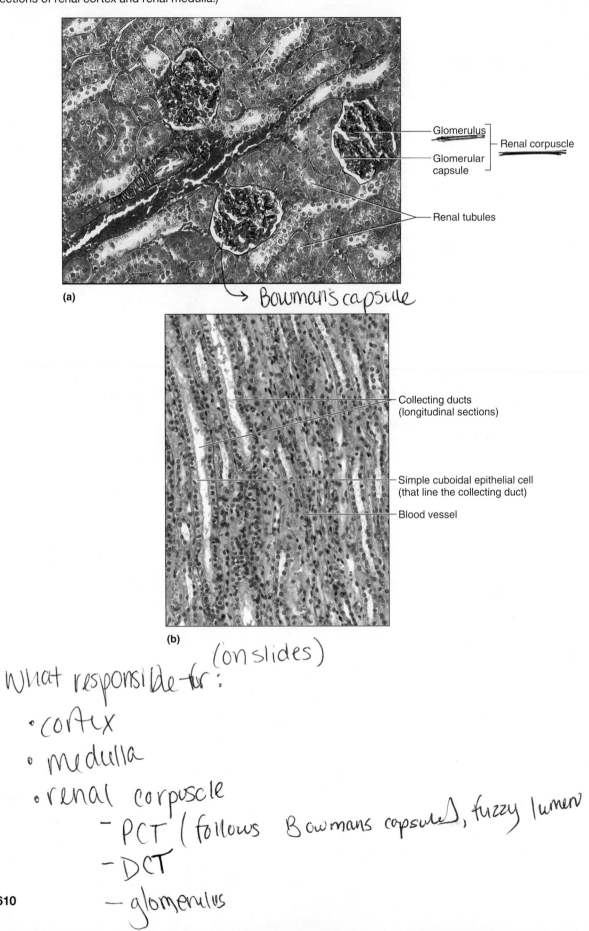

Glomerulus ⎤
                ⎥ Renal corpuscle
Glomerular ⎦
capsule

Renal tubules

(a)

→ Bowman's capsule

Collecting ducts
(longitudinal sections)

Simple cuboidal epithelial cell
(that line the collecting duct)

Blood vessel

(b)

What responsible for: (on slides)
• cortex
• medulla
• renal corpuscle
    – PCT (follows Bowmans capsule), fuzzy lumen
    – DCT
    – glomerulus

PROCEDURE C: Ureter, Urinary Bladder, and Urethra

The *ureters* are paired tubes about 25 cm (10 inches) long extending from the renal pelvis to the *ureteral openings (ureteric orifices)* of the *urinary bladder.* Peristaltic waves, created from rhythmic contractions of smooth muscle, occur about every 30 seconds, transporting the urine toward the urinary bladder. The bladder possesses mucosal folds called *rugae* that enable it to distend for temporary urine storage. Coarse bundle layers of smooth muscle compose the somewhat broad *detrusor muscle.* The triangular floor of the bladder, the *trigone,* is bordered by the two ureteral openings and the opening into the urethra, which is surrounded by the *internal urethral sphincter.* The internal urethral sphincter is formed by a thickened region of the detrusor muscle (fig. 56.6).

Although the bladder can hold about 600 mL of urine, the desire to urinate from the pressure of a stretched bladder occurs when the bladder has about 150–200 mL of urine. During *micturition (urination)* the powerful detrusor muscle contracts, and the internal urethral sphincter is forced open. The *external urethral sphincter,* at the level of the urogenital diaphragm, is under somatic control. Because the external urethral sphincter is skeletal muscle, there is considerable voluntary control over the voiding of urine.

The *urethra* extends from the exit of the urinary bladder to the *external urethral orifice.* In the female the urethra is about 4 cm (1.5 inches) long. In the male, the urethra includes passageways for both urine and semen and extends the length of the penis. The male urethra can be divided into three sections: prostatic urethra, membranous (intermediate) urethra, and spongy urethra (fig. 56.6*b*). A total length

FIGURE 56.6 AP|R Ureters and frontal sections of urinary bladder and urethra of (a) female and (b) male.

Ureter
Detrusor muscle
Rugae
Ureteral openings
Trigone
Internal urethral sphincter
Urethra
Urogenital diaphragm
External urethral orifice
External urethral sphincter

(a) Female

Prostate gland
Prostatic urethra
Membranous (intermediate) urethra
Spongy urethra
Penis
External urethral orifice

(b) Male

of about 20 cm (8 inches) would be characteristic for the urethra of a male.

1. Obtain a microscope slide of a cross section of a ureter and examine it using low-power magnification. Locate the *mucous coat* layer next to the lumen. Examine the middle *muscular coat* composed of longitudinal and circular smooth muscle cells responsible for the peristaltic waves that propel urine from the kidneys to the urinary bladder. The outer *adventitia,* composed of connective tissue, secures the ureter in the retroperitoneal position (fig. 56.7).

2. Examine the mucous coat using high-power magnification. The specialized tissue is transitional epithelium, which allows changes in its thickness when unstretched and stretched.

3. Prepare a labeled sketch of a ureter in Part D of the laboratory assessment.

4. Obtain a microscope slide of a segment of the wall of a urinary bladder and examine it using low-power magnification. Examine the *mucous coat* next to the lumen and the *submucous coat* composed of connective tissue just beneath the mucous coat. Examine the *muscular coat* composed of bundles of smooth muscle fibers interlaced in many directions. This thick, muscular layer is called the *detrusor muscle* and functions in the elimination of urine. Also note the outer *adventitia* of connective tissue (fig. 56.8).

5. Examine the mucous coat using high-power magnification. The tissue is transitional epithelium, which allows changes in its thickness from unstretched when the bladder is empty to stretched when the bladder distends with urine.

6. Prepare a labeled sketch of a segment of the urinary bladder wall in Part D of the laboratory assessment.

FIGURE 56.7 AP|R Micrograph of a cross section of a ureter (40×).

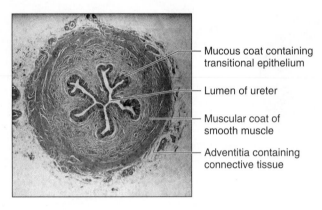

- Mucous coat containing transitional epithelium
- Lumen of ureter
- Muscular coat of smooth muscle
- Adventitia containing connective tissue

FIGURE 56.8 AP|R Micrograph of a segment of the human urinary bladder wall (40×). (*Note:* A serous layer is on the superior surface of the urinary bladder wall instead of the adventitia as the outermost layer.)

- Adventitia of connective tissue
- Muscular coat of smooth muscle (detrusor muscle)
- Submucous coat of connective tissue
- Mucous coat contains transitional epithelium next to lumen
- Mucosal fold
- Lumen of urinary bladder

7. Obtain a microscope slide of a cross section of a urethra and examine it using low-power magnification. Locate the muscular coat composed of smooth muscle fibers. Locate groups of mucous glands, called *urethral glands,* in the urethral wall that secrete protective mucus into the lumen of the urethra. Examine the mucous coat using high-power magnification. The mucous membrane is composed of a type of stratified epithelium. The specific type of stratified epithelium varies from transitional, to stratified columnar, to stratified squamous epithelium between the urinary bladder and the external urethral orifice. Depending upon where the section of the urethra was taken for the preparation of the microscope slide, the type of stratified epithelial tissue represented could vary (fig. 56.9).

8. Complete Part E of the laboratory assessment.

FIGURE 56.9 AP|R Cross section through the urethra (10×).

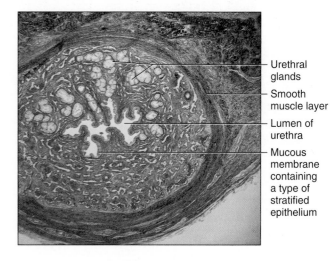

- Urethral glands
- Smooth muscle layer
- Lumen of urethra
- Mucous membrane containing a type of stratified epithelium

56

Urinary Organs

Name _____

Date _____

Section _____

The (A) corresponds to the indicated Learning Outcome(s) (O) found at the beginning of the Laboratory Exercise.

PART A: Assessments

1. Label the features indicated in figure 56.10 of a kidney (frontal section).

FIGURE 56.10   Label the structures in the frontal section of a kidney.

Renal cortex
Renal medulla
Renal sinus
1
2
3
4
5
6
7
8
9

2. Match the terms in column A with the descriptions in column B. Place the letter of your choice in the space provided.

Column A	Column B
a. Calyces	_____ **1.** Superficial region around the renal medulla
b. Hilum of kidney	_____ **2.** Extensions of renal pelvis containing urine
c. Nephron	_____ **3.** Conical mass of tissue within renal medulla
d. Renal column	_____ **4.** Projection with tiny openings into a minor calyx
e. Renal cortex	_____ **5.** Hollow chamber within kidney
f. Renal papilla	_____ **6.** Microscopic functional subunit of kidney
g. Renal pelvis	_____ **7.** Located between renal pyramids
h. Renal pyramid	_____ **8.** Superior, funnel-shaped sac at end of ureter inside the renal sinus
i. Renal sinus	_____ **9.** Medial depression for blood vessels and ureter to enter kidney chamber

PART B: Assessments

Sketch a representative section of the renal cortex and the renal medulla. Label the glomerulus, the glomerular capsule, and sections of renal tubules in the renal cortex. Label a longitudinal section and cross section of a collecting duct in the renal medulla. A2

Renal cortex (_____x)	Renal medulla (_____x)

PART C: Assessments

Complete the following:

1. Distinguish between a renal corpuscle and a renal tubule. A2 _____

_____

_____

2. Number the following structures to indicate their respective positions in relation to the nephron. Assign the number 1 to the structure nearest the glomerulus. A3

_____ Ascending limb of nephron loop

_____ Collecting duct

_____ Descending limb of nephron loop

_____ Distal convoluted tubule

_____ Glomerular capsule

_____ Proximal convoluted tubule

_____ Papillary duct in renal papilla

616

3. Number the following structures to indicate their respective positions in the blood pathway within the kidney. Assign the number 1 to the vessel nearest the abdominal aorta. (A4)

_____ Afferent arteriole

_____ Arcuate artery

_____ Arcuate vein

_____ Cortical radiate artery

_____ Cortical radiate vein

_____ Efferent arteriole

_____ Glomerulus

_____ Interlobar artery

_____ Interlobar vein

_____ Peritubular capillary (or vasa recta)

_____ Renal artery

_____ Renal vein

_____ Segmental artery

PART D: Assessments

Sketch a cross section of a ureter and label the three layers and the lumen. Sketch a segment of a urinary bladder and label the four layers and the lumen. (A5)

Ureter (_____ ×)	Urinary bladder (_____ ×)

PART E: Assessments

Number the following structures to indicate respective positions in the pathway of urine flow in a male. Assign the number 1 to the structure nearest the papillary duct in a renal papilla. **A6**

_____ External urethral orifice

_____ Major calyx

_____ Membranous urethra

_____ Minor calyx

_____ Prostatic urethra

_____ Renal pelvis

_____ Spongy urethra

_____ Ureter

_____ Ureteral opening

_____ Urinary bladder

LABORATORY EXERCISE 57

Urinalysis

Purpose of the Exercise

To perform the observations and tests commonly used to determine the characteristics and composition of urine and to analyze the clinical implications of abnormal results.

Learning Outcomes

After completing this exercise, you should be able to:

O1 Evaluate the physical characteristics of a urine sample: color, transparency, odor, specific gravity.

O2 Test, record, and interpret results of the urine sample, using individual and/or combination test strips, for the presence of substances such as: specific gravity, pH, glucose, protein, ketones, bilirubin, urobilinogen, hemoglobin/occult blood, nitrites, and leukocytes.

O3 Analyze the constituents of the urine sample, and provide reasonable clinical implications for abnormal results.

O4 Perform a microscopic study of urine sediment.

O5 Summarize the results of these observations and tests.

The **O** corresponds to the assessments **A** indicated in the Laboratory Assessment for this Exercise.

Pre-Lab

Carefully read the introductory material and examine the entire lab. Be familiar with normal urine components from lecture or the textbook. Answer the pre-lab questions.

Pre-Lab Questions Select the correct answer for each of the following questions:

1. Urine consists of about ____ percent water and ____ percent solutes.
   a. 100; 0
   (b.) 95; 5
   c. 90; 10
   d. 50; 50

2. A urinalysis consists of all of the following *except*
   a. physical characteristics.
   b. chemical analysis.
   c. microscopic examination.
   (d.) body temperature analysis.

3. Microscopic solids are stained with _____ in order to make identifications possible.
   a. reagents
   b. iodine
   c. methylene blue
   (d.) Sedi-stain

4. Which of the following would be the most typical urinary output in a day?
   a. 0.5 liters
   b. 0.8 liters
   (c.) 1.2 liters
   d. 2.2 liters

5. Which of the following would represent a normal pH range of urine?
   (a) 4.6–8.0
   b. 0–14
   c. 7.35–7.45
   d. 5–10

619

Urine is the product of three processes of the nephrons within the kidneys: *glomerular filtration, tubular reabsorption,* and *tubular secretion.* As a result of these processes, various waste substances are removed from the blood, and body fluid and electrolyte balance are maintained. Consequently, the composition of urine varies considerably because of differences in dietary intake and physical activity from day to day and person to person. The normal urinary output ranges from 1.0–1.8 liters per day. Typically, urine consists of 95% water and 5% solutes. The volume of urine produced by the kidneys varies with such factors as fluid intake, environmental temperature, relative humidity, respiratory rate, and body temperature.

An analysis of urine composition and volume often is used to evaluate the functions of the kidneys and other organs. This procedure, called *urinalysis,* is a clinical assessment and a diagnostic tool for certain pathological conditions and general overall health. A urinalysis and a complete blood analysis complement each other for an evaluation of certain diseases, such as diabetes mellitus, and general health.

A urinalysis involves three aspects: *physical characteristics, chemical analysis,* and a *microscopic examination.* Physical characteristics of urine that are noted include volume, color, transparency, and odor. The chemical analysis of solutes in urine addresses urea and other nitrogenous wastes; electrolytes; pigments; and possible glucose, protein, ketones, bilirubin, urobilinogen, and hemoglobin. The specific gravity and pH of urine are greatly influenced by the components and amounts of solutes. An increase in the viscosity of blood due to dehydration, polycythemia, kidney inflammation, or other causes will result in an increase in urine specific gravity. A diet high in protein and whole wheat products lowers urine pH, while bacterial infection may increase urine pH; thus diet and pathological conditions have an effect on urinalysis. An examination of microscopic solids, including cells, casts, and crystals, assists in the diagnosis of injury, various diseases, and urinary infections.

 WARNING

While performing the following tests, you should wear disposable latex gloves so that skin contact with urine is avoided. Observe all safety procedures listed for this lab. (Normal and abnormal simulated urine specimens can be used instead of real urine for this lab.)

PROCEDURE A: Physical and Chemical Analysis

1. Proceed to the restroom with a clean, disposable container if collected urine is to be used for this lab exercise. The first small volume of urine should not be collected because it contains abnormally high levels of microorganisms from the urethra. Collect a midstream sample of about 50 mL of urine. The best collections are the first specimen in the morning or one taken 3 hours after a meal. Refrigerate samples if they are not used immediately.

2. Place a sample of urine in a clean, transparent container. Describe the *color* of the urine. Normal urine varies from light yellow to amber, depending on the presence of urochromes, end-product pigments produced during the decomposition of hemoglobin.

   Abnormal urine colors include yellow-brown or green, due to elevated concentrations of bile pigments, and red to dark brown, due to the presence of blood. Certain foods, such as beets or carrots, and various drug substances also can cause color changes in urine, but in such cases the colors have no clinical significance. Enter the results of this and the following tests in Part A of Laboratory Assessment 57.

3. Evaluate the *transparency* of the urine sample (judge whether the urine is clear, slightly cloudy, or very cloudy). Normal urine is clear enough to see through. You can read newsprint through slightly cloudy urine; you can no longer read newsprint through cloudy urine. Cloudy urine indicates the presence of various substances, including mucus, bacteria, epithelial cells, fat droplets, and inorganic salts.

4. Gently wave your hand over the urine sample toward your nose to detect the *odor.* Normal urine should have a slight ammonia-like odor due to the nitrogenous wastes. Some vegetables (like asparagus), drugs (like certain vitamins), and diseases (like diabetes mellitus) will influence the odor of urine.

5. Determine the *specific gravity* of the urine sample, which indicates the solute concentration. Specific gravity is the ratio of the weight of something to the weight of an equal volume of pure water. For example, mercury (at 15°C) weighs 13.6 times as much as an equal volume of water; thus, it has a specific gravity of 13.6. Although urine is mostly water, it has substances dissolved in it and is slightly heavier than an equal volume of water. The specific gravity of pure (distilled) water is 1.000, while the specific gravity of urine is higher, varying from 1.003 to 1.035 under normal circumstances. If the specific gravity is too low, the urine contains few solutes and represents dilute urine, a likely result of excessive fluid intake or the use of diuretics. A specific gravity above the normal range represents a higher concentration of solutes, likely from a limited fluid intake. Concentrated urine over an extended time increases the risk of the formation of kidney stones (renal calculi).

   To determine the specific gravity of a urine sample, follow these steps:
   a. Pour enough urine into a clean urinometer cylinder to fill it about three-fourths full. Any foam that appears should be removed with a paper towel.
   b. Use a laboratory thermometer to measure the temperature of the urine.
   c. Gently place the urinometer hydrometer into the urine, *and make sure that the float is not touching the sides or the bottom of the cylinder* (fig. 57.1).
   d. Position your eye at the level of the urine surface. Determine which line on the stem of the hydrometer

Hydrometer

Urine Specimen

Urine in urinometer cylinder (jar)

intersects the lowest level of the concave surface (meniscus) of the urine.

e. Liquids tend to contract and become denser as they are cooled, or to expand and become less dense as they are heated, so it may be necessary to make a temperature correction to obtain an accurate specific gravity measurement. To do this, add 0.001 to the hydrometer reading for each 3 degrees of urine temperature above 25°C or subtract 0.001 for each 3 degrees below 25°C. Enter this calculated value in the table in Part A of the laboratory assessment as the test result.

6. Individual or combination reagent strips can be used to perform a variety of urine tests (fig. 57.2). In each case, directions for using the strips are found on the strip container. *Be sure to read them.*

   To perform each test, follow these steps:

   a. Obtain a urine sample and the proper reagent test strip.

   b. Read the directions on the strip container.

   c. Dip the strip in the urine sample as directed on the container.

   d. Remove the strip at an angle and let it touch the inside rim of the urine container to remove any excess liquid.

   e. Wait for the length of time indicated by the directions on the container before you compare the color of the test strip with the standard color scale provided with the container. The value or amount represented by the matching color should be used as the test result.

7. Record your urine test results in Part A of the laboratory assessement.

8. Your urine sample might have been tested using individual test strips, combination test strips, or some from each. Use the following information about any of the tests you conducted to help you interpret the findings of your sample. (Your test results might have been done and recorded in a different sequence than those listed in this section.)

   ▪ *Specific gravity:* Compare your results of the specific gravity using a reagent test strip to that obtained using the urinometer. Given that you used the same urine, your results should be similar. Uncontrolled diabetes mellitus will result in urine that has a higher than normal specific gravity due to the excess solutes (glucose, ketones), while diabetes insipidus will result in urine that has a lower than normal specific gravity due to the diluted solutes.

   ▪ *pH test:* The pH of normal urine varies from 4.6 to 8.0, but most commonly, it is near 6.0 (slightly acidic). A vegetarian diet may make the urine pH more alkaline, while a diet high in protein may make urine pH more acidic. Significant daily variations within the broad normal range are a result of dietary excesses of certain foods and liquids consumed.

   ▪ *Glucose:* Glucose is normally absent in urine, as it is completely reabsorbed from the renal tubules into circulation. However, glucose may appear in the urine temporarily following a meal high in carbohydrates. Glucose in the urine (called *glucosuria*) is a sign of uncontrolled diabetes mellitus.

   ▪ *Protein:* Normally, proteins are not present in urine. However, those of a small molecular size, such as albumins, may appear in trace amounts, particularly after strenuous exercise. Increased amounts of proteins may appear as a result of uncontrolled diabetes mellitus, kidney disease in which the glomeruli are damaged, or high blood pressure (hypertension).

   ▪ *Ketones:* Ketones are organic acids that are products of fat catabolism. They normally only appear in urine in very small amounts, if at all. When ketone production exceeds tissue use, *ketoacidosis* results.

The most common cause of ketoacidosis is uncontrolled diabetes mellitus—due to insulin deficiency, the cells cannot obtain carbohydrates so they rely on fatty acids for fuels. Fasting or starvation may also cause ketoacidosis.

- *Bilirubin:* Bilirubin, which results from hemoglobin decomposition in the liver, normally is absent in urine. Urochrome, a normal yellow component of urine, is a result of additional breakdown of bilirubin. Elevated bilirubin levels, the result of liver disorders that cause obstruction of the biliary tract and excessive red blood cell destruction (*hemolytic anemia*), make the urine appear dark because of the bilirubin in the urine.
- *Urobilinogen:* Urobilinogen is a colorless breakdown product of bilirubin from the action of intestinal bacteria. Urobilinogen is either reabsorbed into the portal circulation to be re-excreted into the bile by the liver or excreted in the feces; however, a small amount travels to the kidneys in the general circulation to be excreted in the urine. Low levels may indicate inadequate bile production or biliary obstruction, while high levels may indicate hemolytic anemia with infectious hepatitis or cirrhosis.
- *Hemoglobin/occult blood:* The appearance of red blood cells in urine is indicative of pathology because erythrocytes are too large to pass through the filtration membrane; it also may be indicative of menses in women. Hemoglobin resides in the red blood cells, and since red blood cells are not normally able to pass through renal tubules, hemoglobin is not normally found in the urine. Its presence in urine usually indicates a renal disease process, a transfusion reaction, an injury to the urinary organs, or menstrual blood.
- *Nitrites:* Normally, nitrites are not found in the urine. Bacteria in the urinary tract make an enzyme that changes urinary nitrates to nitrites, so when nitrites appear in urine, it is a sign of a urinary tract infection (UTI).
- *Leukocytes:* White blood cells in the urine are indicative of an infection of the urinary tract, urinary bladder, or kidney; normally they are not present.

The clinical significance or conclusive diagnosis of a disease is often difficult to achieve. Most often, several blood and urine characteristics along with other symptoms are assessed collectively in order to make a determination of the abnormality and its potential cause. Any test that is showing abnormal results should be validated by a health care professional.

9. Complete Part A of the laboratory assessment.

PROCEDURE B: Microscopic Sediment Analysis

1. A urinalysis usually includes an analysis of urine sediment—the microscopic solids present in a urine sample. This sediment normally includes mucus, certain crystals, and a variety of cells, such as the epithelial cells that line the urinary tubes and an occasional white blood cell. Other types of solids, such as casts or red blood cells, may indicate a disease or injury if they are present in excess. (Casts are cylindrical masses of cells or other substances that form in the renal tubules and are flushed out by the flow of urine.)

   To observe urine sediment, follow these steps:
   a. Thoroughly stir or shake your urine sample to suspend the sediment, which tends to settle to the bottom of the container.
   b. Pour 10 mL of urine into a clean centrifuge tube and centrifuge it for 5 minutes at slow speed (1,500 rpm). Be sure to balance the centrifuge with an even number of tubes filled to the same levels.
   c. Carefully decant 9 mL (leave 1 mL) of the liquid from the sediment in the bottom of the centrifuge tube, as directed by your laboratory instructor. Resuspend the 1 mL of sediment.
   d. Use a medicine dropper to remove some of the sediment and place it on a clean microscope slide.
   e. Add a drop of Sedi-stain to the sample, and add a coverslip.
   f. Examine the sediment with low-power (reduce the light when using low power) and high-power magnifications.
   g. With the aid of figure 57.3, identify the types of solids present.

2. In Part B of the laboratory assessment, make a sketch of each type of sediment that you observed.

3. Complete Part B of the laboratory assessment.

FIGURE 57.3 Types of urine sediment. Healthy individuals lack many of these sediments and possess only occasional to trace amounts of others. (*Note:* Shades of white to purple sediments are most characteristic when using Sedi-stain.)

Crystals

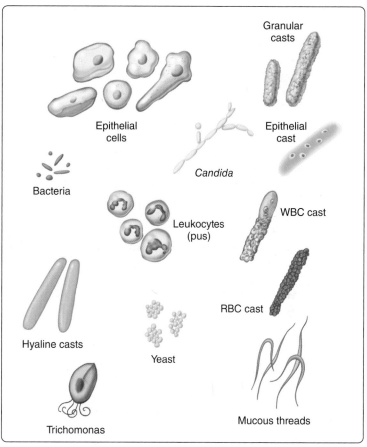

Cells and casts

NOTES

Name _____

Date _____

Section _____

The corresponds to the indicated Learning Outcome(s)
found at the beginning of the Laboratory Exercise.

Urinalysis

PART A: Assessments

1. Enter your observations, test results, and evaluations/clinical implications in the following table: **A1** **A2** **A3** **A5**

Urine Characteristics	Normal Values	Observations and Test Results	Evaluations/ Clinical Implications
Color	Light yellow to amber		
Transparency	Clear		
Odor	Slight ammonia-like		
Specific gravity (corrected for temperature)	1.003–1.035		
pH	4.6–8.0		
Glucose	Negative (0)		
Protein	Negative to trace		
Ketones	Negative (0)		
Bilirubin	Negative (0)		
Urobilinogen	0.1–1.0 mg/dL		
Hemoglobin/occult blood	Negative (0)		
Nitrites	Negative (0)		
Leukocytes	Negative (0)		

2. Summarize the results and clinical implications of the chemical analysis of urine. (A5)

_____

_____

_____

_____

_____

_____

CRITICAL THINKING ASSESSMENT

Joey complained to his physician that he was constantly thirsty, hungry, and had to urinate frequently. His physician ordered a blood and urine test. The results of his urinalysis showed the following abnormal values:

- pH: 3.5
- glucose: 250 mg/100 mL
- ketones: 10 gm/100 mL
- albumin: 30 mg/100 mL
- specific gravity: 1.045

Given Joey's symptoms and urinalysis results, what do you predict is his diagnosis? Explain and provide reasoning. (A3)

_____

_____

_____

_____

PART B: Assessments

1. Make a sketch for each type of sediment you observed. Label any from those shown in figure 57.3. (A4)

2. Summarize the results of the microscopic sediment analysis of urine. (A5)

_____

_____

_____

_____

LABORATORY EXERCISE 58

Male Reproductive System

Purpose of the Exercise

To review the structures and functions of the male reproductive organs and to examine some of their features.

MATERIALS NEEDED

Human torso model

Model of the male reproductive system

Anatomical chart of the male reproductive system

Compound light microscope

Prepared microscope slides of the following:

  Testis section

  Epididymis, cross section

  Ductus deferens, cross section

  Penis, cross section

Learning Outcomes AP|R

After completing this exercise, you should be able to:

O1 Locate and identify the organs of the male reproductive system.

O2 Describe the structures and functions of these organs.

O3 Sketch and label the major features of microscopic sections of the testis, epididymis, ductus deferens, and penis.

The **O** corresponds to the assessments **A** indicated in the Laboratory Assessment for this Exercise.

Pre-Lab

Carefully read the introductory material and examine the entire lab. Be familiar with the structures and functions of the male reproductive organs from lecture or the textbook. Answer the pre-lab questions.

Pre-Lab Questions Select the correct answer for each of the following questions:

1. Sperm are produced in the
   a. seminiferous tubules.
   b. interstitial cells.
   c. epididymis.
   d. ductus deferens.

2. The ductus deferens is located within the
   a. testis.
   b. epididymis.
   c. spermatic cord.
   d. penis.

3. The _____ is the common tube for the passage of urine and semen.
   a. ductus deferens
   b. urethra
   c. ureter
   d. ejaculatory duct

4. Which of the following would *not* be visible in a cross section of a penis?
   a. urethra
   b. corpus spongiosum
   c. corpora cavernosa
   d. epididymis

5. The interstitial cells produce
   a. sperm.
   b. testosterone.
   c. nutrients.
   d. semen.

6. The following glands are paired *except* the
   a. prostate.
   b. testis.
   c. seminal vesicle.
   d. bulbourethral.

7. The smooth muscle represents the thickest layer of the wall of the ductus deferens.
   a. True _____
   b. False _____

8. The rete testis connects directly to the tail of the epididymis.
   a. True _____
   b. False _____

The organs of the male reproductive system are specialized to produce and maintain the male sex cells, to transport these cells together with supporting fluids to the female reproductive tract, and to produce and secrete male sex hormones. These organs include sets of internal and external genitalia.

The *testes* originate within the abdominopelvic cavity, but they descend through an inguinal canal prior to birth to a position in the *scrotum,* which provides a lower temperature necessary for sperm production and storage. Within the testes, numerous *seminiferous tubules* produce *sperm* (*spermatozoa*) by spermatogenesis, and *interstitial cells* between seminiferous tubules produce testosterone. The process of spermatogenesis (meiosis) is described in Laboratory Exercise 60. The interstitial cells then release the testosterone for transport in the bloodstream. Testosterone influences the sex drive and development of the secondary sex characteristics.

The sperm produced in the seminiferous tubules enter a tubular network, the *rete testis,* which joins the tightly coiled *epididymis* through *efferent ductules.* Within head, body, and tail of the epididymis, immature sperm are stored and nourished during their maturation. A muscular *ductus deferens (vas deferens),* part of the *spermatic cord,* passes through the inguinal canal into the pelvic cavity and posterior to the urinary bladder, where it unites with ducts from *seminal vesicles* to form the ejaculatory ducts. An alkaline fluid containing nutrients and prostaglandins is secreted by the seminal vesicles. The ejaculatory ducts unite with the urethra within the *prostate gland.* Additional secretions are added from the prostate gland and two *bulbourethral glands.* The combination of sperm and the secretions from the accessory glands is *semen.*

The urethra, a common tube to convey semen and urine, passes through the *penis* within the *corpus spongiosum* to the external urethral orifice. The corpus spongiosum plus two *corpora cavernosa* provide vascular spaces that become engorged with blood during an erection. A loose fold of skin, forming a cuff over the glans penis, is called the *prepuce,* and is frequently removed in a procedure called a circumcision shortly after birth.

PROCEDURE A: Male Reproductive Organs

1. Study figures 58.1 and 58.2 and the list of structures and descriptions.

FIGURE 58.1 The major structures of the male reproductive system, sagittal view.

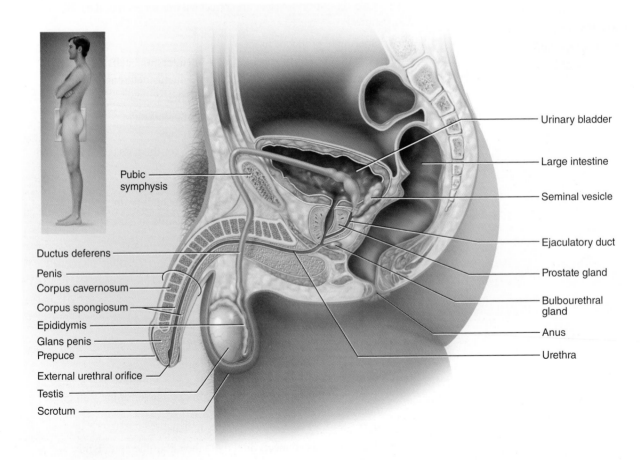

Pubic symphysis

Ductus deferens

Penis

Corpus cavernosum

Corpus spongiosum

Epididymis

Glans penis

Prepuce

External urethral orifice

Testis

Scrotum

Urinary bladder

Large intestine

Seminal vesicle

Ejaculatory duct

Prostate gland

Bulbourethral gland

Anus

Urethra

Spermatic cord

Ductus deferens

Head of epididymis

Efferent ductule

Rete testis

Body of epididymis

Testis, covered by tunica albuginea

Tail of epididymis

Scrotum (reflected)

Spermatic cord

Blood vessels and nerves

Seminiferous tubules

Septum

Lobule

(a)

2 cm

(b)

2. Observe the human torso model, the model of the male reproductive system, and the anatomical chart of the male reproductive system. Locate the following features:

   testes—paired organs suspended by spermatic cord within scrotum

   - lobules—subdivisions of testes
   - seminiferous tubules—long, tiny coiled tubes; site of spermatogenesis
   - interstitial cells (Leydig cells)—endocrine cells between seminiferous tubules; produce and secrete testosterone
   - rete testis—network of channels from seminiferous tubules to efferent ductules
   - efferent ductules—several ducts from rete testis to epididymis

   inguinal canal—tubelike passageway between abdominopelvic cavity and scrotum; contains the spermatic cord

   spermatic cord—contains ductus deferens, blood vessels, nerves, and lymphatic vessels

   epididymis—site of sperm maturation and storage; tube about 6 meters (18 feet) if uncoiled

   ductus deferens (vas deferens)—muscular tube that transports and stores sperm

   ejaculatory duct—tube within prostate gland from ductus deferens to urethra

   seminal vesicles (seminal glands)—paired glands; produce most of seminal fluid that contains fructose and prostaglandins; fructose nourishes sperm and prostaglandins enhance sperm entrance into uterus

   prostate gland—produces seminal fluid that helps to activate sperm; single gland

   bulbourethral glands—tiny paired glands; secretions neutralize any residual acidic urine

   scrotum—protective covering of testes and epididymis

   penis—male copulatory organ

   - corpora cavernosa—paired erectile tissue cylinders; engorge with blood during an erection
   - corpus spongiosum—single erectile tissue cylinder; engorges with blood during an erection
   - tunica albuginea—fibrous capsule around testes and corpora erectile tissues
   - glans penis—expanded distal end of penis
   - external urethral orifice (external urethral meatus)—external opening for urine and semen
   - prepuce (foreskin)—fold of skin around glans penis; removed if a surgical circumcision is performed

3. Complete Part A of Laboratory Assessment 58.

PROCEDURE B: Microscopic Anatomy

1. Obtain a microscope slide of a human testis section and examine it using low-power magnification (fig. 58.3). Locate the thick *tunica albuginea (fibrous capsule)* on the surface and the numerous sections of *seminiferous tubules* inside.

2. Focus on some of the seminiferous tubules using high-power magnification (fig. 58.4). Locate the *basement membrane* and the layer of *spermatogonia* near the basement membrane. Identify some *sustentacular cells* (supporting cells; Sertoli cells), which have pale, oval-shaped nuclei, and some *spermatogenic cells,* which have smaller, round nuclei. Spermatogonia give rise to spermatogenic cells that are in various stages of spermatogenesis as they are forced toward the lumen. Near the lumen of the tube, find some darkly stained, elongated heads of developing sperm (spermatozoa). In the spaces between adjacent seminiferous tubules, locate some isolated *interstitial cells* (Leydig cells) of the endocrine system. Interstitial cells produce the hormone testosterone, transported by the blood.

3. Prepare a labeled sketch of a representative section of the testis in Part B of the laboratory assessment.

4. Obtain a microscope slide of a cross section of *epididymis* (fig. 58.5). Examine its wall using high-power magnification. Note the elongated, *pseudostratified columnar epithelial cells* that compose most of the inner lining. These cells have nonmotile stereocilia (elongated microvilli) on their free surfaces that absorb excess fluid secreted by the testes. Also note the thin layer of smooth muscle and connective tissue surrounding the tube.

5. Prepare a labeled sketch of the epididymis wall in Part B of the laboratory assessment.

6. Obtain a microscope slide of a cross section of *ductus deferens,* and examine it with scan and low-power magnifications (fig. 58.6a). Note the small lumen lined with an epithelium and the thick *smooth muscle* layer composing most of the structure of the duct. Three layers of smooth muscle fibers are distinctive: a middle circular layer between two layers of longitudinal muscle fibers. The peristaltic contractions of the muscular layer transport sperm along the passageway toward the ejaculatory duct and urethra. Sperm storage occurs within the lumen of the ductus deferens and the epididymis. An outer, connective tissue covering of the ductus deferens is called the *adventitia.* Use high-power magnification and examine the epithelium layer composed of pseudostratified columnar epithelial cells (fig. 58.6b).

7. Prepare a labeled sketch of a representative section of a ductus deferens in Part B of the laboratory assessment.

FIGURE 58.3 Micrograph of a human testis (1.7×).

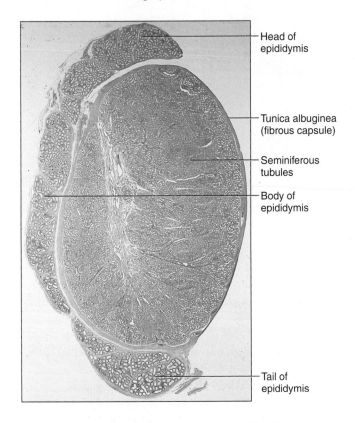

- Head of epididymis
- Tunica albuginea (fibrous capsule)
- Seminiferous tubules
- Body of epididymis
- Tail of epididymis

FIGURE 58.4 **AP|R** Micrograph of a seminiferous tubule (250×).

- Basement membrane
- Spermatogenic cells
- Spermatogonia
- Sperm cells
- Interstitial cells (Leydig cells)
- Seminiferous tubule

FIGURE 58.5   Micrograph of a cross section of a human epididymis (200×).

Pseudostratified columnar epithelial cells

Nonmotile stereocilia

Sperm cells

Smooth muscle cells

Lumen

FIGURE 58.6   AP|R   Ductus deferens. (a) Micrograph of a cross section of the ductus deferens (40×). (b) Micrograph of the wall of the ductus deferens (400×).

Adventitia

Lumen

Epithelium

Smooth muscle

Sperm in lumen of ductus deferens

Smooth muscle

Pseudostratified columnar epithelium

(a)

(b)

8. Study figure 58.7 and the list of structures and descriptions. Obtain a microscope slide of a *penis* cross section, and examine it with scan and low-power magnifications. Identify the following features:

corpora cavernosa—two dorsal erectile tissue cylinders; engorge with blood during an erection

corpus spongiosum—single erectile tissue cylinder around urethra; engorges with blood during an erection

tunica albuginea—fibrous capsule around corpora

urethra—passageway for urine and semen

skin—surrounds entire penis

9. Prepare a labeled sketch of a penis cross section in Part B of the laboratory assessment.

10. Complete Part B of the laboratory assessment.

FIGURE 58.7 Micrograph of a cross section of the body of the penis (5×).

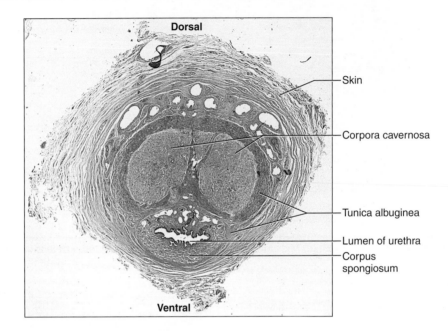

Name _____

Date _____

Section _____

The Ⓐ corresponds to the indicated Learning Outcome(s) Ⓞ found at the beginning of the Laboratory Exercise.

Male Reproductive System

PART A: Assessments

1. Label the structures in figures 58.8 and 58.9.

FIGURE 58.8   Label the major structures of the male reproductive system in this sagittal view.

FIGURE 58.9 Label the diagram of (a) the sagittal section of a testis and (b) a cross section of a seminiferous tubule by placing the correct numbers in the spaces provided. **A1**

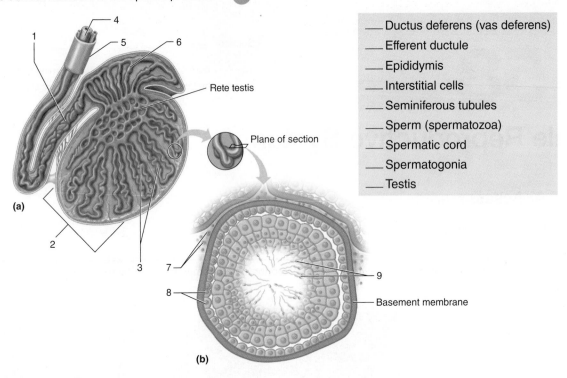

____ Ductus deferens (vas deferens)
____ Efferent ductule
____ Epididymis
____ Interstitial cells
____ Seminiferous tubules
____ Sperm (spermatozoa)
____ Spermatic cord
____ Spermatogonia
____ Testis

(a)

Rete testis

Plane of section

(b)

Basement membrane

2. Match the terms in column A with the descriptions in column B. Place the letter of your choice in the space provided. **A2**

Column A	Column B
a. Bulbourethral glands	_____ 1. Process by which sperm are formed
b. Ductus deferens	_____ 2. Common tube for urine and semen
c. Ejaculatory duct	_____ 3. Sperm duct located within spermatic cord
d. Epididymis	_____ 4. Produce male hormones
e. Interstitial cells	_____ 5. Location of spermatogenesis
f. Penis	_____ 6. Located within prostate gland and can propel semen into urethra
g. Prepuce	_____ 7. Contains sperm and secretions from accessory glands
h. Prostate	_____ 8. Tiny paired glands that secrete into urethra
i. Semen	_____ 9. Maturation and storage site within a coiled tube along testis
j. Seminiferous tubules	_____ 10. Surrounds the urethra and secretes a seminal fluid
k. Spermatogenesis	_____ 11. Contains three columns of erectile tissue
l. Urethra	_____ 12. Removed in a surgical circumcision

PART B: Assessments

1. In the space that follows, sketch a representative area of the organ indicated. Label any of the structures that were observed, and indicate the magnification used for each sketch. **A3**

Testis (_____×)	Epididymis (_____×)
Ductus deferens (_____×)	Penis (_____×)

2. Describe the function of each of the following: **A2**

   a. Scrotum

   _____

   _____

   b. Spermatogenic cell

   _____

   _____

   c. Interstitial cell (Leydig cell)

   _____

   _____

   d. Epididymis

   _____

   _____

   e. Corpora cavernosa and corpus spongiosum

   _____

   _____

NOTES

Female Reproductive System

Purpose of the Exercise

To examine the microscopic and gross anatomy of the female reproductive organs and to describe their basic physiological functions.

MATERIALS NEEDED

Human torso model

Model of the female reproductive system

Anatomical chart of the female reproductive system

Compound light microscope

Prepared microscope slides of the following:

   Ovary section with maturing follicles

   Uterine tube, cross section

   Uterine wall section

For Demonstration Activity:

Prepared microscope slides of the following:

   Uterine wall, early proliferative phase

   Uterine wall, secretory phase

   Uterine wall, early menstrual phase

Learning Outcomes

After completing this exercise, you should be able to:

   01 Locate and identify the organs of the female reproductive system.

   02 Describe the structures and functions of these organs.

   03 Sketch and label the major features of microscopic sections of the ovary, uterine tube, and uterine wall.

The O corresponds to the assessments A indicated in the Laboratory Assessment for this Exercise.

Pre-Lab

Carefully read the introductory material and examine the entire lab. Be familiar with the structures and functions of the female reproductive system from lecture or the textbook. Answer the pre-lab questions.

Pre-Lab Questions Select the correct answer for each of the following questions:

1. Oogenesis originates within the
   a. follicle.   b. uterine tube.
   c. uterus.   d. corpus albicans.

2. Early cleavage division occurs within the
   a. follicle.   b. corpus luteum.
   c. uterine tube.   d. uterus.

3. Which of the following is *not* a structure of the uterus?
   a. cervix   b. fimbriae
   c. endometrium   d. myometrium

4. The _____ are the milk-producing parts of the mammary glands.
   a. areolar glands   b. lactiferous ducts
   c. adipose cells   d. alveolar glands

5. How many openings are located on a human nipple?
   a. 1   b. 2–3
   c. 5–10   d. 15–20

6. Which of the following structures occupies the most anterior position of the external genitalia?
   a. hymen   b. clitoris
   c. external urethral orifice   d. vaginal orifice

7. The broad ligament is the largest ligament involved in holding female reproductive organs in position.
   a. True _____   b. False _____

8. Primordial follicles are concentrated in the medulla region of an ovary.
   a. True _____   b. False _____

The organs of the female reproductive system have a large variety of functions. They are specialized to produce and maintain the female sex cells, to transport these cells to the site of fertilization, to provide a favorable environment for a developing offspring, to move the offspring to the outside, and to produce female sex hormones.

The internal female reproductive organs include the *ovaries,* which produce the egg cells (oocytes) and female sex hormones, the uterine tubes, uterus, and vagina. The external genitalia include the labia majora, labia minora, clitoris, and vestibular glands.

The cortex region of an ovary contains numerous follicles in a variety of developmental stages to produce *ova (eggs)* by the process called *oogenesis (meiosis).* Oogenesis is described in Laboratory Exercise 60. Numerous *primordial follicles* are formed during prenatal development. During the period from puberty to menopause, reproductive cycles occur as individual follicles complete maturation. A *primary follicle* contains a *primary oocyte* that undergoes oogenesis. The primary follicle develops into a *secondary follicle,* and finally a *mature tertiary follicle* shortly before ovulation of a *secondary oocyte.* The ruptured follicle becomes a *corpus luteum,* which secretes estrogens and progesterone, and eventually degenerates to form a *corpus albicans,* composed of connective tissue.

The *uterine tube (oviduct; fallopian tube)* has fingerlike fimbriae that partially envelop the ovary. After ovulation, the secondary oocyte is conveyed through the uterine tube by the current produced by the cilia and peristaltic waves. If fertilization occurs, early cleavage divisions take place during the passage to the *uterus* during the next several days. The uterus has three regions: *fundus, body,* and *cervix.* There are three layers to its wall: *perimetrium, myometrium,* and *endometrium.* Following implantation of an embryo in the endometrium, development continues in the uterus until birth.

The *vagina* has a mucosal lining, and near the vaginal orifice a membrane (hymen) partially closes the orifice unless broken, often during the first intercourse. This is the tube that sperm entered to fertilize the egg, and the fetus, once mature, will exit through the same canal.

The *external genitalia (vulva)* are within a diamond-shaped perineum with boundaries established by the pubis, ischial tuberosities, and coccyx. Larger rounded folds of skin, *labia majora,* protect the *labia minora* and other external genitalia. Between the labia minora is a space (*vestibule*) that encloses the urethral and vaginal openings. The *clitoris* at the anterior vulva has structures similar to those of a penis, including a prepuce, glans, and corpora cavernosa.

The *breasts* contain the mammary glands, which are comprised of many lobes, ducts, ligaments, and adipose tissue. At puberty, estrogens stimulate breast development, and at the birth of a baby, various hormones promote alveolar gland development and milk production (lactation). Lactiferous ducts terminate in 15–20 openings on the nipple in the center of the pigmented areola.

PROCEDURE A: Female Reproductive Organs

1. Study figures 59.1, 59.2, 59.3, and 59.4 of the internal reproductive organs and the list of structures and descriptions.
2. Observe the human torso model, the model of the female reproductive system, and the anatomical chart of the female reproductive system. Locate the following features:

ovaries—paired organs that produce ova and sex hormones
  - medulla—internal region
  - cortex—outer region

ligaments—hold reproductive organs in place
  - broad ligament—largest ligament attaches to uterine tube, ovary, and uterus
  - suspensory ligament of ovary—holds superior end of ovary to pelvic wall
  - ovarian ligament—attaches ovary to uterus
  - round ligament of uterus—band within broad ligament from lateral uterus to the connective tissue of external genitalia

uterine tubes (oviducts; fallopian tubes)—paired tubes about 10 cm (4 inches) long; site of fertilization and early cleavage development
  - fimbriae—slender extensions on end of infundibulum
  - infundibulum—expanded open end near ovary
  - ampulla—middle section; typical site of oocyte fertilization
  - isthmus—narrow portion near uterus

uterus—houses embryonic and fetal development during pregnancy
  - fundus of uterus—wide superior curvature
  - body of uterus—large middle section
  - cervix of uterus—inferior cylindrical end
    • cervical canal—narrow passage within cervix
    • external os—opening into vagina
  - uterine wall—thick and composed of three layers
    • endometrium—inner mucosal layer with simple columnar epithelium and numerous tubular glands
    • myometrium—thick middle layer of smooth muscle; contracts during childbirth
    • perimetrium (serosa; serous coat)—outer layer

rectouterine pouch—inferior recess of peritoneal cavity between rectum and uterus

FIGURE 59.1 AP|R The structures of the female reproductive system.

Uterine tube
Ovary
Rectum
Peritoneum

Fundus of uterus
Body of uterus

Rectouterine pouch
Cervix of uterus

Pubis
Urinary bladder

Urethra
Clitoris
Labium minus
Labium majus

External urethral orifice

Vagina

Anus
Vaginal orifice

FIGURE 59.2 AP|R The female reproductive organs (posterior view). The left-side organs are shown in section and the ligaments are shown on the right side.

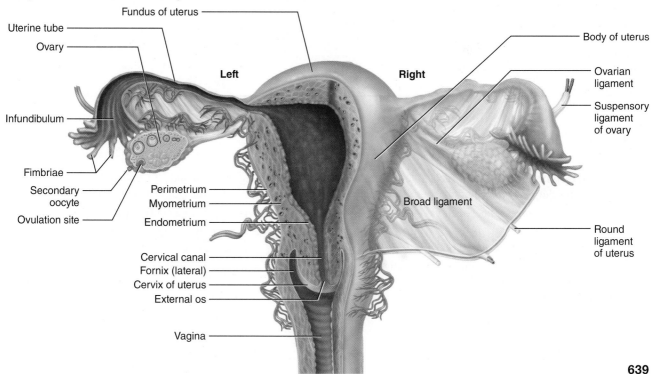

Fundus of uterus

Uterine tube
Ovary

Left   **Right**

Body of uterus

Ovarian
ligament

Suspensory
ligament
of ovary

Infundibulum

Fimbriae

Secondary
oocyte

Ovulation site

Perimetrium
Myometrium
Endometrium

Cervical canal
Fornix (lateral)
Cervix of uterus
External os

Vagina

Broad ligament

Round
ligament
of uterus

FIGURE 59.3 AP|R The female reproductive organs of a cadaver (anterior view).

Suspensory ligament
Fimbriae
Uterine tube
Ovary
Ovarian ligament
Round ligament
Uterus
Vagina

FIGURE 59.4 AP|R The structures associated with the female perineum. The external genitalia (vulva) are within the perineum.

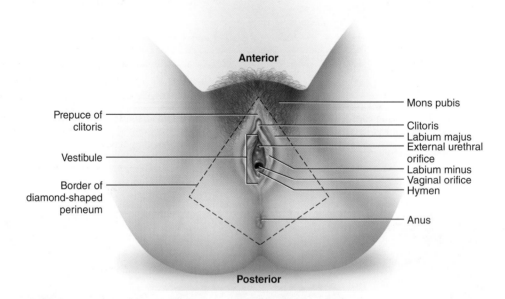

Anterior

Prepuce of clitoris
Vestibule
Border of diamond-shaped perineum

Mons pubis
Clitoris
Labium majus
External urethral orifice
Labium minus
Vaginal orifice
Hymen
Anus

Posterior

vagina—tube about 10 cm (4 inches) long; receives penis during sexual intercourse; birth canal

- fornices—recesses between vaginal wall and cervix
- vaginal orifice—external opening
- hymen—thin fold of mucosa that partially closes vaginal orifice
- mucosal layer—inner vaginal wall of stratified squamous epithelium

3. Study figures 59.4 and 59.5 and the list of structures and descriptions of the external genitalia and the breasts.

external genitalia (vulva; pudendum)

- mons pubis—rounded elevation with adipose tissue anterior to pubic symphysis
- labia majora—folds of skin with hair that enclose and protect structures between them
- labia minora—lesser hair-free folds between labia majora

FIGURE 59.5   AP|R   The structures of a lactating breast (anterior view).

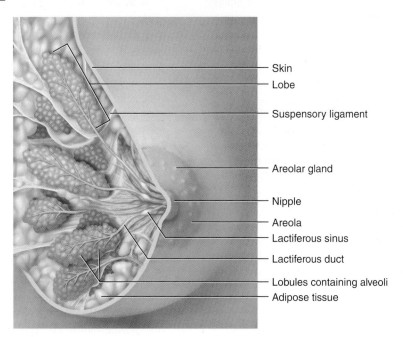

- Skin
- Lobe
- Suspensory ligament
- Areolar gland
- Nipple
- Areola
- Lactiferous sinus
- Lactiferous duct
- Lobules containing alveoli
- Adipose tissue

- vestibule—region between labia minora
- vestibular glands—located on each side of vagina; secrete a lubricating fluid
- clitoris—small projection at anterior end of vestibule; homologous to penis

breasts—paired elevations containing mammary glands

- nipple—protruding region in center of areola containing openings of the lactiferous ducts
- areola—circular pigmented skin around nipple
- alveolar glands—compose the milk-producing parts of mammary glands that develop during pregnancy

- lactiferous ducts—drain the 15–20 lobes of the breast
- adipose tissue—compose much of internal structure of the breast

4. Complete Part A of Laboratory Assessment 59.

PROCEDURE B: Microscopic Anatomy

1. Obtain a microscope slide of an ovary section with maturing follicles, and examine it with low-power magnification (fig. 59.6). Locate the outer layer, or *cortex,* composed of densely packed cells with developing follicles, and the inner layer, or *medulla,* which largely consists of loose connective tissue.

FIGURE 59.6   AP|R   Micrograph of the ovary (80×).

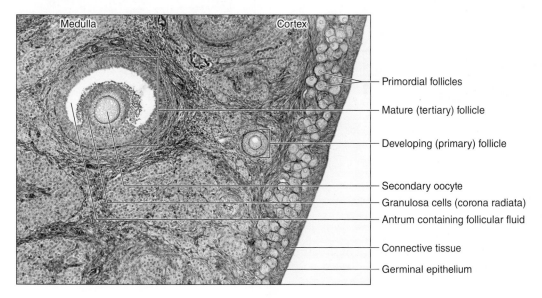

Medulla   Cortex

- Primordial follicles
- Mature (tertiary) follicle
- Developing (primary) follicle
- Secondary oocyte
- Granulosa cells (corona radiata)
- Antrum containing follicular fluid
- Connective tissue
- Germinal epithelium

2. Focus on the cortex of the ovary using high-power magnification (fig. 59.7). Note the thin layer of small cuboidal cells on the free surface. These cells constitute the *germinal epithelium.* Also locate some *primordial follicles* just beneath the germinal epithelium. Each follicle consists of a single, relatively large *primary oocyte* with a prominent nucleus and a layer of simple squamous epithelial cells.

3. Use low-power magnification to search the ovarian cortex for maturing follicles in various stages of development (fig. 59.6). Locate and compare primordial follicles, *primary follicles, secondary follicles,* and *mature (tertiary; vesicular; Graafian) follicles.* Developing follicles have a covering from a single layer to a double layer of folliclular cells. The type of epithelial tissue cells depends upon the stage of development. Mature follicles have a covering of several layers of granulosa cells.

4. Prepare labeled sketches in Part B of the laboratory assessment to illustrate the changes that occur in a follicle as it matures.

5. Obtain a microscope slide of a cross section of a uterine tube. Examine it using low-power magnification. The shape of the lumen is very irregular because of folds of the mucosa layer.

6. Focus on the inner lining of the uterine tube using high-power magnification (fig. 59.8). The lining is composed of *simple columnar epithelium,* and some of the epithelial cells are ciliated on their free surfaces.

7. Prepare a labeled sketch of a representative region of the wall of the uterine tube in Part B of the laboratory assessment.

8. Obtain a microscope slide of the uterine wall section (fig. 59.9). Examine it using low-power magnification, and locate the following:

   endometrium—inner mucosal layer

   myometrium—middle, thick muscular layer

   perimetrium—outer serosal layer

9. Prepare a labeled sketch of a representative section of the uterine wall in Part B of the laboratory assessment.

10. Complete Part C of the laboratory assessment.

FIGURE 59.7 AP|R Micrograph of the ovarian cortex (200×).

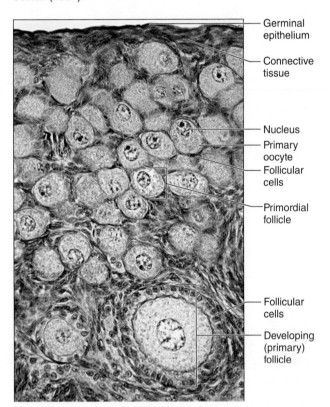

Germinal epithelium

Connective tissue

Nucleus
Primary oocyte
Follicular cells

Primordial follicle

Follicular cells

Developing (primary) follicle

FIGURE 59.8 AP|R Micrograph of a cross section of the uterine tube (800×).

Lumen

Cilia

Cytoplasm

Nucleus

Basement membrane

Connective tissue layer

FIGURE 59.9 AP|R Micrograph of the uterine wall (10×).

Simple columnar epithelium

Lumen

Endometrium

Uterine glands

Myometrium

Perimetrium

DEMONSTRATION ACTIVITY

Observe the slides in the demonstration microscopes. Each slide contains a section of uterine mucosa taken during a different phase in the reproductive cycle. In the *early proliferative phase,* note the simple columnar epithelium on the free surface of the mucosa and the many sections of tubular uterine glands in the tissues beneath the epithelium. In the *secretory phase,* the endometrium is thicker, the uterine glands appear more extensive, and they are coiled. In the *early menstrual phase,* the endometrium is thinner because its surface functional layer has been shed, while retaining the basal layer. Also the uterine glands are less apparent, and the spaces between the glands contain many leukocytes. The basal layer of the endometrium renews the functional layer after menstruation ends.

NOTES

Name _____

Date _____

Section _____

The Ⓐ corresponds to the indicated Learning Outcome(s) Ⓞ found at the beginning of the Laboratory Exercise.

Female Reproductive System

PART A: Assessments

1. Identify the numbered structures in figures 59.10 and 59.11.

FIGURE 59.10 Label the major structures of the female reproductive system in this sagittal view.

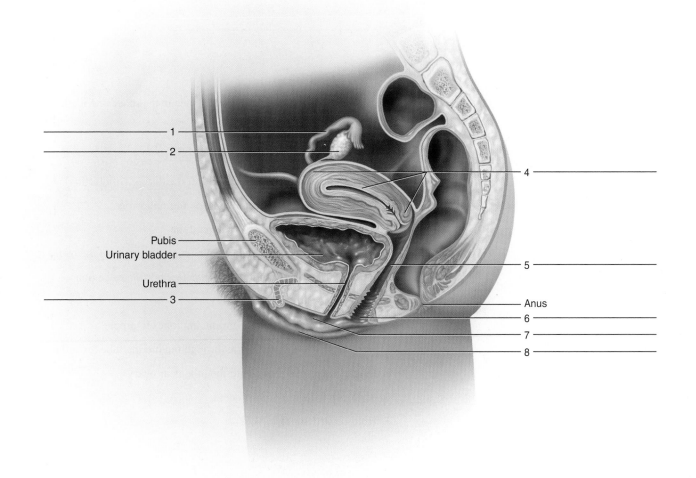

FIGURE 59.11 Label the female reproductive structures in this posterior view.

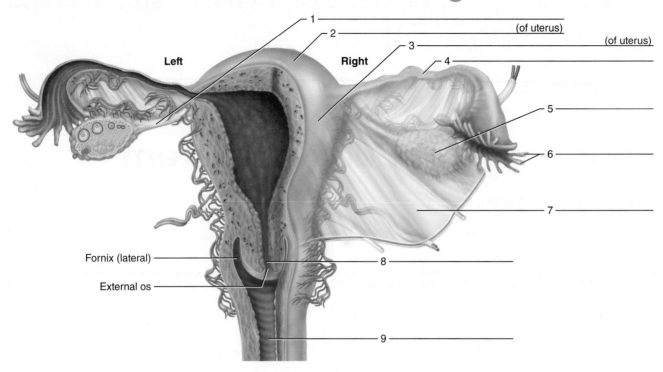

2. Match the terms in column A with the descriptions in column B. Place the letter of your choice in the space provided. **A2**

Column A	Column B
a. Cervix	_____ **1.** Rounded end of uterus near uterine tubes
b. Cilia	_____ **2.** Smooth muscle that contracts with force during childbirth
c. Endometrium	_____ **3.** Thin membrane that partially closes vaginal orifice
d. Fimbriae	_____ **4.** Surface region around external genitalia
e. Fundus	
f. Hymen	_____ **5.** Process by which a secondary oocyte is released from the ovary
g. Lactiferous duct	_____ **6.** Portion of uterus extending into superior portion of vagina
h. Myometrium	_____ **7.** Inner mucosal lining of uterus
i. Ovulation	_____ **8.** Transports milk from milk-producing lobes to nipple
j. Perineum	_____ **9.** Help move secondary oocyte through the uterine tube toward uterus
k. Suspensory ligament	_____ **10.** Fingerlike projections of uterine tube near ovary
l. Vagina	_____ **11.** Birth canal during birth process
	_____ **12.** Holds ovary to pelvic wall

PART B: Assessments

In the space that follows, sketch a representative area of the organ indicated. Label any of the structures observed and indicate the magnification used for each sketch. (A3)

Primordial or primary follicle (_____×)	Secondary or mature follicle (_____×)
Uterine tube (_____×)	Wall of uterus (_____×)

PART C: Assessments

Identify the numbered features in figure 59.12 of a sectioned ovary that represent structures present at some stage during a reproductive cycle.

FIGURE 59.12 Label this ovary by placing the correct numbers in the spaces provided. The arrows indicate development over time; follicles do not migrate through an ovary. **A1**

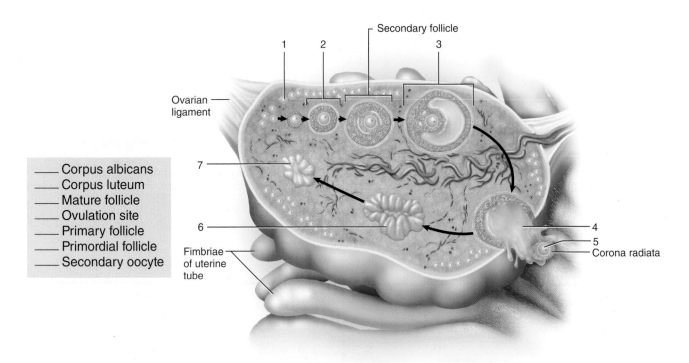

____ Corpus albicans
____ Corpus luteum
____ Mature follicle
____ Ovulation site
____ Primary follicle
____ Primordial follicle
____ Secondary oocyte

Meiosis, Fertilization, and Early Development

Purpose of the Exercise

To describe the process of meiosis and fertilization, to observe sea urchin eggs being fertilized, and to examine various features of the human in early stages of development.

MATERIALS NEEDED

Sea urchin egg suspension*

Sea urchin sperm suspension*

Compound light microscope

Depression microscope slide

Coverslip

Medicine droppers

Models of animal meiosis

Models of stages of human fertilization

Models of human development

Prepared microscope slide of the following:
Sea urchin embryos (cleavage, blastula, and gastrula stages)

For Learning Extension Activity:

Vaseline

Toothpick

For Demonstration Activity:

Preserved mammalian embryos

*See the Instructor's Manual for a source of materials.

Learning Outcomes

After completing this exercise, you should be able to:

O1 Describe meiosis, fertilization, and the early developmental stages of a human.

O2 Identify and sketch the early developmental stages of a sea urchin.

O3 Distinguish the major features of early human development.

The **O** corresponds to the assessments **A** indicated in the Laboratory Assessment for this Exercise.

Pre-Lab

Carefully read the introductory material and examine the entire lab. Be familiar with fertilization and early embryonic development from lecture or the textbook. Answer the pre-lab questions.

Pre-Lab Questions Select the correct answer for each of the following questions:

1. The ovum and sperm unite and form a diploid cell called the
   a. pronucleus.
   b. blastocyst.
   c. morula.
   d. zygote.

2. The _____ is the development stage that implants in the endometrium.
   a. morula
   b. blastocyst
   c. zygote
   d. two-cell stage

3. Which of the following is *not* an extraembryonic membrane?
   a. umbilical cord
   b. amnion
   c. chorion
   d. yolk sac

4. The placenta is comprised of
   a. only the myometrium.
   b. only the decidua basalis.
   c. only the chorionic villi.
   d. combinations of the decidua basalis and chorionic villi.

5. The embryonic stage of human development exists
   a. from fertilization until implantation.
   b. through the first four weeks of development.
   c. through the first eight weeks of development.
   d. from fertilization until birth.

6. The zona pellucida gel layer surrounds the secondary oocyte and early development stages through the morula.
   a. True _____
   b. False _____

7. A single completed oogenesis sequence produces four functional ova (eggs).
   a. True _____   b. False _____

8. A single completed spermatogenesis sequence produces four functional sperm.
   a. True _____   b. False _____

Meiosis is a distinctive process of nuclear division in the testes and ovaries that occurs only during *gametogenesis* and produces sex cells called *gametes.* In males the process is more specifically called *spermatogenesis,* and it produces gametes known as *sperm.* In females the process is more specifically called *oogenesis,* and it produces gametes known as *secondary oocytes.* Meiosis is a double division that includes *meiosis I (first meiotic division* or *reduction division)* followed by *meiosis II (second meiotic division)*, and during this entire process, a single *diploid (2n)* cell is reduced to four *haploid (n)* cells. In males the process of spermatogenesis produces four functional haploid sperm from each completed sequence. In females, oogenesis produces a secondary oocyte (sometimes referred to as the ovum or egg) along with a small *polar body* from each completed sequence of meiosis I. Oogenesis begins prior to puberty and continues through menopause. Meiosis II will occur if a sperm cell penetrates the secondary oocyte. (If the secondary oocyte is not fertilized, the second meiotic division does not happen.)

The secondary oocyte undergoes meiosis II only if sperm penetration occurs, producing a large fertilized ovum and a second tiny polar body that degenerates. The first small polar body from meiosis I will either degenerate or undergo the second meiotic division, resulting in two degenerating polar bodies. Polar bodies degenerate and die because they lack cytoplasmic nutrients. Ovulation takes place when a secondary oocyte, surrounded by the *corona radiata* (granulosa cells) and by a gel layer called the *zona pellucida,* is expelled from the ovary into the uterine tube. A sperm must penetrate the zona pellucida in order to enter the secondary oocyte. Upon fertilization, the pronuclei of the ovum and sperm come together and combine their chromosomes, forming a single diploid cell called the *zygote.*

Recall that mitosis was described as part of the cell cycle in Laboratory Exercise 7. The mitosis phase of the cell cycle included events during interphase, prophase, metaphase, anaphase, telophase, and cytokinesis. There are many similarities between the various phases of mitosis and meiosis. However, meiosis differs from mitosis in some significant ways, including:

- Meiosis has two consecutive nuclear divisions: meiosis I followed by meiosis II. Mitosis is a single division.
- Meiosis results in four genetically different daughter cells; mitosis produces two genetically identical daughter cells.
- Meiosis produces haploid cells; mitosis produces diploid cells.

- Meiosis occasionally includes exchanges of genetic material between homologous chromosomes, called *crossing-over,* which increase genetic variability. Crossing-over does not occur during mitosis.

Shortly after fertilization, the zygote undergoes cell division (mitosis and cytokinesis) to form two cells. These two cells become four, then in turn divide into eight, and so forth. A solid cluster of cells forms, the *morula,* which is still surrounded by the zona pellucida. The cells of the morula continue cell cycles, and the hollow *blastocyst* develops. The blastocyst contains an *inner cell mass,* which develops into the embryo, and an outer *trophoblast* layer, which develops into the extraembryonic membranes and a portion of the *placenta.* These early *cleavage* divisions, resulting in an increasing number of cells that are smaller than the zygote, continue for most of the first week of development. The blastocyst implants into the endometrium about 6–7 days after fertilization. The early stages of development in some animals, such as sea urchins, are similar to those in humans.

During the *embryonic stage* of development, the blastocyst develops into a *gastrula* containing three primary germ layers: the *ectoderm, endoderm,* and *mesoderm,* which give rise to particular tissues and body systems. The extraembryonic membranes include the amnion, yolk sac, allantois, and chorion. The *amnion* extends around the embryo and is filled with amniotic fluid, providing a protective cushion and constant environment. The *yolk sac* forms early blood cells and a portion of the embryonic digestive tube; however, the placental region, not the yolk, provides the nutritive functions. The *allantois* gives rise to the umbilical blood vessels and forms part of the urinary bladder. The *chorion* extends around the embryo and the other membranes, and it eventually fuses with the amnion to form the amniochorionic membrane. In human development, the embryonic stage lasts through the first eight weeks, followed by the *fetal stage* until birth.

The *placenta* is comprised of maternal and fetal components. The maternal component is a region of the *endometrium (decidua basalis);* the fetal portion contains an elaborate series of *chorionic villi* formed from the chorion. An examination of the placenta upon birth (the afterbirth) reveals a rough surface of the maternal portion and a smooth surface with the attached umbilical cord of the fetal part.

PROCEDURE A: Meiosis and Fertilization

1. Review the introductory material related to meiosis and study figure 60.1. Examine the models of animal meiosis and place them in the proper sequence of stages of meiosis. Observe that there are phases during meiosis I and meiosis II that seem similar to some of the events of phases during mitosis. Note that replicated chromosomes are present in prophase I of meiosis, with sister

FIGURE 60.1 AP|R Meiosis phases illustrated for a cell with a diploid (2n) = 4 (2 pairs of homologous chromosomes). Human cells have a diploid (2n) = 46 (23 pairs of homologous chromosomes).

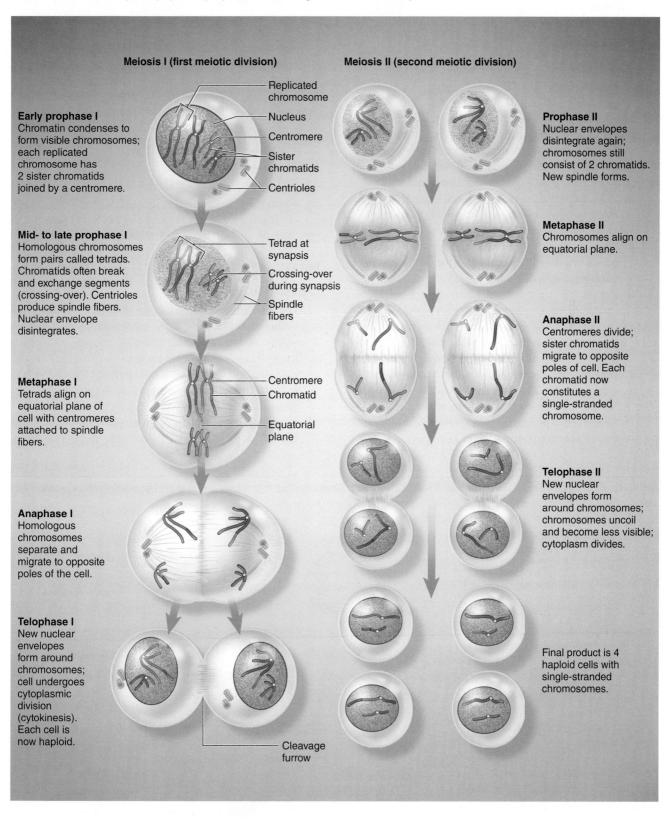

Meiosis I (first meiotic division)

Early prophase I
Chromatin condenses to form visible chromosomes; each replicated chromosome has 2 sister chromatids joined by a centromere.

- Replicated chromosome
- Nucleus
- Centromere
- Sister chromatids
- Centrioles

Mid- to late prophase I
Homologous chromosomes form pairs called tetrads. Chromatids often break and exchange segments (crossing-over). Centrioles produce spindle fibers. Nuclear envelope disintegrates.

- Tetrad at synapsis
- Crossing-over during synapsis
- Spindle fibers

Metaphase I
Tetrads align on equatorial plane of cell with centromeres attached to spindle fibers.

- Centromere
- Chromatid
- Equatorial plane

Anaphase I
Homologous chromosomes separate and migrate to opposite poles of the cell.

Telophase I
New nuclear envelopes form around chromosomes; cell undergoes cytoplasmic division (cytokinesis). Each cell is now haploid.

- Cleavage furrow

Meiosis II (second meiotic division)

Prophase II
Nuclear envelopes disintegrate again; chromosomes still consist of 2 chromatids. New spindle forms.

Metaphase II
Chromosomes align on equatorial plane.

Anaphase II
Centromeres divide; sister chromatids migrate to opposite poles of cell. Each chromatid now constitutes a single-stranded chromosome.

Telophase II
New nuclear envelopes form around chromosomes; chromosomes uncoil and become less visible; cytoplasm divides.

Final product is 4 haploid cells with single-stranded chromosomes.

chromatids joined by a common centromere. Also during prophase I, a *tetrad* of pairs of homologous chromosomes forms during a stage called *synapsis*. It is during this tetrad stage that portions of chromatids may be exchanged. This is called crossing-over, and it results in greater variations of gene combinations. The two cells produced during meiosis I are haploid as the homologous replicated chromosomes separate. DNA replication does not occur during meiosis II. Sister chromatids separate and migrate to opposite poles during anaphase II. The cells that result from meiosis II are haploid. AP|R

2. Review the introductory material related to fertilization and early embryonic development. Study figure 60.2 and models showing various stages of human fertilization. Answer the questions in Part A of Laboratory Assessment 60.

3. Although it is difficult to observe fertilization in mammals since the process occurs internally, it is possible to view forms of external fertilization. For example, egg and sperm cells can be collected from sea urchins, and the process of fertilization can be observed microscopically. To make this observation, follow these steps:

   a. Place a drop of sea urchin egg-cell suspension in the chamber of a depression slide and add a coverslip.

   b. Examine the egg cells using low-power magnification.

   c. Focus on a single unfertilized egg cell with high-power magnification and sketch the cell in Part A of the laboratory assessment.

   d. Remove the coverslip and add a drop of sea urchin sperm-cell suspension to the depression slide. Replace the coverslip and observe the sperm cells with high-power magnification as they cluster around the egg cells. This attraction is stimulated by gamete secretions.

   e. Observe the egg cells with low-power magnification once again and watch for the appearance of *fertilization membranes*. Such a membrane forms as soon as an egg cell is penetrated by a sperm cell; it looks like a clear halo surrounding the egg cell.

   f. Focus on a single fertilized egg cell and sketch it in Part A of the laboratory assessment.

PROCEDURE B: Sea Urchin Early Development

1. Obtain a prepared microscope slide of developing sea urchin embryos. This slide contains embryos in various stages of cleavage. Search the slide using low-power magnification, and locate embryos in two-, four-, eight-cell, and morula stages. Observe that cleavage results in an increase of cell numbers; however, the cells get progressively smaller.

2. Using low-power magnification, locate the blastula and identify its blastocoel (hollow central cavity). Then,

---

FIGURE 60.2 AP|R   Fertilization and early embryonic development in the uterine tube and implantation of blastocyst stage in endometrium of uterus.

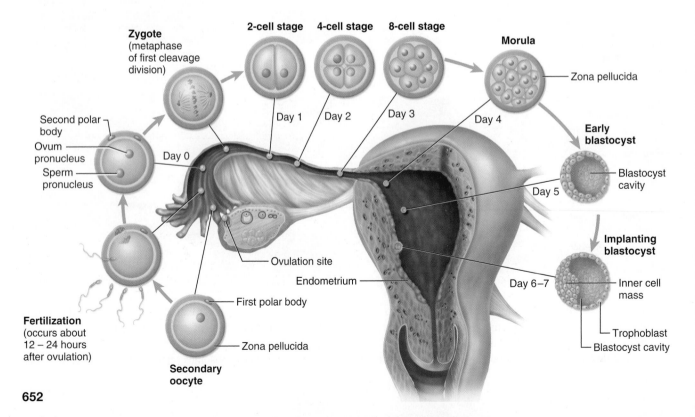

652

locate the gastrula, and identify the ectoderm (outermost germ layer) and endoderm (innermost germ layer).

3. Prepare a sketch of each stage in Part B of the laboratory assessment.

PROCEDURE C: Human Early Development

1. Study figures 60.3 and 60.4 and the list of structures and descriptions.
2. Observe the models of human development and identify the following features:

blastocyst—implantation stage of embryo into the endometrium

- blastocyst cavity (blastocoel)—hollow part
- inner cell mass (embryoblast)—develops into the actual embryo
- trophoblast—develops into the extraembryonic membranes and part of the placenta

primary germ layers—primitive tissues that develop into all organs

- ectoderm—outer layer; develops into epidermal structures and the nervous system
- endoderm—inner layer; gives rise to linings of digestive, respiratory, and urinary organs
- mesoderm—middle layer; develops into bones, muscles, dermis, and connective tissues

chorion—forms fetal portion of placenta

chorionic villi—branched projections of chorion

amnion—membrane that encloses embryo

amniotic fluid—protective fluid between amnion and embryo

yolk sac—produces early blood cells and gonadal tissues of embryo

allantois—forms part of umbilical cord

umbilical cord—suspends embryo in amniotic fluid; attached to the placenta

- umbilical arteries (2)—transport oxygen-poor blood of fetus to maternal portion of placenta
- umbilical vein (1)—transports oxygen-rich blood from maternal portion of placenta to embryo

placenta—site of exchange of nutrients, gases, and wastes between embryonic and maternal blood

3. Complete Part C of the laboratory assessment.

FIGURE 60.3 Structures associated with this 12-week-old fetus: (a) fetus, membranes, and uterus; (b) fetal and maternal vasculature of placenta.

FIGURE 60.4 Human placenta (afterbirth): (a) fetal surface; (b) maternal (uterine) surface. At full term, the placenta is about 18 cm in diameter, 4 cm thick, and weighs about 450 g (one pound).

Umbilical cord

Smooth fetal surface

Amniotic sac

(a)

(b)

DEMONSTRATION ACTIVITY

Observe the preserved mammalian embryos that are on display. In addition to observing the developing external body structures, identify such features as the chorion, chorionic villi, amnion, yolk sac, umbilical cord, and placenta. What special features provide clues to the types of mammals these embryos represent?

_____

_____

_____

Name _____

Date _____

Section _____

The Ⓐ corresponds to the indicated Learning Outcome(s) Ⓞ found at the beginning of the Laboratory Exercise.

LABORATORY ASSESSMENT

60

Meiosis, Fertilization, and Early Development

PART A: Assessments

Complete the following statements:

1. Human gametes contain _____ chromosomes. **A1**

2. The diploid number of human chromosomes found in human body cells is _____. **A1**

3. At the end of meiosis, _____ haploid daughter cells are produced. **A1**

4. To create genetic variation, _____ occasionally occurs during synapsis with homologous chromosomes that are arranged in a tetrad. **A1**

5. The cell resulting from fertilization is called a(n) _____. **A1**

6. Tiny cells that form during meiotic divisions of oogenesis and then degenerate are called _____. **A1**

7. Fertilization and early embryonic development occur within the lumen of the _____. **A1**

8. A solid ball of about sixteen cells is called a _____. **A1**

9. The part of meiosis that is most similar to mitosis is _____. **A1**

10. Meiosis I reduces the chromosome number in half, so it is also called _____ division. **A1**

11. Implantation of the _____ into the endometrium occurs around day 6–7 after fertilization. **A1**

12. The two-cell, four-cell, eight-cell, and morula stages of early embryonic development represent the _____ phase, where cells increase in number but are smaller than the zygote. **A1**

Prepare sketches of the following: **A2**

Single unfertilized sea urchin egg (_____×)	Single fertilized sea urchin egg (_____×)

PART B: Assessments

Prepare sketches of the following:

Two-cell sea urchin embryo (_____×)

Four-cell sea urchin embryo (_____×)

Eight-cell sea urchin embryo (_____×)

Morula sea urchin embryo (_____×)

Blastula of sea urchin (_____×)

Gastrula of sea urchin (_____×)

PART C: Assessments

Match the terms in column A with the descriptions in column B. Place the letter of your choice in the space provided. (A3)

Column A	Column B
a. Allantois	_____ **1.** Gel layer around oocyte and early cleavage stages
b. Amniotic fluid	_____ **2.** Hollow ball of cells
c. Blastocyst	_____ **3.** Develops into extraembryonic membranes
d. Chorionic villi	_____ **4.** Site of exchange of nutrients, gases, and wastes between the embryonic and maternal blood
e. Endometrium	
f. Fetal stage	_____ **5.** Forms fetal portion of placenta
g. Gastrula	_____ **6.** Forms maternal portion of placenta
h. Inner cell mass	_____ **7.** Forms the umbilical cord blood vessels
i. Placenta	_____ **8.** Contains the ectoderm, endoderm, and mesoderm
j. Trophoblast	_____ **9.** Fluid that protects the embryo from jarring movements and provides a watery environment for development
k. Zona pellucida	
	_____ **10.** Part of the blastocyst that develops into the embryo
	_____ **11.** Stage of development that begins at the end of week 8

12. Summarize your observations of any similarities noted between sea urchin and human early development stages. (A2) (A3)

_____

_____

_____

_____

NOTES

LABORATORY EXERCISE 61

Genetics

Purpose of the Exercise

To observe some selected human traits, to use pennies and dice to demonstrate laws of probability, and to solve some genetic problems using a Punnett square.

MATERIALS NEEDED

Pennies (or other coins)

Dice

PTC paper

Astigmatism chart

Ichikawa's or Ishihara's color plates for color blindness test

Many examples of human genetics in this laboratory exercise are basic, external features to observe. Some traits are based upon simple Mendelian genetics. As our knowledge of genetics continues to develop, traits such as tongue roller and free earlobe may no longer be considered examples of simple Mendelian genetics. We may need to continue to abandon some simple Mendelian models. New evidence may involve polygenic inheritance, effects of other genes, or environmental factors as more appropriate explanations. Therefore, the analysis of family genetics is not always feasible or meaningful.

Learning Outcomes

After completing this exercise, you should be able to:

01 Examine and record twelve genotypes and phenotypes of selected human traits.

02 Demonstrate the laws of probability using tossed pennies and dice and interpret the results.

03 Predict genotypes and phenotypes of complete dominance, codominance, and sex-linked problems using Punnett squares.

The **O** corresponds to the assessments **A** indicated in the Laboratory Assessment for this Exercise.

Pre-Lab

Carefully read the introductory material and examine the entire lab. Be familiar with genetic terminology and genetic problems from lecture or the textbook. Answer the pre-lab questions.

Pre-Lab Questions Select the correct answer for each of the following questions:

1. The term _____ is used for a person possessing two identical alleles.
   a. homozygous **b.** heterozygous
   c. genotype **d.** phenotype

2. The term _____ is used for the appearance of the person as a result of the way the genes are expressed.
   a. homozygous **b.** heterozygous
   c. genotype **d.** phenotype

3. The _____ is a genetic tool to simulate all possible genetic combinations of offspring.
   a. genome **b.** Punnett square
   c. gamete formation **d.** gene pool

4. The allele is termed _____ if one allele determines the phenotype.
   a. homozygous **b.** heterozygous
   c. dominant **d.** recessive

5. Which of the following represents a dominant human trait?
   a. straight hairline **b.** attached earlobe
   c. dimples **d.** blood type O

6. An attached earlobe is considered a recessive human trait.
   a. True _____ **b.** False _____

7. A diploid chromosome number of 46 and a haploid number of 23 are normal for humans.
   a. True _____ **b.** False _____

8. The possession of freckles is considered a recessive human trait.
   a. True _____ **b.** False _____

Genetics is the study of the inheritance of characteristics. The *genes* that transmit this information are coded in segments of DNA in chromosomes. Homologous chromosomes possess the same gene at the same *locus*. These genes may exist in variant forms, called *alleles*. If a person possesses two identical alleles, the person is said to be *homozygous* for that particular trait. If a person possesses two different alleles, the person is said to be *heterozygous* for that particular trait. The particular combination of these gene variants (alleles) represents the person's *genotype;* the appearance or physical manifestation of the individual that develops as a result of the way the genes are expressed, represents the person's *phenotype.*

If one allele determines the phenotype by masking the expression of the other allele in a heterozygous individual, the allele is termed *dominant.* The allele whose expression is masked is termed *recessive.* A heterozygous person is known as a *carrier* when a recessive gene for a genetic disease "skips" generations and the condition reappears. If the heterozygous condition determines an intermediate phenotype, the inheritance represents *incomplete dominance.*

If more than two allelic forms of a gene exist in the gene pool, it represents an example of *multiple alleles.* Three alleles exist related to the inheritance of the human ABO blood types. However, different alleles are *codominant* if both are expressed in the heterozygous condition.

Some characteristics inherited on the sex chromosomes result in phenotype frequencies that might be more prevalent in males or females. Such characteristics are called *sex-linked (X-linked or Y-linked) characteristics.*

As a result of meiosis during the formation of ova (eggs) and sperm, a mother and father each transmit an equal number of chromosomes (the haploid number 23) to form the zygote (diploid number 46). An offspring will receive one allele from each parent. These gametes combine randomly in the formation of each offspring. Hence, the *laws of probability* can be used to predict possible genotypes and phenotypes of offspring. A genetic tool called a *Punnett square* simulates all possible combinations (probabilities) in offspring genotypes and resulting phenotypes.

PROCEDURE A: Human Genotypes and Phenotypes

A complete set of genetic instructions in one human cell constitutes one's *genome.* The human genome contains about 2.9 billion base pairs representing approximately 20,500 protein-encoding genes. These instructions represent our genotypes and are expressed as phenotypes sometimes clearly observable on our bodies. Some of these traits are listed in table 61.1 and are discernible in figure 61.1.

A dominant trait might be homozygous or heterozygous; only one capital letter is used along with a blank for the possible second dominant or recessive allele. For a recessive trait, two lowercase letters represent the homozygous recessive genotype for that characteristic. Dominant does not always correlate with the predominance of the allele in the gene pool; dominant means one allele will determine the appearance of the phenotype.

1. **Tongue roller/nonroller:** The dominant allele (*R*) determines the person's ability to roll the tongue into a U-shaped trough. The homozygous recessive condition (*rr*) prevents this tongue rolling (fig. 61.1). Record your results in the table in Part A of Laboratory Assessment 61.
2. **Freckles/no freckles:** The dominant allele (*F*) determines the appearance of freckles. The homozygous recessive condition (*ff*) does not produce freckles (fig. 61.1). Record your results in the table in Part A of the laboratory assessment.

Table 61.1 Examples of Some Common Human Phenotypes

Dominant Traits and Genotypes	Recessive Traits and Genotypes
Tongue roller (*R__*)	Nonroller (*rr*)
Freckles (*F__*)	No freckles (*ff*)
Widow's peak (*W__*)	Straight hairline (*ww*)
Dimples (*D__*)	No dimples (*dd*)
Free earlobe (*E__*)	Attached earlobe (*ee*)
Normal skin coloration (*M__*)	Albinism (*mm*)
Astigmatism (*A__*)	Normal vision (*aa*)
Curly hair (*C__*)	Straight hair (*cc*)
PTC taster (*T__*)	Nontaster (*tt*)
Blood type A (I^A __), B (I^B __), or AB ($I^A I^B$)	Blood type O (*ii*)
Normal color vision ($X^C X^C$), ($X^C X^c$), or ($X^C Y$)	Red-green color blindness ($X^c X^c$) or ($X^c Y$)

FIGURE 61.1 Representative genetic traits comparing dominant and recessive phenotypes: (a) tongue roller; (b) nonroller; (c) freckles; (d) no freckles; (e) widow's peak; (f) straight hairline; (g) dimples; (h) no dimples; (i) free earlobe; (j) attached earlobe.

Dominant Traits

Recessive Traits

(a) Tongue roller

(b) Nonroller

(c) Freckles

(d) No freckles

(e) Widow's peak

(f) Straight hairline

FIGURE 61.1   *Continued.*

Dominant Traits   **Recessive Traits**

(g) Dimples

(h) No dimples

(i) Free earlobe

(j) Attached earlobe

3. **Widow's peak/straight hairline:** The dominant allele (*W*) determines the appearance of a hairline above the forehead that has a distinct downward point in the center called a widow's peak. The homozygous recessive condition (*ww*) produces a straight hairline (fig. 61.1). A receding hairline would prevent this phenotype determination. Record your results in the table in Part A of the laboratory assessment.

4. **Dimples/no dimples:** The dominant allele (*D*) determines the appearance of a distinct dimple in one or both cheeks upon smiling. The homozygous recessive condition (*dd*) results in the absence of dimples (fig. 61.1). Record your results in the table in Part A of the laboratory assessment.

5. **Free earlobe/attached earlobe:** The dominant allele (*E*) codes for the appearance of an inferior earlobe that hangs freely below the attachment to the head. The homozygous recessive condition (*ee*) determines the earlobe attaching directly to the head at its inferior border (fig. 61.1). Record your results in the table in Part A of the laboratory assessment.

6. **Normal skin coloration/albinism:** The dominant allele (*M*) determines the production of some melanin, producing normal skin coloration. The homozygous recessive condition (*mm*) determines albinism due to the inability to produce or use the enzyme tyrosinase in pigment cells. An albino does not produce melanin in the skin, hair, or the middle tunic (choroid, ciliary body, and

iris) of the eye. The absence of melanin in the middle tunic allows the pupil to appear slightly red to nearly black. Remember that the pupil is an opening in the iris filled with transparent aqueous humor. An albino human has pale white skin, flax-white hair, and a pale blue iris. Record your results in the table in Part A of the laboratory report.

7. **Astigmatism/normal vision:** The dominant allele (*A*) results in an abnormal curvature to the cornea or the lens. As a consequence, some portions of the image projected on the retina are sharply focused, and other portions are blurred. The homozygous recessive condition (*aa*) generates normal cornea and lens shapes and normal vision. Use the astigmatism chart and directions to assess this possible defect described in Laboratory Exercise 36. Other eye defects, such as nearsightedness (myopia) and farsightedness (hyperopia), are different genetic traits due to genes at other locations. Record your results in the table in Part A of the laboratory assessment.

8. **Curly hair/straight hair:** The dominant allele (*C*) determines the appearance of curly hair. Curly hair is somewhat flattened in cross section, as the hair follicle of a similar shape served as a mold for the root of the hair during its formation. The homozygous recessive condition (*cc*) produces straight hair. Straight hair is nearly round in cross section from being molded into this shape in the hair follicle. In some populations (Caucasians) the heterozygous condition (*Cc*) expresses the intermediate wavy hair phenotype (incomplete dominance). This trait determination assumes no permanents or hair straightening procedures have been performed. Such hair alterations do not change the hair follicle shape, and future hair growth results in original genetic hair conditions. Record your unaltered hair appearance in the table in Part A of the laboratory assessment.

9. **PTC taster/nontaster:** The dominant allele (*T*) determines the ability to experience a bitter sensation when PTC paper is placed on the tongue. About 70% of people possess this dominant gene. The homozygous recessive condition (*tt*) makes a person unable to notice the substance. Place a piece of PTC (phenylthiocarbamide) paper on the upper tongue surface and chew it slightly to see if you notice a bitter sensation from this harmless chemical. The nontaster of the PTC paper does not detect any taste at all from this substance. Record your results in the table in Part A of the laboratory assessment.

10. **Blood type A, B, or AB/blood type O:** Blood type inheritance is an example of multiple-allele inheritance with codominant alleles. There are three alleles (I^A, I^B, and *i*) in the human population affecting RBC membrane structure. These alleles are located on a single pair of homologous chromosomes, so a person possesses either two of the three alleles or two of the same allele. All of the possible combinations of these alleles of genotypes and the resulting phenotypes are depicted in table 61.2. The expression of the blood type AB is a result of both

Table 61.2 Genotypes and Phenotypes (Blood Types)

Genotypes	Phenotypes (Blood Types)
$I^A I^A$ or $I^A i$	A
$I^B I^B$ or $I^B i$	B
$I^A I^B$ (codominant)	AB
ii	O

codominant alleles located in the same individual. Possibly you have already determined your blood type in Laboratory Exercise 43 or have it recorded on a blood donor card. (If simulated blood-typing kits were used for Laboratory Exercise 43, those results would not be valid for your genetic factors.) Record your results in the table in Part A of the laboratory assessment.

11. **Sex determination:** A person with sex chromosomes XX displays a female phenotype. A person with sex chromosomes XY displays a male phenotype. Record your results in the table in Part A of the laboratory assessment.

12. **Normal color vision/red-green color blindness:** This condition is a sex-linked (X-linked) characteristic. The alleles for color vision are also on the X chromosome, but absent on the Y chromosome. As a result, a female might possess both alleles (*C* and *c*), one on each of the X chromosomes. The dominant allele (*C*) determines normal color vision; the homozygous recessive condition (*cc*) results in red-green color blindness. However, a male would possess only one of the two alleles for color vision because there is only a single X chromosome in a male. Hence, a male with even a single recessive gene for color blindness possesses the defect. Note all the possible genotypes and phenotypes for this condition (table 61.1). Review the color vision test in Laboratory Exercise 36 using the color plate in figure 36.6 and color plates in Ichikawa's or Ishihara's book. Record your results in the table in Part A of the laboratory assessment.

13. Complete Part A of the laboratory assessment.

PROCEDURE B: Laws of Probability

The laws of probability provide a mathematical way to determine the likelihood of events occurring by chance. This prediction is often expressed as a ratio of the number of results from experimental events to the number of results considered possible. For example, when tossing a coin there is an equal chance of the results displaying heads or tails. Hence the probability is one-half of obtaining either a heads or a tails (there are two possibilities for each toss). When all of the probabilities of all possible outcomes are considered for the result, they will always add up to 1. To predict the probability of two or more events occurring in succession, multiply the probabilities of each individual event. For example, the probability of tossing a die and displaying a 4 two times in a row is 1/6 × 1/6 = 1/36 (there are six possibilities for each toss). Each toss in a sequence is an *independent event* (chance has no memory). The same laws apply when parents have multiple children (each fertilization is an independent event). Perform the following experiments to demonstrate the laws of probability:

1. Use a single penny (or other coin) and toss it 20 times. Predict the number of heads and tails that would occur from the 20 tosses. Record your prediction and the actual results observed in Part B of the laboratory assessment.

2. Use a single die (*pl.* dice) and toss it 24 times. Predict the number of times a number below 3 (numbers 1 and 2) would occur from the 24 tosses. Record your prediction and the actual results in Part B of the laboratory assessment.

3. Use two pennies and toss them simultaneously 32 times. Predict the number of times two heads, a heads and a tails, and two tails would occur. Record your prediction and the actual results in Part B of the laboratory assessment.

4. Use a pair of dice and toss them simultaneously 32 times. Predict the number of times for both dice coming up with odd numbers, one die an odd and the other an even number, and both dice coming up with even numbers. Record your prediction and the actual results in Part B of the laboratory assessment.

5. Obtain class totals for all of the coins and dice tossed by adding your individual results to a class tally location, as on the blackboard.

6. A Punnett square can be used to visually demonstrate the probable results for two pennies tossed simultaneously. For the purpose of a genetic comparison, an *h* (heads) will represent one "allele" on the coin; a *t* (tails) will represent a different "allele" on the coin.

Coin #1 Possibilities

7. Complete Part B of the laboratory assessment.

PROCEDURE C: Genetic Problems

1. A Punnett square can be constructed to demonstrate a visual display of the predicted offspring from parents with known genotypes. Recall that, in complete dominance, a dominant allele is expressed in the phenotype, as it can mask the other recessive allele on the homologous chromosome pair. During meiosis, the homologous chromosomes with their alleles separate (Mendel's Law of Segregation) into different gametes. An example of such a cross might be a homozygous dominant mother for dimples (*DD*) who has offspring with a father homozygous recessive (*dd*) for the same trait. The results of such

a cross, according to the laws of probability, would be represented by the following Punnett square:

Results: Genotypes: 100% *Dd* (all heterozygous)

Phenotypes: 100% dimples

In another example, assume that both parents are heterozygous (*Dd*) for dimples. The results of such a cross, according to the laws of probability, would be represented by the following Punnett square:

Results: Genotypes: 25% *DD* (homozygous dominant); 50% *Dd* (heterozygous); 25% *dd* (homozygous recessive) (1:2:1 genotypic ratio)

Phenotypes: 75% dimples; 25% no dimples (3:1 phenotypic ratio)

2. Work genetic problems 1 and 2 in Part C of the laboratory assessment.

3. The ABO blood type inheritance represents an example of multiple alleles and codominance. Review table 61.2 for the genotypes and phenotypes for the expression of this trait. A Punnett square can be constructed to predict the offspring of parents of known genotypes. In this example, assume the genotype of the mother is $I^A I^B$, and the father is *ii*. The results of such a cross would be represented by the following Punnett square:

Results: Genotypes: 50% $I^A i$ (heterozygous for A); 50% $I^B i$ (heterozygous for B) (1:1 genotypic ratio)

Phenotypes: 50% blood type A; 50% blood type B (1:1 phenotypic ratio)

Note: In this cross, all of the children would have blood types unlike either parent.

4. Work genetic problems 3 and 4 in Part C of the laboratory assessment.

5. Review the inheritance of red-green color blindness, an X-linked characteristic, in table 61.1. A Punnett square

can be constructed to predict the offspring of parents of known genotypes. In this example, assume the genotype of the mother is heterozygous $X^C X^c$ (normal color vision but a carrier for the color blindness defect), and the father is $X^C Y$ (normal color vision; no allele on the Y chromosome). The results of such a cross would be represented by the following Punnett square:

Female Gametes

		X^C	X^c	
Male Gametes	X^C	$X^C X^C$	$X^C X^c$	Possible Genotypes of Offspring (in boxes)
	Y	$X^C Y$	$X^c Y$	

Results: Genotypes: 25% $X^C X^C$; 25% $X^C X^c$; 25% $X^C Y$; 25% $X^c Y$ (1:1:1:1 genotypic ratio)

Phenotypes for sex determination: 50% females; 50% males (1:1 phenotypic ratio)

Phenotypes for color vision: Females 100% normal color vision (however, 50% are heterozygous carriers for color blindness). Males 50% normal; 50% with red-green color blindness (1:1 phenotypic ratio)

Phenotypes for sex determination and color vision combined: 50% normal females; 25% normal males; 25% males with color blindness (2:1:1 phenotypic ratio)

Note: In X-linked inheritance, males with color blindness received the recessive gene from their mothers.

6. Complete Part C of the laboratory assessment.

NOTES

Name _____

Date _____

Section _____

The (A) corresponds to the indicated Learning Outcome(s) (O) found at the beginning of the Laboratory Exercise.

Genetics

PART A: Assessments

1. Enter your test results for genotypes and phenotypes in the table. Circle your phenotype and genotype for each of the twelve traits. **A1**

Trait	Dominant Phenotype	Genotype	Recessive Phenotype	Genotype
Tongue movement	Roller	$R\_\_$	Nonroller	rr
Freckles	Freckles	$F\_\_$	No freckles	ff
Hairline	Widow's peak	$W\_\_$	Straight	ww
Dimples	Dimples	$D\_\_$	No dimples	dd
Earlobe	Free	$E\_\_$	Attached	ee
Skin coloration	Normal (some melanin)	$M\_\_$	Albinism	mm
Vision	Astigmatism	$A\_\_$	Normal	aa
Hair shape*	Curly	$C\_\_$	Straight	cc
Taste	PTC taster	$T\_\_$	Nontaster for PTC	tt
Blood type	A, B, or AB	$I^A\_\_$; $I^B\_\_$; or I^AI^B	O	ii
Sex		XX or XY		
Color vision	Normal	$X^CX\_\_$ or X^CY	Red-green color blindness	X^CX^c or X^cY

*In some populations (Caucasians) the heterozygous condition (Cc) results in the appearance of wavy hair, which actually represents an example of incomplete dominance for this trait.

2. Choose at least three dominant phenotypes that you circled, and analyze the genotypes for those traits. If it is feasible to observe your biological parents and siblings for any of these traits, are you able to determine if any of your dominant genotypes are homozygous dominant or heterozygous? _____ If so, which ones? _____ Explain the rationale for your response.

_____

_____

_____

_____

_____

_____

PART B: Assessments

1. Single penny tossed 20 times and counting heads and tails: A2

   Probability (prediction): ___/20 heads ___/20 tails

   (Note: Traditionally, probabilities are converted to the lowest fractional representation.)

   Actual results: _____ heads _____ tails

   Class totals: _____ heads _____ tails

2. Single die tossed 24 times and counting the number of times a number below 3 occurs: A2

   Probability: ____/24 number below 3 (numbers 1 and 2)

   Actual results: ____ number below 3

   Class totals: ____ number below 3 ____ total tosses by class members

3. Two pennies tossed simultaneously 32 times and counting the number of two heads, a heads and a tails, and two tails: A2

   Probability: ____/32 of two heads ____/32 of a heads and a tails ____/32 of two tails

   Actual results: ____two heads ____heads and tails ____two tails

   Class totals: ____two heads ____heads and tails ____two tails

4. Two dice tossed simultaneously 32 times and counting the number of two odd numbers, an odd and an even number, and two even numbers: A2

   Probability: ____/32 of two odd numbers ____/32 of an odd and an even number ____/32 of two even numbers

   Actual results: ____ two odd numbers ____ an odd and an even number ____ two even numbers

   Class totals: ____ two odd numbers ____ an odd and an even number ____ two even numbers

5. Use the example of the two dice tossed 32 times, and construct a Punnett square to represent the possible combinations that could be used to determine the probability (prediction) of odd and even numbers for the resulting tosses. Your construction should be similar to the Punnett square for the two coins tossed that is depicted in Procedure B of the laboratory exercise. A2

6. Complete the following:

   a. Are the class totals closer to the predicted probabilities than your results? _____ Explain your response. A2

      _____

      _____

   b. Does the first toss of the penny or the first toss of the die have any influence on the next toss? _____
      Explain your response. A2

      _____

   c. Assume a family has two boys or two girls. They wish to have one more child, but hope for the child to be of the opposite sex from the two they already have. What is the probability that the third child will be of the opposite sex?
      _____ Explain your response. A2

      _____

   d. What is the probability (prediction) that a couple without children will eventually have four children, all girls?
      _____ Explain your response. A2

      _____

      _____

PART C: Assessments

For each of the genetic problems, (a) determine the parents' genotypes, (b) determine the possible gametes for each parent, (c) construct a Punnett square, and (d) record the resulting genotypes and phenotypes as ratios from the cross. Problems 1 and 2 involve examples of complete dominance; problems 3 and 4 are examples of codominance; problem 5 is an example of sex-linked (X-linked) inheritance.

1. Determine the results from a cross of a mother who is heterozygous (*Rr*) for tongue rolling with a father who is homozygous recessive (*rr*).

2. Determine the results from a cross of a mother and a father who are both heterozygous for freckles.

3. Determine the results from a mother who is heterozygous for blood type B and a father who is homozygous dominant for blood type A.

4. Determine the results from a mother who is heterozygous for blood type A and a father who is heterozygous for blood type B. (A3)

5. Color blindness is an example of X-linked inheritance. Hemophilia is another example of X-linked inheritance, also from a recessive allele (*h*). The dominant allele (*H*) determines whether the person possesses normal blood clotting. A person with hemophilia has a permanent tendency for hemorrhaging due to a deficiency of one of the clotting factors (VIII— antihemophilic factor). Determine the offspring from a cross of a mother who is a carrier (heterozygous) for the disease and a father with normal blood coagulation. (A3)

CRITICAL THINKING ASSESSMENT

Assume that the genes for hairline and earlobes are on different pairs of homologous chromosomes. Determine the genotypes and phenotypes of the offspring from a cross if both parents are heterozygous for both traits. (1) First determine the genotypes for each parent. (2) Determine the gametes, but remember each gamete has one allele for each trait (gametes are haploid). (3) Construct a Punnett square, with sixteen boxes, that has four different gametes from each parent along the top and the left edges. (This is an application to demonstrate Mendel's Law of Independent Assortment.) (4) List the results of genotypes and phenotypes as ratios. (A3)

EXERCISE 62

Cat Dissection: Musculature

Purpose of the Exercise

To examine and identify the skeletal musculature of the cat and to make a comparison with the corresponding musculature of the human.

Learning Outcomes AP|R Cat

After completing this exercise, you should be able to:

O1 Name and locate the major skeletal muscles of the cat.

O2 Name the origins, insertions, and actions of the skeletal muscles designated by the laboratory instructor.

O3 Identify the corresponding skeletal muscles of the human.

The O corresponds to the assessments A indicated in the Laboratory Assessment for this Exercise.

Pre-Lab

Carefully read the introductory material and examine the entire lab. Be familiar with the location, origin, insertion, and action of muscles of the human body from lecture, the textbook, or laboratory exercises.

SKELETAL MUSCLE DISSECTION
TECHNIQUES

Dissect:	To expose the entire length of the muscle from origin to insertion. Most of the procedures are accomplished with blunt probes used to separate various connective tissues that hold adjacent structures together. It does not mean to remove or to cut into the muscles or other organs.
Transect:	To cut through the muscle near its midpoint (belly). The cut is perpendicular to the length of the muscle fibers.
Reflect:	To lift a transected muscle aside to expose deeper muscles or other organs.

Although the aim of this exercise is to become more familiar with the human musculature, human cadavers are not always available for dissection. Instead, preserved cats are often used for dissection because they are relatively small and can be purchased from biological suppliers. Because they are mammals, cats have many features in common with humans, including similar skeletal muscles (with similar names).

On the other hand, cats use four limbs for locomotion, whereas humans use only two limbs. Because the musculature of each type of organism is adapted to provide for its special needs, comparisons of the muscles of cats and humans may not be precise.

As you continue your dissection of various systems of the cat, many anatomical similarities between cats and humans will be observed. These fundamental similarities are called homologous structures. Although homologous structures have a similar structure and embryological origin, the functions are sometimes different.

PROCEDURE A: Skin Removal

1. Cats usually are preserved with a solution that prevents microorganisms from decomposing the tissues. However, because fumes from this preserving fluid may be annoying and may irritate skin, be sure to work in a well-ventilated room and wear disposable gloves to protect your hands.
2. Obtain a preserved cat, a dissecting tray, a set of dissecting instruments, a large plastic bag, and an identification tag.
3. If the cat is in a storage bag, dispose of any excess preserving fluid in the bag, as directed by the laboratory instructor. However, you may want to save some of the preserving fluid to keep the cat moist until you have completed your work later in the year.
4. If the cat is too wet with preserving fluid, blot the cat dry with paper towels.
5. Remove the skin from the cat. To do this, follow these steps:
   a. Place the cat in the dissecting tray with its ventral surface down.
   b. Use a sharp scalpel to make a short, shallow incision through the skin in the dorsal midline at the base of the cat's neck.
   c. Insert the pointed blade of a dissecting scissors through the opening in the skin and cut along the dorsal midline, following the tips of the vertebral spinous processes to the base of the tail (fig. 62.1).
   d. Use the scissors to make an incision encircling the region of the tail, anus, and genital organs. Do not remove the skin from this area.
   e. Use bone shears or a bone saw to sever the tail at its base and discard it (optional procedure).
   f. From the incision at the base of the tail, cut along the lateral surface of each hindlimb to the paw and encircle the ankle. Also clip off the claws from each

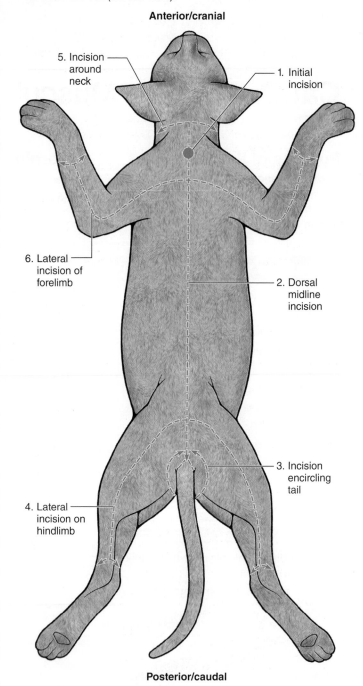

FIGURE 62.1 Incisions to be made for removing the skin of the cat (dorsal view).

Anterior/cranial

5. Incision around neck

1. Initial incision

6. Lateral incision of forelimb

2. Dorsal midline incision

3. Incision encircling tail

4. Lateral incision on hindlimb

Posterior/caudal

paw to prevent them from tearing the plastic storage bag and scratching your skin (optional procedure).
   g. Make an incision from the initial cut at the base of the neck to encircle the neck.
   h. From the incision around the neck, cut along the lateral surface of each forelimb to the paw and encircle the wrist (fig. 62.1).
   i. Grasp the skin on either side of the dorsal midline incision, and carefully pull the skin laterally away from the body. At the same time, use your fingers or a blunt

probe to help remove the skin from the underlying muscles by separating the loose connective tissue (superficial fascia). As you pull the skin away, you may note a thin sheet of muscle tissue attached to it. This muscle is called the *cutaneous maximus,* and it functions to move the cat's skin. Humans lack the cutaneous maximus, but a similar sheet of muscle (platysma) is present in the neck of a human.

j. As you remove the skin, work toward the ventral surface, then work toward the head, and finally work toward the tail. Pull the skin over each limb as if you were removing a glove.

k. If the cat is a female, note the mammary glands on the ventral surface of the thorax and abdomen (fig. 62.2). These glands will adhere to the skin and can be removed from it and discarded.

l. After the skin has been pulled away, carefully remove as much of the remaining connective tissue as possible to expose the underlying skeletal muscles. The muscles should appear light brown and fibrous.

FIGURE 62.2 Mammary glands on the ventral surface of the cat's thorax and abdomen.

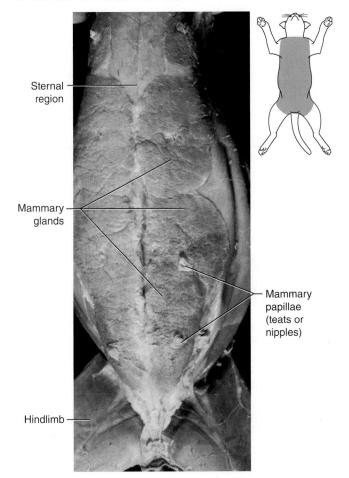

Sternal region

Mammary glands

Mammary papillae (teats or nipples)

Hindlimb

6. After skinning the cat, follow these steps:
   a. Discard the tissues you have removed as directed by the laboratory instructor.
   b. Wrap the skin around the cat to help keep its body moist, and place it in a plastic storage bag.
   c. Write your name in pencil on an identification tag, and tie the tag to the storage bag so that you can identify your specimen.

7. Observe the recommended safety procedures for the conclusion of a laboratory session.

PROCEDURE B: Skeletal Muscle Dissection

1. The purpose of a skeletal muscle dissection is to separate the individual muscles from any surrounding tissues in order to expose the muscles for observation. To do a muscle dissection, follow these steps:
   a. Use the appropriate figure as a guide and locate the muscle to be dissected in the specimen.
   b. Use a blunt probe or a finger to separate the muscle from the surrounding connective tissue along its natural borders. The muscle should separate smoothly. If the border appears ragged, you probably have torn the muscle fibers.

2. If it is necessary to cut through (transect) a superficial muscle to observe a deeper one, use scissors to transect the muscle about halfway between its origin and insertion. Then, lift aside (reflect) the cut ends, leaving their attachments intact.

3. The following procedures will instruct you to dissect some of the larger and more easily identified muscles of the cat. In each case, the procedure will include the names of the muscles to be dissected, figures illustrating their locations, and tables listing the origins, insertions, and actions of the muscles. Compare the muscle origins and insertions to the cat skeleton (fig. 62.3). (More detailed dissection instructions can be obtained by consulting an additional guide or atlas for cat dissection.) As you dissect each muscle, review the muscle figures and tables in the lab manual and identify any corresponding (homologous) muscles of the human body. Also locate these homologous muscles in the human torso model or models of the human upper and lower limbs.

PROCEDURE C: Muscles of the Head and Neck

1. Place the cat in the dissecting tray with its ventral surface up.

2. Remove the skin and underlying connective tissues from each side of the neck, forward to the chin, and up to each ear.

FIGURE 62.3 Skeleton of a cat, lateral view.

FIGURE 62.3 Skeleton of a cat, lateral view.

3. Study figures 62.4 and 62.5, and then locate and dissect the following muscles:

sternomastoid   masseter
sternohyoid   mylohyoid
digastric

4. To find the deep muscles of the cat's neck and chin, carefully remove the sternomastoid, sternohyoid, and mylohyoid muscles from the right side.

5. Study figure 62.6 and locate the following muscles:

hyoglossus   styloglossus   geniohyoid
cleidomastoid   thyrohyoid   sternothyroid

FIGURE 62.4 **AP|R**_{Cat} Muscles on the ventral surface of the cat's head and neck.

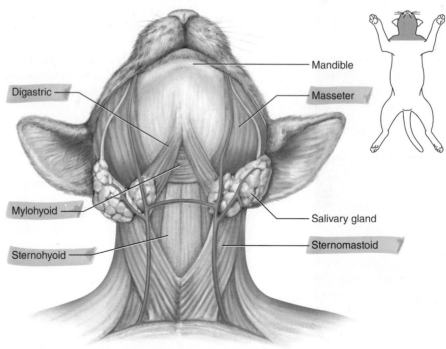

FIGURE 62.5 Deep muscles of the ventral surface of the cat's neck.

Masseter

Digastric

Mylohyoid

Stylohyoid

Sternohyoid

Sternomastoid

FIGURE 62.6 Deep muscles of the cat's neck and chin regions.

Geniohyoid

Styloglossus

Hyoglossus

Thyrohyoid

Cleidomastoid

Sternothyroid

Masseter

Digastric

Mylohyoid

Stylohyoid

Sternohyoid

Sternomastoid

Table 62.1 Muscles of the Head and Neck

Muscle	Origin	Insertion	Action
Sternomastoid	Manubrium of sternum	Mastoid process of temporal bone	Turns and depresses head
Sternohyoid	Costal cartilage	Hyoid bone	Depresses hyoid bone
Digastric	Mastoid process and occipital bone	Ventral border of mandible	Depresses mandible
Mylohyoid	Inner surface of mandible	Hyoid bone	Raises floor of mouth
Masseter	Zygomatic arch	Coronoid fossa of mandible	Elevates mandible
Hyoglossus	Hyoid bone	Tongue	Pulls tongue back
Styloglossus	Styloid process	Tip of tongue	Pulls tongue back
Geniohyoid	Inner surface of mandible	Hyoid bone	Pulls hyoid bone forward
Cleidomastoid	Clavicle	Mastoid process	Turns head to side
Thyrohyoid	Thyroid cartilage	Hyoid bone	Raises larynx
Sternothyroid	Sternum	Thyroid cartilage	Pulls larynx back

6. See table 62.1 for the origins, insertions, and actions of these head and neck muscles.
7. Complete Part A of Laboratory Assessment 62.

PROCEDURE D: Muscles of the Thorax

1. Place the cat in the dissecting tray with its ventral surface up.

2. Remove any remaining fat and connective tissue to expose the muscles in the walls of the thorax and abdomen.
3. Study figures 62.7 and 62.8. Transect the superficial pectoantebrachialis to expose the entire pectoralis major. Locate and dissect the following pectoral muscles:

pectoantebrachialis—well developed in cats
pectoralis major
pectoralis minor
xiphihumeralis

FIGURE 62.7 AP|R Cat Muscles of the cat's thorax and abdomen.

4. To find the deep thoracic muscles, transect the pectoralis minor, pectoralis major, and pectoantebrachialis muscles, and reflect their cut edges to the sides. Remove the underlying fat and connective tissues to expose the following muscles, as shown in figure 62.9:

scalenus	levator scapulae
serratus ventralis	transversus costarum

5. See table 62.2 for the origins, insertions, and actions of these muscles.

FIGURE 62.8 Superficial muscles of the cat's thorax.

- Pectoralis major
- Pectoantebrachialis
- Pectoralis major
- Latissimus dorsi
- Pectoralis minor
- Xiphihumeralis

FIGURE 62.9 Superficial and deep muscles of the cat's thorax.

- Levator scapulae
- Transversus costarum
- Scalenus
- Serratus ventralis
- Rectus abdominis
- Pectoralis major
- Pectoantebrachialis
- Pectoralis minor
- Xiphihumeralis

Deep muscles | Superficial muscles

Table 62.2 Muscles of the Thorax

Muscle	Origin	Insertion	Action
Pectoantebrachialis	Manubrium of sternum	Fascia of forelimb	Adducts forelimb
Pectoralis major	Manubrium of sternum and costal cartilages	Greater tubercle of humerus	Adducts and rotates forelimb
Pectoralis minor	Sternum	Proximal end of humerus	Adducts and rotates forelimb
Xiphihumeralis	Xiphoid process of sternum	Proximal end of humerus	Adducts forelimb
Scalenus	Ribs	Cervical vertebrae	Flexes neck
Serratus ventralis	Ribs	Scapula	Pulls scapula ventrally and posteriorly
Levator scapulae	Cervical vertebrae	Scapula	Pulls scapula ventrally and anteriorly
Transversus costarum	Sternum	First rib	Pulls sternum toward head

FIGURE 62.10 AP|R^{Cat} Muscles of the cat's thorax and abdomen. Some abdominal muscles are partially reflected to show deeper muscles.

Serratus ventralis

Latissimus dorsi

Xiphihumeralis

External oblique

Aponeurosis

Rectus abdominis (beneath aponeurosis)

Internal oblique

External oblique

Transversus abdominis (deepest muscle layer)

PROCEDURE E: Muscles of the Abdominal Wall

1. Study figures 62.7 and 62.10.
2. Locate the *external oblique* muscle in the abdominal wall.
3. Make a shallow, longitudinal incision through the external oblique. Reflect the cut edge and expose the *internal oblique* muscle beneath (fig. 62.7). The fibers of the internal oblique run at a right angle to those of the external oblique.
4. Make a longitudinal incision through the internal oblique. Reflect the cut edge and expose the *transversus abdominis* muscle.
5. Expose the *rectus abdominis* muscle on one side of the midventral line. This muscle lies beneath an aponeurosis.
6. See table 62.3 for the origins, insertions, and actions of these muscles.
7. Complete Part B of the laboratory assessment.

PROCEDURE F: Muscles of the Shoulder and Back

1. Place the cat in the dissecting tray with its dorsal surface up.
2. Remove any remaining fat and connective tissue to expose the muscles of the shoulder and back.
3. Study figures 62.11 and 62.12, and then locate and dissect the following superficial muscles:

   clavotrapezius
   acromiotrapezius
   spinotrapezius
   levator scapulae ventralis
   clavodeltoid
   acromiodeltoid
   spinodeltoid
   latissimus dorsi

Table 62.3 Muscles of the Abdominal Wall

Muscle	Origin	Insertion	Action
External oblique	Ribs and fascia of back	Linea alba	Compresses abdominal wall
Internal oblique	Fascia of back	Linea alba	Compresses abdominal wall
Transversus abdominis	Lower ribs	Linea alba	Compresses abdominal wall
Rectus abdominis	Pubic symphysis	Sternum and costal cartilages	Compresses abdominal wall and flexes trunk

FIGURE 62.11 AP|R^{Cat} Superficial muscles of the cat's shoulder and back (*right side*); deep muscles of the shoulder and back (*left side*).

Note: The clavodeltoid (clavobrachialis) appears as a continuation of the clavotrapezius.

4. Using scissors, transect the latissimus dorsi and the group of trapezius muscles. Reflect their cut edges and remove any underlying fat and connective tissue. Study figures 62.11 and 62.13, and then locate and dissect the following deep muscles of the shoulder and back:

supraspinatus	**infraspinatus**
teres major	**rhomboid**
rhomboid capitis	**splenius**

FIGURE 62.12 Muscles of the cat's shoulder and back.

Clavodeltoid (clavobrachialis)

Spinodeltoid

Acromiotrapezius
Infraspinatus
Teres major

Spinotrapezius

Latissimus dorsi

Clavotrapezius

Acromiotrapezius

Rhomboid capitis

Rhomboid

Spinous process

Lumbodorsal fascia

FIGURE 62.13 Deep muscles of the cat's back. The left forelimb is adducted for photograph.

Clavotrapezius (reflected)

Supraspinatus

Rhomboid capitis

Serratus ventralis

Acromiotrapezius (reflected)
Rhomboid

Table 62.4 Muscles of the Shoulder and Back

Muscle	Origin	Insertion	Action
Clavotrapezius	Occipital bone of skull and spines of cervical vertebrae	Clavicle	Pulls clavicle upward
Acromiotrapezius	Spines of cervical and thoracic vertebrae	Spine and acromion of scapula	Pulls scapula upward
Spinotrapezius	Spines of thoracic vertebrae	Spine of scapula	Pulls scapula upward and back
Levator scapulae ventralis	Occipital bone of skull	Vertebral border of scapula	Pulls scapula toward head
Clavodeltoid	Clavicle	Ulna near semilunar notch	Flexes forelimb
Acromiodeltoid	Acromion of scapula	Proximal end of humerus	Flexes and rotates forelimb
Spinodeltoid	Spine of scapula	Proximal end of humerus	Flexes and rotates forelimb
Latissimus dorsi	Thoracic and lumbar vertebrae	Proximal end of humerus	Pulls forelimb upward and back
Supraspinatus	Fossa above spine of scapula	Greater tubercle of humerus	Extends forelimb
Infraspinatus	Fossa below spine of scapula	Greater tubercle of humerus	Rotates forelimb
Teres major	Posterior border of scapula	Proximal end of humerus	Rotates forelimb
Rhomboid	Spines of cervical and thoracic vertebrae	Medial border of scapula	Pulls scapula back
Rhomboid capitis	Occipital bone of skull	Medial border of scapula	Pulls scapula forward
Splenius	Fascia of neck	Occipital bone	Turns and raises head

5. See table 62.4 for the origins, insertions, and actions of these muscles in the shoulder and back.
6. Review the locations, origins, insertions, and actions designated by the laboratory instructor without the aid of the figures and tables.
7. Complete Part C of the laboratory assessment.

PROCEDURE G: Muscles of the Forelimb

1. Place the cat in the dissecting tray with its ventral surface up.
2. Remove any remaining fat and connective tissue from a forelimb to expose the muscles.
3. Study figure 62.14a, and then locate and dissect the following muscles from the medial surface of the upper forelimb:

    epitrochlearis—well developed in the cat

    triceps brachii

4. Transect the pectoral muscles (pectoantebrachialis, pectoralis major, and pectoralis minor) and reflect their cut edges.
5. Study figures 62.14b and 62.15, and then locate and dissect the following muscles:

    coracobrachialis

    biceps brachii

    long head of the triceps brachii

    medial head of the triceps brachii

6. Study figures 62.16a and 62.17, which show the lateral surface of the upper forelimb.
7. Locate the lateral head of the triceps muscle, transect it, and reflect its cut edges.
8. Study figure 62.16b. Locate and dissect the following deeper muscles:

    anconeus

    brachialis

9. Separate the muscles in the lower forelimb into an *extensor group* (fig. 62.16a) and a *flexor group* (fig. 62.14a). Generally, the extensors are located on the posterior surface of the lower forelimb, and the flexors are located on the anterior surface. A thick, tough layer of fascia (antebrachial fascia) surrounds the forelimb. Remove this fascia to expose and dissect the following muscles:

    brachioradialis

    extensor carpi radialis

    extensor digitorum communis

    extensor digitorum lateralis

    extensor carpi ulnaris

    flexor carpi ulnaris

    palmaris longus

    flexor carpi radialis

    pronator teres

FIGURE 62.14 AP|R^{Cat} (a) Superficial muscles of the cat's left medial shoulder and forelimb. (b) Deep muscles of the left medial shoulder and forelimb.

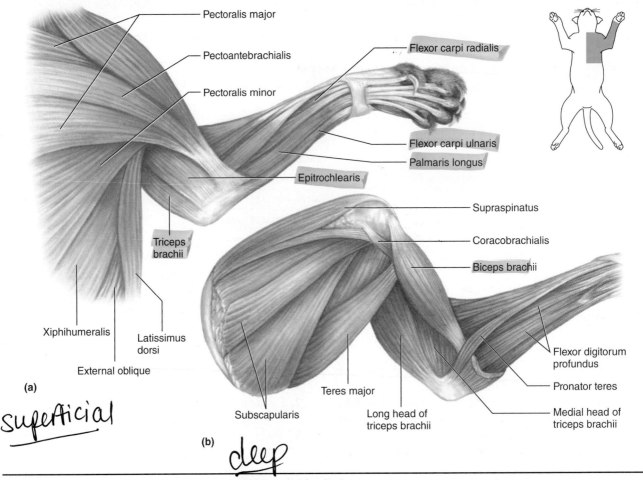

Pectoralis major
Pectoantebrachialis
Pectoralis minor
Flexor carpi radialis
Flexor carpi ulnaris
Palmaris longus
Epitrochlearis
Supraspinatus
Coracobrachialis
Biceps brachii
Triceps brachii
Xiphihumeralis
Latissimus dorsi
External oblique
Flexor digitorum profundus
Pronator teres
Medial head of triceps brachii
Subscapularis
Teres major
Long head of triceps brachii

(a) *superficial*

(b) *deep*

FIGURE 62.15 Deep muscles of the cat's right medial forelimb.

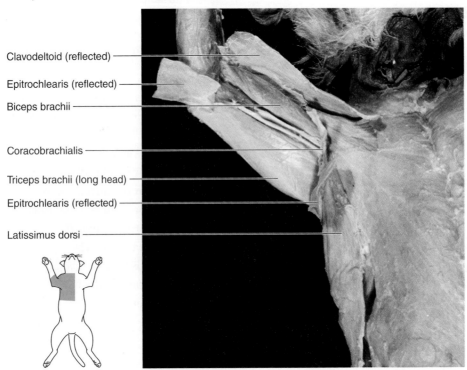

Clavodeltoid (reflected)
Epitrochlearis (reflected)
Biceps brachii
Coracobrachialis
Triceps brachii (long head)
Epitrochlearis (reflected)
Latissimus dorsi

FIGURE 62.16 (a) Superficial muscles of the cat's left lateral shoulder and forelimb. (b) Deep muscles of the left lateral shoulder and forelimb.

superficial

(a)

Lateral head of triceps brachii

Brachioradialis

Clavotrapezius

Clavodeltoid

Acromiodeltoid

Spinodeltoid

Extensor carpi radialis

Extensor digitorum communis

Extensor digitorum lateralis

Extensor carpi ulnaris

Long head of triceps brachii

Teres major

Infraspinatus

Latissimus dorsi

Acromiotrapezius

Spinotrapezius

Lateral head of triceps brachii (cut)

Medial head of triceps brachii

Brachialis

Rhomboid capitis

Extensor carpi ulnaris

Anconeus

Lateral head of triceps brachii (cut)

Teres major

Serratus ventralis

Supraspinatus

Long head of triceps brachii

Infraspinatus

(b)

deep

683

FIGURE 62.17   Superficial muscles of the cat's left lateral shoulder and forelimb, dorsal view.

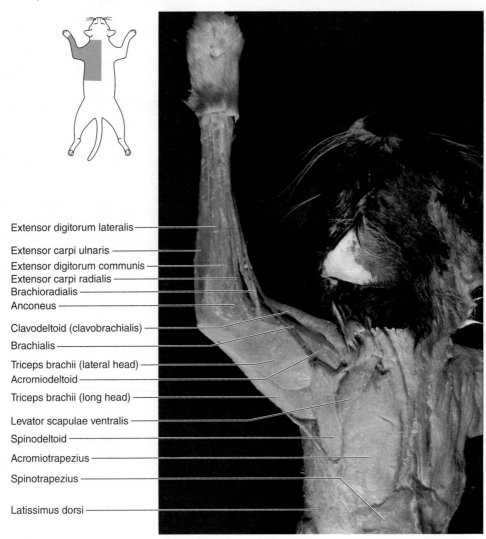

Extensor digitorum lateralis

Extensor carpi ulnaris

Extensor digitorum communis

Extensor carpi radialis

Brachioradialis

Anconeus

Clavodeltoid (clavobrachialis)

Brachialis

Triceps brachii (lateral head)

Acromiodeltoid

Triceps brachii (long head)

Levator scapulae ventralis

Spinodeltoid

Acromiotrapezius

Spinotrapezius

Latissimus dorsi

Table 62.5 Muscles of the Forelimb

Muscle	Origin	Insertion	Action
Coracobrachialis	Coracoid process of scapula	Proximal end of humerus	Adducts forelimb
Epitrochlearis	Fascia of latissimus dorsi	Olecranon process of ulna	Extends forelimb
Biceps brachii	Border of glenoid cavity of scapula	Radial tuberosity of radius	Flexes forelimb

Triceps brachii:
   Lateral head
   Long head
   Medial head
	Deltoid tuberosity of humerus
	Border of glenoid cavity of scapula
	Shaft of humerus
		Olecranon process of ulna
		Olecranon process of ulna
		Olecranon process of ulna
			Extends forelimb
			Extends forelimb
			Extends forelimb

Anconeus	Lateral epicondyle of humerus	Olecranon process of ulna	Extends forelimb
Brachialis	Lateral surface of humerus	Proximal end of ulna	Flexes forelimb
Brachioradialis	Shaft of humerus	Distal end of radius	Supinates lower forelimb
Extensor carpi radialis	Lateral surface of humerus	Second and third metacarpals	Extends wrist
Extensor digitorum communis	Lateral surface of humerus	Digits two to five	Extends digits
Extensor digitorum lateralis	Lateral surface of humerus	Digits two to five	Extends digits
Extensor carpi ulnaris	Lateral epicondyle of humerus and ulna	Fifth metacarpal	Extends wrist
Flexor carpi ulnaris	Medial epicondyle of humerus	Carpals	Flexes wrist
Palmaris longus	Medial epicondyle of humerus	Digits	Flexes digits
Flexor carpi radialis	Medial epicondyle of humerus	Second and third metacarpals	Flexes metacarpals
Pronator teres	Medial epicondyle of humerus	Radius	Pronates lower forelimb
Flexor digitorum profundus	Radius, ulna, and medial epicondyle of humerus	Distal phalanges	Flexes digits

10. Transect the flexor carpi ulnaris, palmaris longus, and flexor carpi radialis to expose the flexor digitorum profundus.

11. See table 62.5 for the origins, insertions, and actions of these muscles of the forelimb.

12. Complete Part D of the laboratory assessment.

PROCEDURE H: Muscles of the Hip and Hindlimb

1. Place the cat in the dissecting tray with its ventral surface up.

2. Remove any remaining fat and connective tissue from the hip and hindlimb to expose the muscles.

3. Study figures 62.18a and 62.19, and then locate and dissect the following muscles from the medial surface of the thigh:

   sartorius

   gracilis

4. Using scissors, transect the sartorius and gracilis, and reflect their cut edges to observe the deeper muscles of the thigh.

FIGURE 62.18 **AP|R^{Cat}** (a) Superficial muscles of the cat's left medial hindlimb. (b) Deep muscles of the left medial hindlimb.

External oblique
Vastus medialis
Sartorius
Rectus abdominis
Adductor longus
Adductor femoris
Gracilis
Semitendinosus

Tensor fasciae latae
Rectus femoris
Adductor longus
Vastus medialis
Adductor femoris
Semimembranosus
Semitendinosus

(a)

(b)

FIGURE 62.19 Superficial muscles of the cat's hindlimb, medial view.

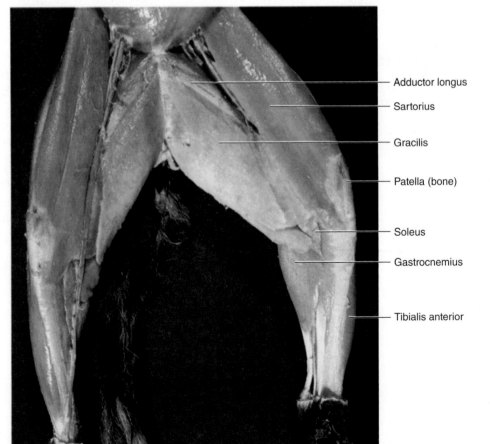

Adductor longus
Sartorius
Gracilis
Patella (bone)
Soleus
Gastrocnemius
Tibialis anterior

5. Study figures 62.18*b* and 62.20, and then locate and dissect the following muscles:

   rectus femoris

   vastus medialis

   adductor longus

   adductor femoris

   semimembranosus

   semitendinosus

   tensor fasciae latae

6. Transect the tensor fasciae latae and rectus femoris muscles and reflect their cut edges. Locate the *vastus intermedius* and *vastus lateralis* muscles beneath (fig. 62.21). (*Note:* In some specimens, the vastus intermedius, vastus lateralis, and vastus medialis muscles are closely united by connective tissue and are difficult to separate.)

7. Study figures 62.22*a* and 62.23, and then locate and dissect the following muscles from the lateral surface of the hip and thigh:

   biceps femoris

   caudofemoralis

   gluteus maximus—small in cats

8. Using scissors, transect the tensor fasciae latae and biceps femoris, and reflect their cut edges to observe the deeper muscles of the thigh.

9. Locate and dissect the following muscles:

   tenuissimus

   gluteus medius

   (*Note:* The tenuissimus muscle and the sciatic nerve may adhere tightly to the dorsal surface of the biceps femoris muscle. Take care to avoid cutting these structures.)

10. On the lateral surface of the lower hindlimb (figs. 62.22 and 62.24), locate and dissect the following muscles:

    gastrocnemius

    soleus

    tibialis anterior

    fibularis (peroneus) group

11. See table 62.6 for the origins, insertions, and actions of these muscles of the hip and hindlimb.

12. Review the locations, origins, insertions, and actions designated by the laboratory instructor without the aid of the figures and tables.

13. Complete Parts E and F of the laboratory assessment.

FIGURE 62.20   Deep muscles of the cat's left hindlimb, medial view.

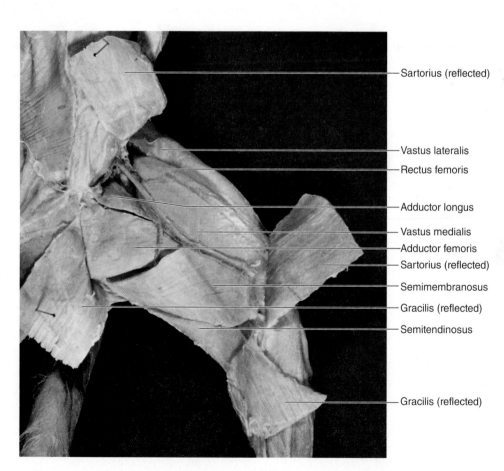

Sartorius (reflected)

Vastus lateralis
Rectus femoris

Adductor longus

Vastus medialis
Adductor femoris
Sartorius (reflected)
Semimembranosus
Gracilis (reflected)
Semitendinosus

Gracilis (reflected)

FIGURE 62.21 Other deep muscles of the cat's left thigh, medial view.

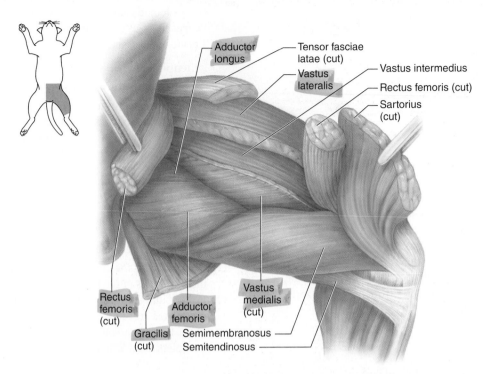

FIGURE 62.22 AP|R Cat (a) Superficial muscles of the cat's left lateral hip and hindlimb. (b) Deep muscles of the left lateral hip and hindlimb.

FIGURE 62.23 Muscles of the cat's left hip and hindlimb, lateral view.

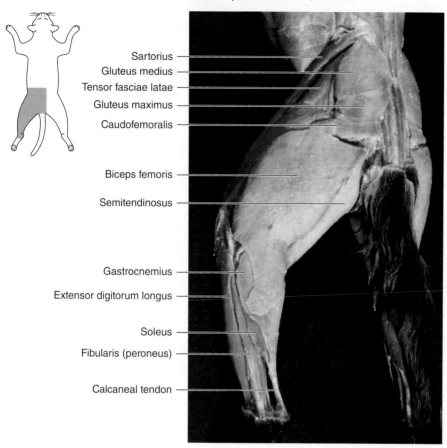

Sartorius
Gluteus medius
Tensor fasciae latae
Gluteus maximus
Caudofemoralis

Biceps femoris

Semitendinosus

Gastrocnemius

Extensor digitorum longus

Soleus

Fibularis (peroneus)

Calcaneal tendon

FIGURE 62.24 Deep muscles of the cat's left hindlimb, lateral view.

Gluteus medius
Tensor fasciae latae
Sartorius
Gluteus maximus

Caudofemoralis (reflected)
Sciatic nerve
Biceps femoris (reflected)
Semimembranosus
Tenuissimus

Semitendinosus

Gastrocnemius

Soleus

Fibularis (peroneus) tertius
Extensor digitorum longus

Table 62.6   Muscles of the Hip and Hindlimb

Muscle	Origin	Insertion	Action
Sartorius	Crest of ilium	Fascia of knee and proximal end of tibia	Rotates and extends hindlimb
Gracilis	Pubic symphysis and ischium	Proximal end of tibia	Adducts hindlimb

Quadriceps femoris:
  Vastus lateralis
  Vastus intermedius
  Rectus femoris
  Vastus medialis
	
	Shaft of femur and greater trochanter
	Shaft of femur
	Ilium
	Shaft of femur
		
		Patella
		Patella
		Patella
		Patella
			
			Extends hindlimb
			Extends hindlimb
			Extends hindlimb
			Extends hindlimb

Adductor longus	Pubis	Proximal end of femur	Adducts hindlimb
Adductor femoris	Pubis and ischium	Shaft of femur	Adducts hindlimb
Semimembranosus	Ischial tuberosity	Medial epicondyle of femur	Extends thigh
Tensor fasciae latae	Ilium	Fascia of thigh	Extends thigh
Biceps femoris	Ischial tuberosity	Patella and tibia	Abducts thigh and flexes lower hindlimb
Tenuissimus	Second caudal vertebra	Tibia and fascia of biceps femoris	Abducts thigh and flexes lower hindlimb
Semitendinosus	Ischial tuberosity	Crest of tibia	Flexes lower hindlimb
Caudofemoralis	Caudal vertebrae	Patella	Abducts thigh and extends lower hindlimb
Gluteus maximus	Sacral and caudal vertebrae	Greater trochanter of femur	Abducts thigh
Gluteus medius	Ilium, and sacral and caudal vertebrae	Greater trochanter of femur	Abducts thigh
Gastrocnemius	Lateral and medial epicondyles of femur	Calcaneus	Extends foot
Soleus	Proximal end of fibula	Calcaneus	Extends foot
Tibialis anterior	Proximal end of tibia and fibula	First metatarsal	Flexes foot
Fibularis (peroneus) group	Shaft of fibula	Metatarsals	Flexes foot

Name _____

Date _____

Section _____

The Ⓐ corresponds to the indicated Learning Outcome(s) Ⓞ found at the beginning of the Laboratory Exercise.

Cat Dissection: Musculature

PART A: Assessments

Complete the following statements:

1. The _____Sternothyroid_____ [*sternocleidomastoid*] muscle of the human is homologous to the sternomastoid muscle of the cat. (A3)

2. The _____masseter_____ muscle elevates the mandible in the human and in the cat. (A2) (A3)

3. Two muscles of the cat that are inserted on the hyoid bone are the _____sternohyoid_____ and the _____mylohyoid_____ . (A1) (A2)

PART B: Assessments

Complete the following:

1. Name two muscles that are found in the thoracic wall of the cat but are absent in the human. (A1) (A3)

   _____   _____

2. Name two muscles that are found in the thoracic wall of the cat and the human. (A3)

   _____   _____

3. Name four muscles that are found in the abdominal wall of the cat and the human. Put an asterisk next to the muscle that is the deepest. (A3)

   _____   _____

   _____   _____

PART C: Assessments

Complete the following:

1. Name three muscles of the cat that together correspond to the trapezius muscle in the human. (A1) (A3)

   _____   _____   _____

2. Name three muscles of the cat that together correspond to the deltoid muscle in the human. (A1) (A3)

   _____   _____   _____

3. Name the muscle in the cat and in the human that occupies the fossa above the spine of the scapula. (A1) (A3)

   _____

4. Name the muscle in the cat and in the human that occupies the fossa below the spine of the scapula. (A1) (A3)

   _____

5. Name two muscles found in the cat and in the human that can rotate the forelimb. (A2) (A3)

   _____   _____

PART D: Assessments

Complete the following:

1. Name two muscles found in the cat and in the human that can flex the forelimb.

_____   _____

2. Name a muscle found in the cat but absent in the human that can extend the forelimb.

_____

3. Name a muscle that has three heads, can extend the forelimb, and is found in the cat and in the human.

_____

PART E: Assessments

Identify the numbered muscles of the cat in figures 62.25 and 62.26.

FIGURE 62.25   Label the numbered muscles of the cat. Superficial muscles are shown on the right side; deep muscles are shown on the left side.

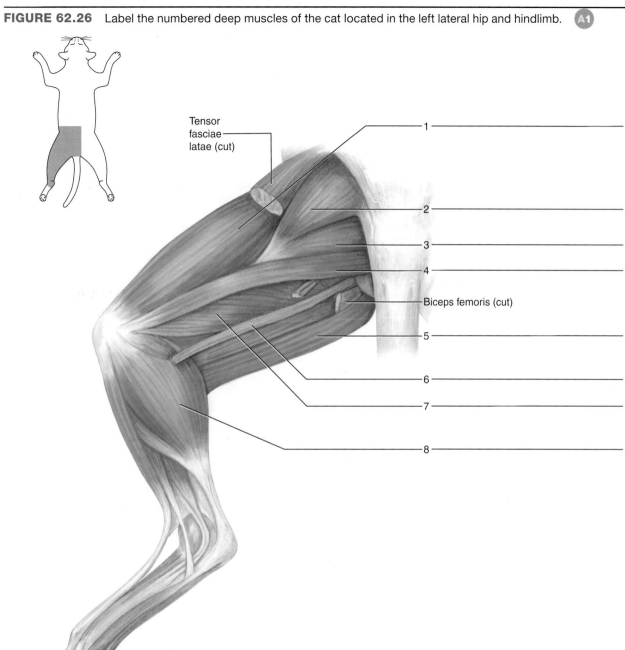

Tensor fasciae latae (cut)

1

2

3

4

Biceps femoris (cut)

5

6

7

8

PART F: Assessments

Observe the muscles on the human torso model and the upper and lower limb models. Each of the muscles in figure 62.27 is found both in the cat and in the human. Identify each of the numbered muscles in the figure by placing its name in the space next to its number.

FIGURE 62.27   Identify the numbered muscles that occur in both the cat and the human.   (A3)

Anterior view   Posterior view

LABORATORY
EXERCISE 63

Cat Dissection: Cardiovascular System

Purpose of the Exercise

To examine and identify the major organs of the cardiovascular system of the cat and to compare them with the corresponding organs of the human.

MATERIALS NEEDED

Preserved cat

Dissecting tray

Dissecting instruments

Disposable gloves

Human torso model

SAFETY

▶ Wear disposable gloves when working on the cat dissection.

▶ Dispose of tissue remnants and gloves as instructed.

▶ Wash the dissecting tray and instruments as instructed.

▶ Wash your laboratory table.

▶ Wash your hands before leaving the laboratory.

Learning Outcomes   AP|R Cat

After completing this exercise, you should be able to:

O1 Locate and identify the major organs of the cardiovascular system of the cat.

O2 Compare the features of the cardiovascular system of the cat with those of the human.

O3 Identify the corresponding cardiovascular organs of the human.

The **O** corresponds to the assessments **A** indicated in the Laboratory Assessment for this Exercise.

Pre-Lab

Carefully read the introductory material and examine the entire lab. Be familiar with the structures and functions of the human cardiovascular system from lecture, the textbook, or laboratory exercises.

In this laboratory exercise, you will dissect the major organs of the cardiovascular system of the cat. As before, while you are examining the organs of the cat, compare them with the corresponding organs of the human torso model.

If the cardiovascular system of the cat has been injected, the systemic arteries will be filled with red latex, and the systemic veins will be filled with blue latex. This will make it easier for you to trace the vessels as you dissect them.

PROCEDURE A: The Arterial System

1. Place the preserved cat in the dissecting tray with its ventral surface up.
2. Open the thoracic cavity and expose its contents. To do this, follow these steps:
   a. Make a longitudinal incision passing anteriorly from the diaphragm along one side of the sternum. Continue the incision through the neck muscles to the mandible. Try to avoid damaging the internal organs as you cut.
   b. Make a lateral cut on each side along the anterior surface of the diaphragm and cut the diaphragm loose from the thoracic wall.
   c. Spread the sides of the thoracic wall outward and use a scalpel to make a longitudinal cut along each side of the inner wall of the rib cage to weaken the ribs. Continue to spread the thoracic wall laterally to break the ribs so that the flaps of the wall will remain open (figs. 63.1 and 63.2).
3. Note the location of the heart and the large blood vessels associated with it. Slit the outer *parietal pericardium* that surrounds the heart by cutting with scissors along the midventral line. Note how this membrane is connected to the *visceral pericardium* that is attached to the surface of the heart. Locate the *pericardial cavity,* the space between the two layers of the pericardium.

FIGURE 63.1 AP|R^Cat Arteries of the cat's thorax, neck, and forelimb.

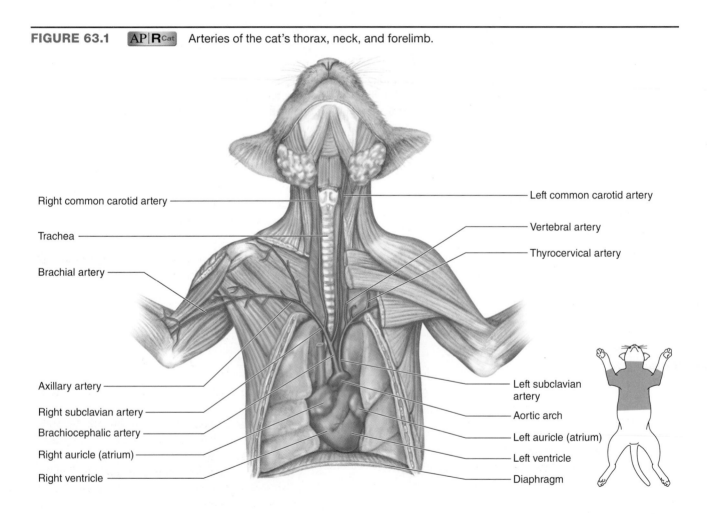

Right common carotid artery

Trachea

Brachial artery

Axillary artery

Right subclavian artery

Brachiocephalic artery

Right auricle (atrium)

Right ventricle

Left common carotid artery

Vertebral artery

Thyrocervical artery

Left subclavian artery

Aortic arch

Left auricle (atrium)

Left ventricle

Diaphragm

FIGURE 63.2 Features of the cat's neck and thoracic cavity.

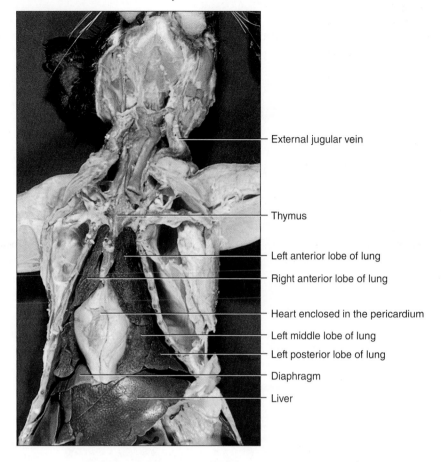

- External jugular vein
- Thymus
- Left anterior lobe of lung
- Right anterior lobe of lung
- Heart enclosed in the pericardium
- Left middle lobe of lung
- Left posterior lobe of lung
- Diaphragm
- Liver

4. Examine the heart (fig. 63.3*a*). The arrangement of blood vessels coming off the aortic arch is different in the cat than in the human body (fig. 63.3*b*). Locate the following:

right atrium

left atrium

right ventricle

left ventricle

pulmonary trunk

aorta

coronary arteries

5. Use a scalpel to open the heart chambers by making a cut along the frontal plane from its apex to its base. Remove any remaining latex from the chambers. Examine the valves between the chambers, and note the relative thicknesses of the chamber walls (fig. 63.4). Do not remove the heart, as it is needed for future examination of the relationship between major blood vessels.

6. Using figures 63.1, 63.4, and 63.5 as guides, locate and dissect the following arteries of the thorax and neck:

aortic arch

brachiocephalic artery—on right only; first branch of the aortic arch

right subclavian artery

left subclavian artery

right common carotid artery

left common carotid artery

7. Trace the right subclavian artery into the forelimb, and locate the following arteries:

vertebral artery

thyrocervical artery

subscapular artery

axillary artery

brachial artery

8. Open the abdominal cavity. To do this, follow these steps:

   a. Use scissors to make an incision through the body wall along the midline from the pubic symphysis to the diaphragm.

   b. Make a lateral incision through the body wall along either side of the inferior border of the diaphragm and along the bases of the thighs.

   c. Reflect the flaps created in the body wall as you would open a book and expose the contents of the abdominal cavity.

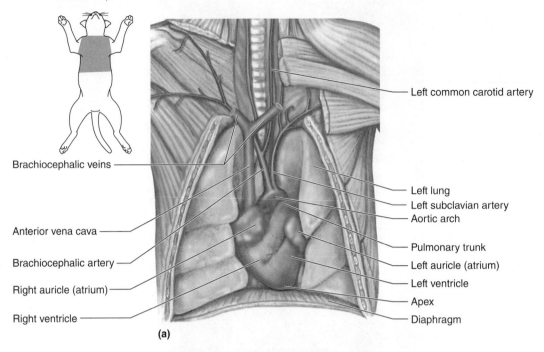

Left common carotid artery

Brachiocephalic veins

Anterior vena cava

Brachiocephalic artery

Right auricle (atrium)

Right ventricle

Left lung
Left subclavian artery
Aortic arch
Pulmonary trunk
Left auricle (atrium)
Left ventricle
Apex
Diaphragm

(a)

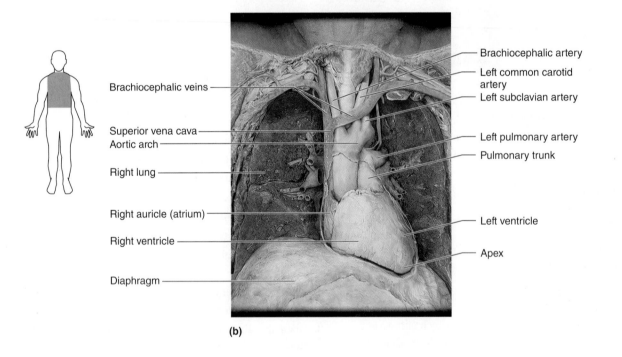

Brachiocephalic veins

Superior vena cava
Aortic arch

Right lung

Right auricle (atrium)

Right ventricle

Diaphragm

Brachiocephalic artery
Left common carotid artery
Left subclavian artery

Left pulmonary artery
Pulmonary trunk

Left ventricle

Apex

(b)

FIGURE 63.4 Blood vessels of the cat's neck and thorax.

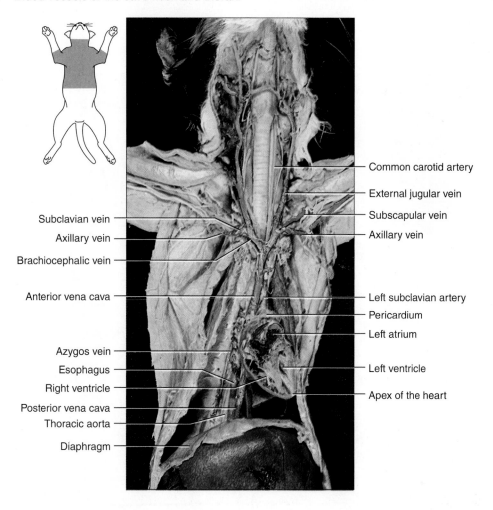

Subclavian vein

Axillary vein

Brachiocephalic vein

Anterior vena cava

Azygos vein

Esophagus

Right ventricle

Posterior vena cava

Thoracic aorta

Diaphragm

Common carotid artery

External jugular vein

Subscapular vein

Axillary vein

Left subclavian artery

Pericardium

Left atrium

Left ventricle

Apex of the heart

FIGURE 63.5 Blood vessels of the cat's upper forelimb and neck.

External jugular vein

Vagus nerve

Nerves of the brachial plexus

Brachiocephalic vein

Lung

Right ventricle

Larynx

Trachea

Left common carotid artery

Brachial vein

External jugular vein

Brachial artery

Subscapular vein

Axillary vein

Subclavian vein

Thymus gland

Parietal pericardium (cut)

Left ventricle

Coronary artery

FIGURE 63.6 Organs of the cat's thoracic and abdominal cavities.

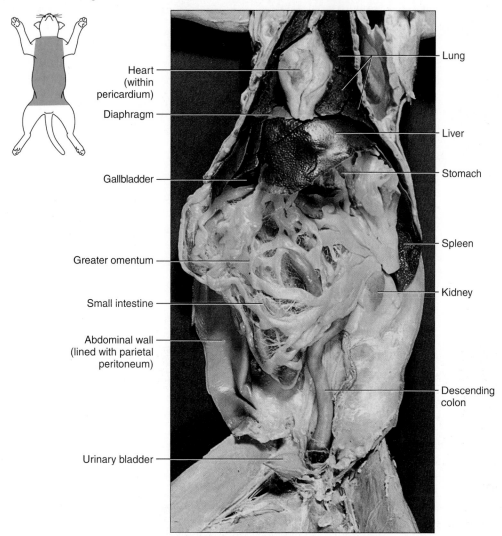

Heart (within pericardium)

Diaphragm

Gallbladder

Greater omentum

Small intestine

Abdominal wall (lined with parietal peritoneum)

Urinary bladder

Lung

Liver

Stomach

Spleen

Kidney

Descending colon

d. Note the *parietal peritoneum* that forms the inner lining of the abdominal wall. Also note the *greater omentum,* a structure composed of a double layer of peritoneum that hangs from the border of the stomach and covers the lower abdominal organs like a fatty apron (fig. 63.6). Reflect the abdominal organs to one side to expose various blood vessels in the abdominopelvic cavity region.

9. As you expose and dissect blood vessels, try not to destroy other visceral organs needed for future studies. Using figures 63.7, 63.8, and 63.9 as guides, locate and dissect the following arteries of the abdomen:

abdominal aorta—unpaired

celiac artery—unpaired

hepatic artery—unpaired

gastric artery—unpaired

splenic artery—unpaired

anterior mesenteric artery—corresponds to superior mesenteric artery; unpaired

renal arteries—paired

adrenolumbar arteries—paired

gonadal arteries—paired

posterior mesenteric artery—corresponds to inferior mesenteric artery; unpaired

iliolumbar arteries—paired

external iliac arteries—paired

internal iliac arteries—paired

caudal artery—unpaired

FIGURE 63.7 AP|R_{Cat}   Arteries of the cat's abdomen and hindlimb.

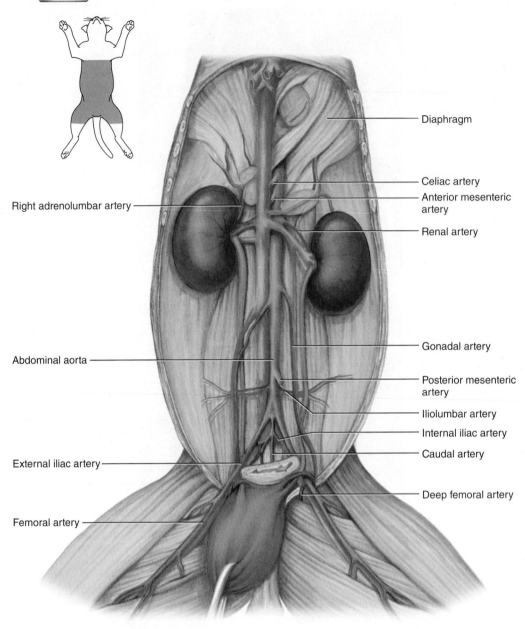

Diaphragm

Celiac artery

Anterior mesenteric artery

Right adrenolumbar artery

Renal artery

Gonadal artery

Abdominal aorta

Posterior mesenteric artery

Iliolumbar artery

Internal iliac artery

Caudal artery

External iliac artery

Deep femoral artery

Femoral artery

FIGURE 63.8 Arteries of the cat's abdomen, left lateral view.

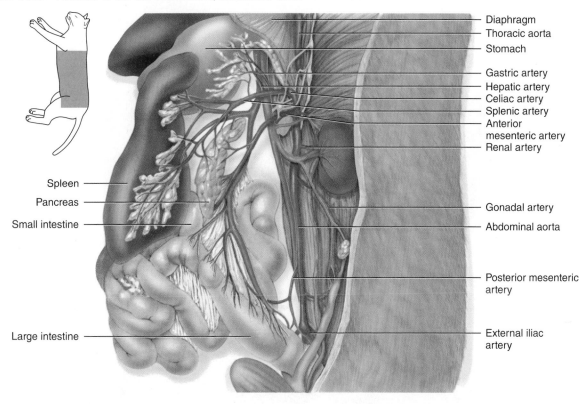

- Diaphragm
- Thoracic aorta
- Stomach
- Gastric artery
- Hepatic artery
- Celiac artery
- Splenic artery
- Anterior mesenteric artery
- Renal artery
- Spleen
- Pancreas
- Small intestine
- Gonadal artery
- Abdominal aorta
- Posterior mesenteric artery
- Large intestine
- External iliac artery

FIGURE 63.9 Blood vessels of the cat's abdominal cavity, with the digestive organs removed.

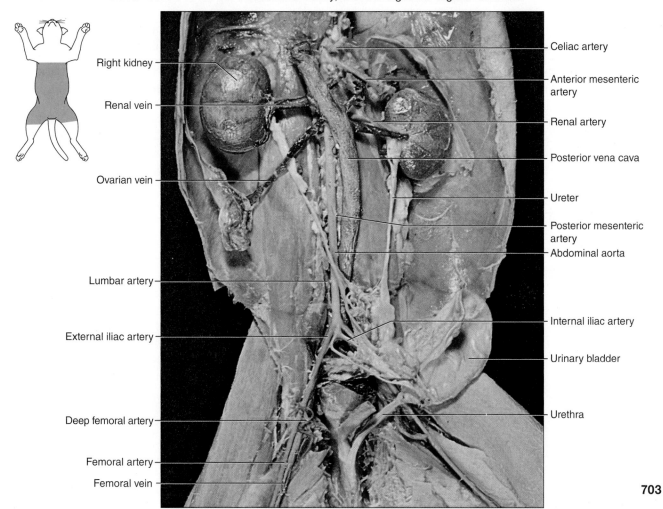

- Right kidney
- Renal vein
- Ovarian vein
- Lumbar artery
- External iliac artery
- Deep femoral artery
- Femoral artery
- Femoral vein
- Celiac artery
- Anterior mesenteric artery
- Renal artery
- Posterior vena cava
- Ureter
- Posterior mesenteric artery
- Abdominal aorta
- Internal iliac artery
- Urinary bladder
- Urethra

FIGURE 63.10   Blood vessels of the cat's upper hindlimb.

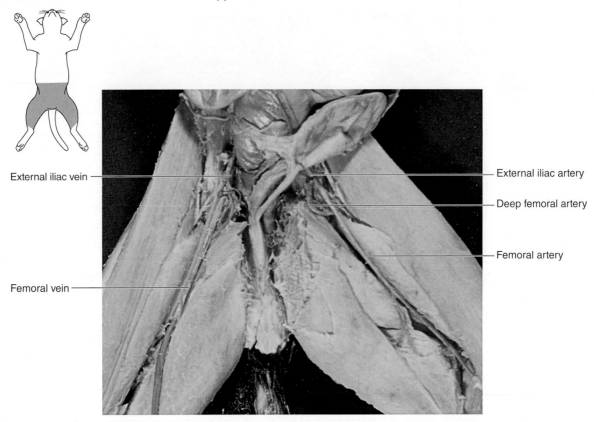

External iliac vein

Femoral vein

External iliac artery

Deep femoral artery

Femoral artery

10. Trace the external iliac artery into the right hindlimb (fig. 63.10), and locate the following:

femoral artery

deep femoral artery

11. Review the locations of the heart structures and arteries without the aid of figures.

12. Complete Part A of Laboratory Assessment 63.

PROCEDURE B: The Venous System

1. Examine the heart again and locate the following veins:

anterior vena cava—corresponds to superior vena cava

posterior vena cava—corresponds to inferior vena cava

pulmonary veins

2. Using figures 63.4, 63.5, and 63.11 as guides, locate and dissect the following veins in the thorax and neck:

right brachiocephalic vein

left brachiocephalic vein

right subclavian vein

left subclavian vein

internal jugular vein

external jugular vein

3. Trace the right subclavian vein into the forelimb and locate the following veins:

axillary vein

subscapular vein

brachial vein

FIGURE 63.11   AP|R_{Cat}   Veins of the cat's thorax, neck, and forelimb.

Right internal jugular vein

Subscapular vein

Brachial vein

Long thoracic vein

Axillary vein

Right subclavian vein

Right brachiocephalic vein

Left external jugular vein

Left subclavian vein

Left brachiocephalic vein

Sternal vein

Anterior vena cava

Posterior vena cava

Diaphragm

FIGURE 63.12 AP|R^{Cat} Veins of the cat's abdomen and hindlimb.

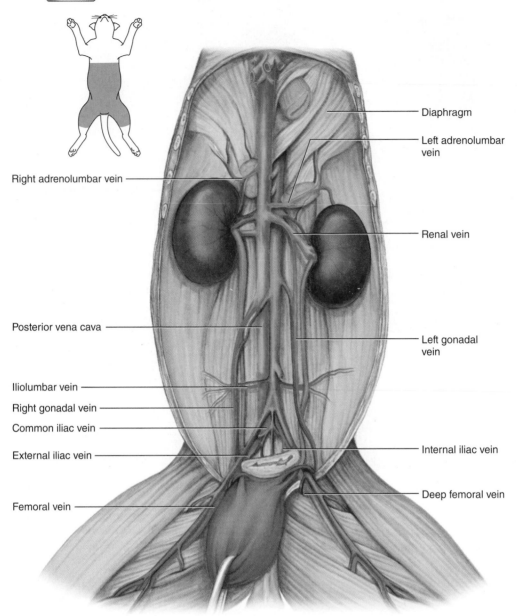

Diaphragm

Left adrenolumbar vein

Right adrenolumbar vein

Renal vein

Posterior vena cava

Left gonadal vein

Iliolumbar vein

Right gonadal vein

Common iliac vein

Internal iliac vein

External iliac vein

Deep femoral vein

Femoral vein

4. Using figures 63.9, 63.12, and 63.13 as guides, locate and dissect the following veins in the abdomen:

posterior vena cava

adrenolumbar vein

anterior mesenteric vein—corresponds to superior mesenteric vein

posterior mesenteric vein—corresponds to inferior mesenteric vein

hepatic portal vein

splenic vein

renal vein

gonadal vein

iliolumbar vein

common iliac vein

internal iliac vein

external iliac vein

5. Trace the external iliac vein into the hindlimb (fig. 63.10) and locate the following veins:

femoral vein

deep femoral vein

6. Review the locations of the veins without the aid of figures.

7. Complete Parts B, C, and D of the laboratory assessment.

FIGURE 63.13 Veins of the cat's abdomen, left lateral view.

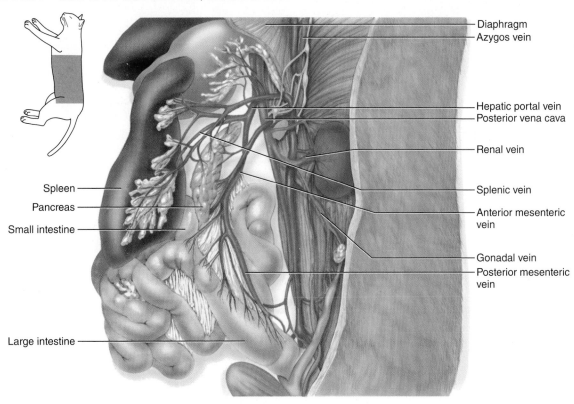

Diaphragm
Azygos vein

Hepatic portal vein
Posterior vena cava

Renal vein

Splenic vein

Anterior mesenteric
vein

Gonadal vein
Posterior mesenteric
vein

Spleen

Pancreas

Small intestine

Large intestine

NOTES

LABORATORY
ASSESSMENT
63

Name _____

Date _____

Section _____

The (A) corresponds to the indicated Learning Outcome(s) (O) found at the beginning of the Laboratory Exercise.

Cat Dissection: Cardiovascular System

PART A: Assessments

Complete the following:

1. Describe the location and position of the heart of the cat and the surrounding attachments of the parietal pericardium. (A1)

_____

_____

2. Describe the relative thicknesses of the walls of the heart chambers of the cat. (A1)

_____

_____

3. Explain how the wall thicknesses are related to the functions of the chambers. (A1)

_____

_____

4. Compare the origins of the common carotid arteries of the cat with those of the human. (A2)

_____

_____

5. Compare the origins of the iliac arteries of the cat with those of the human. (A2)

_____

_____

PART B: Assessments

Complete the following:

1. Compare the origins of the brachiocephalic veins of the cat with those of the human. (A2)

_____

_____

2. Compare the relative sizes of the external and internal jugular veins of the cat with those of the human. (A2)

_____

_____

_____

3. List twelve veins that cats and humans have in common.

_____

_____

_____

PART C: Assessments

Identify the numbered cardiovascular features of the cat in figures 63.14, 63.15, 63.16, and 63.17.

_____

FIGURE 63.14 Label the numbered cardiovascular structures of the cat. A1

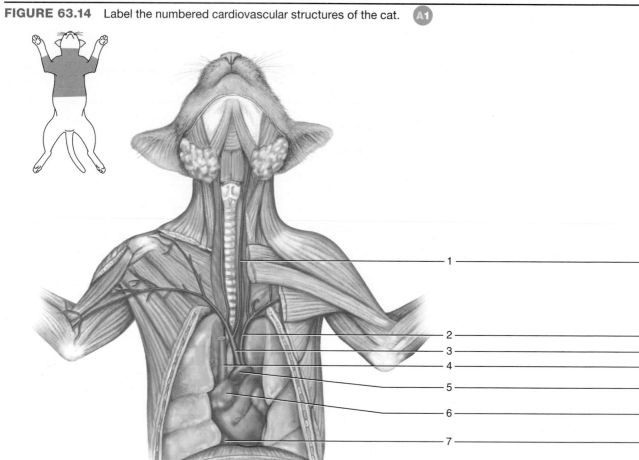

1 ——————————————————

2 ——————————————————
3 ——————————————————
4 ——————————————————
5 ——————————————————
6 ——————————————————
7 ——————————————————

FIGURE 63.15 Label the numbered cardiovascular structures of the cat. **A1**

1 _____

2 _____

3 _____

4 _____

5 _____

6 _____

7 _____

FIGURE 63.16 Label the numbered cardiovascular blood vessels of the cat.

FIGURE 63.17 Label the numbered cardiovascular structures of the cat. A1

PART D: Assessments

Observe the human torso model and figures 63.7, 63.9, and 63.12 of a cat. Locate the labeled features and identify the numbered features in figure 63.18 of the cadaver that were also identified in the cat dissection.

FIGURE 63.18 AP|R^{Cat} Identify the arteries and veins indicated on this anterior view of the abdomen of a cadaver, using the terms provided. A2 A3

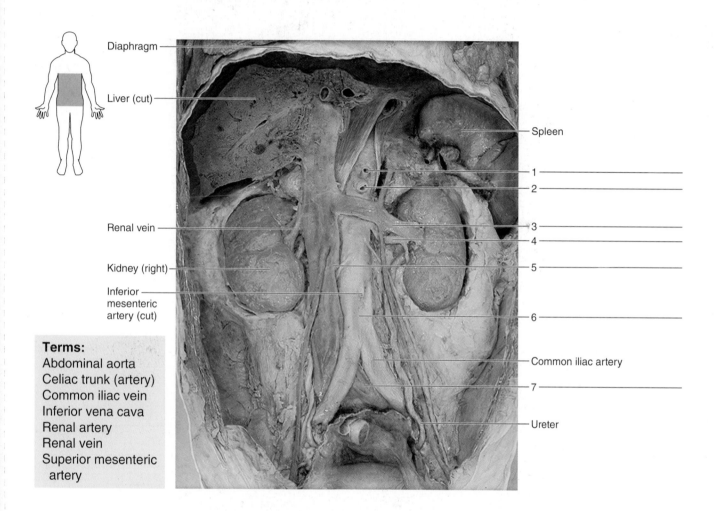

Diaphragm

Liver (cut)

Spleen

1

2

Renal vein

3

4

Kidney (right)

5

Inferior mesenteric artery (cut)

6

Common iliac artery

7

Ureter

Terms:
Abdominal aorta
Celiac trunk (artery)
Common iliac vein
Inferior vena cava
Renal artery
Renal vein
Superior mesenteric artery

NOTES

LABORATORY
EXERCISE 64

Cat Dissection: Respiratory System

Purpose of the Exercise

To examine the major respiratory organs of the cat and to compare and associate them with the organs of the human.

MATERIALS NEEDED

Preserved cat

Dissecting tray

Dissecting instruments

Bone cutter

Disposable gloves

Human torso model

SAFETY

▶ Wear disposable gloves when working on the cat dissection.

▶ Dispose of tissue remnants and gloves as instructed.

▶ Wash the dissecting tray and instruments as instructed.

▶ Wash your laboratory table.

▶ Wash your hands before leaving the laboratory.

Learning Outcomes

After completing this exercise, you should be able to:

01 Locate and identify the major respiratory organs of the cat.

02 Compare the respiratory system of the human with that of the cat.

03 Identify the corresponding organs in the human torso model.

The **O** corresponds to the assessments **A** indicated in the Laboratory Assessment for this Exercise.

Pre-Lab

Carefully read the introductory material and examine the entire lab. Be familiar with the structures and functions of the human respiratory system from lecture, the textbook, or laboratory exercises.

In this laboratory exercise, you will dissect the major respiratory organs of a cat. As you observe these structures in the cat, compare them with those of the human torso model.

PROCEDURE: Respiratory System Dissection

1. Place the preserved cat in a dissecting tray with its ventral surface up.
2. Examine the *external nares* (*nostrils*) and *nasal septum*.

3. Open the oral cavity. To do this, follow these steps:
   a. Use scissors to cut through the soft tissues at the angle of the mouth.
   b. When you reach the bone of the jaw, use a bone cutter to cut through the bone, thus freeing the mandible.
4. Open the oral cavity wide to expose the *hard palate* and the *soft palate* (fig. 64.1). Cut through the tissues of the soft palate and observe the *nasopharynx* above it. Locate the small openings of the *auditory tubes* in the lateral walls of the nasopharynx. Insert a probe into an opening of an auditory tube.

FIGURE 64.1 AP|R_{Cat}   Features of the cat's neck and oral cavity, ventral view.

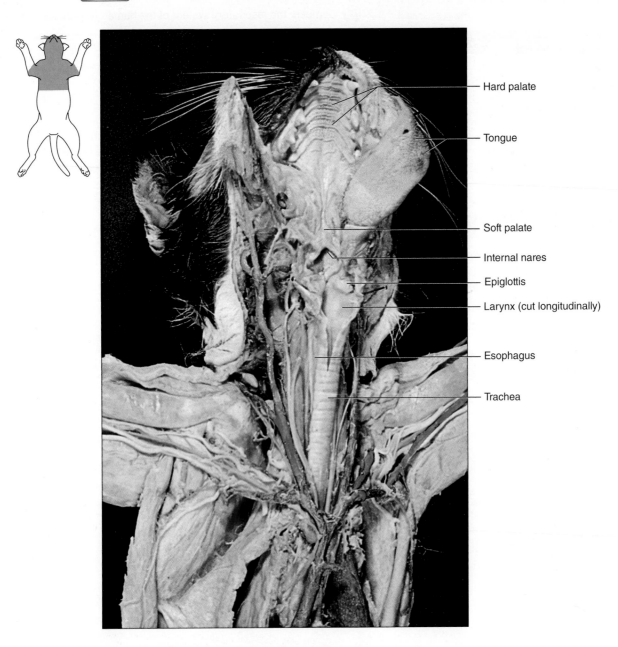

Hard palate

Tongue

Soft palate

Internal nares

Epiglottis

Larynx (cut longitudinally)

Esophagus

Trachea

5. Pull the tongue forward, and locate the *epiglottis* near its base. Also identify the *glottis,* an opening between the *vocal folds (vocal cords)* into the larynx. Locate the *esophagus,* which is dorsal to the larynx (figs. 64.1 and 64.2).

6. Open the thoracic cavity and expose its contents. See figures 64.1, 64.2, and 64.3.

7. Dissect the *trachea* in the neck and expose the *larynx* at the anterior end near the base of the tongue. Note the *tracheal rings* and locate the lobes of the *thyroid gland* on each side of the trachea, just posterior to the larynx. Also note the *thymus gland,* a rather diffuse mass of glandular tissue extending along the ventral surface of the trachea into the thorax (figs. 64.1, 64.2, and 64.3).

8. Examine the larynx, and identify the *thyroid cartilage* and *cricoid cartilage* in its wall (fig. 64.4*a*). Make a longitudinal incision through the ventral wall of the larynx and locate the vocal folds inside. Compare the structures of the larynx of the cat with those of a human (fig. 64.4*b*).

FIGURE 64.2   AP|R^{Cat}   Ventral view of the cat's thoracic cavity.

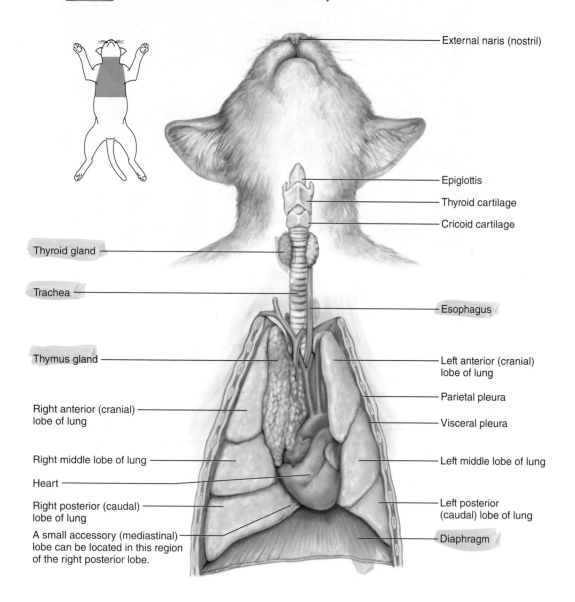

Labels (left side, top to bottom):
- Thyroid gland
- Trachea
- Thymus gland
- Right anterior (cranial) lobe of lung
- Right middle lobe of lung
- Heart
- Right posterior (caudal) lobe of lung
- A small accessory (mediastinal) lobe can be located in this region of the right posterior lobe.

Labels (right side, top to bottom):
- External naris (nostril)
- Epiglottis
- Thyroid cartilage
- Cricoid cartilage
- Esophagus
- Left anterior (cranial) lobe of lung
- Parietal pleura
- Visceral pleura
- Left middle lobe of lung
- Left posterior (caudal) lobe of lung
- Diaphragm

FIGURE 64.3 AP|R^{Cat} Structures of the lower respiratory tract of the cat, ventral view.

Trachea

Left lung

Heart

Right lung

9. Observe a sagittal view of the respiratory tract of the cat (fig. 64.5), and compare this to ventral views of the cat (figs. 64.1, 64.2, and 64.3).

10. Trace the trachea posteriorly to where it divides into *main (primary) bronchi*, which pass into the lungs.

11. Examine the *lungs,* each of which is subdivided into three lobes—an *anterior (cranial),* a *middle,* and a *posterior (caudal) lobe.* The right lung of the cat has a small fourth *accessory (mediastinal; cardiac) lobe* associated with the right posterior lobe. Notice the thin membrane, the *visceral pleura,* on the surface of each lung. Also notice the parietal pleura, which forms the inner lining of the thoracic wall, and locate the spaces of the *pleural cavities* (figs. 64.2 and 64.3).

12. Make an incision through a lobe of a lung and examine its interior. Note the branches of the smaller air passages.

13. Examine the *diaphragm* and note that its central portion is tendinous.

14. Locate the *phrenic nerve.* This nerve appears as a white thread passing along the side of the heart to the diaphragm.

15. Examine the organs located in the *mediastinum.* Note that the mediastinum separates the right and left lungs and pleural cavities within the thorax.

16. Review the locations of the respiratory organs without the aid of figures.

17. Complete Parts A and B of Laboratory Assessment 64.

FIGURE 64.4 AP|R_{Cat} (a) Larynx and associated structures of the cat, ventral view. (b) Larynx structures of a cadaver, anterior view.

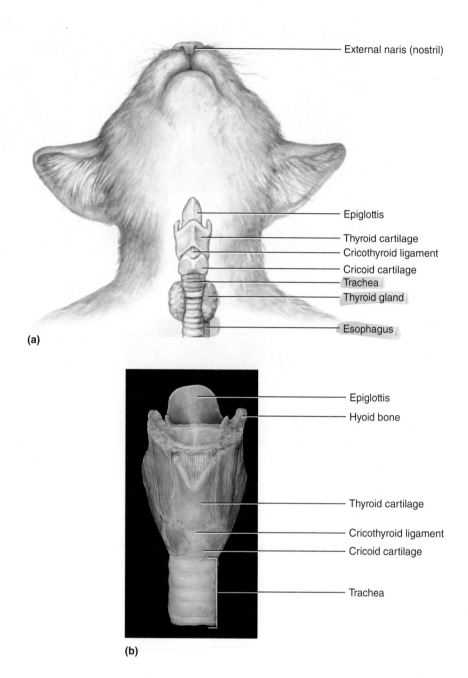

(a)

- External naris (nostril)
- Epiglottis
- Thyroid cartilage
- Cricothyroid ligament
- Cricoid cartilage
- Trachea
- Thyroid gland
- Esophagus

(b)

- Epiglottis
- Hyoid bone
- Thyroid cartilage
- Cricothyroid ligament
- Cricoid cartilage
- Trachea

FIGURE 64.5 AP|R_{Cat} Upper respiratory tract of the cat, sagittal view.

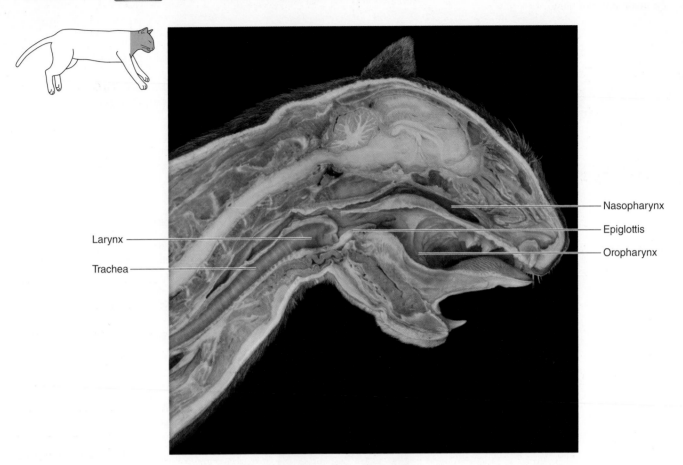

Nasopharynx

Epiglottis

Oropharynx

Larynx

Trachea

Name _____

Date _____

Section _____

The Ⓐ corresponds to the indicated Learning Outcome(s) Ⓞ found at the beginning of the Laboratory Exercise.

LABORATORY
ASSESSMENT
64

Cat Dissection: Respiratory System

PART A: Assessments

Complete the following:

1. What is the purpose of the auditory tubes opening into the nasopharynx? **A1** _____

_____

_____

_____

2. Distinguish between the glottis and the epiglottis. **A1** _____

_____

_____

3. Are the tracheal rings of the cat complete or incomplete? **A1** _____ How does this feature compare with that of the human? **A2**

_____

_____

4. How does the structure of the main (primary) bronchi compare with that of the trachea? **A1** _____

_____

_____

5. Compare the number of lobes in the human lungs with the number of lobes in the cat. **A2** _____

_____

_____

_____

6. Describe the attachments of the diaphragm. **A1** _____

_____

_____

7. What major structures are located within the mediastinum of the cat? **A1** _____

_____

How does this compare to the human mediastinum? **A2** _____

_____

PART B: Assessments

Observe the human torso model and figures 64.4b and 64.6 of a cadaver. Locate the labeled features and identify the numbered features in figure 64.6 that were also identified in the cat dissection.

FIGURE 64.6   **AP|R**^Cat   Identify the respiratory features indicated on this anterior view of the thorax, using the terms provided.   A2   A3

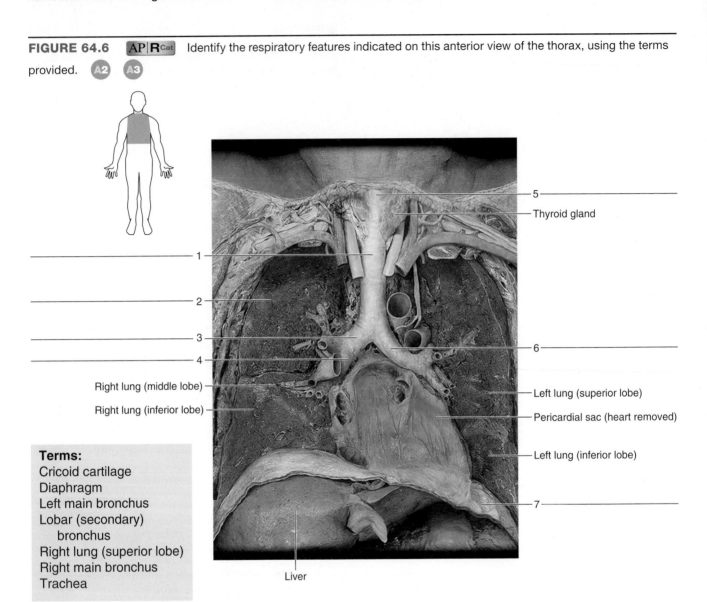

5 ————————
Thyroid gland

1 ————————

2 ————————

3 ————————

4 ————————

6 ————————

Right lung (middle lobe) ————

Right lung (inferior lobe) ————

Left lung (superior lobe) ————

Pericardial sac (heart removed) ————

Left lung (inferior lobe) ————

7 ————————

Liver

Terms:
Cricoid cartilage
Diaphragm
Left main bronchus
Lobar (secondary)
   bronchus
Right lung (superior lobe)
Right main bronchus
Trachea

LABORATORY
EXERCISE 65

Cat Dissection: Digestive System

Purpose of the Exercise

To examine the major digestive organs of the cat and to compare these organs with those of the human.

MATERIALS NEEDED

Preserved cat

Dissecting tray

Dissecting instruments

Bone cutter

Disposable gloves

Hand lens

Human torso model

SAFETY ⚠

▶ Wear disposable gloves when working on the cat dissection.

▶ Dispose of tissue remnants and gloves as instructed.

▶ Wash the dissecting tray and instruments as instructed.

▶ Wash your laboratory table.

▶ Wash your hands before leaving the laboratory.

Learning Outcomes

After completing this exercise, you should be able to:

O1 Locate and identify the major digestive organs of the cat.

O2 Compare the digestive system of the cat with that of the human.

O3 Identify the corresponding organs in the human.

The **O** corresponds to the assessments **A** indicated in the Laboratory Assessment for this Exercise.

Pre-Lab

Carefully read the introductory material and examine the entire lab. Be familiar with the structures and functions of the human digestive system from lecture, the textbook, or laboratory exercises.

In this laboratory exercise, you will dissect the major digestive organs of the cat. As you observe these organs, compare them with those of the human by observing the parts of the human torso model; however, keep in mind that the cat is a carnivore (flesh-eating mammal) and that the organs of its digestive system are adapted to capturing, holding, eating, and digesting the bodies of other animals. In contrast, humans are omnivores (eat both plant and animal substances). Because humans are adapted to eating and digesting a greater variety of foods, comparisons of the cat and the human digestive organs may not be precise.

PROCEDURE: Digestive System Dissection

1. Place the preserved cat in the dissecting tray on its left side.
2. Locate the major salivary glands on one side of the head (fig. 65.1a). To do this, follow these steps:
   a. Clear away any remaining fascia and other connective tissue from the region below the ear and near the joint of the mandible.
   b. Identify the *parotid gland,* a relatively large mass of glandular tissue just below the ear. Note the parotid duct (Stensen's duct) that passes over the surface of the masseter muscle and opens into the oral cavity.

FIGURE 65.1 AP|R_{Cat} Major salivary glands. (a) The cat, lateral view. (b) Cadaver, lateral view.

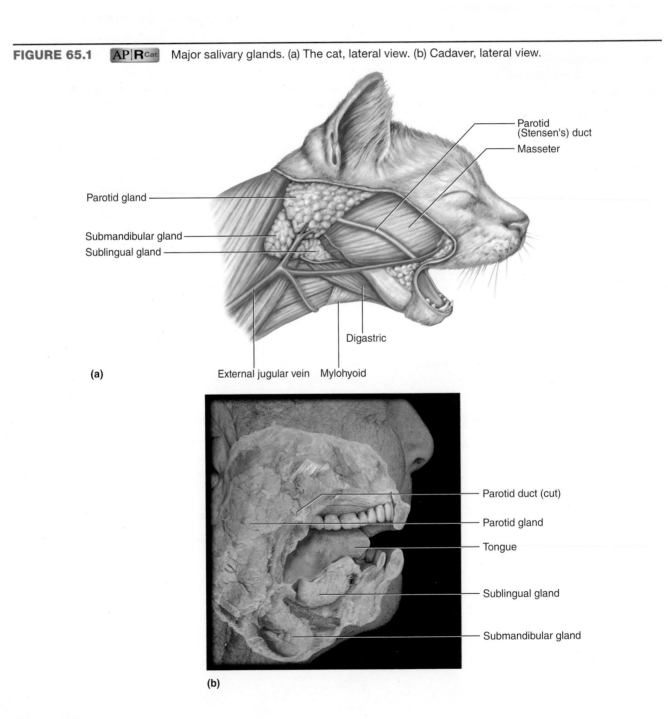

(a)

(b)

c. Look for the *submandibular gland* just below the parotid gland, near the angle of the jaw.

d. Locate the *sublingual gland* that is adjacent and medial to the submandibular gland.

e. Compare the location of the salivary glands of the cat with those of a human (fig. 65.1*b*).

3. Open the oral cavity. To do this, follow these steps:

a. Use scissors to cut through the soft tissues at the angle of the mouth if not previously accomplished.

b. When you reach the bone of the jaw, use a bone cutter to cut through the bone, thus freeing the mandible.

c. Open the mouth wide, and locate the following features:

cheek

lip

vestibule

palate

- hard palate—has transverse ridges (rugae) to help hold food
- soft palate—lacks a uvula

palatine tonsils—small, rounded masses of glandular tissue in the lateral wall of the soft palate

tongue

- frenulum
- papillae—examine with a hand lens to observe spiny projections used to clean fur

4. Examine the teeth of the upper jaw. The adult cat has 6 incisors, 2 canines, 8 premolars, and 2 molars. The lower jaw teeth are similar, except 4 premolars are present.

5. Complete Part A of Laboratory Assessment 65.

6. Examine organs in the abdominal cavity with the cat positioned with its ventral surface up. Review the normal position and function of the greater omentum.

7. Examine the *liver,* which is located just beneath the diaphragm and is attached to the central portion of the diaphragm by the *falciform ligament.* Also, locate the *spleen,* which is posterior to the liver on the left side (fig. 65.2). The liver has five lobes—a right medial, right lateral (which is subdivided into two parts by a deep cleft), left medial, left lateral, and caudate lobe. The caudate lobe is the smallest lobe and is located in the median line, where it projects into the curvature of the stomach. The caudate lobe is covered by a sheet of mesentery, the *lesser omentum,* that connects the liver to the stomach (figs. 65.3 and 65.4). Find the greenish *gallbladder* on the inferior surface of the liver on the right side. Also note the *cystic duct,* by which the gallbladder is attached to the *bile duct,* and the *hepatic duct,* which originates in the liver and attaches to the cystic duct. Trace the bile duct to its connection with the duodenum.

8. Locate the *stomach* in the left side of the abdominal cavity. At its anterior end, note the union with the *esophagus,* which passes through the diaphragm. Identify the *cardia, fundus, body,* and *pyloric part* (listed from entrance to exit) of the stomach. Use scissors to make an incision along the convex border of the stomach from the cardia region to the pyloric part. The lining of the stomach has numerous *gastric folds (rugae).* Examine the *pyloric sphincter,* which creates a constriction between the stomach and small intestine.

9. Locate the *pancreas.* It appears as a grayish, two-lobed, elongated mass of glandular tissue. One lobe lies dorsal to the stomach and extends across to the duodenum. The other lobe is enclosed by the *mesentery,* which supports the duodenum (figs. 65.3 and 65.4).

10. Trace the *small intestine,* beginning at the pyloric sphincter. The first portion, the *duodenum,* travels posteriorly for several centimeters. Then it loops back around a lobe of the pancreas. The proximal half of the remaining portion of the small intestine is the *jejunum,* and the distal half is the *ileum.* Open the duodenum and note the velvety appearance of the villi. Note how the mesentery supports the small intestine from the dorsal body wall. The small intestine terminates on the right side where it joins the large intestine.

11. Locate the *large intestine* and identify the *cecum, ascending colon, transverse colon,* and *descending colon.* Also locate the *rectum,* which extends through the pelvic cavity to the *anus.* Make an incision at the junction between the ileum and cecum and look for the *ileocecal sphincter.* An appendix is absent in cats.

12. Review the locations of the digestive organs without the aid of figures.

13. Complete Parts B, C, and D of the laboratory assessment.

FIGURE 65.2   **AP|R**_{Cat}   Ventral view of abdominal organs of the cat: (a) illustration; (b) dissected specimen.

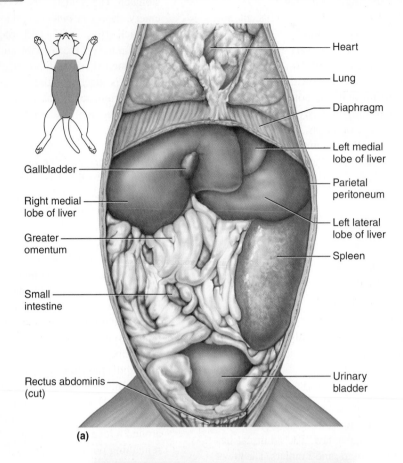

Heart

Lung

Diaphragm

Left medial
lobe of liver

Parietal
peritoneum

Left lateral
lobe of liver

Spleen

Urinary
bladder

Gallbladder

Right medial
lobe of liver

Greater
omentum

Small
intestine

Rectus abdominis
(cut)

(a)

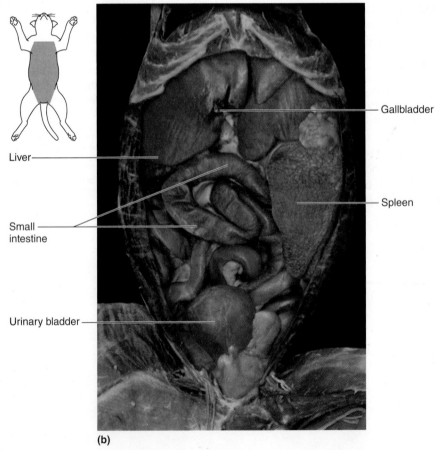

Gallbladder

Spleen

Liver

Small
intestine

Urinary bladder

(b)

FIGURE 65.3 Ventral view of the cat's abdominal organs with intestines reflected to the right side.

Right medial lobe of liver
Bile duct
Lesser omentum
Right lateral lobe of liver
Duodenum
Pancreas
Transverse colon
Small intestine
Mesentery
Descending colon

Gallbladder
Diaphragm
Left medial lobe of liver
Left lateral lobe of liver
Esophagus
Cardia
Fundus
Body
Pyloric part
Stomach
Spleen
Kidney
Parietal peritoneum
Rectum
Urinary bladder

FIGURE 65.4 AP|R^{Cat} Abdominal organs of the cat with the greater omentum removed. The left liver lobes have been reflected to the right side.

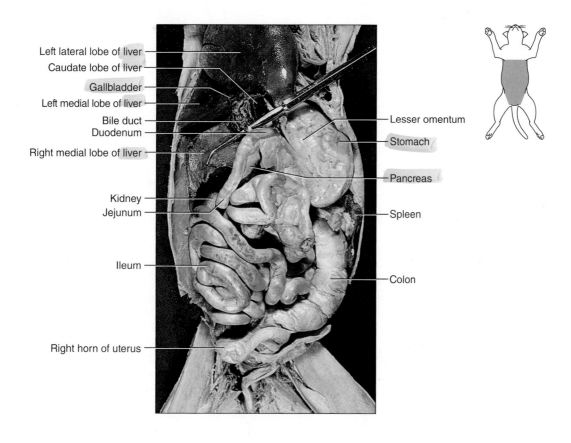

Left lateral lobe of liver
Caudate lobe of liver
Gallbladder
Left medial lobe of liver
Bile duct
Duodenum
Right medial lobe of liver
Kidney
Jejunum
Ileum
Right horn of uterus

Lesser omentum
Stomach
Pancreas
Spleen
Colon

Name _____

Date _____

Section _____

The Ⓐ corresponds to the indicated Learning Outcome(s) Ⓞ found at the beginning of the Laboratory Exercise.

Cat Dissection: Digestive System

PART A: Assessments

Complete the following:

1. Compare the locations of the major salivary glands of the human with those of the cat. A2 _____
_____
_____

2. Compare the types and numbers of teeth present in the cat's upper and lower jaws with those of the human. A2
_____
_____
_____
_____

3. Describe how the cat's teeth are adapted to a special diet. A1 _____
_____
_____

4. What part of the human soft palate is lacking in the cat? A2 _____

5. What do you think is the function of the transverse ridges (rugae) in the hard palate of the cat? A1 _____
_____
_____

6. How do the papillae on the surface of the human tongue compare with those of the cat? A2
_____
_____

PART B: Assessments

Complete the following:

1. Describe how the peritoneum and mesenteries are associated with the organs in the abdominal cavity. (A1) _____

   _____

   _____

   _____

   _____

2. Describe the inner lining of the stomach. (A1) _____

   _____

3. Compare the structure of the human liver with that of the cat. (A2) _____

   _____

   _____

   _____

4. Compare the structure and location of the human pancreas with those of the cat. (A2) _____

   _____

   _____

5. What feature of the human cecum is lacking in the cat? (A2) _____

   _____

PART C: Assessments

Identify the numbered digestive features of the cat in figure 65.5.

FIGURE 65.5 Label the numbered digestive features of the cat: (a) illustration view; (b) dissection view. **A1**

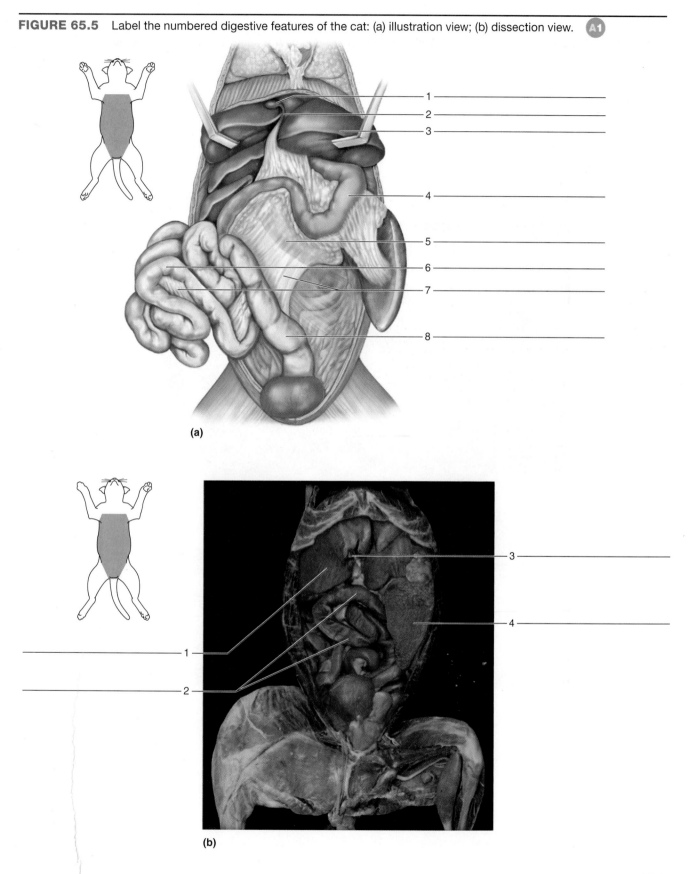

(a)

(b)

Part D   Assessments

Observe the human torso model and figure 65.6 of a cadaver. Locate the labeled features, and identify the numbered features in figure 65.6 that were also identified in the cat dissection.

FIGURE 65.6   **AP|R**Cat   Identify the digestive features indicated on this anterior view of the abdomen of a cadaver, using the terms provided.   A2   A3

Diaphragm

Falciform ligament

Appendix

Urinary bladder

1

2

3

4

5

6

7

8

Terms:
Ascending colon
Gallbladder
Lesser omentum
Liver
Mesentery
Small intestine
Stomach
Transverse colon

Cat Dissection: Urinary System

Purpose of the Exercise

To examine and identify the urinary organs of the cat and to compare them with those of the human.

Preserved cat

Dissecting tray

Dissecting instruments

Disposable gloves

Human torso model

▶ Wear disposable gloves when working on the cat dissection.

▶ Dispose of tissue remnants and gloves as instructed.

▶ Wash the dissecting tray and instruments as instructed.

▶ Wash your laboratory table.

▶ Wash your hands before leaving the laboratory.

Learning Outcomes AP|R Cat

After completing this exercise, you should be able to:

○1 Locate and identify the urinary organs of the cat.

○2 Compare the urinary organs of the cat with those of the human.

○3 Identify the corresponding organs in the human torso model and a cadaver.

The ○ corresponds to the assessments A indicated in the Laboratory Assessment for this Exercise.

Pre-Lab

Carefully read the introductory material and examine the entire lab. Be familiar with the structures and functions of the human urinary system from lecture, the textbook, or laboratory exercises.

In this laboratory exercise, you will dissect the urinary organs of the cat. As you observe these structures, compare them with the corresponding human organs by observing the parts of the human torso model.

PROCEDURE: Urinary System Dissection

1. Place the preserved cat in a dissecting tray with its ventral surface up.
2. Open the abdominal cavity and remove the liver, stomach, and spleen (figs. 66.1 and 66.2).
3. Push the intestines to one side and locate the *kidneys* in the dorsal abdominal wall on either side of the vertebral column. The kidneys are located dorsal to the *parietal peritoneum* (retroperitoneal).
4. Carefully remove the parietal peritoneum and the adipose tissue surrounding the kidneys. Locate the following, using figures 66.1, 66.2, and 66.3 as guides:

   ureters
   renal arteries
   renal veins

5. Locate the *adrenal glands,* which lie medially and anteriorly to the kidneys and are surrounded by connective tissues. The adrenal glands are attached to blood vessels and are separated by the abdominal aorta and posterior vena cava.
6. Expose the ureters by cleaning away the connective tissues along their lengths. They enter the *urinary bladder* on the dorsal surface.
7. Examine the urinary bladder, attached to the abdominal wall by folds of peritoneum that form ligaments. Locate the *medial ligament* on the ventral surface of the bladder that connects it with the linea alba and the *two lateral ligaments* on the dorsal surface.
8. Remove the connective tissue from around the urinary bladder. Use a sharp scalpel to open the bladder, and examine its interior. Locate the openings of the ureters and the urethra on the inside.
9. Expose the *urethra* at the posterior of the urinary bladder (figs. 66.2, 66.3, and 66.4). The urethra terminates with the vagina in a urogenital sinus in the female cat; the urethra enters the penis in the male cat.

FIGURE 66.1 AP|R^{Cat} Urinary system of the female cat's abdominal cavity, with most of the digestive and reproductive organs removed.

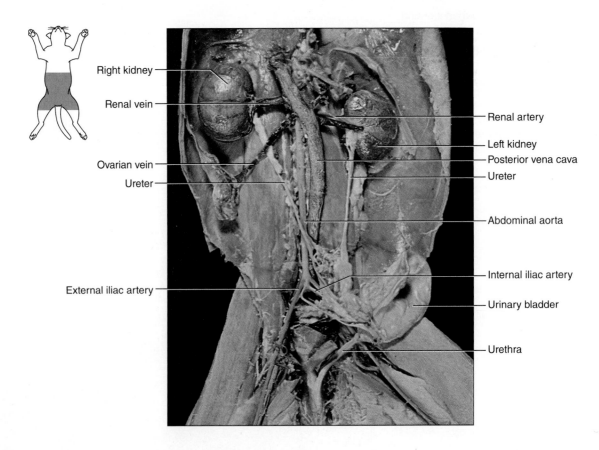

Right kidney
Renal vein
Ovarian vein
Ureter
External iliac artery

Renal artery
Left kidney
Posterior vena cava
Ureter
Abdominal aorta
Internal iliac artery
Urinary bladder
Urethra

FIGURE 66.2 Urinary system of the male cat's abdominal cavity, with most of the digestive and reproductive organs removed.

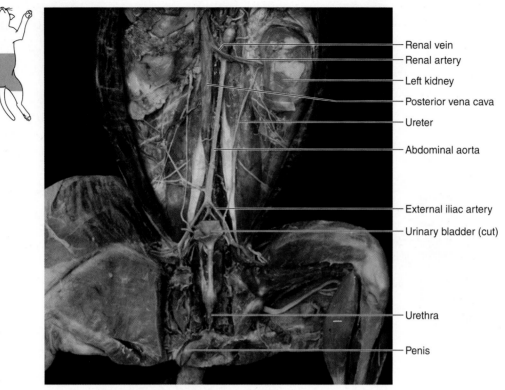

- Renal vein
- Renal artery
- Left kidney
- Posterior vena cava
- Ureter
- Abdominal aorta
- External iliac artery
- Urinary bladder (cut)
- Urethra
- Penis

FIGURE 66.3 The urinary organs of a female cat.

- Adrenal glands
- Kidney
- Renal vein
- Abdominal aorta
- Rectum
- Urethra

- Renal artery
- Posterior vena cava
- Ureters
- Urinary bladder

FIGURE 66.4　AP|R Cat　Female urinary and reproductive systems of a pregnant cat.

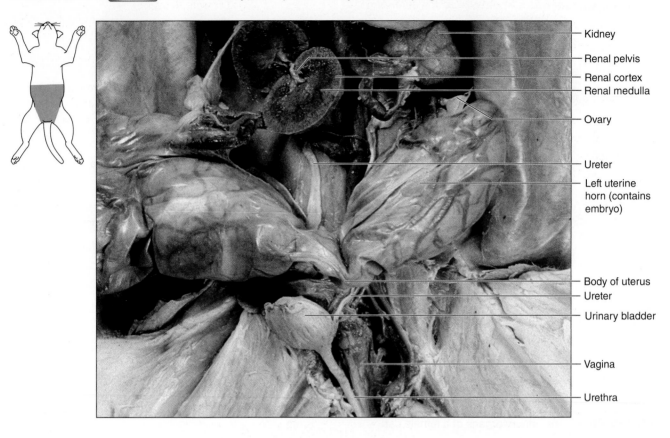

- Kidney
- Renal pelvis
- Renal cortex
- Renal medulla
- Ovary
- Ureter
- Left uterine horn (contains embryo)
- Body of uterus
- Ureter
- Urinary bladder
- Vagina
- Urethra

FIGURE 66.5　Ventral view of the cat kidney.

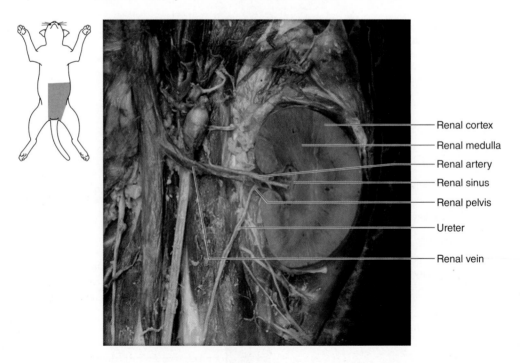

- Renal cortex
- Renal medulla
- Renal artery
- Renal sinus
- Renal pelvis
- Ureter
- Renal vein

10. Remove one kidney and section it longitudinally (figs. 66.4, 66.5, and 66.6a). Identify the following features:

fibrous capsule—encases a kidney

renal cortex—superficial, lighter region

renal medulla—deeper, darker region with striations

renal pyramid—single in cat

renal papilla—single in cat; opens into renal pelvis

renal sinus—hollow chamber of kidney

renal pelvis—funnel-shaped sac at kidney end of ureter

hilum of kidney—indented region containing renal blood vessels and ureter

11. Compare the structures of the kidney of the cat with those of a human (fig. 66.6b).
12. Review the locations of the urinary organs without the aid of figures.
13. Discard the organs and tissues that were removed from the cat, as directed by the laboratory instructor.
14. Complete Parts A and B of Laboratory Assessment 66.

FIGURE 66.6 AP|R^Cat Longitudinal section of a kidney. (a) The cat, (b) cadaver.

(a)

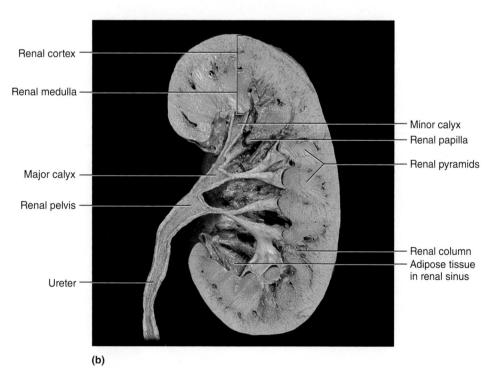

(b)

NOTES

Name _____

Date _____

Section _____

The Ⓐ corresponds to the indicated Learning Outcome(s) Ⓞ found at the beginning of the Laboratory Exercise.

Cat Dissection: Urinary System

PART A: Assessments

Complete the following:

1. Compare the positions of the kidneys in the cat with those in the human. Ⓐ2 _____

_____

_____

_____

2. Compare the locations of the adrenal glands in the cat with those in the human. Ⓐ2 _____

_____

_____

_____

_____

_____

3. What structures of the cat urinary system are retroperitoneal? Ⓐ1 _____

_____

_____

_____

4. Describe the wall of the urinary bladder of the cat. Ⓐ1 _____

_____

_____

_____

5. Compare the renal pyramids and renal papillae of the human kidney with those of the cat. Ⓐ2 _____

_____

_____

_____

PART B: Assessments

Observe the human torso model and figures 66.6*b* and 66.7 of a cadaver. Locate the labeled features and identify the numbered features in figure 66.7 that were also identified in the cat dissection.

FIGURE 66.7 AP|R^{Cat} Identify the urinary features indicated on this anterior view of the abdomen of a cadaver, using the terms provided. A2 A3

Liver
Diaphragm
Spleen
Inferior vena cava (cut)
Abdominal aorta
Major calyx
1
Adipose tissue
4
5
2
6
3

Terms:
Kidney
Renal artery
Renal pelvis
Renal pyramid
Ureter
Urinary bladder

Cat Dissection: Reproductive Systems

Purpose of the Exercise

To examine and identify the reproductive organs of the cat and to compare them with the corresponding organs of the human.

Preserved cat

Dissecting tray

Dissecting instruments

Bone cutters

Disposable gloves

Models of human reproductive systems

▶ Wear disposable gloves when working on the cat dissection.

▶ Dispose of tissue remnants and gloves as instructed.

▶ Wash the dissecting tray and instruments as instructed.

▶ Wash your laboratory table.

▶ Wash your hands before leaving the laboratory.

Learning Outcomes   AP|R^{Cat}

After completing this exercise, you should be able to:

 O1 Locate and identify the reproductive organs of a cat.

O2 Compare the reproductive organs of the cat with those of the human.

O3 Identify the corresponding organs in models of the human reproductive systems and a cadaver.

The **O** corresponds to the assessments **A** indicated in the Laboratory Assessment for this Exercise.

Pre-Lab

Carefully read the introductory material and examine the entire lab. Be familiar with the structures and functions of the human reproductive system from lecture, the textbook, or laboratory exercises.

In this laboratory exercise, you will dissect the reproductive system of the cat. If you have a female cat, begin with Procedure A. If you have a male cat, begin with Procedure B. After completing the dissection, exchange cats with someone who has dissected one of the opposite sex, and examine its reproductive organs.

As you observe the cat reproductive organs, compare them with the corresponding human organs by examining the models of the human reproductive systems.

PROCEDURE A: Female Reproductive System

1. Place the preserved female cat in a dissecting tray with its ventral surface up and open its abdominal cavity.
2. Locate the small, oval *ovaries* just posterior to the kidneys (figs. 67.1 and 67.2*a* and *b*).
3. Examine an ovary, suspended from the dorsal body wall by a fold of peritoneum called the *mesovarium.* Another attachment, the *ovarian ligament,* connects the ovary to the tubular *uterine horn* (fig. 67.1).

4. Near the anterior end of the ovary, locate the funnel-shaped *infundibulum,* which is at the end of a very small *uterine tube* (*oviduct*). Note the tiny projections, or *fimbriae,* that create a fringe around the edge of the infundibulum. Trace the uterine tube around the ovary to its nearby connection with the uterine horn near the attachment of the ovarian ligament.
5. Examine the uterine horn. It is suspended from the body wall by a fold of peritoneum, the *broad ligament,* that is continuous with the mesovarium of the ovary. Also note the fibrous *round ligament,* which extends from the uterine horn laterally and posteriorly to the body wall (figs. 67.1 and 67.2*a* and *b*).
6. To observe the remaining organs of the reproductive system more easily, remove the left hindlimb by severing it near the hip with bone cutters. Then use the bone cutters to remove the left anterior portion of the pelvic girdle. Also remove the necessary muscles and connective tissue to expose the structures shown in figure 67.3.
7. Trace the relatively long uterine horns posteriorly. They unite to form the *uterine body,* which is located between

FIGURE 67.1 Ventral view of the female cat's reproductive system.

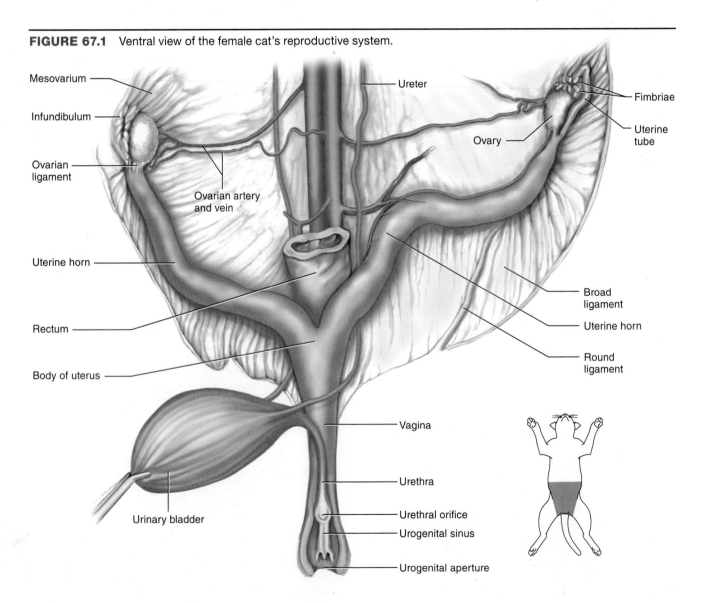

FIGURE 67.2 AP|RCat (a) Female urinary and reproductive systems of a pregnant cat. (b) Ventral view of the female cat pelvis.

(a)

Kidney

Kidney (cut)

Ovary

Ureter

Left uterine horn
(contains embryo)

Body of uterus

Urinary bladder

Vagina

Urethra

Left kidney

Uterine tube

Broad ligament

Uterine horn

Urinary bladder (cut)

Mesovarium

Ovary

Round
ligament
of uterus

Uterine horn

Body of uterus

(b)

FIGURE 67.3 Lateral view of the female cat's reproductive system.

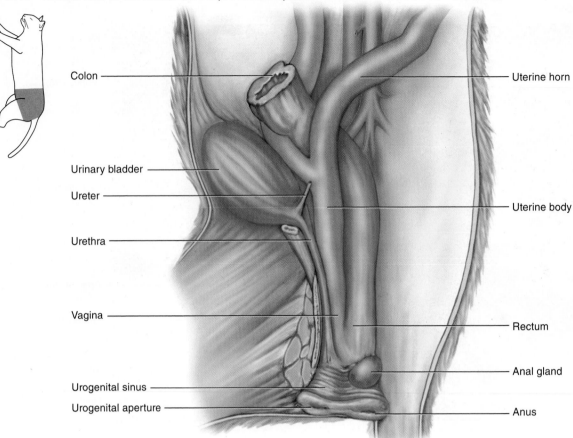

Colon

Urinary bladder

Ureter

Urethra

Vagina

Urogenital sinus

Urogenital aperture

Uterine horn

Uterine body

Rectum

Anal gland

Anus

the urethra and rectum. This Y-shaped uterus allows ample space for several offspring to develop at the same time in the uterine horns (fig. 67.2a and b). The uterine body is continuous with the *vagina,* which leads to the outside.

8. Trace the *urethra* from the urinary bladder posteriorly. The urethra and the vagina open into a common chamber called the *urogenital sinus.* The opening of this chamber, ventral to the anus, is called the *urogenital aperture.* Locate the *labia majora,* which are folds of skin on either side of the urogenital aperture.

9. Use scissors to open the vagina along its lateral wall, beginning at the urogenital aperture and continuing to the body of the uterus. Note the *urethral orifice* in the ventral wall of the urogenital sinus and locate the small, rounded *cervix* of the uterus, which projects into the vagina at its deep end. The *clitoris* is located in the ventral wall of the urogenital sinus near its opening to the outside.

10. Review the locations of the reproductive organs without the aid of figures.

11. Complete Part A of Laboratory Assessment 67.

PROCEDURE B: Male Reproductive System

1. Place the preserved male cat in a dissecting tray with its ventral surface up.

2. Locate the *scrotum* between the hindlimbs. Make an incision along the midline of the scrotum and expose the *testes* inside (fig. 67.4a and b). The testes are separated by a septum, and each testis is enclosed in a sheath of connective tissue.

3. Remove the sheath surrounding the testes and locate the convoluted *epididymis* on the dorsal surface of each testis (fig. 67.5).

FIGURE 67.4 AP|R_{Cat} (a) Male reproductive system of the cat. (b) Ventral view of the male cat pelvis.

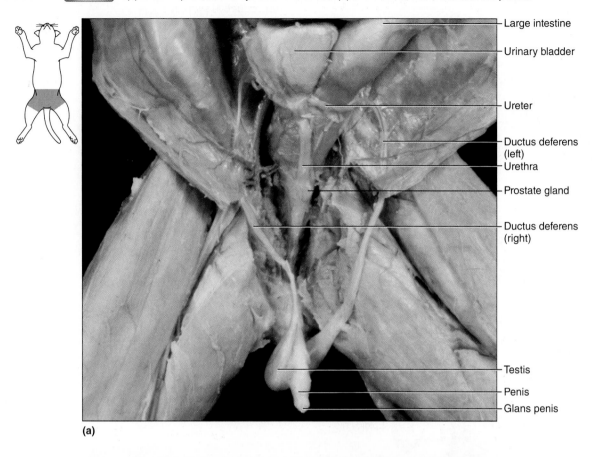

Large intestine
Urinary bladder
Ureter
Ductus deferens (left)
Urethra
Prostate gland
Ductus deferens (right)
Testis
Penis
Glans penis

(a)

Ductus (vas) deferens
Body of the penis
Glans penis
Testis
Epididymis

(b)

FIGURE 67.5 Ventral view of the male cat's urinary and reproductive systems.

- Adrenal gland
- Kidney
- Ureter
- Urinary bladder
- Ductus deferens
- Inguinal canal
- Prostate gland
- Rectum
- Urethra
- Spermatic cord
- Crus of penis
- Bulbourethral gland
- Testis
- Epididymis
- Penis
- Glans penis

FIGURE 67.6 Lateral view of the male cat's reproductive system.

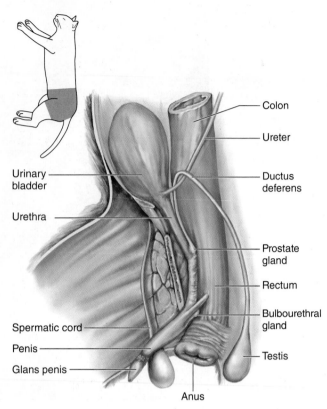

- Colon
- Ureter
- Urinary bladder
- Ductus deferens
- Urethra
- Prostate gland
- Rectum
- Bulbourethral gland
- Spermatic cord
- Penis
- Testis
- Glans penis
- Anus

4. Locate the *spermatic cord* on the right side, leading away from the testis. This spermatic cord contains the *ductus (vas) deferens,* which is continuous with the epididymis, as well as with the nerves and blood vessels that supply the testis on that side. Trace the spermatic cord to the body wall, where its contents pass through the *inguinal canal* and enter the pelvic cavity (fig. 67.5).

5. Locate the *penis,* and identify the *prepuce,* which forms a sheath around the entire penis. Make an incision through the skin of the prepuce and expose the *glans penis.* The glans penis has minute spines on its surface that assist copulation.

6. To observe the remaining structures of the reproductive system more easily, remove the left hindlimb by severing it near the hip with bone cutters. Then use the bone cutters to remove the left anterior portion of the pelvic girdle. Also remove the necessary muscles and connective tissues to expose the organs shown in figure 67.6.

7. Trace the ductus deferens from the inguinal canal to the penis. Note that the ductus deferens loops over the ureter within the pelvic cavity and passes downward behind the urinary bladder to join the urethra. Locate the *prostate gland,* which appears as an enlargement at the junction of the ductus deferens and urethra. The prostate gland is relatively small compared to that in humans and positioned along the urethra some distance from the urinary bladder.

8. Trace the urethra to the penis. Locate the *bulbourethral glands,* which form small swellings on either side at the proximal end of the penis. Seminal vesicles are absent in cats.

9. Use a sharp scalpel to cut a transverse section of the penis. Identify the *corpora cavernosa,* each of which contains many blood spaces surrounded by a sheath of connective tissue.

10. Review the locations of the reproductive organs without the aid of figures.

11. Complete Parts B, C, and D of the laboratory assessment.

Name _____

Date _____

Section _____

The Ⓐ corresponds to the indicated Learning Outcome(s) Ⓞ found at the beginning of the Laboratory Exercise.

LABORATORY ASSESSMENT
67

Cat Dissection: Reproductive Systems

PART A: Assessments

Complete the following:

1. Compare the relative lengths and paths of the uterine tubes (oviducts) of the cat and the human. Ⓐ1 _____

_____

_____

2. How do the shape and structure of the uterus of the cat compare with that of the human? Ⓐ2 _____

_____

_____

3. Assess the function of the uterine horns of the cat. Ⓐ1 _____

_____

_____

4. Compare the relationship of the urethra and the vagina in the cat and in the human. Ⓐ2 _____

_____

_____

PART B: Assessments

Complete the following:

1. Compare the glans penis in the cat and in the human. Ⓐ2 _____

_____

_____

2. How do the location and the relative size of the prostate gland of the cat compare with that of the human? Ⓐ2

_____

_____

3. What glands associated with the human ductus deferens are missing in the cat? Ⓐ2 _____

_____

4. Compare the prepuce in the cat and in the human. Ⓐ2 _____

_____

_____

_____

PART C: Assessments

Identify the numbered reproductive structures of the female cat in figure 67.7 and the male cat in figure 67.8.

FIGURE 67.7   Label the numbered reproductive structures of the female cat. A1

1 ⎯⎯⎯⎯⎯⎯⎯⎯⎯⎯

2 ⎯⎯⎯⎯⎯⎯⎯⎯⎯⎯

3 ⎯⎯⎯⎯⎯⎯⎯⎯⎯⎯

4 ⎯⎯⎯⎯⎯⎯⎯⎯⎯⎯

5 ⎯⎯⎯⎯⎯⎯⎯⎯⎯⎯

6 ⎯⎯⎯⎯⎯⎯⎯⎯⎯⎯

FIGURE 67.8 Label the numbered reproductive structures of the male cat. A1

1 —————————————————

2 —————————————————

3 —————————————————

4 —————————————————

5 —————————————————

6 —————————————————

7 —————————————————

8 —————————————————

Part D Assessments

Observe the models of the human reproductive system and figure 67.9 of a female and a male cadaver. Locate the labeled features and identify the numbered features in figure 67.9 that were also identified in the cat dissection.

FIGURE 67.9 `AP|R`Cat Sagittal (median; midsagittal) plane of a cadaver, (a) female, (b) male. Identify the reproductive features, using the terms provided. A2 A3

Terms:
Body of uterus
Cervix of uterus
Ovary
Uterine tube
Vagina

(a) Posterior ←——→ Anterior

Terms:
Glans penis
Penis
Prostate gland
Scrotum
Testis

(b) Posterior ←——→ Anterior

APPENDIX 1

Preparation of Solutions

Amylase solution, 0.5%

Place 0.5 g of bacterial amylase in a graduated cylinder or volumetric flask. Add distilled water to the 100 mL level. Stir until dissolved. The amylase should be free of sugar for best results; a low-maltose solution of amylase yields good results. (Store amylase powder in a freezer until mixing this solution.)

Benedict's solution

Prepared solution is available from various suppliers.

Caffeine, 0.2%

Place 0.2 g of caffeine in a graduated cylinder or volumetric flask. Add distilled water to the 100 mL level. Stir until dissolved.

Calcium chloride, 2.0%

Place 2.0 g of calcium chloride in a graduated cylinder or volumetric flask. Add distilled water to the 100 mL level. Stir until dissolved.

Calcium hydroxide solution (limewater)

Add an excess of calcium hydroxide to 1 L of distilled water. Stopper the bottle and shake thoroughly. Allow the solution to stand for 24 hours. Pour the supernatant fluid through a filter. Store the clear filtrate in a stoppered container.

Epsom salt solution, 0.1%

Place 0.5 g of Epsom salt in a graduated cylinder or volumetric flask. Add distilled water to the 500 mL level. Stir until dissolved.

Glucose solutions

1. *1.0% solution.* Place 1 g of glucose in a graduated cylinder or volumetric flask. Add distilled water to the 100 mL level. Stir until dissolved.
2. *10% solution.* Place 10 g of glucose in a graduated cylinder or volumetric flask. Add distilled water to the 100 mL level. Stir until dissolved.

Iodine-potassium-iodide (IKI solution)

Add 20 g of potassium iodide to 1 L of distilled water, and stir until dissolved. Then add 4.0 g of iodine, and stir again until dissolved. Solution should be stored in a dark, stoppered bottle.

Methylene blue

Dissolve 0.3 g of methylene blue powder in 30 mL of 95% ethyl alcohol. In a separate container, dissolve 0.01 g of potassium hydroxide in 100 mL of distilled water. Mix the two solutions. (Prepared solution is available from various suppliers.)

Monosodium glutamate (MSG) solution, 1%

Place 1 g of monosodium glutamate in a graduated cylinder or volumetric flask. Add distilled water to the 100 mL level. Stir until dissolved.

Physiological saline solution

Place 0.9 g of sodium chloride in a graduated cylinder or volumetric flask. Add distilled water to the 100 mL level. Stir until dissolved.

Potassium chloride, 5%

Place 5.0 g of potassium chloride in a graduated cylinder or volumetric flask. Add distilled water to the 100 mL level. Stir until dissolved.

Quinine sulfate, 0.5%

Place 0.5 g of quinine sulfate in a graduated cylinder or volumetric flask. Add distilled water to the 100 mL level. Stir until dissolved.

Ringer's solution (frog)

Dissolve the following salts in 1 L of distilled water:

6.50 g sodium chloride
0.20 g sodium bicarbonate
0.14 g potassium chloride
0.12 g calcium chloride

Sodium chloride solutions

1. *0.9% solution.* Place 0.9 g of sodium chloride in a graduated cylinder or volumetric flask. Add distilled water to the 100 mL level. Stir until dissolved.
2. *1.0% solution.* Place 1.0 g of sodium chloride in a graduated cylinder or volumetric flask. Add distilled water to the 100 mL level. Stir until dissolved.
3. *3.0% solution.* Place 3.0 g of sodium chloride in a graduated cylinder or volumetric flask. Add distilled water to the 100 mL level. Stir until dissolved.
4. *5.0% solution.* Place 5.0 g of sodium chloride in a graduated cylinder or volumetric flask. Add distilled water to the 100 mL level. Stir until dissolved.

Starch solutions

1. *0.5% solution.* Add 5 g of cornstarch to 1 L of distilled water. Heat until the mixture boils. Cool the liquid, and pour it through a filter. Store the filtrate in a refrigerator.

2. *1.0% solution.* Add 10 g of cornstarch to 1 L of distilled water. Heat until the mixture boils. Cool the liquid, and pour it through a filter. Store the filtrate in a refrigerator.
3. *10% solution.* Add 100 g of cornstarch to 1 L of distilled water. Heat until the mixture boils. Cool the liquid, and pour it through a filter. Store the filtrate in a refrigerator.

Sucrose, 5% solution

Place 5.0 g of sucrose in a graduated cylinder or volumetric flask. Add distilled water to the 100 mL level. Stir until dissolved.

Wright's stain

Prepared solution is available from various suppliers.

Assessment of Laboratory Assessments

Many assessment models can be used for laboratory reports. A rubric, which can be used for performance assessments, contains a description of the elements (requirements or criteria) of success to various degrees. The term rubric originated from *rubrica terra,* which is Latin for the application of red earth to indicate anything of importance. A rubric used for assessment contains elements for judging student performance, with points awarded for varying degrees of success in meeting the objectives. The content and the quality level necessary to attain certain points are indicated in the rubric. It is effective if the assessment tool is shared with the students before the laboratory exercise is performed.

Following are two sample rubrics that can easily be modified to meet the needs of a specific course. Some of the elements for these sample rubrics may not be necessary for every laboratory exercise. The generalized rubric needs to contain the possible assessment points that correspond to objectives for a specific course. The point value for each element may vary. The specific rubric example contains performance levels for laboratory assessments. The elements and the point values can easily be altered to meet the value placed on laboratory reports for a specific course.

ASSESSMENT: Generalized Laboratory Report Rubric

Element	Assessment Points Possible	Assessment Points Earned
1. Figures are completely and accurately lableled.		
2. Sketches are accurate, contain proper labels, and are of sufficient detail.		
3. Colored pencils were used extensively to differentatiate structures on illustrations.		
4. Matching and fill-in-the-blank answers are completed and accurate.		
5. Short-answer/discussion questions contain complete, thorough, and accurate answers. Some elaboration is evident for some answers.		
6. Data collected are complete, are accurately displayed, and contain a valid explanation.		

TOTAL POINTS: POSSIBLE _____ EARNED _____

ASSESSMENT: Specific Laboratory Report Rubric

Element	Excellent Performance (4 points)	Proficient Performance (3 points)	Marginal Performance (2 points)	Novice Performance (1 point)	Points Earned
Figure labels	Labels completed with ≥ 90% accuracy.	Labels completed with 80%–89% accuracy.	Labels completed with 70%–79% accuracy.	Labels <70% accurate.	
Sketches	Accurate use of scale, details illustrated, and all structures labeled accurately.	Minor errors in sketches. Missing or inaccurate labels on one or more structures.	Sketch not realistic. Missing or inaccurate labels on two or more structures.	Several missing or inaccurate labels.	
Matching and fill-in-the-blanks	All completed and accurate.	One or two errors or omissions.	Three or four errors or omissions.	Five or more errors or omissions.	
Short-answer and discussion questions	Answers complete, valid, and contain some elaboration. No misinterpretations noted.	Answers generally complete and valid. Only minor inaccuracies noted. Minimal elaboration.	Marginal answers to the questions and contains inaccurate information.	Many answers incorrect or fail to address the topic. May be misinterpretations.	
Data collection and analysis	Data complete and displayed with a valid interpretation.	Only minor data missing or a slight misinterpretation.	Some omissions. Not displayed or interpreted accurately	Data incomplete or show serious misinterpretations	

TOTAL POINTS EARNED _____

Correlation of Laboratory Exercises and Ph.I.L.S. 4.0 Lab Simulations

LABORATORY EXERCISE	PH.I.L.S. 4.0
Fundamentals of Human Anatomy and Physiology	
1 Scientific Method and Measurements **2** Body Organization, Membranes, and Terminology **3** Chemistry of Life **4** Care and Use of the Microscope	
Cells	

5 Cell Structure and Function
   Ph.I.L.S. 4.0 Lesson 2 Metabolism: Size and Basal Metabolic Rate
6 Movements Through Cell Membranes
   Ph.I.L.S. 4.0 Lesson 1 Osmosis and Diffusion: Varying Extracellular Concentration
7 Cell Cycle
	4 Cyanide and Electron Transport
	2 Size and Basal Metabolic Rate
	1 Varying Extracellular Concentration

Tissues	
8 Epithelial Tissues **9** Connective Tissues **10** Muscle and Nervous Tissues	
Integumentary System	
11 Integumentary System	
Skeletal System	
12 Bone Structure and Classification **13** Organization of the Skeleton **14** Skull **15** Vertebral Column and Thoracic Cage **16** Pectoral Girdle and Upper Limb **17** Pelvic Girdle and Lower Limb **18** Fetal Skeleton **19** Joint Structure and Movements	
Muscular System	
20 Skeletal Muscle Structure and Function Ph.I.L.S. 4.0 Lesson 5 Skeletal Muscle Function: Stimulus-Dependent Force Generation **21** Electromyography: BIOPAC Exercise **22** Muscles of the Head and Neck **23** Muscles of the Chest, Shoulder, and Upper Limb **24** Muscles of the Vertebral Column, Abdominal Wall, and Pelvic Floor **25** Muscles of the Hip and Lower Limb	**5** Stimulus-Dependent Force Generation **6** Weight and Contraction **7** The Length-Tension Relationship **8** Principles of Summation and Tetanus **9** EMG and Twitch Amplitude

LABORATORY EXERCISE	PH.I.L.S. 4.0
Lymphatic System	
49 Lymphatic System	
Respiratory System	
50 Respiratory Organs	
51 Breathing and Respiratory Volumes	**37** Altering Body Position
Ph.I.L.S. 4.0 Lesson 38 Respiration: Altering Airway Volume	**38** Altering Airway Volume
52 Spirometry: BIOPAC Exercise	
53 Control of Breathing	**3** Respiratory Quotient
	39 Exercise-Induced Changes
	40 Deep Breathing and Cardiac Function
Digestive System	
54 Digestive Organs	**41** Glucose Transport
55 Action of a Digestive Enzyme	
Urinary System	
56 Urinary Organs	**42** Anti-Diuretic Hormone
57 Urinalysis	
Reproductive Systems and Development	
58 Male Reproductive System	
59 Female Reproductive System	
60 Meiosis, Fertilization, and Early Development	
61 Genetics	
Fetal Pig Dissection Exercises (similar for cat version)	
62 Fetal Pig Dissection: Musculature	
63 Fetal Pig Dissection: Cardiovascular System	
64 Fetal Pig Dissection: Respiratory System	
65 Fetal Pig Dissection: Digestive System	
66 Fetal Pig Dissection: Urinary System	
67 Fetal Pig Dissection: Reproductive Systems	
Supplemental Laboratory Exercises*	
S-1 Skeletal Muscle Contraction	**6** Weight and Contraction
Ph.I.L.S. 4.0 Lesson 8 Skeletal Muscle Function: Principles of Summation and Tetanus	**8** Principles of Summation and Tetanus
S-2 Nerve Impulse Stimulation	**12** The Compound Action Potential
Ph.I.L.S. 4.0 Lesson 12 Action Potentials: The Compound Action Potential	
S-3 Factors Affecting the Cardiac Cycle	
Ph.I.L.S. 4.0 Lesson 21 Frog Heart Function: Thermal and Chemical Effects	**21** Thermal and Chemical Effects

*These exercises are available in the eBook via Connect Anatomy & Physiology and also online for instructor distribution; see Instructor Resources via Connect Library tab.

CREDITS

Photo Credits

Front Matter
Dedication: © Laura Chiavini, Creative Communication Specialist, Kishwaukee College; About the Author, Terry R. Martin: © J & J Photography; Brian H. Kipp: © Brian Kipp; Cynthia Prentice-Craver: © Bill Craver; Phillip D. Snider: © Leslee Shae Snider.

Lab Exercise 1
Figure 1.2, 1.3: © J & J Photography.

Lab Exercise 2
Figure 2.4, 2.5: © J & J Photography; 2.6: © McGraw-Hill Education/Joe DeGrandis, photographer; 2.10 (1): © McGraw-Hill Education/Karl Rubin, photographer; 2.10 (2): © Living Art Enterprises/Science Source; 2.10 (middle): © McGraw-Hill Education/Joe DeGrandis, photographer; 2.10 (3): © McGraw-Hill Education/Karl Rubin, photographer.

Lab Exercise 4
Figure 4.1: © J & J Photography; 4.2: © Stockbyte/Getty Images RF; 4.3: © Nikon Instruments, Inc.; 4.4a-b, 4.5, 4.8, 4.9: © J & J Photography.

Lab Exercise 5
Figure 5.3: © Ed Reschke; 5.5a-b: © Dr. Gopal Murti/SPL/Science Source.

Lab Exercise 6
Figure 6.1: © J & J Photography; 6.5a-c: © David M. Phillips/Science Source.

Lab Exercise 7
Figure 7.2, 7.3, 7.5: © Ed Reschke/Getty Images; 7.6a: © James Cavallini/Science Source; 7.6b: © CNRI/Science Source; 7.7a-d: © Ed Reschke/Getty Images.

Lab Exercise 8
Figure 8.1a-b: © Ed Reschke/Getty Images; 8.1c-d: © Victor P. Eroschenko; 8.1e: © McGraw-Hill Education/Dennis Strete, photographer; 8.1f: © McGraw-Hill Education/Al Telser, photographer; 8.1g: © McGraw-Hill Education/Dennis Strete, photographer; 8.1h-i: © McGraw-Hill Education/Al Telser, photographer; 8.1j-k: © Ed Reschke; 8.2a-c: © J & J Photography.

Lab Exercise 9
Figure 9.1a-c: © McGraw-Hill Education/Al Telser, photographer; 9.1d-e: © McGraw-Hill Education/Dennis Strete, photographer; 9.1f: © Visuals Unlimited/Corbis; 9.1g: © McGraw-Hill Education/Al Telser, photographer; 9.1h: © Victor P. Eroschenko; 9.1i: © McGraw-Hill Education/Al Telser, photographer; 9.1j: © McGraw-Hill Education/Dennis Strete, photographer; 9.1k: © McGraw-Hill Education/Al Telser, photographer.

Lab Exercise 10
Figure 10.1a: © Ed Reschke/Getty Images; 10.1b: © McGraw-Hill Education/Al Telser, photographer; 10.1c: © Ed Reschke/Getty Images; 10.2: © McGraw-Hill Education/Al Telser, photographer.

Lab Exercise 11
Figure 11.1b: © The University of Toledo; 11.3: © McGraw-Hill Education/Dennis Strete, photographer; 11.5a: © Per Kjeldsen; 11.5b: © Image Source/Getty Images RF; 11.5c-d: © McGraw-Hill Education/Dennis Strete, photographer; 11.5e: © Carolina Biological/Visuals Unlimited/Corbis.

Lab Exercise 12
Figure 12.2: Courtesy of Utah Valley Regional Medical Center, Department of Radiology; 12.4a: © Ed Reschke/Getty Images; 12.4b-c: Courtesy of John W. Hole, Jr.; 12.6: © McGraw-Hill Education/Dennis Strete, photographer; 12.8a-b, 12.10a-b: © J & J Photography.

Lab Exercise 13
Figure 13.4: © J & J Photography.

Lab Exercise 14
Figure 14.6, 14.9-14.12: © J & J Photography.

Lab Exercise 15
Figure 15.2a-b, 15.3a-c: © J & J Photography; 15.8: © Stockbyte/PunchStock RF.

Lab Exercise 16
Figure 16.8: Courtesy Dale Butler; 16.9: © Magán-Domingo/agefotostock.com; 16.10: Courtesy Dale Butler; 16.12: © J & J Photography.

Lab Exercise 17
Figure 17.6-17.8: Courtesy Dale Butler; 17.9: © J & J Photography.

Lab Exercise 18
Figure 18.1-18.6, 18.8, 18.9: © J & J Photography.

Lab Exercise 19
Figure 19.3b: © McGraw-Hill Education/Rebecca Gray, photographer/Don Kincaid, dissections; 19.4a-h, 19.5a-g: © J & J Photography.

Lab Exercise 20
Figure 20.2a: © Victor B. Eichler, Ph.D.; 20.3: © Ed Reschke/Getty Images; 20.5a, 20.7: © H.E. Huxley.

Lab Exercise 22
Figure 22.8: © McGraw-Hill Education/APR; 22.9a-c: © J & J Photography.

Lab Exercise 23
Figure 23.9, 23.11: © McGraw-Hill Education/APR; 23.12a-c: © J & J Photography.

Lab Exercise 24
Figure 24.5: © McGraw-Hill Education/APR.

Lab Exercise 25
Figure 25.10: © McGraw-Hill Education/APR; 25.11: © McGraw-Hill Education/Photo and Dissection by Christine Eckel; 25.12a-b: © J & J Photography.

Lab Exercise 26
Figure 26.1: © Suza Scalora/Getty Images RF; 26.2a: © McGraw-Hill Education/Joe DeGrandis, photographer; 26.2b-c: © McGraw-Hill Education/Eric Wise, photographer; 26.3a: © McGraw-Hill Education/Joe DeGrandis, photographer; 26.3b: © Suza Scalora/Getty Images RF; 26.4, 26.5a-b: © McGraw-Hill Education/Joe DeGrandis, photographer; 26.6: © McGraw-Hill Education/Eric Wise, photographer; 26.7, 26.8a-c: © McGraw-Hill Education/Joe DeGrandis, photographer; 26.9a-b: © McGraw-Hill Education/Eric Wise, photographer.

Lab Exercise 27
Figure 27.4: © McGraw-Hill Education/Al Telser, photographer; 27.5, 27.7: © Ed Reschke/Getty Images; 27.8: © David Becker/Science Source; 27.10: © Ed Reschke; 27.11: © Ed Reschke/Getty Images.

Lab Exercise 28
Figure 28.2: From: A Stereoscopic Atlas of Anatomy by David L. Basett. Courtesy of Dr. Robert A. Chase, MD; 28.8: © McGraw-Hill Education/APR; 28.10: © McGraw-Hill Education/Photo and Dissection by Christine Eckel; 28.14: © Ed Reschke/Getty Images.

Lab Exercise 29
Figure 29.3a-e: © J & J Photography.

Lab Exercise 30
Figure 30.2: © McGraw-Hill Education/APR; 30.4: © McGraw-Hill Education/Rebecca Gray, photographer/Don Kincaid, dissections; 30.9: © McGraw-Hill Education/APR; 30.12b: © McGraw-Hill Education/Joe DeGrandis, photographer; 30.13: © McGraw-Hill Education/Dennis Strete, photographer.

Lab Exercise 32
Figure 32.2, 32.5: © J & J Photography; 32.7: © McGraw-Hill Education/APR; 32.8: © J & J Photography.

Lab Exercise 33
Figure 33.1, 33.2: © Ed Reschke/Getty Images.

Lab Exercise 34
Figure 34.1: © Biophoto Associates/Science Source; 34.3: © Victor B. Eichler, Ph.D.; 34.4: © Ed Reschke/Getty Images.

Lab Exercise 35
Figure 35.4: © Per Kjeldsen; 35.5: © J & J Photography; 35.7: © Lisa Klancher; 35.9, 35.10: © J & J Photography; 35.12: © Kage Mikrofotografie/Phototake.

Lab Exercise 36
Figure 36.5: © J & J Photography; 36.6: © Steve Allen/Getty Images RF.

Lab Exercise 37
Figure 37.5: © J & J Photography; 37.6: © Biophoto Associates/Science Source; 37.7a-b, 37.8, 37.10: © J & J Photography; 37.11: © Dr. John D. Cunningham/Visuals Unlimited/Corbis.

Lab Exercise 38

Figure 38.3, 38.5: © J & J Photography; 38.6: © Science VU/Visuals Unlimited/Corbis.

Lab Exercise 39

Figure 39.2: © Biophoto Associates/Science Source; 39.3: © Science VU/Visuals Unlimited/Corbis; 39.4: © Biophoto Associates/Science Source; 39.6: © Visuals Unlimited/Corbis; 39.8: © Biophoto Associates/Science Source; 39.10: © Ed Reschke/Getty Images; 39.12: © Visuals Unlimited/Corbis.

Lab Exercise 40

Figure 40.1: © Visuals Unlimited/Corbis; 40.2, 40.3: © McGraw-Hill Education/Al Telser, photographer; 40.4: © Dr. Frederick Skvara/Visuals Unlimited/Corbis.

Lab Exercise 41

Figure 41.2: © J & J Photography; 41.4, 41.5(all), 41.7: © McGraw-Hill Education/Al Telser, photographer.

Lab Exercise 42

Figure 42.2: © Kemtec Science; 42.3a-d: © McGraw-Hill Education/James Schaffer, photographer; 42.4: © SPL/Science Source; 42.5a-b, p. 453a-b: © J & J Photography.

Lab Exercise 43

Figure 43.3: © Lester V. Bergman/Corbis; 43.5: © Jean-Claude Revy/ISM/Phototake; 43.7: © Image Source/Getty Images RF.

Lab Exercise 44

Figure 44.7: © Ed Reschke/Getty Images; 44.9-44.12: © J & J Photography; 44.13: © McGraw-Hill Education/APR; 44.14: © J & J Photography.

Lab Exercise 47

Figure 47.2: © Biophoto Associates/Science Source.

Lab Exercise 48

Figure 48.2: © Image Source/Getty Images RF; 48.5: © Johner/SuperStock RF.

Lab Exercise 49

Figure 49.3: © McGraw-Hill Education/Dennis Strete, photographer; 49.8a: © McGraw-Hill Education/Rebecca Gray, photographer/Don Kincaid, dissections; 49.8b: © J & J Photography; 49.8c: © John Watney/Science Source; 49.9: © Microscape/SPL/Science Source; 49.10, 49.11: © Biophoto Associates/Science Source.

Lab Exercise 50

Figure 50.2b: © McGraw-Hill Education/Joe DeGrandis, photographer; 50.4: © CNRI/Science Source; 50.6: © Victor P. Eroschenko; 50.7: © Biophoto Associates/Science Source; 50.9: © Dr. Fred Hossler/Visuals Unlimited/Corbis; 50.10: © Ed Reschke/Getty Images; 50.13: © McGraw-Hill Education/APR.

Lab Exercise 51

Figure 51.4-51.6: © J & J Photography.

Lab Exercise 54

Figure 54.4: © McGraw-Hill Education/Dennis Strete, photographer; 54.5: © Ed Reschke/Getty Images; 54.7: © Ed Reschke; 54.10: © McGraw-Hill Education/Dennis Strete, photographer; 54.11b: © Victor P. Eroschenko; 54.14: © J & J Photography; 54.15:

© Dennis Strete; 54.16: © Ed Reschke/Getty Images; 54.17, 54.19: © McGraw-Hill Education/APR.

Lab Exercise 56

Figure 56.2b: © J & J Photography; 56.5a: © Biophoto Associates/Science Source; 56.5b: © McGraw-Hill Education/Al Telser, photographer; 56.7, 56.8: © Biophoto Associates/Science Source; 56.9: © Ed Reschke/Getty Images.

Lab Exercise 57

Figure 57.2: © McGraw-Hill Education.

Lab Exercise 58

Figure 58.2a: © McGraw-Hill Education/Dennis Strete, photographer; 58.3: © Biophoto Associates/Science Source; 58.4, 58.5: © Ed Reschke; 58.6a-b: © McGraw-Hill Education/Al Telser, photographer; 58.7: © Michael Peres.

Lab Exercise 59

Figure 59.3: © McGraw-Hill Education/Rebecca Gray, photographer/Don Kincaid, dissections; 59.6, 59.7: © Ed Reschke/Getty Images; 59.8: © Michael Peres; 59.9: © McGraw-Hill Education/Carol D. Jacobson, Ph.D., Dept. of Veterinary Anatomy, Iowa State University.

Lab Exercise 60

Figure 60.4a-b: © Dr. Kurt Benirschke.

Lab Exercise 61

Figure 61.1a-j: © J & J Photography.

Lab Exercise 62

Figure 62.2: Courtesy of Dr. Sheril Burton; 62.3: © McGraw-Hill Education/Photo and Dissection by

Christine Eckel; 62.5: © McGraw-Hill Education/APR; 62.8, 62.10, 62.12, 62.13, 62.15, 62.17, 62.19, 62.20, 62.23, 62.24: Courtesy of Dr. Sheril Burton.

Lab Exercise 63

Figure 63.2: Courtesy of Dr. Sheril Burton; 63.3b: © McGraw-Hill Education/APR; 63.4-63.6, 63.9, 63.10: Courtesy of Dr. Sheril Burton; 63.16-63.18: © McGraw-Hill Education/APR.

Lab Exercise 64

Figure 64.1: Courtesy of Dr. Sheril Burton; 64.3, 64.4b-64.6: © McGraw-Hill Education/APR.

Lab Exercise 65

Figure 65.1b, 65.2b: © McGraw-Hill Education/APR; 65.4: Courtesy of Dr. Sheril Burton; 65.5b, 65.6: © McGraw-Hill Education/APR.

Lab Exercise 66

Figure 66.1: Courtesy of Dr. Sheril Burton; 66.2: © McGraw-Hill Education/APR; 66.4: © McGraw-Hill Education/Ralph W. Stevens III, Ph.D., photographer; 66.5: © McGraw-Hill Education/APR; 66.6b: © J & J Photography; 66.7: © McGraw-Hill Education/APR.

Lab Exercise 67

Figure 67.2a: © McGraw-Hill Education/Ralph W. Stevens III, Ph.D., photographer; 67.2b: © McGraw-Hill Education/APR; 67.4a: © McGraw-Hill Education/Ralph W. Stevens III, Ph.D., photographer; 67.4b, 67.9a-b: © McGraw-Hill Education/APR.

INDEX

Articulations in bones, 123, 125*f*
Arytenoid cartilage, 542*f*, 544
Ascending aorta, 466*f*–68*f*, 507*f*, 509
Ascending colon, 510*f*, 589*f*, 590
Ascending tracts, 292
Association areas, 314, 318*f*, 319
Association neurons, 282
Asters, 69*f*, 70*f*
Astigmatism, 386, 387, 660*t*, 663
Astigmatism test, 387, 387*f*
Astrocytes, 282, 283*t*, 286*f*–87*f*
Atherosclerosis, 450
Atlas, 146, 146*f*–147*f*, 192*f*
Atoms, 22
ATP, 56
Atria, 466
Atrioventricular bundle, 480, 482*f*, 494, 494*f*
Atrioventricular node, 480, 482*f*, 494, 494*f*
Atrioventricular orifices, 469
Atrioventricular sulcus, sheep, 471*f*
Atrioventricular valves, 466, 468*f*, 469
Audiometers, 400
Auditory acuity test, 396
Auditory areas, 319
Auditory association area, 318*f*
Auditory ossicles, 132, 396, 397*f*
Auditory stimuli, reaction time, 331–33
Auditory tubes, 396, 397*f*
   cat, 716
Auricles
   ear, 396, 397*f*
   heart, 466*f*, 469
   sheep, 471*f*
Autocrine secretions, 414
Autonomic reflexes, 306, 390
Axial skeleton, 122
Axilla, 270*f*
Axillary artery, 509, 509*f*
   cat, 696*f*
Axillary fold, 270*f*–71*f*
Axillary lymph nodes, 532*f*
Axillary nerve, 298*f*
Axillary region, 16*f*
Axillary vein, 511*f*, 512
   cat, 699*f*–700*f*, 705*f*
Axis, 146, 146*f*–147*f*, 192*f*
Axon collateral, 284*f*
Axon hillock, 284*f*, 285
Axons, 282, 282*f*–85*f*, 286–88
Azygos vein, cat, 699*f*, 707*f*

B

Babinski reflex, 308
Bacteria in urine, 623*f*
Balance, 396, 405–9, 406*f*–8*f*
Ball-and-socket joints, 184*t*, 185*f*
Banana sectioning activity, 81, 81*f*
Bárány test, 409
Basal cells, 364
Basal metabolic rate, 51
Basal nuclei, 316*f*, 318
Basal surface, epithelial tissue, 78
Basement membrane

in blood vessels, 504*f*
of composite cell, 43*f*
in epithelial tissues, 78, 79*f*, 105*f*
seminiferous tubule, 630*f*, 633*f*
of uterine tubes, 642*f*
Base of heart, 466, 469
Base of microscope, 30, 30*f*
Basic solutions, 22
Basilar artery, 508*f*
Basilar membrane, 398*f*
Basilic vein, 511*f*, 512
Basophil cells, 415–16, 416*f*, 438*f*, 439, 439*t*
B blood type, 458*t*, 459*f*
Benedict's test, 23–24, 59
Best-fit lines, 6
Beta cells, 418, 419*f*, 428, 429–430, 429*f*, 430*f*
Beta waves, 342, 342*f*, 343*t*
Biaxial joints, 185
Biceps brachii muscle, 199*f*, 233*f*, 237, 237*t*, 241*f*, 257*f*, 270*f*–71*f*
   cat, 682*f*, 685*t*
Biceps femoris muscle, 258, 260*t*, 261*f*–62*f*
   cat, 688*f*–89*f*, 690*t*, 693*f*
Biceps reflex, 308, 309*f*
Bile, 582, 622
Bile duct, 419*f*, 586, 586*f*, 587*f*, 595*f*
   cat, 725, 727*f*
Bilirubin, 622
Binocular microscopes, 31*f*
BIOPAC exercises
   display tools, 209*t*
   electrocardiography, 493–98
   electroencephalography, 341–45
   electromyography, 207–15
   reaction time, 329–333
   software and hardware, 208
   spirometry, 565–570
Bipolar leads, 484
Bipolar neurons, 282*t*, 283*f*
Bipolar recording method, 342
Bitter taste, 364
Biuret test, 23
Bladder. *See* Urinary bladder
Blastocoel, 653
Blastocyst, 650, 652*f*, 653
Blastula, 69, 75*f*
Blind spot, 374, 375*f*, 378*f*, 379
Blind-spot demonstration, 389
Blood
   cell structure, 89*f*, 90*t*, 435–39, 437*f*–38*f*
   locations, 90*t*
   osmosis exercise, 65–66
   testing, 445–450, 453–55
   typing, 457–461, 660*t*, 663
   in urine, 622, 623*f*
Blood glucose, 428
Blood pressure, 517–521, 526
Blood smear, 436, 436*f*
Blood vessels. *See* Arteries; Veins
Blood volume, 519
Body (vertebra), 147, 148*f*
Body cavities, 10–11, 12*f*

Body fluids, pH, 22
Body of stomach, 584, 585*f*
Body of uterus, 638
Body regions, 14, 15*f*–16*f*
Body tube of microscope, 30*f*, 31
Bohr effect, 455
Bone extracellular matrix, 112, 114, 115*f*, 116
Bone marrow, 532*f*
Bones. *See also* Skeleton
   articulations, 123, 125*f*
   cat skeleton, 674*f*
   cell structure, 89*f*, 90*t*
   indications of skeletal size from, 4
   locations, 90*t*
   lower limbs, 122, 123*f*, 166, 168, 169*f*
   major features, 112
   as major organs, 13
   middle ear, 122, 132, 396, 397*f*–98*f*, 399*f*
   pelvic girdle, 122, 123*f*
   skull, 122, 123*f*, 131–36, 132*f*, 133*f*–34*f*, 135*f*, 136*f*, 137*f*, 137*t*
   structure and classification, 111–16
   thoracic cage, 122, 123*f*
   upper limbs, 122, 123*f*
   vertebral column, 122, 123*f*
Bony labyrinth, 396
Bony landmarks, 267–275, 268*f*–274*f*
Borders of scapula, 157
Bowman's capsule, 609
Boyle's law, 553–54, 553*f*
Brachial artery, 509, 509*f*, 518*f*
   cat, 696*f*, 700*f*
Brachialis muscle, 233*f*, 237, 237*t*
   cat, 683*f*–84*f*, 685*t*
Brachial plexus, 295*f*, 296, 298*f*
   cat, 700*f*
Brachial pulse, 520
Brachial region, 16*f*
Brachial vein, 511*f*, 512
   cat, 700*f*, 705*f*
Brachiocephalic arteries, 477*f*
   cat, 696*f*, 698*f*
   sheep, 471*f*–73*f*
Brachiocephalic trunk, 466*f*, 508*f*, 509, 509*f*
Brachiocephalic veins, 475*f*, 477*f*, 511, 511*f*
   cat, 698*f*–700*f*, 705*f*
Brachioradialis muscle, 233*f*, 236*f*–37*f*, 237, 237*t*
   cat, 683*f*–84*f*, 685*t*
Brain
   blood vessels, 508*f*, 509, 511
   as major organ, 13
   respiratory centers, 574, 575*f*
   sheep, 349–353, 350*f*–54*f*
   structures and functions, 313–19, 314*f*–19*f*
Brainstem, 317*f*, 318*t*, 319, 319*f*
Brain waves, 342–45, 342*f*, 343*t*
Breast, 638, 641, 641*f*

Breath holding, 578
Breathing, 551–56, 565–570. *See also* Respiratory system
Breathing control, 573–78
Broad epicranial aponeurosis, 222*f*
Broad ligament, 638, 639*f*
   cat, 742, 742*f*
Broca's area, 318*f*, 319
Bronchi, 13, 540, 540*f*, 543*f*, 544
   cat, 718
Bronchial tree, 544
Bronchioles, 544, 546*f*
Bronchomediastinal trunk, 531, 532*f*
Buccal branch of facial nerve, 323*f*
Buccal region, 16*f*
Buccinator muscle, 222, 222*f*, 223*t*
Buffers, 22
Buffy coat, 446
Bulbospongiosus muscle, 249, 249*t*, 250*f*
Bulbourethral gland, 628, 628*f*, 629
   cat, 746, 746*f*
Bundle branches, 480, 482*f*
Bundle of His, 494

C

Cabbage water tests, 22–23
Calcaneal reflex, 308, 308*f*
Calcaneal region, 16*f*
Calcaneal tendon, 261*f*–62*f*, 274*f*
Calcaneus, 169, 169*f*, 274*f*
   cat, 674*f*
Calcitonin, 417, 417*f*
Calcium hydroxide, 577
Calcium oxalate, 623*f*
Calcium phosphate, 112, 623*f*
Calyx, 740*f*
   cat, 737*f*
Canaliculi, 112, 114, 115*f*, 374*f*, 375, 588*f*
Canals in bone, 123, 125*f*
Cancellous bone. *See* Spongy bone
Candida, 623*f*
Canines, 582*f*, 583
Capillaries, 504, 507*f*
   lymphatic, 530, 530*f*
   peritubular, 608*f*
Capillary tubes, 447–48, 447*f*, 450, 450*f*
Capitate bone, 159, 159*f*
Capitulum, 158, 158*f*
Carbohydrates, 22
Carbon dioxide in blood, 574
Carbon dioxide in expired air, 577
Cardiac conduction system, 480, 481, 482*f*, 494
Cardiac cycle, 479–484, 482*f*–83*f*, 489–491, 494
Cardiac muscle, 96, 96*f*, 97*t*, 208, 469*f*
Cardiac output, 519, 527
Cardiac veins, 466*f*–67*f*, 471
   sheep, 471
Cardia of stomach, 584, 585*f*
Cardiopulmonary resuscitation, 149
Cardiovascular system. *See also* Arteries; Heart; Veins

Hip muscles, 255–58, 256f–58f, 258t
Hippuric acid, 623f
Histology, 78. See also Tissues
Homeostasis, 22, 51, 330, 423–25
Homologous chromosomes, 71f
Homozygous individuals, 660
Hormones
    diabetic physiology, 427–430
    role in metabolism, 423–25
    sites of production, 414–18
Hot plates, 59f
Human chromosomes, 71, 71f
Humerus, 235f, 237f
    basic features, 158, 158f
    cat, 674f
    in elbow joint, 192f
    estimating skeletal size from, 4
    location in human, 122, 123f, 124f, 156f
    surface features, 271f
Hyaline cartilage, 89f, 90t, 112, 186, 186f, 540–44, 545f
Hyaline casts in urine, 623f
Hydrochloric acid, 582, 585
Hydrogen ion concentration, 22, 22f, 574
Hydrometer, 620–21, 621f
Hydrostatic pressure, 56, 58
Hydroxyapatite, 112
Hymen, 640, 640f
Hyoglossus muscle, cat, 675f, 676t
Hyoid bone, 122, 132, 224, 226f, 269f, 541f, 542f
    cat, 719f
Hyperextension, 188f, 190t
Hyperopia, 386, 386f
Hypertension, 56
Hypertonic solutions, 56, 58, 65–66
Hyperventilation, 578
Hypodermis, 88f, 102, 102f, 103f
Hypogastric region, 15f
Hypoglossal canal, 134f, 137t
Hypoglossal cranial nerve, 297f
Hypoglossal nerve, 320, 320f, 321t, 322t
    sheep, 351, 352f
Hypoglycemia, 428
Hypothalamus, 317f, 318t, 319, 414, 415f–16f
    sheep, 353, 353f
Hypotheses, 2
Hypotonic solutions, 56, 58, 65–66
H zones, 199, 199f

I

I bands, 199, 199f
IC. See Inspiratory capacity
Ileocecal junction, 589f, 596f
Ileocecal sphincter, cat, 725
Ileocecal valve, 589, 589f
Ileum, 510f, 589f
    cat, 725, 727f
Iliac crest, 166, 166f–167f, 268f, 270f
Iliac fossa, 166, 166f–167f
Iliac spine, 167f, 256f
Iliacus muscle, 256, 256f–57f, 258t

Iliocostalis muscles, 246, 247f
Iliolumbar artery, cat, 702f
Iliolumbar vein, cat, 706f
Iliopsoas muscles, 256, 257f
Iliotibial tract, 256f–57f, 274f
Ilium, 122, 166
Incisions for cat skin removal, 672f
Incisive foramen, 133f, 137t
Incisors, 582f, 583
Incomplete dominance, 660
Incus, 132, 396, 397f–98f, 399f
Independent assortment, law of, 670
Independent events, 664
Independent variables, 2
Inferior, illustrated, 14f
Inferior articular processes, 147
Inferior canaliculi, 374f
Inferior colliculus, 319f
    sheep, 351f, 353
Inferior lobe, lung, 544
Inferior mesenteric artery, 509, 510f
Inferior mesenteric vein, 512, 512f
Inferior nasal concha, 132f, 135f, 136
Inferior oblique muscle, 375, 375f, 376t
Inferior orbital fissure, 132f, 136f, 137t
Inferior pubic ramus, 167f, 168
Inferior rectus muscle, 374f–75f, 375, 376t
Inferior vena cava, 466f–68f, 470, 470f, 476f, 507f, 512, 512f, 606f
    sheep, 472, 472f
Inferior vertebral notch, 147, 148f
Inflation reflex, 574
Infraglenoid tubercle, 157f, 158
Infrahyoid muscles, 224
Infraorbital foramen, 132f–133f, 137t
Infraspinatus muscle, 234f–235f, 235t, 236
    cat, 679f–680f, 681t, 683f
Infraspinous fossa, 157f, 158
Infundibulum, 414, 638
    cat, 742, 742f
    sheep, 352f, 353
Inguinal canal, 629, 746f
    cat, 746f
Inguinal ligament, 248f, 252f, 511f, 512f
Inguinal lymph nodes, 532f, 534f
Inguinal region, 16f
Inner cell mass, 650
Inner ear, 396, 397f, 398f
Innominate bone, 122, 166
Insertions of joints, 187, 198
Inspiration, 552, 566, 574, 577f
Inspiratory capacity, 555f, 556, 566, 569
Inspiratory muscles, 526
Inspiratory reserve volume, 555f, 556, 566, 569
Insula, 315, 317f, 318f, 364
Insulin, 428–29

Insulin-dependent diabetes mellitus, 428
Insulin resistance, 428
Insulin shock, 428–29
Integration center, 306
Integumentary system, 13, 101–6, 102f, 103f, 105f
Interatrial septum, 466, 468f, 476f, 494f
Intercalated discs, 96f, 97t, 469f
Intercostal muscles
    location in human, 225f, 232f–33f, 252f
    origin and insertion, 233t, 553t, 576t
    role in ventilation, 232–33, 552f, 553t, 576t, 577f
Intercostal spaces, 480, 481f
Interlobar artery, 607f
Interlobar vein, 607f
Intermediate cuneiform, 169, 169f
Internal acoustic meatus, 134f, 135, 137t
Internal carotid artery, 508f, 509
Internal elastic lamina, 504f
Internal iliac artery, 509, 510f–11f
    cat, 702f–3f, 734f
Internal iliac vein, 512, 512f
    cat, 706f
Internal intercostal muscles
    location in human, 232f–33f, 252f
    origin and insertion, 233t, 553t
    role in ventilation, 233, 552f, 553t, 576t, 577f
Internal jugular vein, 511, 511f, 531, 532f
    cat, 705f
Internal nares, cat, 716f
Internal oblique muscle, 246, 247f–48f, 248f, 249t
    cat, 676f, 678, 678f, 679t
Internal respiration, 540
Internal urethral sphincter, 611, 611f
Interneurons, 282, 282t, 306, 330, 330f, 331
Interosseous membrane, 159f
Interphase, 68, 68f, 70f
Interstitial cells, 628, 629, 630, 630f
Intertubercular sulcus, 158, 158f
Intervals, ECG, 495, 498
Interventricular foramen, 316f
Interventricular septum, 466, 468f, 473f, 477f, 494f
Interventricular sulcus, 469
Intervertebral discs, 89f, 146f, 147
Intervertebral foramina, 146f, 147
Intestinal lymphatic trunk, 531, 532f
Intestines, 12f, 13. See also Large intestine; Small intestine
Intramembranous bones, 122
Intramembranous ossification, 176
Inversion, 189f, 190t
Iodine-potassium-iodide solution, 59

Iodine test, 24, 59
IPMAT acronym, 68
Iris, 375f, 376, 376t, 378, 379f
Iris diaphragm of microscope, 30f, 31, 32, 35
Irregular bones, 112
Irritant receptors, 574
IRV. See Inspiratory reserve volume
Ischial ramus, 167f, 168
Ischial spine, 166f, 167f, 168
Ischial tuberosity, 167f, 168, 279f
Ischiocavernosus muscle, 249, 249t, 250f
Ischium, 122, 166, 168
Islets of Langerhans, 418, 419f, 429, 429f
Isometric contraction, 203
Isotonic solutions, 56, 58, 65–66

J

Jaw movements, 222
Jejunum, 586f, 589f
    cat, 725, 727f
Joints
    capsule, 186f
    cavity, 186f
    types of, 184–85, 184t, 185f, 186, 186f–187f
Journal, BIOPAC, 208
Jugular foramen, 133f–134f, 135, 137t, 142f
Jugular notch, 149, 270f, 481f
Jugular trunk, 531
Juxtamedullary nephrons, 606

K

Karyotypes, 71f
Keratin, 78, 81t
Ketoacidosis, 621–22
Ketones, 621–22
Kidneys
    adrenal glands on, 418f
    cat, 701f, 703f, 727f, 734, 734f–37f, 737, 743f, 746f
    epithelial tissues from, 79f
    location in human, 606f, 713f
    as major organs, 13
    structure and functions, 606
Kilograms, 3t
Kilometers, 3t
Kinetochores, 69f
Knee-jerk reflex, 306f, 308, 308f
Knee joint, 186, 187f
Korotkoff sounds, 520
Kyphotic curve, 271f

L

Labia majora, 638, 639f, 640, 640f, 750f
    cat, 744
Labia minora, 638, 639f, 640, 640f
Lacrimal bone, 132f–133f, 136, 136f, 140f
Lacrimal fluid, 374
Lacrimal glands, 374, 374f, 375
Lacrimal punctum, 374f, 375
Lacrimal sacs, 374f, 375

Medial ligament, cat, 734
Medial longitudinal arch, 273*f*
Medial malleolus, 168, 169*f*, 273*f*–74*f*
Medial meniscus, 187*f*
Medial pterygoid muscle, 222, 223*f*, 223*t*
Medial rectus muscle, 375, 375*f*, 376*t*
Medial rotation, 188*f*, 190*t*
Medial sacral crest, 146
Median aperture, 316*f*
Median cubital vein, 511*f*, 512
Median nerve, 298*f*
Median sacral artery, 509, 510*f*
Median sacral crest, 149, 149*f*
Mediastinum, 10*f*, 11
Medulla oblongata, 317*f*, 318*t*, 319, 319*f*, 574, 575*f*–76*f*
    sheep, 350, 350*f*–53*f*, 353
Medullary cavities, 114*f*, 116
Meiosis, 68, 650–52, 651*f*
Meissner's corpuscles, 358, 358*t*, 359*f*
Melanin, 105*f*, 106, 662–63
Melanocytes, 106
Membranous labyrinth, 396
Membranous urethra, 611, 611*f*
Mendelian genetics, 659
Mendel's law of independent assortment, 670
Mendel's law of segregation, 664
Meningeal branch, 294
Meninges
    cranial, 314–15, 314*f*, 315*f*
    spinal, 292–93, 292–94, 292*f*, 300
Menisci, 186, 187*f*
Mental foramen, 132*f*–133*f*, 136, 137*t*
Mental protuberance, 132*f*, 136
Mental region, 16*f*
Merocrine sweat glands, 102, 103*f*, 105*f*, 106*t*
Mesentery, 589, 589*f*
    cat, 725, 727*f*
Mesoderm, 650, 653
Mesovarium, 742
    cat, 742*f*
Metabolism, 42, 51–54, 423–25
Metacarpal bones, 122, 123*f*, 124*f*, 159, 159*f*, 192*f*
    cat, 674*f*
Metacarpophalangeal joint (MP), 272*f*
Metaphase, 68, 70*f*, 651*f*
Metatarsal bones, 119*f*, 122, 123*f*, 169, 169*f*, 262*f*
    cat, 674*f*
Metatarsal head, 273*f*
Meters, 3*t*
Metric system, 2, 3*t*
Mice, metabolism, 51–54
Microglia, 282, 283*t*, 286*f*
Micrometers, 3*t*, 34, 37
Micrometer scales in microscopes, 34
Micrometer slides, 34*f*

Microscopes, 29–36
Microtubules, 43*f*, 44*t*, 69*f*
Microvilli, 43*f*, 44*t*, 79*f*, 591*f*
Micturition, 611
Midbrain, 317*f*, 318*t*, 319, 319*f*
    sheep, 351*f*–53*f*, 353
Middle cardiac vein, 467*f*, 471
Middle cerebral artery, 508*f*, 509
Middle deltoid muscle, 271*f*
Middle ear, 11, 122, 396, 397*f*, 398*f*, 399*f*
Middle layer of eye, 374, 376
Middle lobe, lung, 544
Middle nasal concha, 132*f*, 136
Middle scalene muscle, 225*t*
Middle tunic, 662–63
Milligrams, 3*t*
Milliliters, 3*t*
Millimeter rulers, 32, 33, 56
Millimeters, 3*t*
Milliseconds, 3*t*
Minor calyx, 606, 607*f*, 608
Minor duodenal papilla, 586*f*, 595*f*
Minute respiratory volume, 556
Mitochondria, 42, 43*f*, 44*t*, 197*f*, 198*f*
Mitosis, 68–69, 68*f*, 70*f*, 650
Mitral area, 481*f*
Mitral valve, 466, 468*f*, 469, 473*f*, 480, 507*f*
Mixed nerves, 294
M lines, 199, 199*f*
Mnemonics, 68
Molars, 582*f*, 583
Molecules, 22
Monocytes, 438*f*, 439, 439*t*
Monosaccharides. *See* Sugars
Mons pubis, 640, 640*f*
Morula, 650, 652*f*
Motor areas, 314, 319
Motor association area, 318*f*, 319
Motor neurons
    role in reflexes, 306, 330, 330*f*, 331
    in spinal nerves, 292
    structures and functions, 196, 282, 282*t*, 284*f*
Motor units
    defined, 196, 208
    recruitment, 203–4, 210, 211*f*, 213–15
    structure, 203*f*
Mouth, 13
Movement, joint, 183–87, 188*f*–89*f*, 190*t*
MP, 272*f*
M phase, 68
Mucous coat of ureter, 612, 612*f*
Mucous threads in urine, 623*f*
Multiaxial joints, 185
Multifidus muscle, 246, 247*f*
Multiple alleles, 660
Multipolar neurons, 282, 282*t*, 283*f*
Muscle fibers, 96*f*, 196, 198, 198*f*, 208
Muscles (cat)
    dissection guidelines, 673

forelimb, 681–85, 682*f*–84*f*, 685*t*
head and neck, 673–76, 674*f*–75*f*, 676*t*
hip and hindlimb, 685–87, 686*f*–89*f*, 690*t*
shoulder and back, 678–681, 679*f*–680*f*
thoracic and abdominal wall, 676–78, 676*f*–78*f*, 678*t*, 679*t*
Muscles (human)
    abdominal wall, 245–49, 247*f*–48*f*
    antagonistic pairs, 199*f*
    cardiac, 96, 96*f*, 97*f*, 469*f*
    chest and shoulder, 231–36, 232*f*–35*f*, 233*t*–35*t*
    eye, 374*f*, 375*f*, 376*t*, 388*f*
    eyelids, 376*t*
    fatigue, 210, 214*f*, 215
    head and neck, 221–24, 222*f*–26*f*
    hip, 255–58, 256*f*–58*f*, 258*t*
    inspiratory and expiratory, 526
    lower limb, 255–261, 256*f*–262*f*, 260*t*, 262*t*
    naming, 196
    pelvic floor, 246, 249, 249*t*, 250*f*
    respiratory, 552, 552*f*, 553, 553*t*, 574, 576*f*, 577*f*
    roles of, 199*t*, 200
    skeletal muscle overview, 195–200
    tissue examination, 78, 95–97, 96*f*, 469*f*
    upper limb, 236–38, 236*f*–37*f*, 237*t*–38*t*
    vertebral column, 246, 246*t*, 247*f*
Muscle spindles, 307
Muscle strains, 246
Muscle twitches, 203, 208
Muscular coat of ureter, 612, 612*f*
Muscular system, 13
Musculocutaneous nerve, 298*f*
Myelin sheath, 283*t*, 285*f*, 286*f*, 287*f*, 288, 288*f*
Mylohyoid muscle, 224, 226*t*
    cat, 674*f*–75*f*, 676*t*, 724*f*
Myocardium, 468*f*, 469, 477*f*
    sheep, 471
Myofibrils, 196*f*–197*f*, 198, 198*f*
Myofilaments, 197*f*
Myometrium, 638, 639*f*, 643*f*, 653*f*
Myopia, 386, 386*f*
Myosin filaments, 199*f*

N

Nail bed, 104, 104*f*
Nail body, 104, 104*f*
Nail matrix, 104, 104*f*
Nail plate, 104, 104*f*
Nail root, 104, 104*f*
Nails, 102, 104, 104*f*
Nares, 540*f*–41*f*
    cat, 717*f*
Nasal bone, 132*f*–134*f*, 136, 140*f*
Nasal cavity, 365*f*, 540, 540*f*

location, 11
    as part of respiratory system, 13
Nasal conchae, 132*f*, 135*f*, 136, 365*f*, 540, 541*f*
Nasalis muscle, 222, 222*f*, 223*t*
Nasal meatuses, 540, 541*f*
Nasal mucosa, 322*f*
Nasal region, 16*f*
Nasal septum, 540
Nasolacrimal ducts, 374*f*, 375
Nasopharynx, 541*f*
    cat, 716, 720*f*
Navicular, 169, 169*f*
Near point of accommodation, 388, 388*f*, 389*t*
Nearsightedness, 386, 386*f*
Neck muscles, 221–24, 224*f*–26*f*
Neck of rib, 149
Negative feedback systems, 414
Nephron loop, 608*f*–9*f*
Nephrons, 606, 608–9, 608*f*–9*f*
Nerve deafness, 399, 400
Nerve impulses, recording, 342–45
Nerves
    cranial, 282, 314, 320, 320*f*, 321*t*, 322*f*–23*f*, 322*t*
    as major organs, 13
    reflexes, 305–9
    spinal, 282, 287*f*, 294–96, 295*f*–300*f*
    structure, 282, 286–88
Nervous system
    major organs, 13
    as major tissue type, 78
    reflexes, 305–9, 330–31
    tissue structure, 95–97, 97*f*, 281–86, 282*t*–83*t*, 283*f*–88*f*
Neurilemma, 282, 285*f*, 288, 288*f*
Neurofibrils, 285, 285*f*
Neuroglia, 96, 97*f*, 282, 285*f*–86*f*
Neuromuscular junctions, 197, 197*f*, 203*f*, 211*f*
Neuron processes, 97, 103*f*, 197
Neurons. *See also* Nerves; Nervous system
    basic features, 282
    examining, 96, 97
    reflex arcs, 306, 330–31
    of retina, 376, 377*f*
    secretions, 414
    in skeletal muscle, 196
Neurotransmitters, 414
Neutral solutions, 22
Neutrophils, 437*f*, 438*f*, 439, 439*t*
Night vision, 379
Nissl bodies, 285, 285*f*
Nitrites in urine, 622
Nociceptors, 358, 358*t*
Node of Ranvier, 284*f*, 288, 288*f*
Nodules
    lymphatic, 533, 533*f*, 534*f*
    spleen, 535
Non-insulin-dependent diabetes mellitus, 428
Norepinephrine, 418
Normal breathing, 578, 578*f*
Normal vision, 386, 386*f*

Nosepiece of microscope, 30f, 31–32, 32f, 36f
Nostrils, 374f, 540f–41f
  cat, 717f
Notches in bone, 125f
Nuclear envelopes, 42, 44t, 68f, 70f
Nuclear pores, 43f
Nuclei (cell)
  basic features, 42, 43f, 44t
  connective tissues, 88f, 89f
  division, 70f
  in epithelial tissues, 79f–80f
  muscle, 96f, 196, 198f, 469f
  nervous tissue, 97f, 284f, 285
  photoreceptors, 376, 377f
  in taste buds, 366f
Nucleoli, 42, 43f, 44t, 284f, 285
Nucleoplasm, 68f
Numerical aperture, 32f
Nystagmus, 406, 409

O

Obesity, 428
Objective lens systems, 30f, 31–32, 32f
Oblique sections, 81, 81f
O blood type, 458t, 459f
Observations in science, 2
Obturator foramen, 166, 167f, 168
Obturator nerve, 299f
Occipital bone, 132, 133f–134f, 137t, 153f, 176f
Occipital condyle, 133f–134f, 135
Occipital lobe, 315, 317f
  sheep, 350, 350f
Occipital region, 16f
Occipital ridge, 269f
Occult blood, 622
Oculomotor nerve, 320, 320f, 321t, 322t, 390
  sheep, 351, 352f
Odontoid process, 147, 147f
Odor of urine, 620
Oil immersion lenses, 32, 32t, 35–36
Olecranon fossa, 158, 158f
Olecranon process, 158, 159f, 271f–72f
  cat, 674f
Olfaction, 364–65, 365f
Olfactory association area, 318f, 319
Olfactory bulb, 322f, 365f, 508f
  sheep, 352f–53f, 353
Olfactory epithelium, 365f
Olfactory foramina, 134f, 142f
Olfactory hairs, 364, 365f
Olfactory nerve, 320, 320f, 322f, 322t, 364, 365f
Olfactory receptors, 364, 365f
Olfactory tract, 321t, 322f, 365f
  sheep, 352f
Oligodendrocytes, 282, 283t, 286f
Omentum. See Greater omentum; Lesser omentum
Omohyoid muscle, 224, 226f, 226t
Oocyte, 638, 639f
Oogenesis, 638, 650

Openings in bones, 123. See also Foramina; Sinuses
Ophthalmic branch of trigeminal, 323f
Ophthalmic division, 321t
Ophthalmoscope, 377, 377f, 378f
Opposition, 189f, 190t
Optic canal, 137t
Optic chiasma, 317f, 319, 320f, 508f
  sheep, 352f–53f, 353
Optic disc, 374, 375f, 376, 377, 378f, 379
Optic foramen, 134f, 136f, 137t, 142f
Optic nerve, 320, 320f, 321t, 322t, 375f, 378, 390
  sheep, 352f
Optic tract, sheep, 352f
Oral cavity, 11, 582, 582f
  cat, 716, 725
Oral region, 16f
Ora serrata, 375f, 376
Orbicularis oculi muscle, 222, 222f, 223t, 374, 374f, 376t
Orbicularis oris muscle, 222, 223t
Orbital cavity, 11
Orbital region, 16f
Orbit of eye, 136f
Organic molecules, 22, 23–24
Organ of Corti, 396, 398f, 399f
Organ systems, 13
Origins, joint, 187, 198
Oropharynx, 541f, 582f
  cat, 720f
Osmosis, 56, 57, 58, 65–66
Ossification, 122, 176
Osteons, 89f, 112, 115f
Otic region, 16f
Otolithic membrane, 406, 407f
Otoliths, 406, 407f
Outer ear, 396, 397f
Outer layer of eye, 374, 376
Ova, 638
Oval window, 396, 397f–98f
Ovarian cortex, 641–42, 642f
Ovarian ligament, 638, 639f, 640f, 648f
  cat, 742f
Ovarian vein, cat, 703f, 734f
Ovaries, 415f, 638, 639f–641f
  cat, 736f, 742, 742f–43f
  as major organs, 13
Oviduct, 638
Ovulation, 650
Ovulation site, 639f
Oxygen consumption, 51–53
Oxygen dissociation curve, 446
Oxyhemoglobin, 446
Oxyphil cells, 417, 417f
Oxytocin, 416

P

Pacemaker, 480
Pacemaker cells, 494
Pacinian corpuscles, 358, 358t, 359f
Pain, referred, 19

Pain receptors, 331, 358, 358t
Palate
  cat, 725
  hard, 133f, 540f, 582f, 716f
  soft, 540f, 582f, 716f
  structure, 583
Palatine bone, 133f–134f, 136, 136f, 137t
Palatine process, 133f, 136
Palatine tonsil, 535f, 582f, 583
  cat, 725
Palmar aponeurosis, 236f, 240f
Palmaris longus muscle, 236f, 238, 238t
  cat, 682f, 685t
Palmar region, 16f
Palpation, 122, 268
Palpebra, 374
Pancreas
  with associated structures, 419f
  blood vessels, 510f
  cat, 703f, 707f, 725, 727f
  location in human, 12f, 415, 415f, 595f
  as major organ, 13
  structure, 586, 586f, 595f
  tissue examination, 418, 429–430, 429f, 430f, 587f
Pancreatic amylase, 600
Pancreatic ducts, 419f, 586, 586f, 595f
Pancreatic islets, 418, 419f, 429–430, 429f
Pancreatic juice, 582
Pancreatic polypeptide, 429
Papillae
  cat tongue, 725
  dermal, 103f, 105f
  hair, 102, 103f, 105f
  tongue, 367f
Papillary duct, 609f
Papillary muscles, 468f, 469, 473f, 476f, 477f
Paracrine secretions, 414
Parafollicular cells, 417, 417f
Paramecium, 45
Paranasal sinuses, 132, 540
Parasympathetic division (ANS), 390, 480
Parathyroid gland, 13, 415f, 417, 417f
Parfocal objectives, 33
Parietal bones, 113f, 132, 132f–135f, 139f, 176f
Parietal cells, 585, 585f
Parietal lobe, 315, 315f, 317f
  sheep, 350, 350f, 352f
Parietal pericardium, 11, 11f, 466, 468f, 469
  cat, 696
Parietal peritoneum, 10, 11, 12f, 248f
  cat, 701, 701f, 726f–27f, 734
Parietal pleura, 11, 11f, 540, 544
  cat, 717f
Parieto-occipital sulcus, 315, 317f
Parotid duct, 583, 583f
  cat, 724, 724f

Parotid salivary glands, 583, 583f
  cat, 724, 724f
Passive processes, 56
Patella, 112, 113f, 122, 123f, 168, 169f, 187f, 259f, 262f, 273f–74f
  cat, 674f, 686f
Patellar ligament, 187f, 256f, 259f, 260f, 263f
Patellar reflex, 306f, 308, 308f
Patellar region, 16f
PCL. See Posterior cruciate ligament (PCL)
Pectinate muscles, 468f, 473f
Pectineus muscle, 256, 257f, 258t
Pectoantebrachialis muscle, cat, 676f–77f, 678t, 682f
Pectoral girdle, 155–59
  bones of, 122, 123f
Pectoralis major muscle, 233f, 235t, 236, 270f
  cat, 676f–77f, 678t, 682f
Pectoralis minor muscle, 233f, 234t, 252f, 553t, 576t, 577f
  cat, 676f–77f, 678t, 682f
Pectoral region, 16f
Pedal region, 16f
Pedicles, 147, 148f
Pelvic cavity, 10, 10f, 11
Pelvic diaphragm, 250f
Pelvic floor muscles, 249, 249t, 250f
Pelvic girdle, 122, 123f, 165–170
Pelvic inlet, 166f
Pelvic lymph nodes, 532f
Pelvic outlet, 166
Pelvis, 166, 166f, 168t
  cat, 674f
Penis
  cat, 745f–46f, 746
  as major organ, 13
  muscles surrounding, 250f
  structure, 628f, 629, 632, 632f
  as urinary organ, 611f
Pepsin, 582
Pepsinogen, 585
Percent saturation of hemoglobin, 446
Perception, 358
Perforating canals, 115f
Pericardial cavity, 10f, 11, 11f, 468f, 469, 475f
  cat, 696
Pericardial sac, 722f
  sheep, 471
Pericardium, 10, 466, 469, 475f
  cat, 697f, 699f–701f
Perilymph, 396, 398f
Perimetrium, 638, 639f, 643f
Perimysium, 196, 196f–197f, 198
Perineal raphe, 250f
Perineal region, 16f
Perineum, 246, 249
Perineurium, 282, 287f–88f
Periodontal ligament, 583f
Periosteum, 114f, 115, 115f
Peripheral chemoreceptors, 574, 576f